Renewable and Alternative Energy: Concepts, Design and Applications

Renewable and Alternative Energy: Concepts, Design and Applications

Edited by George Thomson

SYRAWOOD
PUBLISHING HOUSE
New York

Published by Syrawood Publishing House,
750 Third Avenue, 9th Floor,
New York, NY 10017, USA
www.syrawoodpublishinghouse.com

Renewable and Alternative Energy: Concepts, Design and Applications
Edited by George Thomson

International Standard Book Number: 978-1-68286-612-2 (Hardback)

Cataloging-in-Publication Data

Renewable and alternative energy : concepts, design and applications / edited by George Thomson.
    p. cm.
Includes bibliographical references and index.
ISBN 978-1-68286-612-2
1. Renewable energy sources. 2. Power resources. I. Thomson, George.
TJ808 .R46 2018
333.794--dc23

TABLE OF CONTENTS

Permissions

List of Contributors

Index

PREFACE

Renewable energy refers to energy that is derived from sources that are replenishable, such as geothermal heat, solar, tidal and wind energy. These energy sources are environment-friendly, and can help reduce pollution on a global scale. Renewable and alternative energy are mainly used in electricity power generation, transportation and water heating at present. This book studies, analyses and upholds the pillars of renewable energy and its utmost significance in modern times. It will serve as a valuable source of reference for graduate and post graduate students.

This book unites the global concepts and researches in an organized manner for a comprehensive understanding of the subject. It is a ripe text for all researchers, students, scientists or anyone else who is interested in acquiring a better knowledge of this dynamic field.

I extend my sincere thanks to the contributors for such eloquent research chapters. Finally, I thank my family for being a source of support and help.

Editor

The Potential Wind Power Resource in Australia: A New Perspective

1

Willow Hallgren*, Udaya Bhaskar Gunturu, Adam Schlosser

The MIT Joint Program on the Science and Policy of Global Change, Massachusetts Institute of Technology, Cambridge, Massachusetts, United States of America

Abstract

Australia's wind resource is considered to be very good, and the utilization of this renewable energy resource is increasing rapidly: wind power installed capacity increased by 35% from 2006 to 2011 and is predicted to account for over 12% of Australia's electricity generation in 2030. Due to this growth in the utilization of the wind resource and the increasing importance of wind power in Australia's energy mix, this study sets out to analyze and interpret the nature of Australia's wind resources using robust metrics of the abundance, variability and intermittency of wind power density, and analyzes the variation of these characteristics with current and potential wind turbine hub heights. We also assess the extent to which wind intermittency, on hourly or greater timescales, can potentially be mitigated by the aggregation of geographically dispersed wind farms, and in so doing, lessen the severe impact on wind power economic viability of long lulls in wind and power generated. Our results suggest that over much of Australia, areas that have high wind intermittency coincide with large expanses in which the aggregation of turbine output does not mitigate variability. These areas are also geographically remote, some are disconnected from the east coast's electricity grid and large population centers, which are factors that could decrease the potential economic viability of wind farms in these locations. However, on the eastern seaboard, even though the wind resource is weaker, it is less variable, much closer to large population centers, and there exists more potential to mitigate it's intermittency through aggregation. This study forms a necessary precursor to the analysis of the impact of large-scale circulations and oscillations on the wind resource at the mesoscale.

Editor: Francois G. Schmitt, CNRS, France

Funding: The authors gratefully acknowledge the financial support for this work provided by the MIT Joint Program on the Science and Policy of Global Change through a number of federal agencies and industrial sponsors (for the complete list, see http://globalchange.mit.edu/sponsors/current.html). The funders of the Joint Program had no role in study design, data collection and analysis, decision to publish, or preparation of the manuscript.

Competing Interests: The authors have declared that no competing interests exist.

* Email: hallgren@mit.edu

Introduction

The general climatology of the winds in Australia has been documented on a national basis [1,2,3] and at the state level [4,5,6,7], using a variety of methodologies [8]. Such climatologies indicate that Australia has wind resources that are in places comparable to those in northern Europe, and indicate that the location of the strongest winds is in western, southwestern, and southern Australia, and southeastern coastal regions [8].

The physical quantity conventionally used to describe the wind energy potential in Australia is wind speed in m/s, whereas in the USA, wind atlases show maps of wind power density (WPD) to describe the quality of the wind resource. Most previous published studies use the mean to characterize the central tendency of the wind resource, however histograms of the wind resource measured using wind power density are characteristically skewed with long-tailed distributions [9] (Figure S1). Therefore, wind power studies based only on the total mean WPD do not give a representative picture of the central tendency of the wind power potential and also omit valuable information in terms of wind intermittency, variability and the temporal distribution of power generation [10], which would affect estimates of power production and required backup [11].

Variability in the wind resource has major ramifications for the economics and therefore the feasibility of wind power generation

and distribution, and hence measures of variability are useful for wind energy policy makers. Yet, very few atlases show maps of wind variability [9], and when they do it is typically in terms of the standard deviation of the wind speed or WPD. However, the economic viability of wind power as an alternative energy source strongly depends on how reliable the resource is, in terms of its availability and persistence, as well as other factors such as proximity to high-capacity power transmission lines, and how remote it is from population centers and the electricity grid. The reliability of wind power can in theory be increased by mitigating the natural intermittency of the wind resource, by aggregating power from wind farms that are geographically dispersed, with the aim of achieving a more continuous wind resource over large areas, and there have been several studies trying to address this issue [12,13].

Wind power production doubled in the 5 years to 2012, and has grown 340% since 1997, to meet 3.4% of Australia's total electricity demand and 26% of total renewable energy generated, which is a bit less than half that generated by hydropower [14]. Wind power is likely to become economically competitive in the coming decades, and is projected to grow by 350% when wind power projects currently in development come online in the next few years [15]. This projected expansion of wind energy conforms to national policies that were designed to lower carbon emissions, including legislation that was introduced to put a price on carbon, and the Renewable Energy Target of 20% by 2020 [15]. In light of this policy directive, there is a need to increase the accuracy and practical relevance of the assessment of Australia's wind power resource.

We assess Australia's potential wind power resource with alternative metrics of abundance, variability and intermittency that provide deeper insights about the stability of the wind resource at a widespread deployment scale [9,11] over long time periods, using a robust, multi-decadal dataset.

Several authors explore the variability and intermittency of the wind resource at many scales [16,17]. There are fewer studies at the mesoscale scale range than at smaller scale ranges, despite the fact that knowledge of variability at this scale is important to the management and control of wind power generation [16]. Our study focuses on variability and intermittency at the hourly scale and above -the mesoscale- and addresses the type of scenario, to take just one example, in which long wind lulls spanning weeks, during sustained periods of high pressure, have been known to occur in countries such as the UK and Germany (Oswald et al 2008 [18], telegraph article [19]). These instances have implications for the reliability of power generated, as well as the potential backup and storage required to sustain power delivery. The goal of the present paper is to characterize the wind resource in Australia and its inherent variability, as a necessary precursor to studies of the impact of large-scale climate oscillations on the variability of the wind resource at different scales.

Questions our study asks include: (1) What is the geographical distribution of the abundance, variability, availability, and persistence of wind power density (WPD), and do these differ with higher turbine hub heights? (2) Where can wind intermittency be mitigated by the aggregation of geographically dispersed wind farms?

Methods

2.1. Data

We have sought to address some of the limitations of previous wind resource studies that used data that had a coarse spatial and temporal resolution, a relatively short record length, and sparse and uneven coverage [20,11]. We used 31 years of hourly $1/2° \times 2/3°$ resolution MERRA (Modern Era Retrospective Analysis for Research and Applications [21] data (from 0030 on January 1st, 1979 to 2330 on 31st December, 2009) to reconstruct the wind field at several turbine hub heights 50 m, 80 m, and 150 m, since the MERRA dataset does not provide wind speeds at different hub heights. These heights were chosen to represent the recent 1990's (US) 50 m standard wind turbine hub height [22,23], and the 80 m hub height, which has become more common as technology develops, and the potentially much higher hub heights in the future.

Wind speed and then wind power density were computed at these different heights using boundary layer flux data (consisting of such parameters as surface roughness, displacement height and friction velocity) and similarity theory of the atmospheric boundary layer [9]. By doing this, we sought to improve on previous wind resource constructions that used a constant scaling exponent (irrespective of surface roughness) to scale the wind speed from a lower altitude (usually 10 m) to that of the turbine hub height. We use WPD (W m^{-2}) to describe the wind resource as it is a function of not only wind speed but also density, which also varies in space and time. It indicates how much wind energy can be harvested at a location by a wind turbine but is independent of wind turbine characteristics. In a recent study, Farkas [24] found that non-consideration of air density causes an

root mean square (RMS) error of 16% in wind potential, which is a considerable difference, and therefore air density should be an important consideration in estimating the wind resource potential. The domain considered for our study spans the entire Australian continent plus Tasmania, between 10°S and 45°S latitudes and 110°E and 155°E longitudes.

While the resolution of the data used in this study is lower than the mesoscale, there have also been many studies that establish the utility of data at the GCM resolution (e.g. Schwartz and George, 1999) [25] for understanding the variability and impact of large-scale circulations at a regional scale. Several studies have used a similar dataset, although with a shorter record length, to estimate the potential wind resource in China [26] and also globally [27]. However, for studying inter-decadal variability, we argue that the longer record length of the data is as essential an attribute. This is because, according to sampling theorem, a dataset has to have at least 20 years of data for understanding inter-decadal variability. Hence this construction was designed, and is most appropriate, for such studies.

All other constructions that span only a few years fail to represent such variability. Moreover, studies such as those by Pryor, Barthelmie and Schoof (2006) [28], Chadee and Clarke (2013) [29], use data with a lower resolution than that used here, to study similar issues (inter-annual variability of wind indices across Europe, large-scale wind energy potential of the Caribbean, etc.). This would indicate that our data resolution is suitable for the purpose of our research, and represents an improvement to the resolution of a number of prior studies [22,26,27,28,29,30,31,32].

Since Gunturu and Schlosser (2012) [9] have already done a thorough evaluation of the lowest model layer wind speed data taken directly from MERRA, and since this study uses the same data, as it is a continuation of theirs, it is unnecessary to reproduce this validation of the MERRA data here. As the original MERRA wind data that this study employs is in the public domain, the description of the methodology will enable others to construct the wind power density dataset that is used in this study.

2.2. Comparison with Existing Wind Climatologies in Australia

Here, the wind resource is constructed for a hub height of 80 m, and was compared with a publicly available map of 80 m wind speed, since a publicly accessible wind power density map was not available. This was done in order to understand the ability of this constructed dataset to reproduce large-scale spatial features. This map was originally published by the Australian government and also used by various state governments [15]. We also use publicly available maps of the location of wind farms in South Australia and New South Wales to validate our wind resource construction [33,34].

2.3. Wind resource metrics

The metrics we use in our study are wind abundance, variability and intermittency in the form of availability and persistence [9,11]. Most previously published studies use the mean to characterize the central tendency of the wind resource. Since the mean is not a robust measure of the central tendency for distributions with long tails, we use the median, which is immune to the extreme values in the distribution, as a robust measure of the central tendency of the wind resource, and provides a better evaluation of it's abundance. As Pryor and Barthelmie (2011) [20] point out "there is a need for accurate data pertaining to metrics of the wind climate beyond the central tendency, and trends in annual mean wind speeds have little bearing on the viability of wind energy."

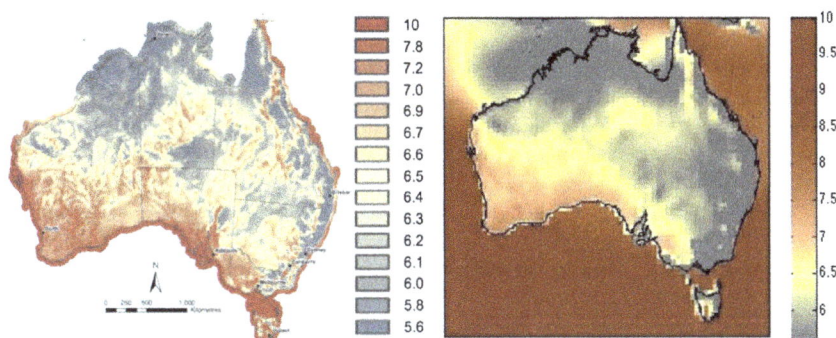

Figure 1. Comparison of mean wind speed (m/s) at an 80 m turbine hub height across Australia. (Left) Map developed by the Australian Government Department of the Environment, Water, Heritage and the Arts in 2008, and (right) the map constructed from MERRA data.

(a) Mean WPD Wm-2

(d) Median WPD Wm-2

(b) 80m – 50m

(e) 80m – 50m

(c) 150m -50m

(f) 150m -50m

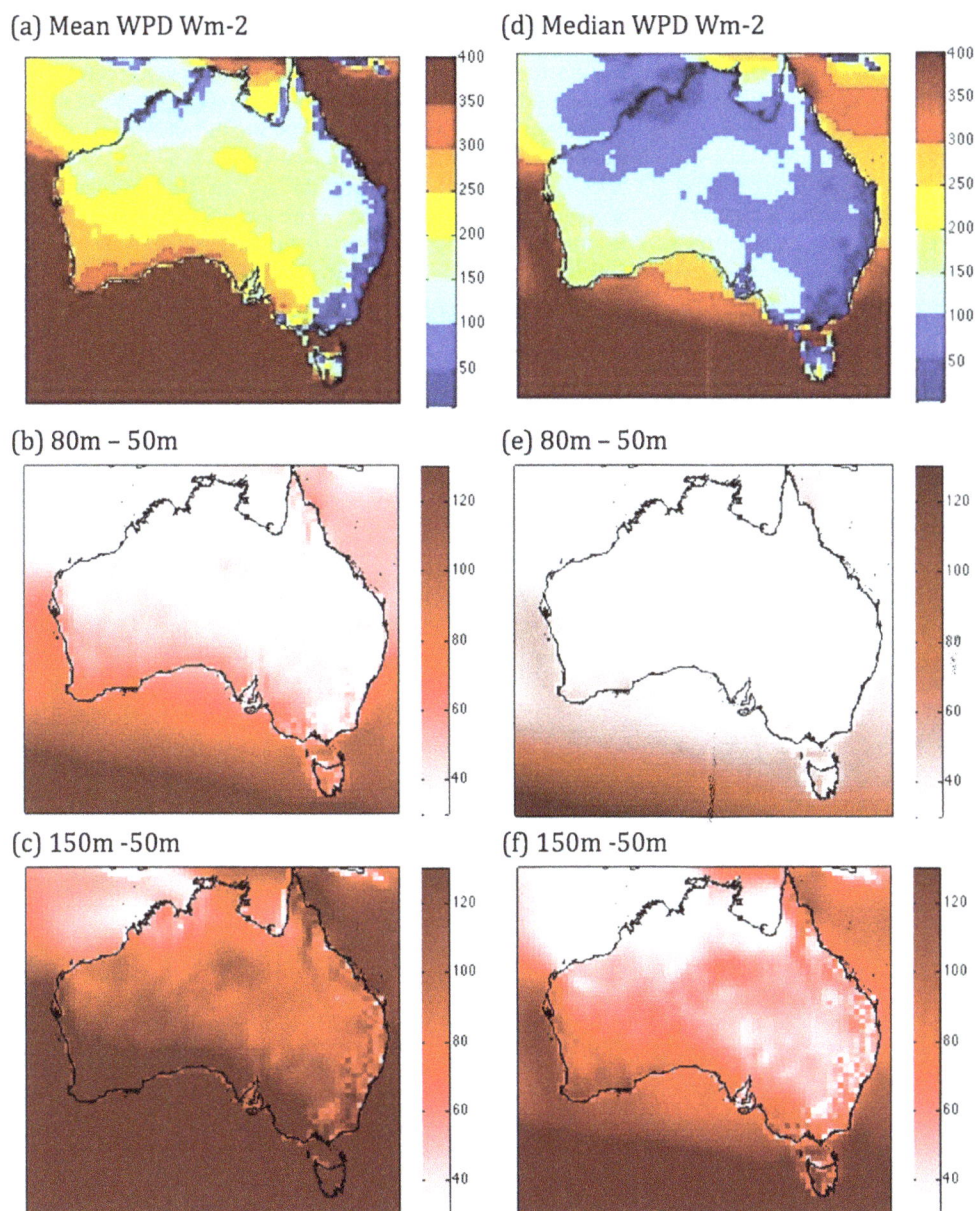

Figure 2. Measures of abundance. (a) The mean WPD at 50 m, (b) the change in the mean from 50 m to 80 m and from (c) 50 m to 150 m, (d) median wind power density at 50 m, (e) the change in the median from 50 m to 80 m and from (f) 50 m to 150 m. All units are W m^{-2}.

Instead of using the standard deviation to represent the variability of the wind resource, we argue that the variability of the wind resource is better captured in terms of the robust coefficient of variation (RCoV), since it is calculated using the median, which we argue is a more accurate representation of the wind power at a given site than the mean. We also use the Inter-quartile range (IQR) as a measure of the statistical dispersion, higher values of which can indicate the greater possibility of 'swings' of the WPD at a location, and therefore the amount of backup power that needs to be maintained.

In addition to these measures of variability, we also look at two measures of the intermittency of the wind - availability (or lack of) and persistence - since these are important indicators of intermittency, which is recognized as one of the key limitations to large-scale installation of wind power. We apply the reliability theory concept of availability to wind power, as a measure of the temporal distribution of the wind resource, and therefore of the reliability of a wind power generation system. We calculate the percentage of hours in our time series where WPD$>$200 W m^{-2}, and use the inverse of this - unavailability - of non-useful WPD (i.e. proportion of hours where $<$200 W m^{-2}), to characterize the geographic distribution of the reliability of the wind resource [11], and as one measure of intermittency. Our rationale for choosing the 200 W m^{-2} cutoff is the same as Gunturu and Schlosser (2011) [11], and incorporates a number of contributing arguments, which are detailed in Text S1. Mean episode length (i.e. number of hours of WPD above 200 W m^{-2}) was calculated as a measure of the persistence of the WPD, which is important in the planning and development of a robust deployment strategy for harvesting wind power.

We use Gunturu and Schlosser's (2011) [11] technique to analyze the potential value of aggregating the power generated by geographically dispersed wind farms in a roughly 1000×1000 km box (19×19 grid cells), in order to mitigate intermittency in the wind resource. Values of anticoincidence [35], and null-antic-oincidence were calculated for each grid cell (see Fig. 5 in Gunturu and Schlosser, 2011) by converting the time series of WPD at each grid point into a binary sequence of 1 s and 0 s depending on if the WPD is greater or less than the 200 W m^{-2} we use as the cutoff useful for viable commercial generation. We base our analysis of anticoincidence on these binary sequences. Two grid points are said to be anticoincident when the hourly time series of WPD is greater than 200 W m^{-2} at one of the two points, but not both, for 50% of the total length of the time series. We also calculate the null-anticoincidence, which offers a somewhat more relaxed criterion. Null-anticoincidence refers to the number of grid points in a roughly 1000×1000 km area surrounding a central point which have usable wind power ($>$200 W m^{-2}), when the central point does not, for at least 50% of the time when there's no wind at the central point [11]. If the region within this analysis area shows higher values of anticoincidence then this means that there will be fewer coincident lulls in the wind resource across the region, and that aggregating power from geographically dispersed wind farms will be more likely to mitigate the intermittency of the wind resource across the region as a whole.

Our choice for using a box this size for the anticoincidence analysis was based on the fact we are looking at the wind resource at a regional scale, hence, this is the scale at which we studied anticoincidence: the mesoscale. For more information on the rationale for the box size, refer to the Supplementary Information. In terms of the temporal scale used in this study, while there have been several methods and technologies to mitigate intermittency at the operational scale, such intermittency for the grid operations occurs at micro-to-hundreds of seconds. But for the scale that this

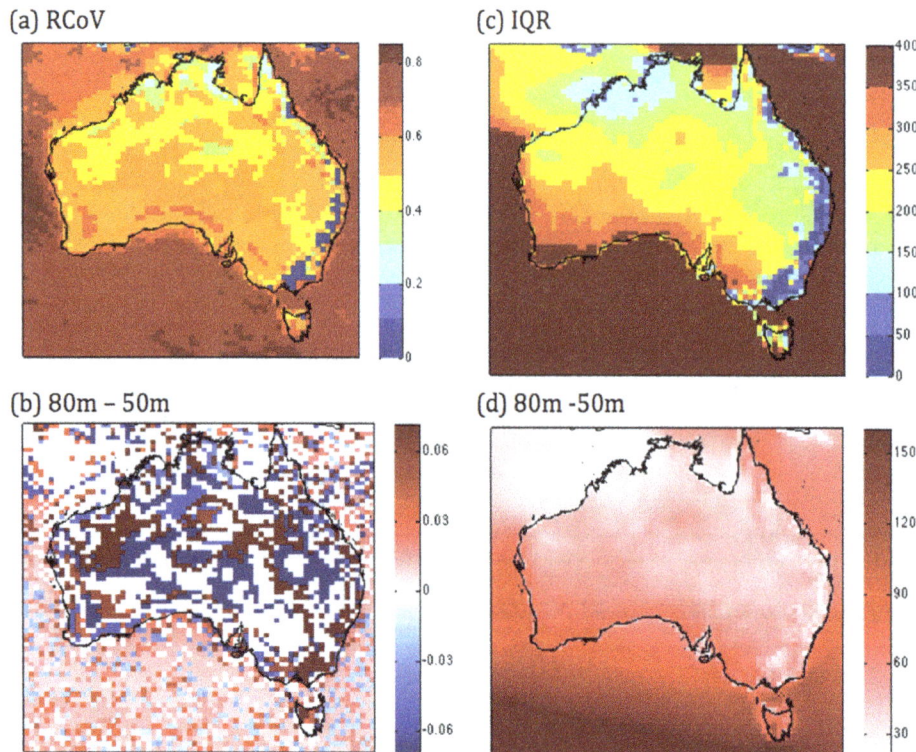

Figure 3. Measures of variation. (a) The robust coefficient of variation (RCoV -unitless) of WPD at 50 m, (b) the change in the RCoV from 50 m to 80 m, (c) inter-quartile range (IQR, W m^{-2}) at 50 m, (d) the change in the IQR from 50 m to 80 m.

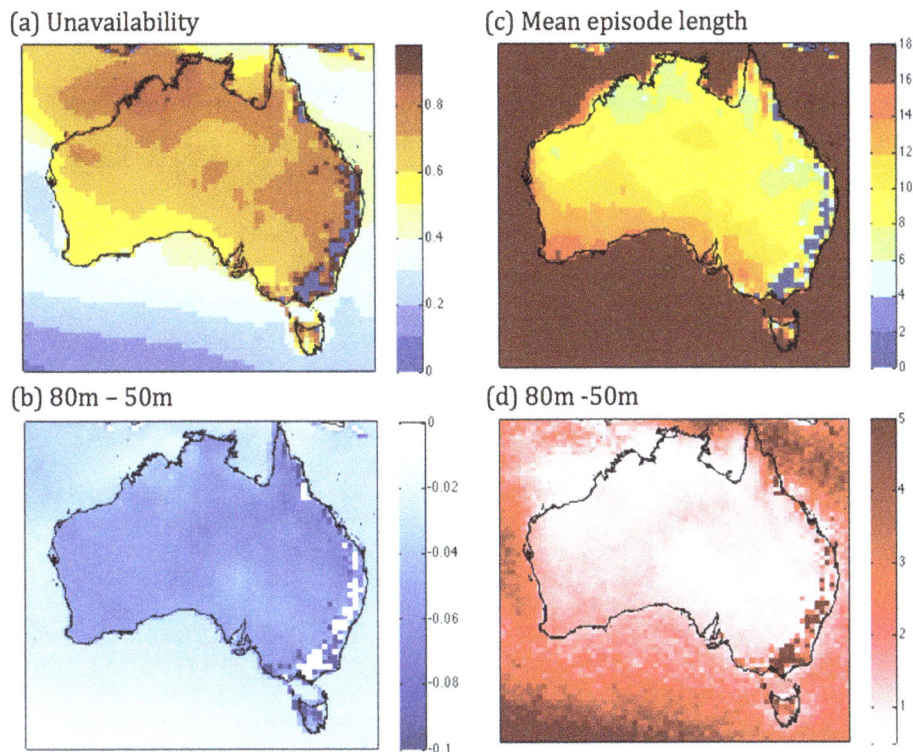

(a) Unavailability

(c) Mean episode length

(b) 80m – 50m

(d) 80m -50m

Figure 4. Measures of intermittency. (a) The unavailability of WPD at 50 m (fraction of time), (b) the change in the unavailability from 50 m to 80 m, (c) the mean episode length at 50 m (hours) (d) the change in the mean episode length from 50 m to 80 m.

study pertains to, no methods or technologies have yet been developed to deal with intermittency, to the knowledge of the authors, at the scale of one hour or more, in which case, the issues of back up and resource adequacy become important.

Results and Discussion

Wind speed and wind power density were computed at several wind turbine hub heights using boundary layer flux data from the Modern Era Retrospective-analysis for Research and Applications (MERRA) [21] and similarity theory of the atmospheric boundary layer [9]. We use wind speed to compare our results to existing wind atlases (as the reference atlas for Australia uses wind speed instead of wind power density to measure wind power potential), as well as a range of metrics to analyze wind power density, including wind abundance, variability, and intermittency in the form of availability and persistence [11,9]. Detailed descriptions of the data and methodology are described in the Methods section.

3.1. Comparison of MERRA and Australian Government maps of wind speed at 80 m

Our approximately 50 km×67 km (½ degree×⅔ degree) map of 80 m above ground level wind speed (Fig. 1) is quantitatively and geographically similar to the 9 km×9 km resolution map of wind speed at the same height produced by the Australian Government Department of the Environment, Water, Heritage and the Arts (hereafter referred to as AGD) [15]. This map was created by WindLab (www.windlab.com) for the AGD and is derived from observed weather station data taken from Bureau of Meteorology weather stations for the years 1995–2005, for the entire continent, and supplemented with commercially produced meteorological datasets, which are then assimilated into a high

resolution broad-area wind mapping model called WindScape [36]. WindScape uses a regional scale weather model (The Air Pollution Model (TAPM [37]) to improve the resolution of the observed data, and also a fine scale computational fluid dynamics model Raptor and/or Raptor-NL to create fine scale resolution maps of the wind resource over broad areas. The maps created are validated and adjusted to achieve consistency with observational data at ground level [38].

While our construction of the wind resource matches qualitatively and quantitatively very well with that of the AGD map overall, there are differences between the two maps in some regions. Our results mostly show slightly lower values for most areas compared to corresponding areas on the AGD map (Table 1). For example, a comparison of our map with the maps of NSW [33] and South Australian [34] wind farms, indicates regions where areas of better wind resources, as shown on our constructed map, coincide with existing wind farm deployments on [33] and [34], particularly in NSW, even though wind speed values in these areas might be slightly lower on our constructed map, than on the AGD map, as shown in Fig. 1. On that map, this is not always the case - it shows even better wind resources outside of these regions of wind farm deployment. This indicates that our map does actually capture areas of good wind resource in areas where there are existing wind farms.

Furthermore, our coarser-resolution map of the wind resource shows fewer orographic effects of the Great Dividing Range than the AGD map. Nevertheless, our map captures precisely the areas where there are existing wind farms on the New South Wales Southern Highlands and Blue Mountains [33,34]. So although our map has a resolution which does not capture as much topographical detail as the government map, it captures precisely the areas where there are existing wind farms, for instance, our

Table 1. Comparison of the range of values (m/s) in many areas of the 80 m wind speed map constructed from MERRA data to the one produced by the Australian Government.

Regions of similarity	MERRA data map	Australian government map
East coast, Tasmania	5.6–7.0	6.5–7.8
Western Victoria	6.5–7.0	Mostly >7.0
SE South Australia	6.4–7.2	Up to 7.8
Central Australia	5.6–7.0	5.8–6.6

The first region encompasses much of the East coast, and includes southeast and northeast QLD, and Tasmania.

map shows two small regions on the eastern seaboard of good wind resource, which is where all but one of the existing wind farms are currently located.

Reasons for the differences seen in these two maps could be due to the lower spatial resolution of our constructed map and the lower temporal record length of the AGD map. Since the AGD wind resource map has been constructed by running a mesoscale model (TAPM) for 11 years (and all other constructions also span only a few years), the record length of the construction is short compared to the record length of our construction, which represents an average over 31 years, that includes many years of low and high wind. Short record lengths do not represent interannual variability and climate scale (i.e. more than a few years) oscillations like the El Nino Southern Oscillation (ENSO) robustly.

3.2. Measures of abundance and variability

Reflecting the wind speed patterns of previous Australian wind atlases, our constructed map of mean WPD at 50 m (Fig. 2a) shows that the strongest wind resources occur in southwest Western Australia, southern South Australia, and Tasmania, and south-western Victoria. It is lowest in mountainous areas along the Great Dividing Range in eastern Australia, in northwest Australia, and northwest QLD. Most of the continent has mean WPD values below 300, and most of the populated east coast of the country has values below 200 W m^{-2} at this resolution, which is the cutoff for the production of usable power that turbines can produce, the rationale for which is detailed in Text S1. As turbine hub height increases to 80 (Fig. 2b) and 150 m (Fig. 2c), there is an increase in mean WPD of up to about 40 and 100 W m^{-2} in the northern two-thirds of Australia and 80 and 160 W m^{-2} (and higher in

Tasmania) in the south respectively. While the mean WPD construction reflects the other known datasets that illustrate wind speed, we extend the analysis that has historically been done, and look at other metrics of the resource that could be useful for assessing the economics of wind power generation and also for operational stability.

The map of median WPD at 50 m (Fig. 2d) indicates that a greater part of the continent has WPD below the 200 W m^{-2} value. Compared with the mean WPD in Fig. 2a, the median values are almost half of the mean values throughout much of the country. This implies that the distribution is very skewed, and hence we argue that the median is a much more robust measure of central tendency and therefore a more appropriate metric to represent WPD. As turbine hub height increases to 80 m (Fig. 2e) and then 150 m (Fig. 2f), there is less of an increase in median WPD compared to mean; up to about 30 and 80 W m^{-2} in the northern half of Australia, and up to about 50 and 120 W m^{-2} (and higher in Tasmania) along the southern part of the country. This scenario implies that the number of hours which show an increase in WPD are about the same as those which show a decrease, however the increase of WPD in those hours which show an increase, is greater than the decrease of WPD in the hours which show a decrease. We infer from this that variability and intermittency of the resource are increasing while the median resource is increasing.

Most maps of the variability of the wind resource use the standard deviation. We do not use the normal standard deviation. In line with our argument that the median is a better metric, being non-parametric, we use the 'robust coefficient of variation' (RCoV) that is the ratio of median deviation about the median to the median. Our results show that the highest RCoV values occur in southwest Tasmania and WA, and in southern South

Anti-coincidence   Null-anti-coincidence

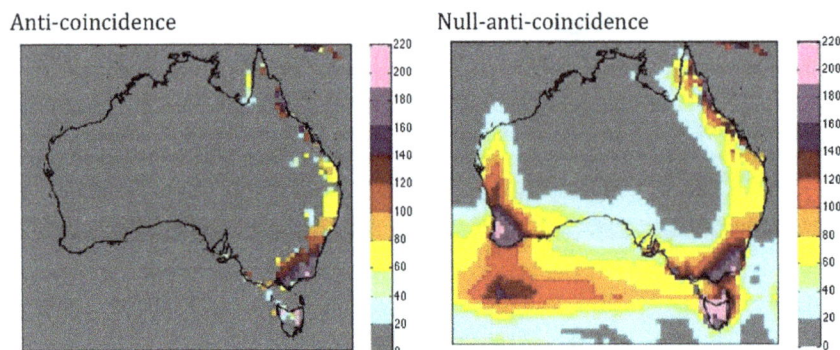

Figure 5. Anticoincidence (left) and Null-anticoincidence (right) of wind power density, at 50 m. Units indicate the number of grid points in a ~1000×1000 km box surrounding the gridpoint in question which are anticoincident to the central gridpoint, which is when the hourly time series of WPD is greater than 200 W m-2 at one of the two points, but not both, for 50% of the total length of the time series.

Australia, but inland from the coastline, which indicates these areas have relatively higher variability compared to the abundance in terms of the median (Fig. 3a). The lowest values, indicating a less variable, more reliable wind resource, occur along the southeastern seaboard and in parts of northern Australia near the coast.

RCoV increases with hub height in some areas (e.g. southeastern Australia), and decreases in others (much of inland SA) (Figs. 3b, S2 (a)). This is because although the median increases with height everywhere, variability decreases in some regions and increases in others. When the median increases with the hub height, and the variability also increases as much or more, RCoV (which is the ratio of deviation to central tendency) also increases. The RCoV decreases when the median increases but the variation does not increase so much (i.e. the ratio decreases). A scenario where RCoV decreases with height indicates that raising the hub height would better harvest the greater wind resources at higher hub heights, with lowered variability and intermittency. With greater surface friction, the standard deviation of the wind in the boundary layer increases [39]. Therefore, the boundary layer roughness predominantly determines the impact that raising the hub height has on the RCoV of the wind resource.

The interquartile range (Fig. 3c) is a measure of an important measure of dispersion in the wind resource since it is immune from the effect of outlying extreme values. Thus it is one of the robust measures of dispersion. As such, it can provide an insight as to the possibility of swings in the wind resource and therefore the amount of backup power that needs to be maintained. At 50 m, the areas that show high IQR (Fig. 3c) tend to coincide with areas that have the highest mean and median WPD (Figs. 2a and 2d, southwest and southern parts of the continent), and increases more with turbine hub height in these areas (Fig. 3d, S2 (b)). The regions that have low mean WPD also have the lowest IQR (e.g. east coast). IQR increases with turbine hub height across the country (Fig. 3d, S2 (b)).

If we consider just abundance and variability, regions that have high WPD and low variability (as shown by IQR) are areas where the wind resource could potentially be harnessed economically. Unfortunately, in Australia, our analysis indicates that at the resolution of this study, the areas which have mean WPD> 200 W m^{-2} also have an IQR of at least the same magnitude if not greater, though undoubtedly there are isolated areas where this would not be the case – but our relatively coarse dataset is unable to show this. However, an additional, very important consideration for harnessing wind power economically at a widespread deployment scale is the extent of its episodic nature - or intermittency.

3.3. Measures of intermittency and the potential for its mitigation

To explicitly gauge the intermittency of WPD, we first consider a metric of unavailability (given as fraction of time WPD is less than a minimum threshold - see Data section). We find that unavailability, which decreases with height, is generally highest in the areas where mean (or median) WPD is low (far northwest Australia, northern Tasmania, and just west of the Great Dividing Range on the eastern seaboard). The lowest values are seen along the eastern seaboard, indicating more reliable winds in these areas. Large areas scattered throughout northern and eastern Australia exhibit relatively high values (above 0.65), with the southwestern third of the country exhibiting moderate values (Fig. 4a). Unavailability decreases with height, as might be expected (WPD increases, so given the 200 W m^{-2} threshold of availability, it also increases), except for the areas which have the lowest mean

WPD values – higher altitude areas along the eastern seaboard - which show a negligible change in unavailability with a change in height (Figs. 4b, S3 (a)).

The availability of WPD as a continuous resource over time is also considered. The spatial pattern of mean episode length (defined as the average time that WPD is continuously above the same threshold) closely resembles that of the mean WPD. We found that the mean episode length at 50 m hub height (c) is lowest in parts of the Great Dividing Range in the east of the country, where WPD is low, and highest in the southern Australia, south-west Western Australia, and Tasmania, where WPD is highest. Mean episode length increases with height most where the mean WPD is lowest, along the Great dividing range in the east (Figs 4d, S3 (b)). Conversely, areas where mean episode length is highest show only small increases (<2 hours) with increasing hub height to 80 m (Fig. 4d), and raising the hub height to 150 m results in a near linear response in terms of additional episode length (Fig. S3(b)).

The coincidence (or lack thereof - see Methods) of intermittent wind power in different places sets the scope of installed backup generation capacity required to maintain a steady power supply, as well as the benefits of the aggregation of wind resources. The areas with the lowest unavailability (suggesting low wind intermittency, or more reliable, steady winds) coincide with areas of moderate to high anticoincidence at 50 m, such as along the eastern seaboard. 'Anticoincidence' denotes the occurrence of one event without the simultaneous occurrence of another [35]. The greatest intensity of anticoincident points is in the southeast of the continent, including northeast Tasmania (Fig. 5). However, these areas also have a small episode length (suggesting less persistent winds), which suggests that the aggregation of wind farms may indeed help mitigate wind intermittency in the more densely populated southeast of Australia.

Davy and Coppin (2003) [40] found that the variability in the total wind power output in south east Australia can be reduced to some extent by wider distribution of numerous wind farms, but remains substantial, thus their analysis suggests some degree of anticoincidence of southeastern Australia's wind resource. Their analysis spanned 4 years from March 1999 to March 2003, and also used hourly automatic weather station data from nine sites located on the SE Australian coast. It is useful to note that this period includes a marked La Nina episode and so the wind record may contain some anomalies, and may not be suitable for more general inferences. The record length used in our study, by contrast, is much longer and spans many ENSO cycles, and therefore can be used to infer the mean picture more robustly.

There are areas in Australia with relatively high intermittency - high unavailability and quite low mean episode length - such as northern and northwest Australia, that overlap a vast swathe of the continent west of the Great Dividing Range that shows little anticoincidence of WPD. These are the areas where aggregating turbines would be least effective, at the spatial and temporal scales analyzed.

However, an analysis of the null-anticoincidence (Figure 5) across Australia suggests that there may be some merit in linking wind farms across large areas to increase the reliability of the power supply in areas which show low anticoincidence and moderate to high intermittency, such as parts of the northern QLD coast, inland NSW, and parts of western Victoria and Tasmania, all of which show high values of null-anticoincidence. This may improve the reliability of wind power in these areas. These results agree well with previous research that has shown the coexistence of higher values of anticoincidence with regions that have high topographical inhomogeneity (i.e. mountain ranges) and proximity to the sea. This research has also co-located low anticoincidence areas to low surface roughness (flat terrain), semi-arid climate and terrains, with climate characterized by anti-cyclones which occur over large areas, leading to a large coincidence of low wind states across these high pressure systems [11].

Summary and Conclusions

Our study suggests that many areas with the strongest widespread wind resource, in terms of both mean and median WPD (SW Western Australia, southern South Australia and Tasmania, and SW Victoria) also score relatively highly on measures of variability (IQR, RCoV) and exhibit moderate levels of intermittency, in terms of reliability (i.e. unavailability) and persistence (mean episode length). Much of the areas which have moderate to high wind intermittency also have very low antic-coincidence, as defined in the Methods section, suggesting that there are large expanses of the continent in which aggregating turbines would be less effective, based on our study, at the spatial and temporal scales analyzed (keeping in mind the limitations of this study, described below). These areas also tend to be geographically remote from the bulk of the Australian population on the east coast (certainly in Western Australia, Northern Territory and South Australia), disconnected from the east coast's electricity grid (Western Australia, Northern Territory), and often are not connected or located near enough high capacity electricity infrastructure (parts of South Australia) [41], all of which would decrease the potential economic viability of wind farms in these locations.

However, in eastern Australia (along the Great Dividing Range and the eastern seaboard), many areas exhibit a comparatively poorer wind resource (in terms of the mean and median), and the broad scale mean WPD is below the 200 W m^{-2} cutoff. However, the variability is also lower in these areas, the reliability is better, and the potential to mitigate intermittency (in the form of relatively low persistence) by the aggregation of wind farms, is larger; these areas tend to have higher values of anticoincidence, and null-anticoincidence. Our results broadly agree with those of Davy and Coppin (2003) [40] who demonstrated that variability in the total wind power output in south east Australia can be reduced to some extent by wider distribution of numerous wind farms.

There are several assumptions and limitations of our study which require articulating, the most important being the mapping scale issues that this study raises, whereby coarser resolution maps can overestimate the area available at a given wind speed, and will also potentially fail to depict many areas with good resources which occur at a scale smaller than the resolution our study employs (1/2×2/3 degree, or about 55×73 km square) [8]. Therefore, we acknowledge that our results are at least partly scale and resolution dependent. That being said, the continuous assimilation of observations to run the model enhances the efficacy of the MERRA data, i.e. if there are many sites that have good subgrid scale wind resources, this will be taken into consideration because the observations at these point locations are fed into the data assimilation cycle.

We assumed a neutral boundary layer, as do most of the wind resource assessments, including that by National Renewable Energy Laboratory (NREL) [22]. The wind energy atlas of the United States [22] justifies the neutral boundary condition as a first approximation, because the wind speeds (4–25 m/s) at which much of the power is produced in turbines occur at neutral stability. Parameterization of boundary layer stability into wind

resource estimation is still a much researched area and we are working towards one such improvement.

The temporal resolution of the MERRA dataset is one hour, and as such, sub-hourly wind intermittency cannot be studied, even though this type of shorter scale intermittency can impact the voltage and frequency stability of a power grid [11]. Also, the MERRA data is created from the assimilation of observational data and satellite remote sensed data into a global model, and will reflect any imperfections of the model and the assimilation procedure, and will have an influence on the results presented here.

These limitations notwithstanding, we note that our data and results are not meant to be used for assessments of the deployability of wind farms at individual sites. Our wind resource construction is a tool to understand the geophysical nature of the resource at a regional scale and its variability, and the impact of large-scale atmospheric circulations and phenomena on the resource and its variability.

For this purpose, the multi-decade span of the MERRA data provides a more robust assessment of the temporal characteristics (i.e. mean, median, availability, intermittency, etc.) of wind power than that used in other studies. As described previously, while the data sets that exist have high spatial and temporal resolution, they do not have the record length required to assess the variability of the resource at the regional scale over longer time scales.

On the other hand, the constructed wind resource data described here uses a much longer record length, and this will allow future studies to utilize it to analyze the variability of the resource at different time scales (like the intra-seasonal and ENSO cycle time scales) and in response to different atmospheric oscillations like the El Nino Southern Oscillation and the Madden Julian Oscillation. This data will also be useful for analyzing the economic viability and the levelized costs of wind power compared to other energy sources, as well as for developing strategies for deployment such as the best pattern for aggregation. Studies such as this can conceivably delineate how far intermittency can be mitigated by aggregation and could play a role in the faster deployment of wind farms.

Supporting Information

Figure S1 An example of a histogram of wind power density that shows a typical skewed distribution.

Figure S2 Measures of variation. (a) the change in the RCoV from 50 m to 150 m, (b) the change in the IQR from 50 m to 150 m.

Figure S3 Measures of intermittency. (a) the change in the unavailability from 50 m to 150 m, (b) the change in the mean episode length from 50 m to 150 m.

Text S1 Rational for the cut-off employed to calculate the intermittency metrics.

Author Contributions

Conceived and designed the experiments: UBG AS. Performed the experiments: WH. Analyzed the data: WH UBG. Contributed reagents/materials/analysis tools: WH UBG. Wrote the paper: WH UBG.

References

1. Gentilli J (1971) The World Survey of Climatology. Volume 13: Climates of Australia and New Zealand. Amsterdam: Elsevier. 405 p.
2. Parkinson G (1986) Atlas of Australian Resources: Third series. Volume 4: Climate. Canberra: Division of National Mapping.
3. Mills D (2001) Assessing the potential contribution of renewable energy to electricity supply in Australia. PhD Thesis. Department of Geographical Sciences and Planning, University of Queensland.
4. Dear SJ (1991) Victorian Coastal Wind Atlas. Renewable Energy Authority Victoria and State Electricity Commission of Victoria.
5. Dear SJ, Lyons TJ, Bell MF (1990) Western Australian Wind Atlas. Minerals and Energy Research Institute of Western Australia.
6. Electricity Trust of South Australia (1989) South Australian Wind Energy Program 1984–1988. Technical Report. A joint project of ETSA, SADME and SENRAC. 83 p.
7. Blakers A, Crawford T, Diesendork M, Hill G, Outhred H (1991) The role of wind energy in reducing greenhouse gas emissions. Report R9317, Australian Government Department of Arts, Sport and Environment, Tourism and Territories (DASETT).
8. Coppin PA, Ayotte KA, Steggel N (2003) Wind Resource Assessment in Australia - A Planners Guide. Report by the Wind Energy Research Unit, CSIRO Land and Water. 96 p.
9. Gunturu UB, Schlosser CA (2012) Characterization of wind power resource in the United States. Atmos Chem Phys 12: 9687–9702.
10. Hennessey J Jr (1997) Some Aspects of Wind Power Statistics. Journal of Applied Meteorology 16: 119–128.
11. Gunturu UB, Schlosser CA (2011) Characterization of Wind Power Resource in the United States and its Intermittency. MIT Joint Program Report No. 209. 65 p.
12. Kahn E (1979) The Reliability of Distributed Wind Generators. Electric Power Systems Research 2: 1–14.
13. Archer CL, Jacobson MZ (2007) Supplying baseload power and reducing transmission requirements by interconnecting wind farms. Journal of Applied Meteorology and Climatology 46: 1701–1717.
14. Clean Energy Council (2012) Clean Energy Australia Report. Online report. Available: http://www.cleanenergycouncil.org.au/dms/cec/reports/2013/CleanEnergyAustraliaReport2012_web_final300513/Clean%20Energy%20Australia%20Report%202012.pdf. Accessed 2012 Jul 22.
15. Sinclair Knight Merz (2010) Renewable Resourceful Victoria, The renewable energy potential of Victoria, Part 2 - Energy Resources. Online report. Available: http://www.dpi.vic.gov.au/_data/assets/pdf_file/0006/38841/SKM-DPI-Renewable-Energy-Part2-v5_Part1.pdf. Accessed 2013 Jul 20.
16. Calif R, Schmitt F (2014) Multiscaling and joint multiscaling description of the atmospheric wind speed and the aggregate power output from a wind farm. Nonlinear Processes in Geophysics 21: 379–392.
17. Calif R, Schmitt F, Huang Y (2013) Multifractal description of the wind power fluctuations using arbitrary order Hilbert spectral analysis. Physica A: Statistical Mechanics and Its Applications 392(18): 4106–4120.
18. Oswald J, Raine M, Ashraf-Ball H (2008) Will British weather provide reliable electricity? Energy Policy 36(8): 3212–3225.
19. Mason R (2010) Wind farms produced 'practically no electricity' during Britain's cold snap. The Telegraph. Available: http://www.telegraph.co.uk/finance/newsbysector/energy/6957501/Wind-farms-produced-practically-no-electricity-during-Britains-cold-snap.html. Accessed 2013 Jan 7.
20. Pryor SC, Barthelmie RJ (2011) Assessing climate change impacts on the near-term stability of the wind energy resource over the United States. Proceedings of the National Academy of Sciences 108: 8167–8171.
21. Rienecker MM, Suarez MJ, Gelaro R, Todling R, Bacmeister J, et al. (2011) MERRA – NASA's Modern-Era Retrospective Analysis for Research and Applications, Journal of Climate 24: 3624–3648.
22. Elliott DL, Holladay CG, Barchet WR, Foote HP, Sandusky WF (1987) Wind Energy Resource Atlas of the United States. NASA STI/Recon Technical Report N, 87: 24819.
23. Elliott DL, Wendell LL, Gower GL (1991) An Assessment of the Available Windy Land Area and Wind Energy Potential in the Contiguous United States, Technical Report, Pacific Northwest Laboratories, Richland, WA, USA. 7306–7321.
24. Farkas Z (2011) Considering Air Density in Wind Power Production. arXiv preprint arXiv: 1103.2198. Available: http://arxiv.org/pdf/1103.2198.pdf. Accessed 2014 Apr 21.
25. Schwartz M, George R (1999) On the use of reanalysis data for wind resource assessment. National Renewable Energy Laboratory, Golden, Colorado. Presented at the 11th Applied Climatology Conference, American Meteorological Society, Dallas, Texas. January 10–15, 1999. Available: http://www.nrel.gov/docs/fy99osti/25610.pdf.
26. McElroy MB, Lu X, Nielson CP, Wang Y (2009) Potential for wind-generated electricity in China. Science 325: 1378–1380.
27. Lu X, McElroy MB, Kiviluoma J (2009) Global potential for wind-generated electricity. Proceedings of the National Academy of Sciences 106(27): 10933–10938.
28. Pryor SC, Schoof JT, Barthelmie RJ (2006) Winds of Change? Projections of Near-Surface Winds Under Climate Change Scenarios. Geophysical Research Letters, 33, L11702, doi:10.1029/2006GL026000.
29. Chadee XT, Clarke RM (2014) Large-scale wind energy potential of the Caribbean region using near-surface reanalysis data. Renewable and Sustainable Energy Reviews 30: 45–58.
30. Archer CL, Jacobson MZ (2003) Spatial and temporal distributions of US winds and wind power at 80 m derived from measurements. Journal of Geophysical Research 108. D9: 4289.
31. Archer CL, Jacobson MZ (2005) Evaluation of global wind power. Journal of Geophysical Research 110(D12): D12110.
32. Archer CL, Jacobson MZ (2007) Supplying baseload power and reducing transmission requirements by interconnecting wind farms. Journal of Applied Meteorology and Climatology 46(11): 1701–1717.
33. Clarke D (2013) Wind farms in New South Wales. Available: https://maps.google.com/maps/ms?ie = UTF&msa = 0&msid = 201184947181960624999.0004b83097a9d77f0c0f4&dg = feature. Accessed 2013 Dec 20.
34. Clarke D (2013) Wind farms in South Australia. Available: http://maps.google.com/maps/ms?ie = UTF&msa = 0&msid = 201184947181960624999.0004b827f26385521d5b3&dg = feature. Accessed 2013 Dec 20.
35. Wiktionary (2013) Available: http://en.wiktionary.org/wiki/anticoincidence.Accessed 2013 Sept 1.
36. Steggle N, Ayotte K, Davy R, Coppin P (2002) Wind prospecting in Australia with WINDSCAPE. Paris: Proceedings of Global Wind Power Conference.
37. Hurley PJ, Blockley A, Rayner K (2001) Verification of a prognostic meteorological and air pollution model for year-long predictions in the Kwinana industrial region of Western Australia. Atmospheric Environment 35: 1871–1880.
38. International Renewable Energy Agency (2012) IRENA Case Study 2013: Renewable Energy resource Mapping in Australia, International Renewable Energy Agency Report. 4 p.
39. Panofsky HA, Dutton JA (1984) Atmospheric turbulence: Models and methods for engineering applications. Wiley. 397 p.
40. Davy R, Coppin P (2003) South East Australia Wind Power Study. Report by the Wind Energy Research Unit, CSIRO Atmospheric Research. 24 p.
41. Geoscience Australia, ABARE (2010) Canberra: Australian Energy Resource Assessment. 358 p.

2

Resilience of Natural Gas Networks during Conflicts, Crises and Disruptions

Rui Carvalho[1]*, Lubos Buzna[2], Flavio Bono[3], Marcelo Masera[4], David K. Arrowsmith[1], Dirk Helbing[5,6]

1 School of Mathematical Sciences, Queen Mary University of London, London, United Kingdom, 2 University of Zilina, Univerzitna 8215/1, Zilina, Slovakia, 3 European Laboratory for Structural Assessment, Institute for the Protection and Security of the Citizen (IPSC), Joint Research Centre, Ispra(VA), Italy, 4 Energy Security Unit, Institute for Energy and Transport, Joint Research Centre, Petten, The Netherlands, 5 ETH Zurich, Zurich, Switzerland, 6 Risk Center, ETH Zurich, Swiss Federal Institute of Technology, Zurich, Switzerland

Abstract

Human conflict, geopolitical crises, terrorist attacks, and natural disasters can turn large parts of energy distribution networks offline. Europe's current gas supply network is largely dependent on deliveries from Russia and North Africa, creating vulnerabilities to social and political instabilities. During crises, less delivery may mean greater congestion, as the pipeline network is used in ways it has not been designed for. Given the importance of the security of natural gas supply, we develop a model to handle network congestion on various geographical scales. We offer a resilient response strategy to energy shortages and quantify its effectiveness for a variety of relevant scenarios. In essence, Europe's gas supply can be made robust even to major supply disruptions, if a fair distribution strategy is applied.

Editor: Matjaž Perc, University of Maribor, Slovenia

Funding: This work was supported by the Engineering and Physical Sciences Research Council under grant number EP/H04812X/1, by APVV (project APVV-0760-11) and by VEGA (project 1/0339/13). The funders had no role in study design, data collection and analysis, decision to publish, or preparation of the manuscript.

Competing Interests: The authors have declared that no competing interests exist.

* E-mail: r.carvalho@qmul.ac.uk

Introduction

Almost everything we do in the course of a day involves the use of energy. Yet, history has taught us that the threats to the security of supply come in unexpected ways [1,2]. Examples of unforeseen energy crises include the recent disputes between Russia and Ukraine over the price of natural gas (2005–2006, 2007–2008, 2008–2009) [3], the disruption of the oil and gas production industry in the US following Hurricanes Katrina and Rita (2005) [4], the terrorist attack on the Amenas gas plant that affected more than 10% of Algerian production of natural gas (2013) [5], and the supply shortage in March 2013, when the UK had only 6 hours worth of gas left in storage as a buffer [6]. New vulnerabilities could come from cyber attacks to the infrastructure [1], particularly in the case of state-driven attacks [2]; be the result of prolonged uncertainty or inaction on energy security in the US or Europe [7]; or derive from an extended period of extremely volatile prices due to intense international conflict [2].

Natural gas, a fossil fuel that accounts for 24% of energy consumption in OECD-Europe [8], has been at the heart of these crises. Gas is expensive to transport, and this is done mainly over a pipeline network. The investments are large and are made with long-term horizons, often of decades, and the costs are covered by locking buyers into long-term contracts [9]. Moreover, current infrastructure investments in Europe still derive from a historical dependency on supply from Russia and North Africa [10]. This dependency leaves the European continent exposed to both a pipeline network that was not designed to transport large quantities of gas imported via Liquefied Natural Gas (LNG) terminals, and to the effects of political and social instabilities in countries that are heavily dependent either on the export of natural gas (e.g., Algeria, Libya, Qatar or Russia) or its transit (e.g., Ukraine). Hence, it is challenging to build infrastructure that will be resilient to a wide range of possible crisis scenarios [11].

In a crisis, less delivery may mean greater congestion. This is due to the breakdown of major transit routes or production losses in affected areas, which cause the supply network to be used in different ways from what it was designed for. Hence, the available resources cannot be distributed well with the remaining transport capacities [12] [13,14]. This is why we need a method to handle congestion.

To manage the gas pipeline network during crises, we propose a decentralized model of congestion control that distributes the available network capacity to each route, without sacrificing network throughput [15–17]. A central controller makes the system vulnerable both to attacks on the control centre and to delays and failures of the lines of communication through the network. In contrast, a decentralized method is more resilient to failures because damage to the network has only a local effect and the need for communication is reduced [15,18]. To illustrate our model, we analyse the throughput of the present and planned pipeline networks across a range of different crisis scenarios at European, country and urban levels. The most challenging scenario corresponds to a hypothetical crisis with Russia with a complete cut-off of supply to Europe. We analyse how to alleviate the impact of such scenarios, by the identification of country groups with similar interests, which should cooperate closely to manage congestion on the network. This acknowledges that many of the 21st century challenges, such as the management of energy grids and infrastructure networks [19–21], cannot be solved by technology alone, but do have a relevant behavioural or social component [22–24].

Results

Data set and model

Our data set is organized in four layers (see "Databases" in File S1), three of which are shown in Figure 1. The first layer is the population density, which we compute from the 2012 Landscan global population data set. The second layer is the European gas pipeline network and Liquefied Natural Gas (LNG) terminals, which we extract from the Platts 2011 geospatial data set. This infrastructure is a spatial network, where nodes and links are geographically located, and links have capacity and length attributes. The third layer is defined by the urban areas in Europe with 100,000 or more inhabitants, and we compile it from the *European Environment Agency* and *Natural Earth*. The fourth layer is the network of annual movements of gas via pipelines and of Liquefied Natural Gas (LNG) via shipping routes (see Figure 2). We represent gas flowing from an exporting country m (including LNG) to an importing country n, by a directed network with weighted adjacency matrix T_{mn}.

Gas enters the network at source nodes, is transported over long distances on the pipeline transmission network, and then passed to the distribution network that delivers it to consumers. Here we model only the transport of gas on the transmission network. To model consumption spatially, we first need a tessellation of each country into disjoint sets of urban and non-urban areas, such that the pipeline network in an area is associated with the population it serves. Urban areas are naturally defined by the boundary of their spatial polygons. We partition non-urban areas by a Voronoi tessellation with the gas pipeline nodes as generators, respecting country borders and excluding all urban areas (see "The Model" in File S1).

We assume that the flow of gas on each pipeline intersecting an urban polygon (i.e., the border of the urban area) is directed towards the centre of the urban area. For simplicity, we also assume that such pipelines supply the urban area from the closest node to the urban polygon that is located inside the urban area. Moreover, each non-urban area is defined by a Voronoi cell, and

we assume that it is supplied by the cell generator node (see "The Model" in File S1).

To connect sink to source nodes with paths (see Contract Paths in Methods), we first go through each non-zero entry in the T_{mn} transport matrix and link each sink node in an importing country n to the $\Phi_{mn} = \min(10, s_m)$ closest nodes in an exporting country m, if m is a country, or to all LNG terminals in country n, if m is LNG, where s_m is the number of gas pipeline nodes in an exporting country m.

To allocate demand to individual paths, we start with the assumption that the demand T_{mn} of an importing country n from an exporting country m is distributed proportionally to the population of country n [25]. We next split the demand T_{mn} among all source to sink paths between countries m and n, proportionally to the population served by each sink node. We now have a value of demand associated with each path, and therefore with each sink node. Finally, we replace each path by a set of identical paths, each having the minimum demand on the network. This implies that all paths have the same demand, while doubling the demand on a path is equivalent to creating two identical paths with the original demand (see "The Model" in File S1).

To begin integrating routing and congestion control, we first consider how to distribute the capacity c_i of one single congested link over the $b_i = \sum_{j=1}^{\rho} B_{ij}$ paths that pass through the link, where B is the link-path incidence matrix ($B_{ij} = 1$ if link i belongs to the path r_j and $B_{ij} = 0$ otherwise), and where ρ is the number of paths on the network (see Table 1 of the File S1). To find the exact routing for these paths, we apply an iterative algorithm that, for each source-sink pair, finds the path with minimum effective path length, where the effective link length is given by $\tilde{l}_i = (\langle h_i \rangle / h_i)^{\alpha} l_i$, l_i is the length of link i, $h_i = c_i/(1 + b_i)$, and $\alpha = 0.03$ (see "The Model" in File S1).

We consider two baseline scenarios: the present and future networks. The present baseline scenario is the network that has been operational since 2011; the future baseline scenario extends the present network by the planned and under construction pipelines. To determine the network effects of crises, we analyse a range of scenarios that consist in hypothetically removing exporting (e.g., Russia) or transit (e.g., Ukraine) countries from the baseline scenarios. The scenarios are, thus, identified by the baseline (present or future) and the hypothetically removed country. For example, the present Russia scenario is given by the present network excluding Russia, that is removing all entries in the transport matrix T_{mn} that are movements of gas originating in Russia. Similarly, the future Ukraine scenario is determined by removing all Ukrainian nodes and links from the future network.

Broadly, there are three strategies to manage congestion [26]. First, expanding the network capacity is the most obvious way to lower congestion. The EU has a plan to build major pipelines crossing the continent, that should lower European dependency on Russia (see planned pipelines in Figure S1 of the File S1). Here, we include these planned pipelines in the future scenarios, but make no suggestions for extra infrastructure because the costs of expanding network capacity are high, and thus our focus is on how to best manage the existing and planned network capacity. Second, implementing congestion pricing is a way to cap the consumption of heavy users that cause network bottlenecks. Finally, by identifying groups of importing countries that have similar dependencies on exporting countries, we map a vast number of consumers to a relatively small number of parties that may be able to cooperate during crises [27].

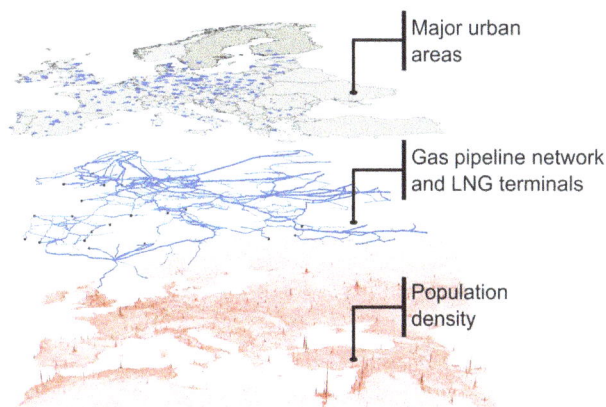

Figure 1. Spatial data layers involved in our analysis: population density (source: Landscan 2012); gas pipeline network and Liquefied Natural Gas (LNG) terminals (source: Platts 2011); and major urban areas (sources: European Environment Agency and Natural Earth). Map composed in ESRI ArcGIS.

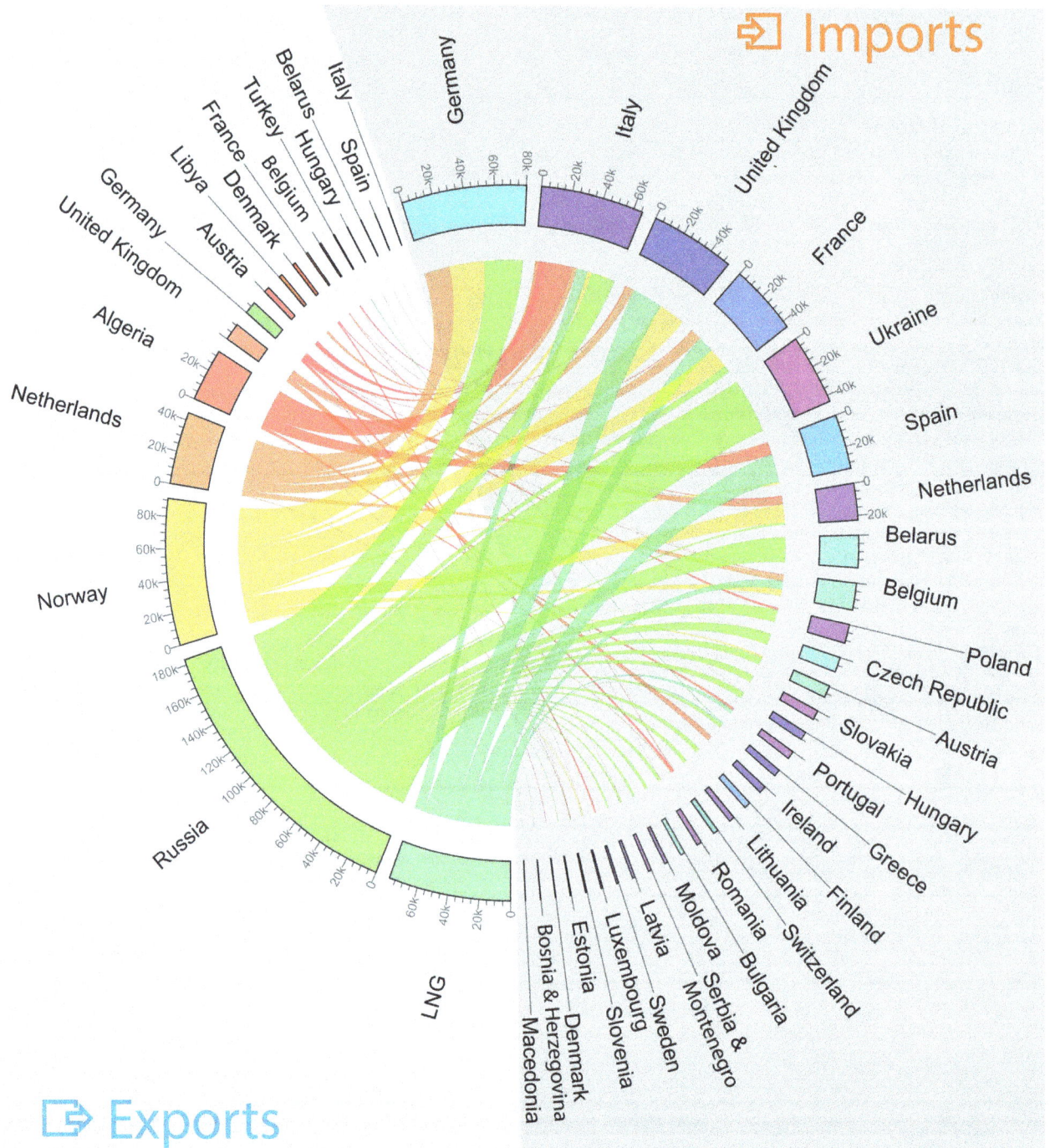

➲ Imports

➯ Exports

Figure 2. Natural gas imports by pipeline and via Liquefied Natural Gas (LNG) terminals in Europe during 2011 (million cubic meters). Gas exporting (importing) countries are on the left (right) of the image. For each exporting country, we show the breakdown of the volumes of gas exported annually, together with the importing countries served. For each importing country, we show the volumes of gas imported annually, together with the diversity of supply.

We are aiming at controlling congestion in situations where the network has to perform a function for which it was not designed. For congestion control, we are using the *proportional fairness* algorithm (see Methods), which is inspired by the way capacity is managed on the Internet [15,28,29]. This approach could be adjusted to other types of critical infrastructures, such as the power grid and road networks. The main idea behind proportional fairness is to use pricing on the links in order to control congestion

(see Methods and "Congestion Control" in File S1). Use of non-congested links is free up to a threshold, above which the cost that a path incurs for using a link increases linearly, but steeply, with the difference between link capacity and link utilization. Hence, paths that traverse many congested links pay a high cost for contributing to congestion, and thus get a smaller flow allocation than paths that avoid congestion. A flow is proportionally fair if, to increase a path flow by a percentage ε, we have to decrease a set of other path flows, such that the sum of the percentage decreases is larger or equal to ε. We view the network as an optimizer and the proportional fairness policy as a distributed solution to a global optimization problem [30,31].

Simulation Results

For each scenario, we hypothetically remove the scenario country from the network and, if m is an exporting country, remove row m in the T_{mn} transport matrix. Since the network topology and the flow network T_{mn} depend on the scenario, we then re-compute the source-sink pairs, the demand of each pair, and we also replace every source-sink path with a number of identical paths, each having the minimum demand in the network. Finally, we apply the proportional fairness congestion control algorithm to the resulting network and paths. We assume that all countries are willing to cooperate, that is, adhere to the rules of the congestion control policy. To assess the effect of the range of scenarios, we then analyse the throughput at the scales of the European continent, countries, and of urban areas.

We compute the global network throughput, which is the sum of the throughput at all sinks (urban and non-urban), for all the scenarios. Our model reproduces successfully the expected consequences of removing the major source and transit countries from the network: the largest decreases in global network throughput are caused by hypothetically removing Russia, Ukraine, the Netherlands, LNG and Norway (see Figure 3).

We say that a country is resilient to crises if it combines high throughput per capita across scenarios with a low coefficient of variation of throughput. In addition, the network is considered resilient to a scenario if the vectors of country throughput per capita for the scenario and the baseline scenario are similar. To start addressing the resilience of countries and the network to supply and transit crises, we study the signatures in the scenario space given by the country throughput per capita in each of the 20 scenarios. Similarly, a scenario can be seen as a point in the 32-dimensional space of country throughput. The heat-map in Figure 4A shows the throughput per capita for each pair of countries and scenarios [32].

The country groups, determined by dendrograms and highlighted in gray, reflect a similar level of throughput per capita achieved across the scenarios. Countries belong to the high throughput per capita groups (highlighted in dark gray in the figure) due to a combination of effects: diversity of supply; good access to network capacity (strategic geographical location); and a relatively small population (see "Results" in File S1). The coefficient of variation, shown in Figure 4B for present and future scenarios, measures the normalized dispersion of country throughput per capita using the mean as a measure of scale. Larger values indicate that the throughput accessible to a country varies across scenarios. Figure 4B shows that countries in Eastern Europe have high coefficient of variation of throughput per capita in the scenarios where we hypothetically remove Russia or Ukraine. In other words, countries in Eastern Europe are still very much dependent on one single source country (Russia) and one major transit country (Ukraine). Unexpectedly, we observe a spillover effect from countries, such as Germany, which make large investments in infrastructure. These countries themselves seem to benefit less from such investments than some of their smaller neighbours. The reason behind this spillover is that countries with plentiful access to network capacity provide routes for neighbouring countries to also access such capacity.

Figure 4A can be read from left to right: the scenarios that cause the largest disruption appear on the left, and the most benign scenarios are on the right. The present and future scenarios are

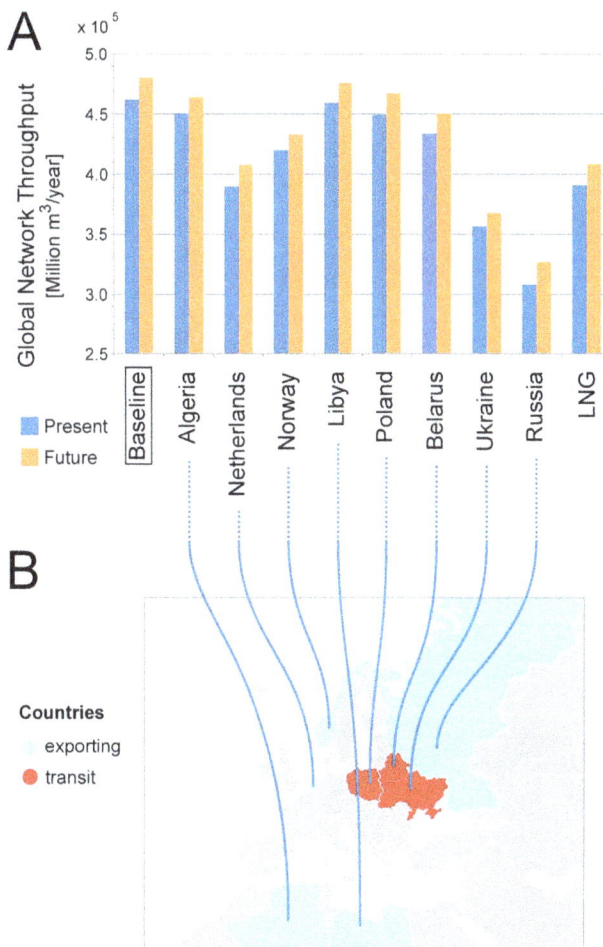

Figure 3. Global network throughput by scenario. (A) A scenario is named after the country that is hypothetically removed from the network, and coloured in blue (orange) if the country is removed from the present (future) baseline scenario. (B) The country removed per scenario is coloured cyan (red) on the map, if it is an exporting (transit) country. The total network throughput increases by 6.3% from the present baseline to the future baseline scenario (i.e., when the future and planned pipelines are added to the present network). The most challenging scenarios are the hypothetical removal of Russia, followed by Ukraine, the Netherlands and LNG. When Russia is removed from the network, the global network throughput falls by 32.7% relative to the present baseline and by 28.1% in relation to the future baseline. Figure created from authors' data with ESRI ArcGIS.

clustered together when either Russia, Ukraine, the Netherlands, or Belarus are removed from the network, demonstrating that the new pipelines being built will only improve slightly the consequences of a hypothetical crisis with one of the major exporting countries (Russia or the Netherlands), or with a critical transit country (Ukraine or Belarus). It is thus very hard to change the consequences of such scenarios even by building new pipelines.

We illustrate our model at a fine geographical scale in the heat-map of Figure 4C, where we show the throughput for urban areas in Europe with 1.5 million inhabitants or more, as the scenarios vary. The figure suggests possible classifications of cities into groups, highlighted in gray. We observe in Figure 4D that the coefficient of variation is larger for cities in Eastern Europe than for cities elsewhere (except Berlin, Vienna and Dusseldorf). Note that Dublin is resilient to all scenarios because it is supplied from the UK, which we never removed from the network. Observe also

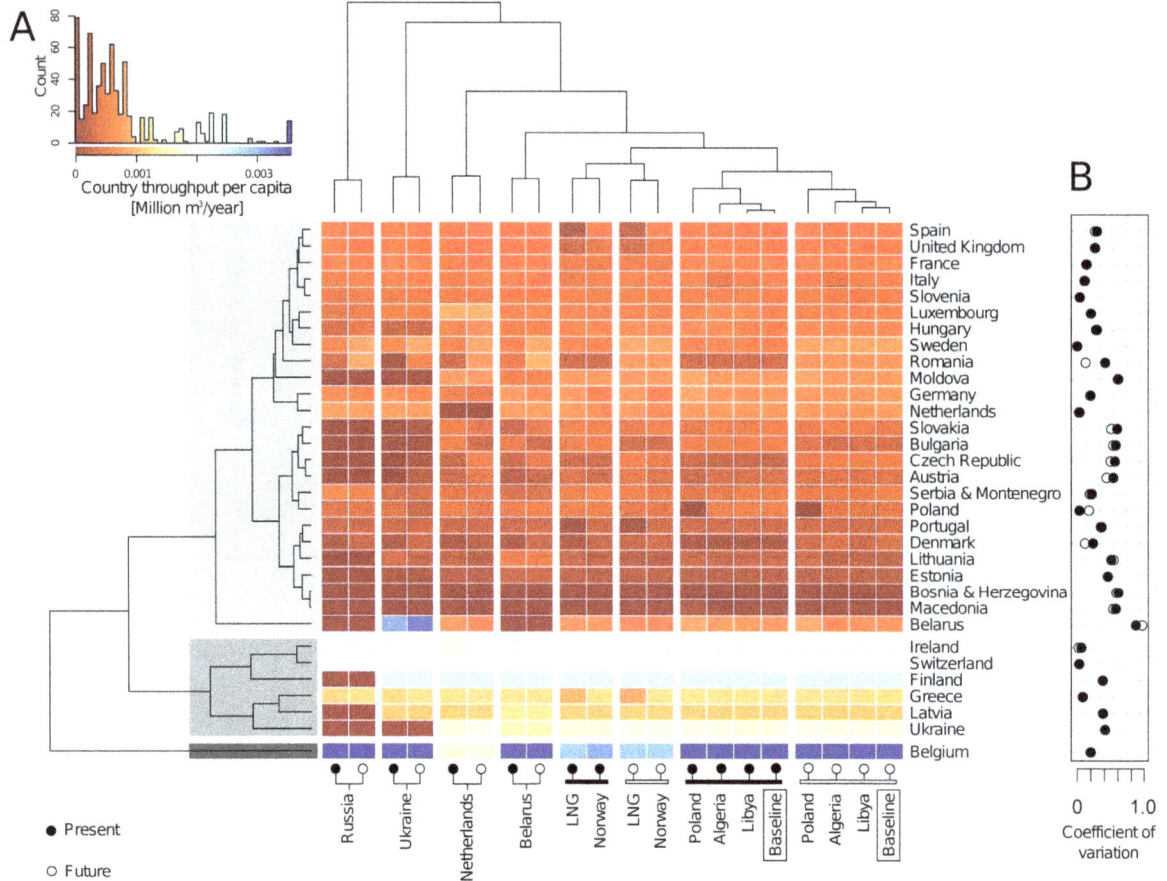

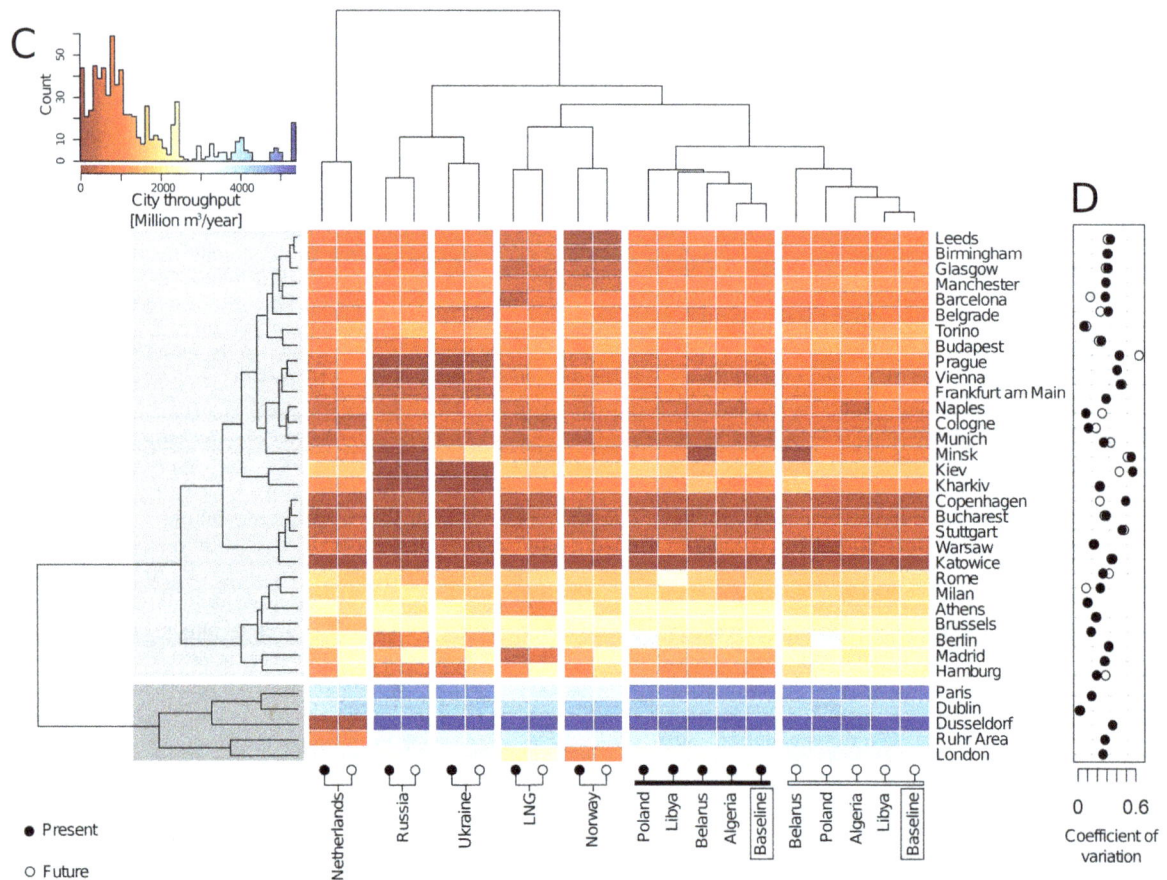

Figure 4. Heat-map [32], illustrating the variation of throughput across various scenarios and the effect of a scenario on the network. The dendrograms are computed using a hierarchical clustering algorithm with the Euclidean norm and average linkage clustering. (A) Heat-map of throughput at country level across various scenarios, allowing for a comparative analysis of the present versus future baseline scenarios, as well as of crises versus baseline scenarios; (B) Coefficient of variation of throughput per capita of a country; (C) Heat-map of throughput at urban level; (D) Coefficient of variation of throughput at urban scale. The gray areas denote groups of countries and urban areas that share common patterns of throughput across scenarios.

that Austria gets most of its gas from Russia, and only a little from Norway, so Vienna is in a similar situation to Eastern European cities.

Taken together, Figures 4A–D illustrate the resilience of countries, urban areas and the network to the scenarios, by showing how countries and urban areas with similar reactions to different types of crises are grouped together by throughput or by its coefficient of variation, and how different scenarios are clustered by their effect on the countries and urban areas.

The most challenging scenario is a hypothetical crisis that would cut-off supply from Russia to Europe. To investigate how Europe could make use of its internal gas production to minimize the impact of such a crisis, we simulate and quantify the effect of replacing gas supply from Russia with supply from Norway and the Netherlands. To do this, we start by creating a cut of European countries into two groups. Group I is made of the geographical cluster of countries that are heavily dependent on Russian gas, and is defined by Eastern Europe (http://eurovoc.europa.eu/100277) together with Estonia, Finland, Greece, Latvia and Lithuania. Group II is defined by all other countries in our study (see "Databases" in File S1). We consider a new scenario where Russia is removed from the network and the demand of countries in group I is rerouted to the Netherlands and Norway. To do this, we first create new paths linking each importing country in group I to Norway and the Netherlands (see "The Model" in File S1) and we update the matrix of gas flows to T'_{mn} (see Methods). Next, we apply a prefactor $0 \leq \beta \leq 1$ to the values of the demand T'_{mn} of countries in group II. The effect of β is to lower the utilization of the network by countries of group II that do not depend heavily on Russia. These countries typically have a high value of demand, and hence by curtailing their demand, there will be more capacity available to transport gas from Norway and the Netherlands to group I countries. In Figure 5, we observe that group I countries increase their access to network capacity as β decreases. Group II countries, such as Austria, that are geographically on the main routes that link Norway and the Netherlands to group I countries, decrease their throughput as β decreases. These countries are crucial: their throughput decreases as they share their network to benefit the more populous group I countries. In contrast, access to network capacity in routes supplying group II countries, such as Germany and Italy, is broadly unaffected, even as β is lowered considerably, because routes from Norway and the Netherlands to group I countries use little network capacity from these group II countries. Despite the increase in throughput for countries in group I as β decreases, Figure 5 shows the difficulty in replacing Russia by the Netherlands and Norway. Although we can hope to recover between 40 and 50% of the baseline throughput for the Czech Republic and Slovakia, we will only recover up to 5% of the Russian supply to Ukraine and up to 20% of the Austrian supply.

Discussion

Agreed political management processes are needed for crises scenarios, to guarantee supply to the most affected countries and urban areas and minimize the loss of gas by populations. Here, we propose a decentralized algorithm inspired by congestion control

on the Internet, which would eliminate the need of improvisation and complicated, lengthy negotiations every time a crisis occurs. Such mechanism has a stabilizing effect because it lowers the resource deficiency of the most affected countries [11,27]. We demonstrate how a wide range of scenarios impacts network throughput at global, country and urban levels, and how countries and urban areas react to scenarios of hypothetical crises. We show and quantify how countries that are heavily dependent on Russian supply can lower the impact of a crisis, if other countries accept to reduce their demand. Finally, our model tries to systematically compare alternative policy options during energy crises, using complex system models [33].

In summary, Europe is not necessarily trapped and helpless during energy crises. The long-term interest in the sustainability of the gas industry makes governments and the industry likely to invest in rules and norms to enhance reciprocity and collective efforts during crises. Because the number of governments and companies ultimately involved in taking the decisions in Europe is relatively high, governments could implement decentralized solutions similar to the one we propose here, perhaps with a centralized control solution as backup. At its heart, energy security, like preparedness for future pandemics [34], is about cooperation among nations [1]. To avoid European-wide crises, nations must cooperate to share access to their critical infrastructure networks.

Methods

Let $G = (V,E,c,l)$ be an undirected and connected weighted graph with no loops, node-set V and link-set $E = \{1, \ldots, \eta\}$. Each link i has a capacity c_i and a length l_i. The network has a set of ρ paths connecting source to sink nodes. All links of a path transport the same *path flow*. Different paths can share a link, even to perform transport in different directions (e.g., , during distinct time intervals).

The relationship between links and paths can be described by the *link-path incidence matrix* B as follows. Set $B_{ij} = 1$ if the link i belongs to the path r_j, and set $B_{ij} = 0$ otherwise. Matrix B has dimensions $\eta \times \rho$, and maps paths to the links contained in these paths. When B is applied to a vector of path flows, the resulting vector with components $(Bf)_i = \sum_{j=1}^{\rho} B_{ij} f_j$ is the total flow on the links, or link throughput. We say that a link is a *bottleneck* if the sum of the path flows of paths that pass through it is equal to the link capacity. We assume that flows are elastic, that is path flows are not fixed by demand, but can be adjusted according to the available network capacity.

Contract paths

The pipeline contracts are for physical point-to-point transport on a given system over a *contract path* [35,36]. The contract path is a route between a pair of source and sink nodes, such that gas flows from source to sink along that path and the transport costs are only incurred on links along that route.

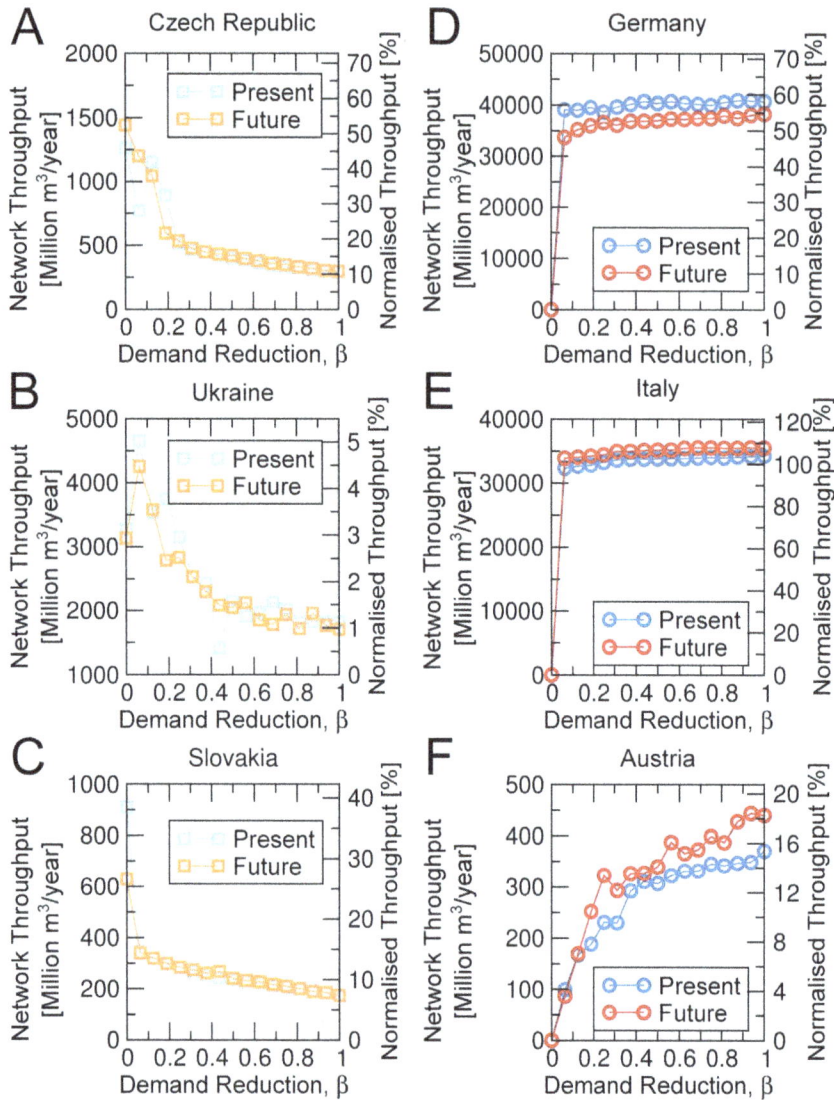

Figure 5. Network throughput of selected countries in a hypothetical crisis with Russia. The right axis shows the country throughput relative to the present baseline scenario. To minimize the impact of the loss of Russian supply, we re-allocate paths that originate in Russia to Norway and the Netherlands (see Methods). We then partition countries into two groups: group I is composed of Eastern Europe (http://eurovoc.europa.eu/100277) together with Estonia, Finland, Greece, Latvia and Lithuania, and group II includes all other countries in our study. Group II countries have a demand of $\beta T'_{mn}$, where $0 \leq \beta \leq 1$. Panels (A)–(C) show the throughput for selected group I countries (open squares), whereas panels (D)–(F) illustrate the throughput for group II countries (open circles). Panels (A)–(C) demonstrate that countries in group I benefit from curtailing the demand of countries in group II. In contrast, panels (D)–(E) show that some countries in group II are largely unaffected even when their own demand is curtailed considerably. Finally, panel (F) demonstrates that supply to Austria is dominated by the demand reduction prefactor, β. Indeed, Austria is crossed by routes from Norway and the Netherlands to group I countries, and these routes get a higher allocation of available capacity as Austrian demand decreases (i.e., as β decreases).

Proportional fairness congestion control

The key to a decentralized algorithm for proportional fairness is to translate the optimization problem into an autonomous system of coupled differential equation, with a fixed point equivalent to the optimal solution of the optimization problem. To do this, we use the result that the stable fixed point of a system of differential equations is the maximum of the equations' Lyapunov function.

A primal algorithm. A decentralized algorithm for congestion control (see "Congestion Control" in File S1) solves the system of coupled ODEs:

$$\frac{d}{dt}f_j(t) = 1 - f_j(t)\sum_{i=1}^{\eta} B_{ij}\mu_i(t), \tag{1}$$

where the price on link i is

$$\mu_i(t) = p_i\left(\sum_{j=1}^{\rho} B_{ij}f_j(t)\right), \tag{2}$$

and the price function is given by

$$p_i(y) = \frac{\max(0, y - c_i + \epsilon)}{\epsilon^2}. \qquad (3)$$

A dual algorithm. Consider a system where the shadow prices vary gradually as a function of the path flows (see "Congestion Control" in File S1):

$$\frac{d}{dt}\mu_i(t) = \sum_{j=1}^{\rho} B_{ij}f_j(t) - q_i(\mu_i(t)), \qquad (4)$$

where

$$f_j(t) = \frac{1}{\sum_{i=1}^{\eta} B_{ij}\mu_i(t)}, \qquad (5)$$

and $q(\cdot)$ is the inverse of $p(\cdot)$. As $\epsilon \to 0$, the dual and primal algorithms become equivalent. The rate of convergence to the stable point is a function of the link-path incidence matrix B and of the derivatives of p_i (primal algorithm) and q_i (dual algorithm), and increases with the magnitude of the latter [15].

Rerouting the demand from Russia to the Netherlands and Norway

When Russia is removed from the network, we reroute paths between group I countries and Russia to paths between group I countries and the Netherlands and Norway. To do this, we pair the new source and sink nodes as described in Section "The Model" in File S1, but we modify the T_{mn} matrix of gas flows. The new T'_{mn} matrix is found by reallocating the demand from Russia

for group I countries to the Netherlands and Norway, proportionally to the production of gas of these two exporting countries:

$$\begin{aligned} T'_{(NO)n} &= a_{NO}T_{(RU)n} \\ T'_{(NL)n} &= a_{NL}T_{(RU)n}, \end{aligned} \qquad (6)$$

where $a_{NO} = \dfrac{\sum_j T_{(NO)j}}{\sum_j T_{(NO)j} + T_{(NL)j}}$ and $a_{NL} = \dfrac{\sum_j T_{(NL)j}}{\sum_j T_{(NO)j} + T_{(NL)j}}$ are the normalised proportions of supply from Norway and the Netherlands, respectively.

Supporting Information

File S1 Supplementary information file which includes supplementary text, figures, and tables.

Acknowledgments

We thank ETHZ for granting access to the Brutus high-performance cluster. We also thank Tom De Groeve and Luca Vernaccini from the Crisis Monitoring and Response Technologies (CRITECH) Action group of the Global Security and Crisis Management Unit, Institute for the Protection and Security of the Citizen, Joint Research Centre for their support.

Author Contributions

Conceived and designed the experiments: RC LB FB MM DKA DH. Performed the experiments: RC LB. Analyzed the data: RC LB. Contributed reagents/materials/analysis tools: FB MM. Wrote the paper: RC LB DKA DH.

References

1. Yergin D (2012) The Quest: Energy, Security, and the Remaking of the Modern World. New York: Penguin Books.
2. Levi M (2013) The Power Surge: Energy, Opportunity, and the Battle for America's Future. New York: Oxford University Press.
3. (2009 January 5) The customary gas stand-off. The Economist.
4. Mouawad J (2005 September 29) Natural gas prices set record, pointing to costly winter. The New York Times.
5. Chrisafis A, Borger J, McCurry J, Macalister T (2013 January 25) Algeria hostage crisis: the full story of the kidnapping in the desert. The Guardian.
6. Plimmer G, Chazan G (2013 May 23). UK gas supply six hours from running out in March. Financial Times.
7. Schiermeier Q (2013) Germany's energy gamble. Nature 496: 156–158.
8. International Energy Agency (2012) Natural Gas Information. OECD/International Energy Agency.
9. Wright S (2012 July 14) Special report on natural gas. The Economist.
10. Moniz EJ (2011) The future of natural gas. Technical report, MIT.
11. Helbing D (2013) Globally networked risks and how to respond. Nature 497: 51–9.
12. Energy and Environmental Analysis (2005) Hurricane damage to natural gas infrastructure and its effect on the U.S. natural gas market. Technical report, Energy Foundation.
13. Voropai NI, Senderov SM, Edelev AV (2012) Detection of "bottlenecks" and ways to overcome emergency situations in gas transportation networks on the example of the european gas pipeline network. Energy 42: 3–9.
14. Lochner S (2011) Identification of congestion and valuation of transport infrastructures in the european natural gas market. Energy 36: 2483–2492.
15. Kelly FP, Maulloo AK, Tan DKH (1998) Rate control for communication networks: shadow prices, proportional fairness and stability. Journal of the Operational Research Society 49: 237–252.
16. Bertsimas D, Farias VF, Trichakis N (2011) The price of fairness. Operations Research 59: 17–31.
17. Carvalho R, Buzna L, Just W, Helbing D, Arrowsmith DK (2012) Fair sharing of resources in a supply network with constraints. Physical Review E 85: 046101.
18. Quattrociocchi W, Caldarelli G, Scala A (2013) Self-healing networks: redundancy and structure. arXiv:1305.3450
19. Brummitt CD, D'Souza RM, Leicht EA (2012) Suppressing cascades of load in interdependent networks. Proceedings of the National Academy of Sciences of the United States of America 109: E680–E689.

20. Buldyrev SV, Parshani R, Paul G, Stanley HE, Havlin S (2010) Catastrophic cascade of failures in interdependent networks. Nature 464: 1025–1028.
21. Havlin S, Kenett DY, Ben-Jacob E, Bunde A, Cohen R, et al. (2012) Challenges in network science: Applications to infrastructures, climate, social systems and economics. European Physical Journal-Special Topics 214: 273–293.
22. Clark WC, Dickson NM (2003) Sustainability science: The emerging research program. Proceedings of the National Academy of Sciences of the United States of America 100: 8059–8061.
23. Ajmone-Marsan M, Arrowsmith D, Breymann W, Fritz O, Masera M, et al. (2012) The emerging energy web. European Physical Journal-Special Topics 214: 547–569.
24. Levin S (2013) The mathematics of sustainability. Notices of the AMS 60: 392–393.
25. Bettencourt LMA, Lobo J, Helbing D, Kuehnert C, West GB (2007) Growth, innovation, scaling, and the pace of life in cities. Proceedings of the National Academy of Sciences of the United States of America 104: 7301–7306.
26. Frischmann BM (2012) Infrastructure: The Social Value of Shared Resources. New York: Oxford University Press.
27. Dietz T, Ostrom E, Stern PC (2003) The struggle to govern the commons. Science 302: 1907–1912.
28. Kelly F (1997) Charging and rate control for elastic traffic. European Transactions on Telecommunications 8: 33–37.
29. Srikant R (2004) The Mathematics of Internet Congestion Control. Boston: Birkhäuser.
30. Chiang M, Low SH, Calderbank AR, Doyle JC (2007) Layering as optimization decomposition: A mathematical theory of network architectures. Proceedings of the IEEE 95: 255–312.
31. Kelly F, Raina G (2011) Explicit Congestion Control: charging, fairness and admission management, in Next-Generation Internet: Architectures and Protocols (eds Byrav Ramamurthy, George N. Rouskas, and Krishna Moorthy Sivalingam). Cambridge University Press.
32. Eisen MB, Spellman PT, Brown PO, Botstein D (1998) Cluster analysis and display of genomewide expression patterns. Proceedings of the National Academy of Sciences of the United States of America 95: 14863–14868.
33. Lempert RJ (2002) A new decision sciences for complex systems. Proceedings of the National Academy of Sciences of the United States of America 99: 7309–7313.

34. Colizza V, Barrat A, Barthelemy M, Valleron AJ, Vespignani A (2007) Modeling the worldwide spread of pandemic inuenza: Baseline case and containment interventions. PLOS Medicine 4: 95–110.

35. Hunt S (2002) Making Competition Work in Electricity. Wiley Finance.

36. Kirschen DS, Strbac G (2004) Fundamentals of Power System Economics. Wiley-Blackwell.

3

Isolation and Evaluation of Oil-Producing Microalgae from Subtropical Coastal and Brackish Waters

David K. Y. Lim[1], Sourabh Garg[1], Matthew Timmins[1,2], Eugene S. B. Zhang[1], Skye R. Thomas-Hall[1], Holger Schuhmann[1], Yan Li[1], Peer M. Schenk[1]*

1 School of Agriculture and Food Sciences, The University of Queensland, Brisbane, Queensland, Australia, 2 ARC Centre of Excellence in Plant Energy Biology, Centre for Metabolomics, School of Chemistry and Biochemistry, The University of Western Australia, Crawley, Western Australia, Australia

Abstract

Microalgae have been widely reported as a promising source of biofuels, mainly based on their high areal productivity of biomass and lipids as triacylglycerides and the possibility for cultivation on non-arable land. The isolation and selection of suitable strains that are robust and display high growth and lipid accumulation rates is an important prerequisite for their successful cultivation as a bioenergy source, a process that can be compared to the initial selection and domestication of agricultural crops. We developed standard protocols for the isolation and cultivation for a range of marine and brackish microalgae. By comparing growth rates and lipid productivity, we assessed the potential of subtropical coastal and brackish microalgae for the production of biodiesel and other oil-based bioproducts. This study identified *Nannochloropsis* sp., *Dunaniella salina* and new isolates of *Chlorella* sp. and *Tetraselmis* sp. as suitable candidates for a multiple-product algae crop. We conclude that subtropical coastal microalgae display a variety of fatty acid profiles that offer a wide scope for several oil-based bioproducts, including biodiesel and omega-3 fatty acids. A biorefinery approach for microalgae would make economical production more feasible but challenges remain for efficient harvesting and extraction processes for some species.

Editor: Jonathan H. Badger, J. Craig Venter Institute, United States of America

Funding: The authors wish to thank the Australian Research Council (LP0883380, LP0990558), Pacific Seeds, Advanta India, North Queensland & Pacific Biodiesel and Queensland Sea Scallops Trading for financial support of this work. The funders had no role in study design, data collection and analysis, decision to publish, or preparation of the manuscript.

Competing Interests: Funding has been received from the Australian Research Council, Pacific Seeds, Advanta India, North Queensland & Pacific Biodiesel and Queensland Sea Scallops Trading. There are no patents, products in development or marketed products to declare.

* E-mail: p.schenk@uq.edu.au

Introduction

Interest in a renewable source of biofuels has recently intensified due to the increasing cost of petroleum-based fuel and the dangers of rising atmospheric CO_2 levels. Among the various candidates for biofuel crops, photosynthetic microalgae have the advantage that they have high growth rates and can be cultured on non-arable land [1,2,3].

At present, microalgae are commercially grown at scale for fatty acid-derived nutraceuticals and as feed and food supply. Significant interest in microalgae for oil production is based on their ability to efficiently convert solar energy into triacylglycerides (TAGs), which can be converted to biodiesel via transesterification reactions [1,4,5]. Oleaginous microalgae are capable of accumulating 20–50% of their dry cell weight as TAGs and potentially have a productivity superior to terrestrial crops used as first generation biofuel feedstock [6]. Theoretical calculations of microalgal oil production (liter/ha) are 10 to 100-fold greater than traditional biodiesel crops such as palm oil [7], corn and soybeans [6,8,9], although large-scale commercial algal oil production has yet to be established. Another major advantage of microalgae over higher plants as a fuel source is their environmental benefits. Despite having to grow in an aquatic medium, microalgae production may require less water than

terrestrial oleaginous crops and can make use of saline, brackish, and/or coastal seawater [10,11]. This allows the production of microalgae without competing for valuable natural resources such as arable land, biodiverse landscapes and freshwater. Furthermore, a microalgae-based biofuel industry has tremendous potential to capture CO_2. In high efficiency, large microalgae cultivation systems, the potential capture efficiency of CO_2 can be as high as 99% [12], effectively capturing 1.8 kg of CO_2 per kg of dry biomass [13]. Although CO_2 captured this way into biodiesel will eventually be released upon combustion, this would displace the emission of fossil CO_2 and the remaining biomass (e.g. ~70% of dry weight) can be fed into downstream carbon sequestration processes. For example, sequestering carbon into hard C-chips (Agri-char) via pyrolysis can be used to improve soil fertility, mitigating climate change by reintroducing durable carbon back into the soil [14], although it is debatable how long this carbon will actually stay in the soil.

Aside from biodiesel production, microalgae are gaining a reputation as ''biofactories'' due to the varied composition of their biomass. Akin to today's petroleum refinery, which produces a range of fuels and derivative products, a well-managed and equipped microalgal biorefinery can produce biodiesel and other value-add products such as protein, carbohydrates and a range of fatty acids (FAs). High value omega-3 fatty acids (ω-3) such as eicosapentaenoic (EPA), docosahexanoic (DHA), alpha-linolenic acid (ALA) and arachidonic (AA) are not desirable FAs for biodiesel production. Nevertheless, these ω-3 polyunsaturated fatty acids (PUFAs) are highly valued in human nutrition and therapeutics [15] and are linked to a wide range of cardio and circulatory benefits [16]. Ω-3 fatty acids also play an important role in aquaculture, increasing growth performance and reducing mortality in the shellfish industry [17,18,19]. This ability to produce value-adding products in addition to biodiesel is important to reduce production cost and make large-scale production viable.

The inherent advantages of a microalgal fuel source are unfortunately offset by current limitations to economically produce it on a large-scale. For example, the cost for obtaining dry biomass, large hexane requirements and limited hexane recycling capacity are currently hindering economic viability. It was estimated that the current cost of producing 1 tonne of microalgal biomass with an average 55% (w/w$_{DryWeight}$) oil content needs to be reduced by 10-fold in order to be competitive with petroleum diesel [8]. Furthermore, despite estimates that suggest microalgal oil production (US\$9–\$25/gallon in ponds, \$15–\$40 in photobioreactors) could be cheaper than the current price of oil [20], companies commercially producing microalgae have not been able to achieve the predicted yields and production costs. Typical lipid yields of 10 g m^{-2}d^{-1} (Skye Thomas-Hall, personal communication) are still short of achieving the current best case scenarios of 103 to 134 g m^{-2}d^{-1} [21]. The industry is still in its infancy, although recent research and development efforts by large oil companies (e.g. Exxon, BP, Chevron and Shell) would certainly increase production capacity and decrease production costs.

As large variations (10–50%) in lipid content exist between different species of microalgae [22,23], it is necessary to identify strains with high lipid content and suitable lipid composition. The need for high-yielding microalgae is straightforward, as this directly translates to an overall increase in production, although lipid production during normal growth needs to be distinguished from lipid accumulation in response to adverse conditions (e.g. nutrient starvation). Lipid composition is equally important, as quantitative and qualitative differences in the TAG content of a given species will affect the quality of biodiesel and its ability to meet fuel standards. Fuels with high cetane number fatty acids (e.g. myristic acid, palmitic acid, stearic acid) are desirable [24], as higher cetane fuels have better combustion quality and the right cetane number of biodiesel is required to meet an engine's cetane rating [25]. Microalgal lipids are mostly polyunsaturated, which have a low cetane number and are more prone to oxidation. This can create storage problems and are thus preferred to be at a minimum level for biodiesel production. Nevertheless, polyunsaturated fatty acids lower the cold filter plugging point (CFPP) of fuel and are crucial in colder climates to enable the biodiesel to perform at lower temperatures [3]. With these factors in mind, an "ideal composition" of fatty acids would consist of a mix of saturated and monounsaturated short chain fatty acids in order to have a very low oxidative potential whilst retaining a good CFPP rating and cetane number.

To date, research efforts have focused on lipid production of individual species, usually investigating the effects different growth conditions have on lipid production and content [26,27,28,29,30]. Unfortunately, direct comparisons of results between studies are unreliable, given the different growth conditions and experimental parameters of each species and also the different methods used for lipid extraction. There is growing interest to compare lipid content and FA composition of multiple microalgae species [11,31,32,33,34,35]. Several studies have revealed algae genera such as *Tetraselmis, Nannochloropsis* and *Isochrysis* to have highest high lipid content, particularly under nutrient-deprived conditions [11,31].

Nutrient deprivation is regarded as an efficient way to stimulate lipid production in microalgae in several microalgae species [11,29,36,37], especially saturated and monosaturated FAs [6,38,39]. Unfortunately, lipid accumulation is often associated with a reduction in biomass, which reduces overall lipid accumulation. A batch culture strategy can be adopted to obtain maximal biomass productivity as well as induction of lipid accumulation through nutrient deprivation. Although a common research practice, only Rodolfi et al. [11] have published lipid profiles of multiple microalgae species in a batch culture setting.

The target of our work was to identify the most effective microalgal TAG producers for biodiesel production using a basic batch culture strategy. Most studies utilize experimental designs that include aeration of media volumes of 1 L to 10 L in order identify microalgae strains with high lipid content [31,32,33,36,40]. To provide a direct comparison between different species, this study evaluated eleven microalgae strains collected from local Australian coastal waterways and other collections that originate in various places in the world. Strains were first characterized by microscopy and partial 18S ribosomal RNA sequencing and total fatty acid methyl ester (FAME) contents were then analyzed via GC/MS, which quantifies the fatty acids in triacylglycerides in each strain, thus providing the most accurate representation of the substrate available for biodiesel production. Using growth rate, FAME productivity and FA composition as criteria, this study identified several algae strains to be suitable for biodiesel, including *Tetraselmis* sp. and *Nannochloropsis* sp. as highly versatile candidate strains for a multiple-product algal biorefinery.

Materials and Methods

Microalgae strain collection and isolation

Microalgae were collected as 10 mL water samples from coastal rock pools, freshwater lakes and brackish (tidal) riverways. After initial cultivation of the mixed cultures with F medium [41] pure cultures were isolated by performing serial dilutions and the use of a micromanipulator (Leica DMIL with Micromanipulator). Strains *Chlorella* sp. BR2 and *Nannochloropsis* sp. BR2 originated from the same water sample and were collected from the Brisbane river (27°31'21"S 153°0'32"E; high tide at 10 am in August 2007 on a sunny day). Strain *Tetraselmis* sp. M8 was collected in an intertidal rock pool at Maroochydore (26°39'39"S 153°6'18"E; 12 pm on 6 August 2009). Additional, microalgae strains used in this study were obtained from the Australian National Algae Culture Collection (ANACC, CSIRO) and Queensland Sea Scallops Trading Pty Ltd (Bundaberg, Australia) (Table 1). All primary stock cultures were maintained aerobically in 100 mL Erlenmeyer flasks with constant orbital shaking (100 rpm) at 25°C, under a 12:12 h light/dark photoperiod of fluorescent white light (120 μmol photons m^{-2}s^{-1}). All cultures except *Chlorella* sp. were grown in seawater complemented with F medium [41]. *Chlorella* sp. was cultured in freshwater complemented with F medium. Primary stock cultures were sub-cultured every 3 weeks to minimize bacterial growth. Non-sterile cultures were used and maintained, as difficulties in maintaining axenic cultures in real production would arise and axenic cultures had been reported to have low biomass productivity, most likely because algae-associated bacteria may assist in nutrient recycling [42]. However, all microalgae cultures were checked during cell counting to ensure that no contamination with other microalgae occurred.

Table 1. Sources and 18S rRNA sequence accessions of microalgae strains used in this study.

Species	Genbank Accession	Location of Origin
Tetraselmis sp. M8	JQ423158	Maroochydore, Qld, Australia
Tetraselmis chui	JQ423150	East Lagoon, Galveston, TX, USA
Tetraselmis suecica	JQ423151	Brest, France
Nannochloropsis sp. BR2	JQ423160	Brisbane River, Brisbane, Australia
Dunaliella salina	JQ423154	Alice Springs, NT, Australia
Chaetoceros calcitrans	JQ423152	Unknown
Chaetoceros. muelleri	JQ423153	Oceanic Institute, Hawaii, USA
Pavlova salina	JQ423155	Sargasso Sea
Pavlova lutheri	JQ423159	Unknown location, UK
Isochrysis galbana	JQ423157	Unknown location, UK
Chlorella sp. BR2	JQ423156	Brisbane River, Brisbane, Australia

Standard protocol for batch culture growth analysis, lipid induction phase and sampling for lipid analysis

A standard protocol was designed to allow direct comparisons of growth rates and lipid productivity between cultures. To standardize inoculum cell densities, cultures were first grown to late logarithmic phase in F medium. Late-log phase of each culture was determined when daily cell count of the pre-culture revealed a less than 20% increase in cell density. A total of 1 mL of pre-culture in late-log phase was used as inoculum (7 to 9 hours after start of light cycle) for 20 mL seawater (SW) complemented with F medium in 100 mL Erlenmeyer flasks. A minimum of three parallel cultures were grown in conditions as described above. Cell counts were performed on days 0, 2, 4, 6 and 7 post inoculation using a haemocytometer. After day 7, nutrient deprivation to stimulate lipid production was achieved by removal of previous medium by centrifugation (1,200×g, 5 min) and replacement with only SW (without F medium). Cultures were then grown for another 48 h before 4 mL of wet biomass from each replicate was harvested for lipid analyses.

Fatty Acid Methyl Ester (FAME) analyses

Algae cultures (4 mL each) were centrifuged at 16,000× g for 3 min. The supernatant was discarded and lipids present in the algal pellet were hydrolyzed and methyl-esterified by shaking (1,200 rpm) with 300 µL of a 2% H_2SO_4/methanol solution for 2 h at 80°C; 50 µg of heneicosanoic acid (Sigma, USA) was added as internal standard to the pellet prior to the reaction. A total of 300 µL of 0.9% (w/v) NaCl and 300 µL of hexane was then added and the mixture was vortexed for 20 s. Phase separation was performed by centrifugation at 16,000× g for 3 min. A total of 1 µL of the hexane layer was injected splitless into an Agilent 6890 gas chromatograph coupled to a 5975 MSD mass spectrometer. A DB-Wax column (Agilent, 122–7032) was used with running conditions as described for Agilent's RTL DBWax method (Application note: 5988–5871EN). FAMEs were quantified by taking the ratio of the integral of each FAME's total ion current peak to that of the internal standard (50 µg). The molecular mass of each FAME was also factored into the equation. Identification of FAME was based on mass spectral profiles, comparison to standards, and expected retention time from Agilent's RTL DBWax method (Application note: 5988–5871EN).

DNA isolation and sequencing

Genomic DNA was isolated from all algal species via a phenol-chloroform method [43] on a pellet obtained by centrifugation of 10 mL of algal culture at the late-log phase. DNA amplification from genomic DNA containing a partial 18S ribosomal RNA region was performed by PCR using the following primers: Forward: 5′-GCGGTAATTCCAGCTCCAATAGC-3′ and Reverse: 5′-GACCATACTCCCCCCGGAACC-3′. Briefly, DNA was denatured at 94°C for 5 min and amplified by 30 cycles of denaturation at 95°C for 30 s, annealing at 58°C for 30 s, and extension at 72°C for 1 min. There was a final extension period at 72°C for 10 min prior to a 4°C hold. The PCR product was isolated using a Gel PCR Clean-Up Kit (Qiagen). For sequencing reactions, 25 ng of PCR product was used as template with 10 pmol of the above primers in separate reactions in a final volume of 12 µL. The samples were then sent to the Australian Genome Research Facility in Brisbane for sequencing. All new data has been deposited in GenBank (Table 1).

Identification of microalgae and phylogenetic analysis

Nucleotide sequences were obtained from the NCBI database based on the BLAST results of each algae sequenced in this study. When sequences from multiple isolates of a species were available, two nucleotide sequences were chosen: (i) highest max score sequence, (ii) highest max score sequence with identified genus and species. Strains Tetraselmis sp. M8, Chlorella sp. BR2 and Nannochloropsis sp. BR2 were isolated by the authors and other strains were obtained from the Australian National Algae Culture Collection (ANACC), CSIRO and Queensland Sea Scallops Trading Pty Ltd (QSST), Bundaberg (Table 1). In total, 22 sequences from the NCBI database and eleven sequences from algae in this study were aligned with the MAFTT [44]. The resulting alignment was then manually inspected for quality and the end gaps trimmed. Phylogenetic analyses of the sequences was performed with PhyML 3.0 [45] using the ML method. Default settings were used, with the exception that 100 bootstraps were used in a nonparametric bootstrap analysis instead of an approximate likelihood ratio test as this is the more commonly used method in recent reports.

Analytical methods

Measurement of nitrate and phosphate levels in the photo-bioreactor was performed using colorimetric assays (API, Aquar-

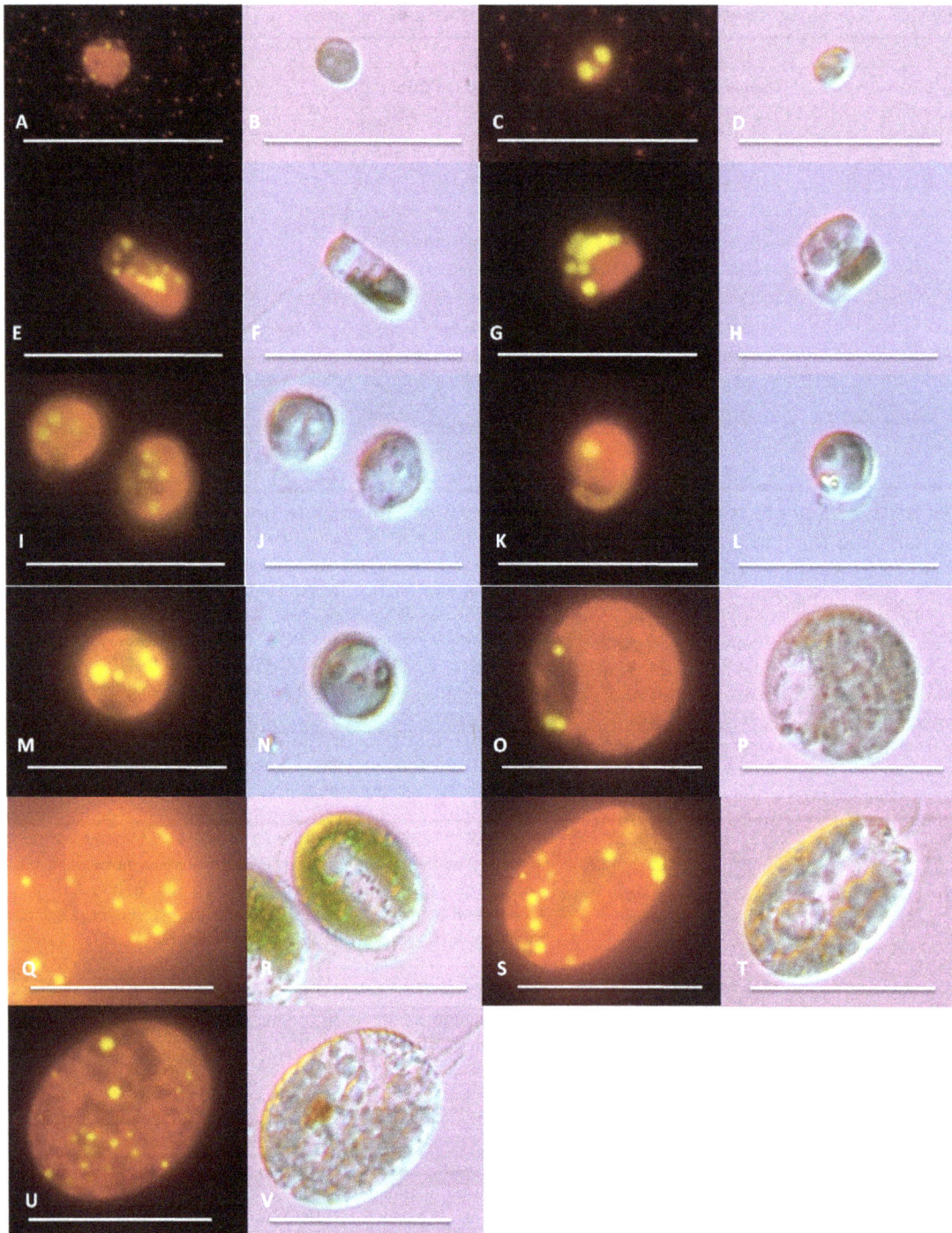

Figure 1. Epifluorescent (A, C, E, G, I, K, M, O, Q, S, U) and differential interference contrast (B, D, F, H, J, L, N, P, R, T, V) images of eleven microalgae used in this study. *Chlorella* sp. BR2 (A, B), *Nannochloropsis* sp. BR2 (C, D), *Chaetoceros muelleri* (E, F), *Chaetoceros calcitrans* (G, H), *Pavlova lutheri* (I, J), *Pavlova salina* (K, L), *Isochrysis* sp. (M, N), *Dunaliella salina* (O, P), *Tetraselmis chui* (Q, R), *Tetraselmis* sp. M8 (S, T) and *Tetraselmis suecica* (U, V). All images were taken at 100x magnification. Bars represent 20 μm.

ium Pharmaceuticals and Nutrafin, respectively). Growth rate, doubling time and lipid productivity were calculated as follows. The average growth rate was calculated using the equation $\mu = Ln(N_y/N_x)/(t_y-t_x)$ with N_y and N_x being the number of cells at the start (t_x) and end (t_y) of the growth phase (7 days). Average doubling time (T_{Ave}) was calculated using the equation $T = (t_y-t_x)/log_2 (N_y/N_x)$ over the growth period of 7 days. The specific growth rate (μ_{Max}) was calculated between the 2 days of maximum slope on the average cell density x-axis time plot [31,46]. Lipid productivity $(\mu g \ mL^{-1} \ day^{-1})$ was calculated as total lipid content $(\mu g/mL)$ over the duration of the entire batch culture (laboratory cultures – 9 days, outdoor culture – 12 days).

Microscopic analyses

After a lipid induction phase, microalgae cells were stained with 2 μg/mL Nile red (dissolved in acetone; Sigma, USA) for 15 minutes and photographed using a fluorescent Olympus BX61 microscope and an Olympus DP10 digital camera. Differential interference contrast (DIC) and epifluorescent (excitation: 510–550 nm, emission: 590 nm) images were obtained at 1000× magnification with oil immersion.

Mid-scale outdoor cultivation

In order to evaluate the growth performance and lipid productivity of microalgae in a medium-scale outdoor setting, *Tetraselmis* sp. was selected and tested in a 1000 L outdoor photobioreactor built by The University of Queensland's Algae Biotechnology Laboratory (www.algaebiotech.org) between 20th May 2011 to 1st June 2011 (sunny conditions 22°C–26.5°C). An initial cell density of 1.3×10^6/mL was cultured in SW + F/2 medium for 10 days (pH 8.8; maintained by the addition of CO_2) followed by 2 days of nutrient starvation (nitrogen measurements were 0 mg/L on day 10). Cell counts were conducted on days 0, 2, 4, 6, 7, 10, 11 and 12 and cultures were checked to ensure that no contamination with other microalgae occurred. To facilitate comparison with laboratory protocols, growth parameters were determined within the first 7 days of culture. At day 10, 4 mL of culture was sampled for lipid analysis.

Statistical analysis

Data for growth rates and lipid productivity was statistically analyzed by one-way analysis of variance (ANOVA) with different microalgae species as the source of variance and growth rate or lipid productivity as dependant variables. This was followed by Bonferroni's multiple comparisons test where appropriate.

Results

Strain collection, isolation and morphological and phylogenetic characterization of candidate microalgal biofuel strains

Over 200 water samples were collected from diverse aquatic habitats from subtropical regions in Queensland, Australia. These included samples from rock pools in coastal areas at the Sunshine Coast, Moreton Bay, Heron Island, Gold Coast and North Stradbroke Island, as well as freshwater samples from Somerset Dam, Wivenhoe Dam and brackish samples from tidal rivers, including the Brisbane and Logan rivers. Additional microalgae strains were obtained from culture collections at ANACC, CSIRO, and two local isolates from QSST, Bundaberg. Visual microscopy (Figure 1) confirmed the isolation of uniclonal cultures. Morphological comparisons to other described microalgae suggested that these strains belonged to the genera *Tetraselmis*, *Chlorella*, *Nannochloropsis*, *Dunaniella*, *Chaetoceros*, *Pavlova* and *Isochrysis*.

Nile red staining and growth analysis (Table 2, Figures 1) revealed eleven candidate strains that met the criteria required for biodiesel production (i.e. easy cultivation with no special nutrient requirements, fast growth rate, seawater-strength (35 ppt) salinity tolerance and high lipid production). One promising freshwater culture (*Chlorella* sp. BR2) was also included. Under nutrient-deprived conditions, lipids produced by microalgal cells were observed as bright yellow globules when stained with Nile red and viewed under epifluorescent light (Figure 1).

To specify the identity of the microalgae strains used in our experiments, a partial 18S region of the ribosomal RNA gene was amplified by PCR and sequenced. The obtained sequences were then compared to existing sequences in the NCBI database by the BLAST algorithm (for Genbank accession numbers see Table 1). Homology (sequence identity) searches confirmed a close relationship of the isolated candidate strains *Chlorella* sp. BR2, *Nannochloropsis* sp. BR2 and *Tetraselmis* sp. M8 with other members of the genera *Chlorella* and *Tetraselmis*. *Chlorella* sp. BR2 had a sequence identity of 99% with *Chlorella* sp. Y9, (Genbank Acc. No. JF950558) and *Chlorella vulgaris* CCAP 211/79 (Acc. No. FR865883). *Tetraselmis* sp. M8 shared a sequence identity of 99% with *Tetraselmis suecica* (CS-187) and *Tetraselmis chui* (CS-26). To characterize the diversity of the 11 microalgae strains and their relationship to other microalgae, the obtained sequences from this study were phylogenetically analyzed. The obtained maximum likelyhood phylogenetic tree (Figure 2) depicts the placement of each microalgae strain used in this study with chosen BLAST results.

BLAST 18S rRNA sequence comparison of eleven strains from this study to each other and the NCBI database (Figure 2) confirmed the taxonomic classification (suggested by microscopic studies or CSIRO/QSST) in all species based on the maximum score, while revealing high similarity within a species.

Comparison of growth rates, doubling times and cell densities of microalgae strains

To determine and compare growth rates, doubling times and cell densities, all microalgae strains were grown as three side-by-side cultures. After inoculation, an initial lag phase was observed in most cultures, except *Chorella* sp. BR2, *C. calcitrans*, *C. muelleri* and *I. galbana*, where exponential growth was observed immediately upon inoculation (Figures 3–4). Exponential growth in all cultures occurred till day 7 but for *D. salina*, *P. lutheri*, *Chlorella* sp. BR2 and *Nannochloropsis* sp. BR2, a lag phase was observed on day 4. *D. salina* culture remained in lag phase till day 7, while *P. lutheri*, *Chlorella* sp. BR2 and *Nannochloropsis* sp. BR2 resumed growth after day 6.

The highest average growth rate (μ_{ave}) was found for *P. lutheri* (0.48 μL^{-1}) and *P. salina* (0.45 μL^{-1}) (Table 2), that were significantly (p<0.05) higher to all other species that had a μ_{ave} of 0.34 μL^{-1}. Specific growth rates (μ_{exp}), were also compared with ANOVA, revealing that *T. chui* had the highest μ_{exp} at 1.03 μL^{-1}, followed by *Tetraselmis* sp. M8 (0.93 μL^{-1}) and *P. salina* (0.88 μL^{-1}). The fastest doubling times that were significantly different to the others were found for *P. lutheri* (1.45 days) and *Tetraselmis* sp. M8 (outdoor) (1.48 days) (Figure 3), while other microalgae strains had an average doubling time of 2.06 days. Maximum growth occurred during day 0 to day 4.

FAME productivity and fatty acid composition

GC/MS analysis revealed *Nannochloropsis* sp. (6.24 μg mL^{-1} day^{-1}) to be the highest FAME producer (ANOVA, P<0.05 in all cases), followed by *D. salina* (4.78 μg mL^{-1} day^{-1}; ANOVA, P<0.05 in all cases except *Chlorella* sp. BR2, 3.9 μg mL^{-1} day^{-1}) (Table 3; Figure 5). On the other hand, *T. chui* (1.5 μg mL^{-1} day^{-1}) and *T. suecica* (1.49 μg mL^{-1} day^{-1}) were the lowest FAME producers. The FA profile of *Nannochloropsis* sp. BR2, *C. calcitrans* and *C. muelleri* consisted predominantly of C16, C16:1 and C20:5 (>70% in total), while *Chaetoceros* strains produced C14 (10.5–11.6%). *Tetraselmis* sp. M8 contained most notably C18:3 (28.9%) and C16 (22.5%), as well as C18:2s (11.7%). *D. salina* and *Chlorella* sp. BR2's FA profile consisted mostly (nearly 90%) of C16, C18 and their unsaturated derivatives. In *T. chui* and *T. suecica*, C16 (35–37%), unsaturated C18s (37–43%) and unsaturated C20s (8–12%) were the main FAs. *I. galbana*'s FA profile was spread across C14 (19%), C16 (16%), C18:1 (22%), C20:3 (22%) and C20:6

Table 2. Growth rate analysis of eleven microalgae strains during growth phase (7 days) of batch culture.

Species	μ_{Ave}	μ_{Exp}	Day of μ_{Exp}	DT_{Ave} [days]	Cell density$_{Max}$ [$\times 10^6$ cells mL^{-1}]	Dry weight (g L^{-1})
Nannochloropsis sp. BR2	0.32	0.62$^{c, d}$	2–4	2.18^c	48.4	0.53
Tetraselmis sp. M8	0.35	0.93$^{a, b}$	2–4	2.00^c	2.07	0.75
T. chui	0.35	1.03^a	2–4	1.98^c	1.56	0.42
T. suecica	0.37	0.5^d	0–2	1.85$^{b, c}$	1.52	0.73
D. salina	0.30	0.76$^{a, b, c, d}$	2–4	2.31^c	2.14	0.37
C. calcitrans[1]	0.34	0.59$^{c, d}$	0–2	2.03^c	4.71	n/a
C. muelleri[1]	0.35	0.71$^{a, b, c, d}$	0–2	1.94$^{b, c}$	4.65	0.50
I. galbana[1]	0.35	0.61$^{b, c, d}$	0–2	1.96$^{b, c}$	4.45	0.45
P. lutheri[1]	0.48^a	0.76$^{a, b, c, d}$	0–2	1.45^a	3.95	0.45
P. salina	0.45^a	0.88$^{a, b, c}$	2–4	1.54$^{a, b}$	5.47	1.68
Chlorella sp. BR2	0.34	0.86$^{a, b, c}$	0–2	2.06^c	13.8	0.59
Tetraselmis sp.M8[3]	0.47	0.48	6–7	1.45	1.61	0.58

[1]Value represents mean of two replicate samples.
[2]Different letter superscripts down a column indicate significant difference at 95% level (ANOVA, Bonferroni's test; $P<0.05$).
[3]Mid-scale outdoor culture.

(12%). Approximately 44% of *P. salina*'s FAs consist of C14 and C16 FAs, with C20:5 and C22:6 FAs accounting for another 26%. *P. lutheri*'s FA profile consisted largely of C16 (25%), C16:1 (29%), C20:5 (22%) and C14 (11%).

On average, saturated FAs accounted for 40% of the total FAs in this study, consisting mostly of C16 (27.2%), C14 (7.2%) and C18 (6%). Similar amounts (37.4%) of FAs were polyunsaturated and included EPA C20:5 (9.6%), ALA C18:3 (10.4%) and DHA C22:6 (3.9%). Monounsaturated FAs accounted for 21% of the total FAs, consisting mostly of C16:1 (11.7%) and C18:1 (8.3%). *P. salina* was found to have the highest saturated FA (53%), *C. calcitrans* the highest monounsaturated FA (40%), and *D. salina* the highest polyunsaturated FA content (60%). C16 was found to be a major FA (17–37%) in all the strains tested, particularly in *T. chui*, *T. suecica* and *Nannochloropsis* sp. BR2. C16:1 FAs were predominantly found in *C. calcitrans*, *C. muelleri* and *Nannochloropsis* sp. BR2, while highest C14 content was found in *P. salina* and *I. galbana*. *I. galbana* also had the highest content of C18:1 FAs, while C18:3 FAs were predominantly found in *D. salina*, *Chlorella* sp. BR2 and *Tetraselmis* sp. M8. *Nannochloropsis* sp. BR2 and *P. lutheri* both had the highest content of EPA C20:5 FAs while DHA C22:6 was predominantly found in *P. salina*. *D. salina* was the only strain found to produce C16:4. It should be noted that due to the small culture volumes in this study certain fatty acids may have remained undetectable.

Outdoor scale-up

The highest lipid productivity for the microalgae strains tested in this study, was measured for *Nannochloropsis* sp. BR2 (Figure 5). However, based on its versatility and resourcefulness of fatty acids, its short doubling times, its ease of handling, and its potentially better lipid extraction efficiency, *Tetraselmis* sp. M8 was identified as a suitable candidate for large-scale cultivation whose FAME profiles would also meet the criteria for a future microalgae biorefinery. To compare laboratory cultivation with larger outdoor cultivation, *Tetraselmis* sp. M8 culture was grown in a 1000 L closed photobioreactor that was inoculated with 20 L of saturated culture. This mid-scale outdoor culture achieved a cell density of 1.6×10^6 cells mL^{-1} on day 7, eventually arriving at

2.3×10^6 cells mL^{-1} on day 10. Maximum growth rate was found between day 4 and 6 (Table 2) and was similar to average growth rates (0.47 μL^{-1} and 0.5 μL^{-1}, respectively). The culture entered stationary phase during starvation (after day 10), and cell count did not increase. The mid-scale, outdoor cultivation of *Tetraselmis* sp. M8 achieved a FAME productivity of 4.8 μL mL^{-1} day^{-1}, consisting mostly of C16 (20.8%), C18 (10.1%) and C18 unsaturated fatty acids (44.6%).

Discussion

In a microalgae-based oil industry, high oil productivity is crucial to achieving commercial feasibility. While growth conditions (e.g. solar radiation and temperature) and culture management are important, the suitable microorganism is fundamental to produce the desired quality and quantity of oil. A suitable microalgae strain must have high lipid productivity, either by possessing a high basal lipid content and/or be inducible to accumulate significant amounts of lipids. The selected strain should also be easily harvested, amenable to efficient oil extraction and flexible enough to adapt to changing physio-chemical conditions in an outdoor environment [11]. Thus, a locally isolated strain would likely adapt better to local changing environmental conditions and provide a more stable and productive culture.

Sampling at local waterways focused on inter-tidal rock pools, where the microclimate alters frequently between optimal growth conditions and unfavorable conditions (e.g. low nutrients, micro-oxic conditions, anaerobiosis, low/high light or dry, hot or cold conditions or rapid changes in salinity). Sampling at such locations was considered advantageous because suboptimal conditions would require the algae there to accumulate photo-assimilates such as starch or lipids that have important storage functions in order to survive, thereby increasing the chances of obtaining high lipid content strains [3]. This was followed by an isolation process targeted to select for high growth rate microalgae strains that could be induced to accumulate lipids under nutrient-deprived conditions. Isolation of uni-clonal microalgae strains by serial dilution and plating in F-supplemented medium was designed to

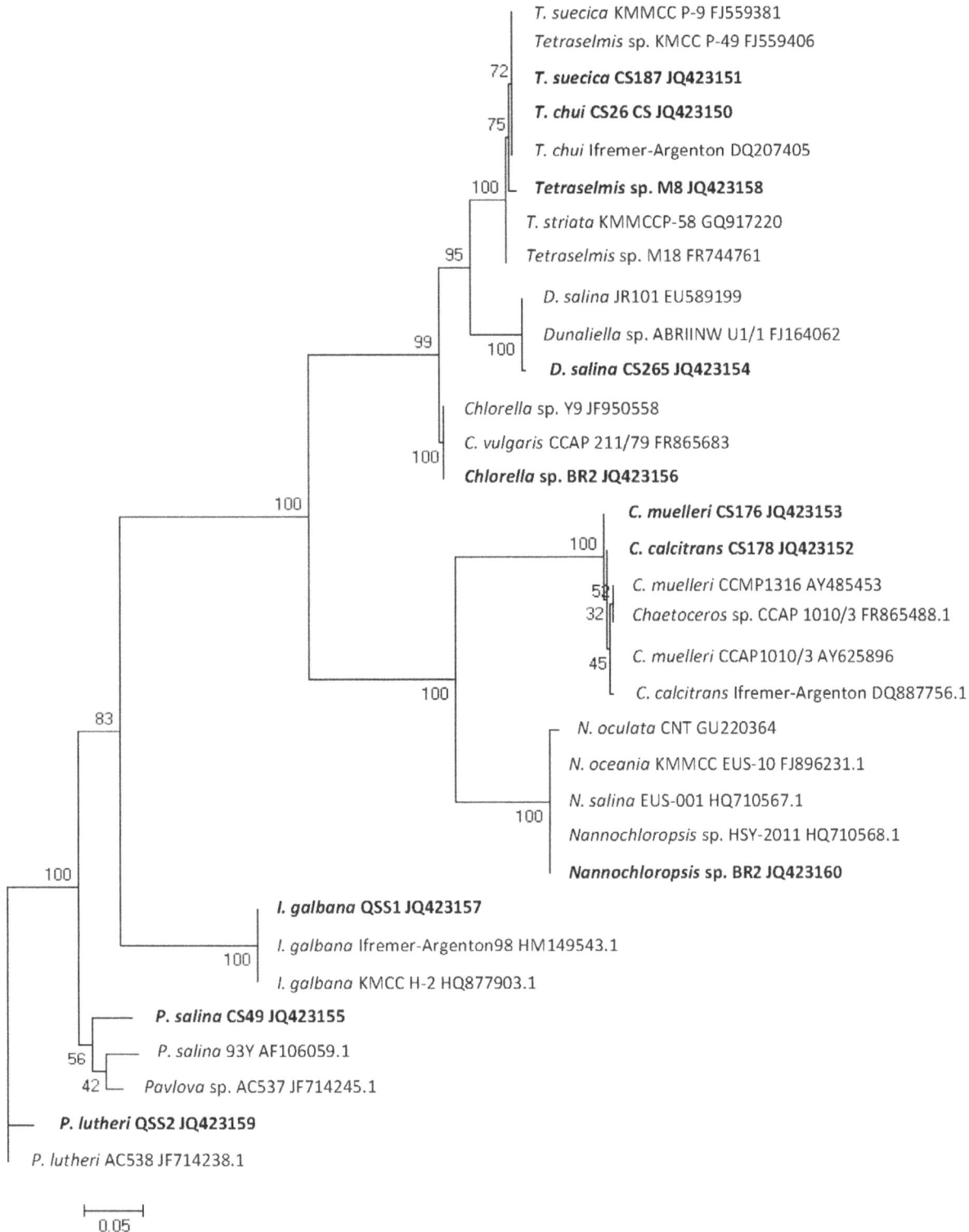

Figure 2. Maximum likelihood phylogenetic tree of 18S rRNA gene sequences from microalgae used in this study. Selected sequences from the NCBI database were also included (see Methods for selection criteria). Microalgae analyzed in this study are shown in bold. Numbers represent the results of 100 bootstrap replicates.

select strains which grew well in F/2 medium, a common nutrient mix used for microalgae culture [31,32,40,41]. Serial dilutions would also select for fast growing strains, which would inevitably dominate a culture. Special attention must be given to ensure that a single fast growing strain does not dominate other potentially high lipid content strains but that may have a slower growth rate. After 48 hours of nutrient deprivation, Nile red staining of the isolated uni-clonal cultures revealed several strains with substantial

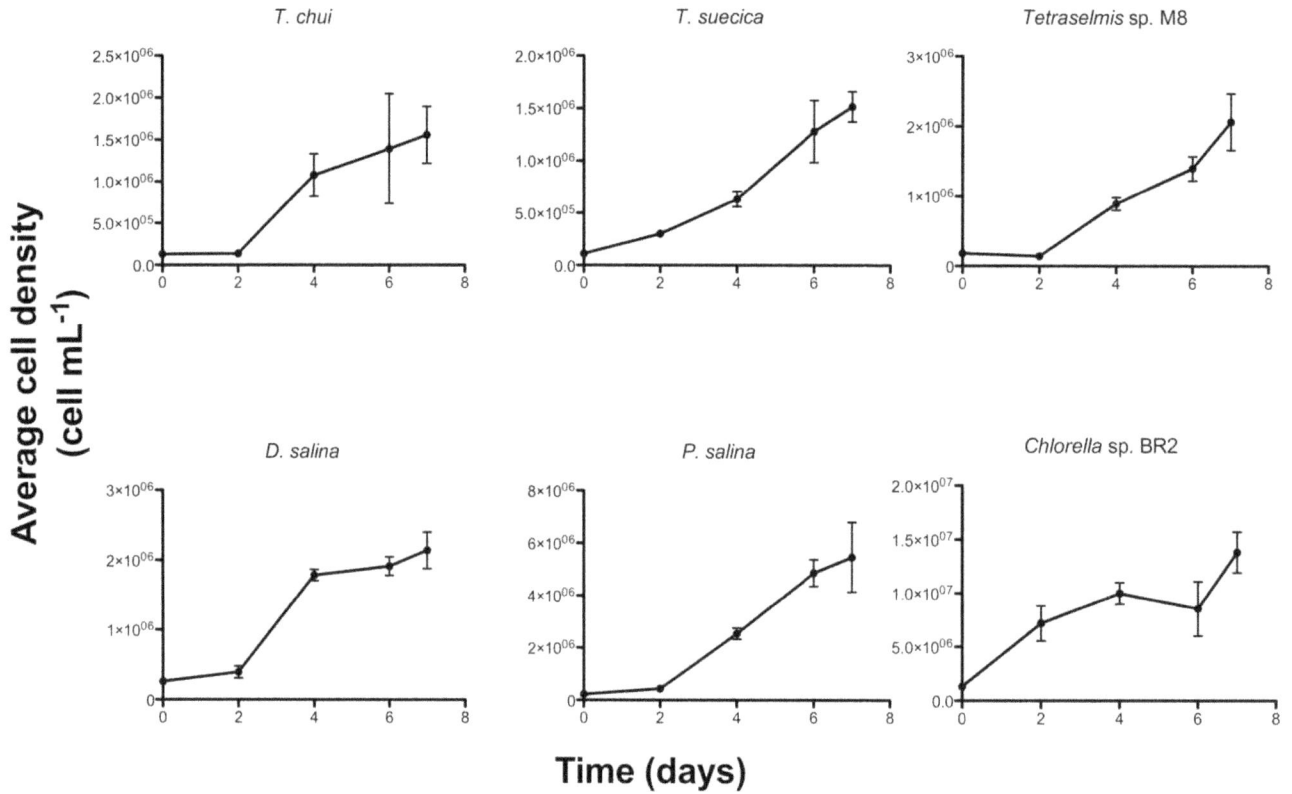

Figure 3. Growth curves of different microalgae in this study. *T. chui, T. suecica, Tetraselmis* sp. M8, *D. salina, P. salina* and *Chlorella* sp. BR2. Shown are average cell densities ± SD from three biological replicates.

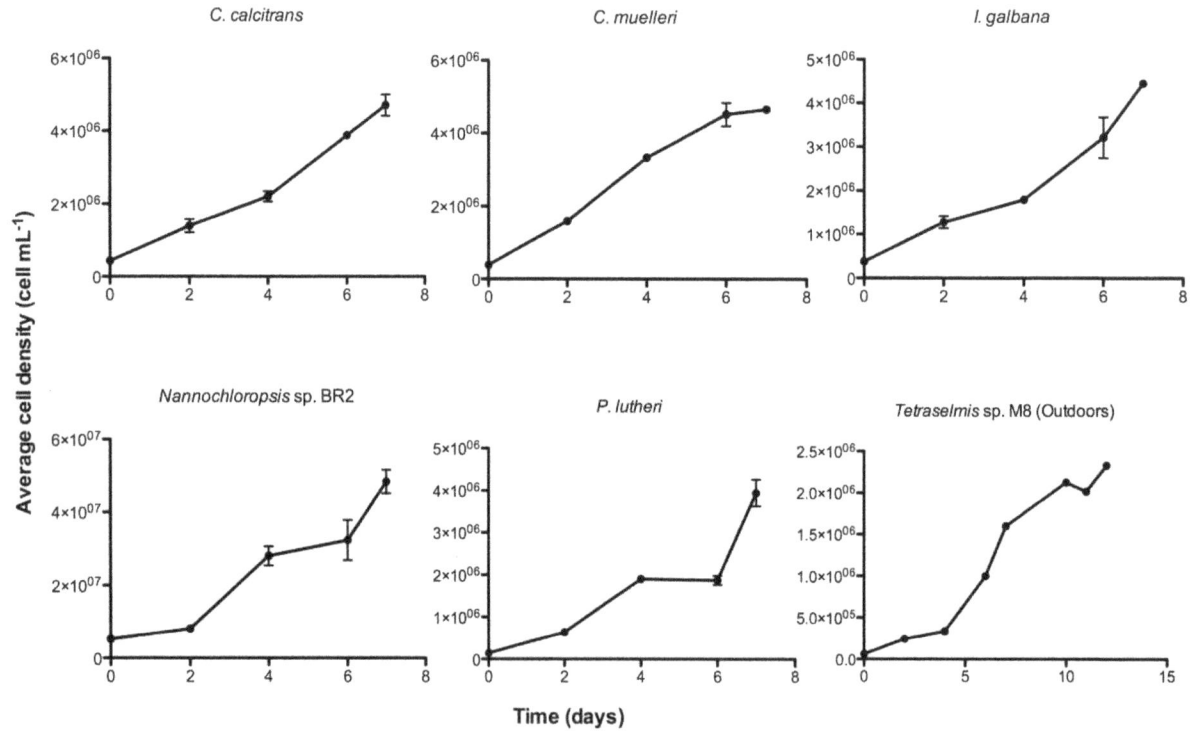

Figure 4. Growth curves of different microalgae in this study. *C. calcitrans, C. muelleri, I. galbana, Nannochloropsis* sp. BR2, *Chlorella* sp. BR2, *P. lutheri* & *Tetraselmis* sp. M8 (Outdoors). Shown are average cell densities ± SD from two biological replicates (3 replicates for *Nannochloropsis* sp. BR2 & 1 for *Tetraselmis* sp. M8 (Outdoors)).

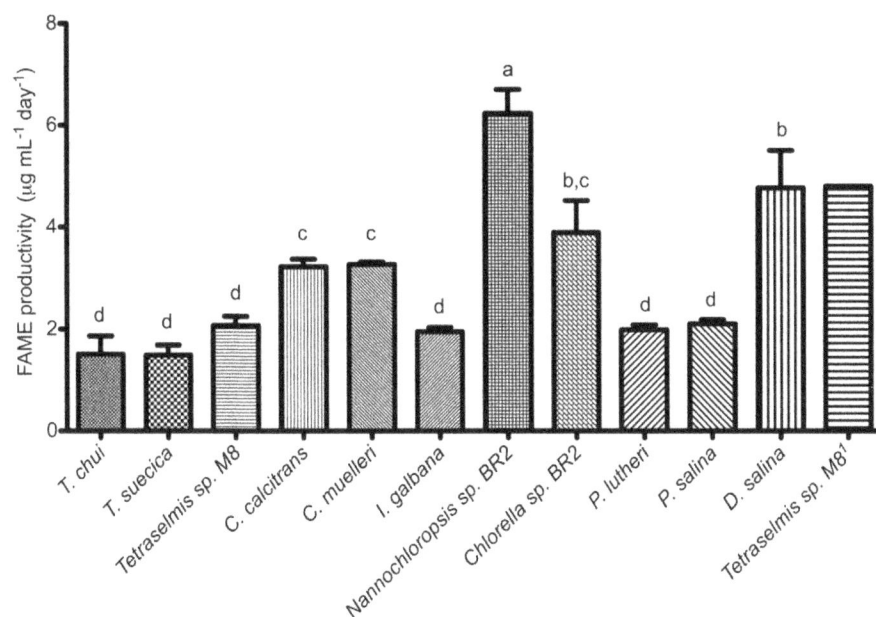

Figure 5. FAME levels of microalgae strains grown in batch culture (7 days growth + 2 days starvation by replacement of medium with seawater). Values shown are the averages of three biological replicates ± SD (except *Tetraselmis* sp.[1]). Different superscripts indicate significant difference at 95% level (ANOVA, Bonferroni's test; P<0.05). [1]Mid-scale outdoors culture.

lipid producing potential. An inherent problem with using Nile red staining was that differences in cell wall structure between species do not allow for equal staining and prevented accurate comparison of lipid productivity between species. For this reason some species with thick cell walls (e.g. some other *Nannochloropsis* species) that were not included in the subsequent analysis may still have a strong potential as future microalgae crops.

A standard protocol was established to identify the top FAME-producing microalgae strains by comparing the growth rates, FAME productivity and composition of the 11 microalgae strains in this study. Growth rate and FAME productivity data was then compared with other literature (Table 4). It is crucial that any comparison must take into consideration the different growth conditions, culture system and lipid analysis methods (available in Table S1). Both average growth rate (μ_{ave}) and specific growth rate (μ_{exp}) of the 11 analyzed microalgae strains were calculated from cell count growth curves (Figures 3–4). Overall, μ_{ave} found in the present study were similar or higher than μ_{ave} published by [36] and [34], aside from [32] which had nearly twice the μ_{ave} (Table 4). The specific growth rate (μ_{exp}) of microalgae is more widely reported in the literature, although many studies only present growth in biomass productivity [11,30,33,35,47]. Comparison with available literature revealed the present study's overall μ_{exp} to be higher than most, with the exception of microalgae from three publications [40,48,49]. The overall high growth rates of this study were observed despite a lack of culture conditions such as air bubbling, CO_2 supplementation and longer photoperiods available in other studies (Table 4; Supplementary Table S1). This could be a result of the increased nutrient availability from the F media in comparison with other studies that utilize F/2 media [31,34,36]. Increase in nutrient availability, particularly nitrogen has been documented to increase growth rate [29,30,50], particularly when the nitrogen source in F/2 media, KNO_3 is low (0.75 mM). A previous study on *Nannochloropsis* discovered light intensity to only have a slight effect on growth rates [47], especially during low cell densities (Skye Thomas-Hall, personal communi-

cation) and growth rate discrepancies may be due to differences in prior culture history [51]. Ultimately, *T. chui* and *Tetraselmis* sp. M8 were found to have the highest μ_{exp}. *Tetraselmis* strains were also the fastest growers in two other studies, [31] and [34]. The growth rate of *Nannochloropsis* sp. in this study was below average, contrary to findings by Huerlimann et al. [31]. FAME analysis by GC/MS revealed *Nannochloropsis* sp. BR2 to be the highest TAG producer, followed by *D. salina* and *Chlorella* sp. BR2. These three strains have been found to also be high lipid producers in other studies. Rodolfi et al. [11] compared the lipid productivity of 30 microalgae strains and found *Nannochloropsis oculata* and *Chlorella* amongst the best producers of lipids, both indoors and outdoors. Likewise, Huerlimann et al. [31] investigated the lipid content of five tropical microalgae and discovered *Nannochloropsis* sp. to be the highest lipid producer. A strain of *Chlorella* was similarly found to be a high lipid producer in an evaluation of ten microalgae strains for oil production [33]. Surprisingly, *Isochrysis* sp., a high lipid producing strain in other studies, [34] and [35], was found to have one of the lowest lipid production rates in this study. Likewise, *Tetraselmis* strains, top lipid producers in other studies, [31] and [11], produced the least amounts of lipids in this study.

Variations in species strains, growth conditions, experimental design and lipid extraction/analysis methods make quantitative comparisons of lipid productivity and FA content between studies very difficult (Supplementary Table S1). Nevertheless, when compared with Patil et al [35], who similarly analyzed FAME productivity by GC/MS, the total FAME/dry weight (%) of *Nannochloropsis* sp. BR2 and *Tetraselmis* sp. M8 was found to be higher, while *I. galbana* produced the same amount of FAME/dry weight. However, GC/MS obtained FAME productivity of this study was found to be lower than other sources (except for [37])(Table 4) that utilized solvent and gravimetric methods to measure total lipids. This was expected as solvent and gravimetric methods would include FFAs, TAGs and other lipid classes such as polar lipids (e.g. phospholipids and glycolipids) [6], wax esters [52], isoprenoid-type lipids, [53], sterols, hydrocarbons and

Table 3. Fatty acid composition in percentage of total FAME of different subtropical Australian microalgae strains after batch culture (7 days growth +2 days starvation).

Fatty acid	Nannochloropsis sp. BR2	T. chui	T. suecica	Tetraselmis sp. M8	D. salina	C. calcitrans	C. muelleri	Isochrysis sp.	P. lutheri	P. salina	Chlorella sp. BR2	Tetraselmis sp. M8 outdoor
C12	0.2	0.1	0.1	-	0.1	-	-	-	-	0.2	0.5	0.8
C14	3.5	0.9	0.9	0.4	0.6	10.5	11.6	19.2	11.4	19.4	0.9	4.2
C15	0.4	0.1	0.2	-	-	-	-	-	-	-	0.2	0.5
C16	33.0	37.3	35.2	22.5	24.7	23.3	26.2	16.4	25.0	24.8	30.9	20.8
C16:1	26.8	2.5	2.3	1.1	2.9	34.1	29.7	2.0	19.1	3.6	4.4	1.3
C16:2	0.4	-	-	5.0	2.5	1.5	2.7	0.9	3.1	-	3.4	-
C16:3	-	0.2	-	-	2.9	4.0	5.5	-	-	-	7.8	0.1
C16:4	-	-	-	-	11.6	-	1.8	-	-	-	-	-
C17	0.4	0.1	-	4.5	-	1.6	1.8	-	-	-	0.4	2.5
C18	3.0	9.0	8.8	3.0	5.8	5.1	4.5	4.4	4.8	8.3	9.7	10.1
C18:1	6.0	13.8	15.3	9.1	5.6	5.8	1.7	21.7	1.3	2.0	9.2	13.6
C18:2	0.9	8.8	19.7	11.7	7.6	0.1	0.2	0.7	-	1.1	7.9	7.0
C18:3	0.4	15.1	8.8	28.9	33.8	0.0	0.4	3.1	0.1	1.3	22.8	11.1
C18:4	-	-	-	-	-	-	-	-	-	6.1	-	12.7
C20	0.2	0.5	0.5	-	0.1	-	-	-	-	-	0.9	-
C20:1	-	1.8	2.1	-	0.1	-	-	5.9	0.1	0.4	0.8	4.6
C20:4	5.9	2.6	3.3	3.4	-	0.9	1.4	13.9	6.1	-	0.1	0.1
C20:5	18.8	7.2	2.9	10.6	1.2	12.7	14.0	0.0	21.8	16.1	-	10.6
C22	-	-	-	-	-	-	-	-	-	-	-	-
C22:4	-	-	-	-	-	-	-	-	-	6.3	-	-
C22:6	-	-	-	-	0.4	0.3	0.4	11.8	7.3	10.5	-	-
Total saturated (%)	40.7	47.9	45.6	30.4	31.4	40.5	44.0	39.9	41.1	53.0	43.6	38.9
Total monounsaturated (%)	32.8	18.2	19.7	10.2	8.6	40.0	31.4	29.6	20.5	5.5	14.4	19.5
Total polyunsaturated (%)	26.5	34.0	34.7	59.5	60.0	19.5	24.6	30.5	38.3	41.4	42.0	41.7
Total FAMEs (µg mL^{-1})	56.1	13.5	13.4	18.7	43.0	29.0	29.5	17.6	17.9	19.0	31.4	57.7
Total FAME/dry weight (%)	10.6	3.2	10.8	2.5	11.4	-	5.9	3.9	4.0	1.2	5.3	9.9

Table 4. Comparison of FAME productivity (μg mL^{-1} day^{-}1) of present study microalgae with lipid productivity of microalgae species from other references.

Species	Lipid productivity [μg mL^{-1} day^{-1}]	References
Nannochloropsis **sp. BR2**	6.2	This study[GCMS, AG]
Nannochloropsis sp.	4.6	Huerlimann et al. (2010)[12h]
Nannochloropsis sp.	48.0	Rodolfi et al. (2009)[24h, CO2]
Nannochloropsis sp.	37.6	Rodolfi et al. (2009)[24h, CO2]
Nannochloropsis sp.	60.9	Rodolfi et al. (2009)[24h, CO2]
Nannochloropsis oculata	10.0	Converti et al. (2009)[24h, CO2]
Tetraselmis **sp. M8**	2.1	This study[GCMS, AG]
Tetraselmis **sp. M8 (outdoor)**	4.8	This study[GCMS]
Tetraselmis sp.	18.6	Huerlimann et al. (2010)[12h]
Tetraselmis sp.	43.4	Rodolfi et al. (2009)[24h, CO2]
Tetraselmis sp.	10.7	Patil et al. (2007)[GCMS, 24h, CO2]
Tetraselmis chui	1.5	This study[GCMS, AG]
Tetraselmis chui	27.0	Rodolfi et al. (2009)[24h, CO2]
Tetraselmis suecica	1.5	This study[GCMS, AG]
Tetraselmis suecica	36.4	Rodolfi et al. (2009)[24h, CO2]
Dunaliella salina	4.8	This study[GCMS, AG]
Dunaliella salina	33.5	Takagi et al. (2006)
Chaetoceros muelleri	3.3	This study[GCMS, AG]
Chaetoceros muelleri	21.8	Rodolfi et al. (2009)[24h, CO2]
Chaetoceros calcitrans	3.2	This study[GCMS, AG]
Chaetoceros calcitrans	17.6	Rodolfi et al. (2009)[24h, CO2]
Chaetoceros sp.	16.8	Renaud et al. (2002)* [12h]
Isochrysis galbana	2.0	This study[GCMS, AG]
Isochrysis sp.	24.9	Renaud et al. (2002)* [12h]
Isochrysis sp.	12.7	Huerlimann et al. (2010)[12h]
Isochrysis sp.	37.7	Rodolfi et al. (2009)[24h, CO2]
I. galbana	12.4	Patil et al. (2007)[GCMS, 24h, CO2]
Pavlova lutheri	2.0	This study[GCMS, AG]
Pavlova lutheri	50.2	Rodolfi et al. (2009)[24h, CO2]
Pavlova salina	2.1	This study[GCMS, AG]
Pavlova salina	49.4	Rodolfi et al. (2009)[24h, CO2]
Pavlova sp.	21.7	Patil et al. (2007)[GCMS, 24h, CO2]
Chlorella **sp.**	3.9	This study[GCMS, AG]
Chlorella sp.	7.1	Chen et al. (2010)[AG]
Chlorella sp.	20.0	Converti et al. (2009)[24h, CO2]
Chlorella sp.	42.1	Rondolfi et al. (2009)[24h, CO2]
Chlorella sorokiana	44.7	Rondolfi et al. (2009)[24h, CO2]
Chlorella sorokiana	1.0	Illman et al. (2000)[24h, CO2]
Chlorella vulgaris	5.3	Illman et al. (2000)[24h, CO2]

*Calculated total lipid content (μg mL^{-1}).
[GCMS]Values obtained by GC/MS.
[24h]Cultures grown with 24 h light and air.
[12h]Cultures grown with 12h light and air.
[CO2]Cultures grown with air supplemented with CO_2.
[AG]Cultures grown with agitation.
For a full comparison of culturing conditions see Table S1.

pigments. Furthermore, different growth conditions in other studies such as growth enrichment with carbon dioxide [48,54], increased photoperiods and light intensity [55], different media volumes and larger initial inoculum would explain for the increased lipid productivity in other studies. This is most evident in the study by Rodolfi et al. [11], where similar strains of *P. salina*

CS-49 and *C. calcitrans* CS-178 were studied under different conditions to reveal significantly different results. It should be noted that the conditions of the current experimental design were not meant to achieve maximum lipid production but to determine the best lipid producing candidates under standard "unoptimized lab conditions", which were *Nannochloropsis* sp. BR2, *D. salina* and *Chlorella* sp. BR2. Higher confidence in the data may be obtained by growing cultures completely independently (i.e. experiments carried out separately at different times with a different culture). Subsequent studies may focus on the comparison of best strains under fully optimized and/or large-scale commercial conditions. In our study, *Tetraselmis* sp M8 was chosen for a scale-up study based on its fast growth rates, culture dominance and ease of harvesting by settling. A comparison of the indoor laboratory conditions to mid-scale (1000 L) outdoor conditions showed that lipid productivity more than doubled under these conditions. Although further long-term studies will be required, these preliminary findings demonstrate the potential for optimization and emphasize that outdoor and large-scale conditions differ strongly from laboratory conditions.

Suitable candidates for biodiesel production require not only high lipid productivity, but also suitable FA content. Recommended FAs for good biodiesel properties include C14:0, C16:0, C16:1, C18:0, C18:1, C18:2 and C18:3 [3,56]. In this study, analyses of FA profiles revealed *Nannochloropsis* sp. BR2, *Chlorella* sp. BR2 and *Chaetoceros* strains (*C. calcitrans* and *C. muelleri*) to be the best candidates (Table 3). In addition to having the highest lipid productivity, the recommended FAs for biodiesel accounted for 73.6% of the total FAs in *Nannochloropsis* sp. BR2, in particular C16 (33%) and C16:1 (26.8%). Huerlimann et al. [31] reported a similar FA composition of *Nannochloropsis* sp. following nutrient deprivation, while Patil et al. [35] also reported *Nannochloropsis* sp. to have the highest C16 and C16:1 content. *Chlorella* sp. BR2 presented slightly lower lipid productivity although having more desired FAs for biodiesel (81.4%). It also had a higher C18 (9.7%) and unsaturated C18 content (39.9%) if compared to *Nannochloropsis* sp. BR2 or the *Chaetoceros* strains; making it more desirable for the production of biodiesel with a higher cold filter plugging point (CFFP) for better performance at low temperatures [3]. Both *C. calcitrans* and *C. muelleri* are good candidates despite only having mediocre lipid productivity due to high levels of C14 FAs (10.5% and 11.6% respectively) and recommended FAs for biodiesel (78.9% and 74.5% respectively). The FA content of *C. calcitrans* was observed in accordance to Lee et al. [34] during low nitrogen conditions, which caused an increase in saturated FAs like C16. *D. salina* was not considered a suitable candidate for biodiesel despite its high lipid productivity due to high levels of PUFAs (C16:4 – 11.6%. C18:3 – 33.8%). Low levels of PUFAs, as evident in *Nannochloropsis* sp. and *C. calcitrans* are desired for biodiesel production as it reduces the need for treatments such as catalytic hydrogenation. *Nannochloropsis* sp. BR2, *C. calcitrans* and *C. muelleri* also exhibited C20:5 (EPA) (18.8%, 12.7% and 14% respectively) that would allow for a biorefinery approach to biodiesel production. It should be noted, however, that microalgal biodiesel is likely to be first used as a drop-in fuel in the future which would allow to achieve blends with the desired fuel properties from most microalgae species.

Commercially feasible production of microalgal biodiesel would require a biorefinery approach to produce biodiesel as well as other value-added products such ω-3 FAs and protein-rich biomass. Microalgae possess the potential to produce high amounts of ω-3 FAs such as EPA (C20:5) and DHA (C22:6) that are used as dietary supplements. The best candidates for EPA and DHA production in this study were found to be *Nannochloropsis* sp. BR2 and the *Pavlova* strains (*P. salina* and *P. lutheri*). Overall, *Nannochloropsis* sp. BR2 produced the highest amounts of ω-3 FAs on account of its high overall lipid and EPA content (18.8%). *P. lutheri* exhibited the highest proportional content of EPA (21.8%), while *Isochrysis* sp. had the highest DHA content (11.8%). The ω-3 FA contents of *Nannochloropsis* sp. and the *Pavlova* strains were comparable to previously published values [31,35,57].

The use of a nutrient starvation phase to improve TAG productivity (particular C16:0 and C16:1) for biodiesel production was successful as C16 and C16:1 FAs were found to be the predominant FAs in the present study. During nutrient limiting conditions, unsaturated FAs are consumed as an energy source and saturated FAs are accumulated [58]. The increase of the % of saturated and monounsaturated FAs during starvation have been well documented in literature for several other species [34,59,60]. While this may prove useful for biodiesel production, the reduction in PUFAs is a problem for ω-3 FA production that has been documented [31,34]. Nevertheless, EPA and DHA contents have been reported to remain consistent despite changes in nutrient level for *T. tetrathele* [40], which may explain the high levels of PUFA observed in *Tetraselmis* sp.

In a 1000 L-outdoor setting, *Tetraselmis* sp. M8 was found to have an increased μ_{Ave} despite a longer lag phase. Cell density achieved by outdoor grown *Tetraselmis* sp. M8 was similar to other large-scale cultures of *Tetraselmis* [61]. FAME productivity and composition were also analyzed, which revealed a near tripling of FAME productivity as well as altered FA composition. High amounts of C16:2, C18:2, C18:3 previously detected in laboratory-grown *Tetraselmis* sp. M8 was found reduced, while higher amounts of recommended FA for biodiesel (particularly C14, C18 & C18:1) were present. The increase in FAME productivity and desirable FA composition of *Tetraselmis* sp. M8 in a mid-scale setting demonstrates that the microalgae isolation and selection technique used in this study can lead to the identification of microalgae strains with potential for large-scale cultivation. Additional factors to be considered for large-scale production include harvesting and oil extraction properties of different microalgae. For example, we noticed that our *Tetraselmis* strains may lose their flagella during stress conditions, resulting in rapid settling that allows easy harvesting/dewatering. Small microalgae, such as *Nannochloropsis* sp., on the other hand may instead be harvested by froth flotation or other techniques, but our results indicate that Nile red staining and lipid extraction may be compromised by thick cell walls in this strain.

Supporting Information

Table S1 Comparison of FAME productivity ($\mu g\ mL^{-1}\ day^{-1}$) of present study microalgae with lipid productivity of microalgae species from other references (including a full comparison of culturing conditions).

Acknowledgments

We wish to thank Tania Catalina Adarme-Vega, Kalpesh Sharma, Felicitas Vernen, Holger Schuhmann, Bart Nijland, Priyanka Nayak, Yamini Kashimshetty, Ekaterina Novak, Miklos Deme and Bernie Degnan for technical assistance and useful discussions. We are also grateful to QSST and CSIRO for provision of additional microalgae strains.

Author Contributions

Conceived and designed the experiments: DKYL SG MT ESBZ SRTH YL PMS. Performed the experiments: DKYL SG MT ESBZ SRTH. Analyzed the data: DKYL SG MT ESBZ SRTH HS YL. Contributed reagents/materials/analysis tools: MT HS PMS. Wrote the paper: DKYL HS MT YL PMS.

References

1. Chisti Y (2007) Biodiesel from microalgae. Biotechnol Adv 25: 294–306.
2. Malcata FX (2011) Microalgae and biofuels: A promising partnership? Trends Biotechnol 29: 542–549.
3. Schenk PM, Thomas-Hall SR, Stephens E, Marx UC, Mussgnug JH, et al. (2008) Second generation biofuels: high-efficiency microalgae for biodiesel production. Bioenerg Res 1: 20–43.
4. Dermirbas A (2009) Potential resources of non-edible oils for biodiesel. Energy Sources Part B – Economics Planning & Policy 4: 310–314.
5. Durret T, Benning C, Ohlrogge J (2008) Plant triacylglycerols as feedstocks for the production of biofuels. Plant J 54: 593–607.
6. Hu Q, Sommerfeld M, Jarvis E, Ghirardi M, Posewitz M, et al. (2008) Microalgal triacylglycerols as feedstocks for biofuel production: perspectives and advances. Plant J 54: 621–639.
7. Ahmad AL, Mat Yasin NH, Derek CJC, Lim JK (2011) Microalgae as a sustainable energy source for biodiesel production: A review. Renew Sustain Energ Rev 15: 584–593.
8. Chisti Y (2008) Biodiesel from microalgae beats bioethanol. Trends Biotechnol 26: 126–131.
9. Gouveia L, Oliveira A (2009) Microalgae as a raw material for biofuels production. J Ind Microbiol Biotechnol 36: 269–274.
10. Kliphus AMJ, Lenneke DW, Vejrazka C (2010) Photosynthetic Efficiency of Chlorella sorokiana in a turbulently mixed short light-path photobioreactor. Biotechnol Progr 26: 687–696.
11. Rodolfi L, Zittelli GC, Bassi N, Padovani G, Biondi N, et al. (2009) Microalgae for oil: strain selection, induction of lipid synthesis and outdoor mass cultivation in a low-cost photobioreactor. Biotechnol Bioeng 102: 100–112.
12. Zeiler KG, Heacox DA, Toon ST, Kadam KL, Brown LM (1995) The use of microalgae for assimilation and utlization of carbon dioxide from fossil fuel-fired power plant flue gas. Energ Convers Manag 36: 707–712.
13. Wang B, Li YQ, Wu N, Lan CQ (2008) CO₂ bio-mitigation using microalgae. Appl Microbiol Biotechnol 79: 707–718.
14. Bridgewater A, Maniatis K (2004) The production of biofuels by thermal chemical processing of biomass. In: Archer M, Barber J, editors. Molecular to global photosynthesis. London: Imperial College Press.
15. Pulz O, Gross W (2004) Valuable products from biotechnology of microalgae. Appl Microbiol Biotechnol 65: 635–648.
16. Ruxton CHS, Reed SC, Simpson MJA, Millington KJ (2004) The health benefits of omega-3 polyunsaturated fatty acids: a review of the evidence. J Human Nutrition Dietetics 17: 449–459.
17. Cavalli RO, Lavens P, Sorgeloos P (1999) Performance of Macrobrachium rosenbergii broodstock fed diets with different fatty acid composition. Aquaculture 179: 387–402.
18. Doroudi MS, Southgate PC, Mayer RJ (1999) Growth and survival of blacklip pearl oyster larvae fed different densities of microalgae. Aquaculture Internat 7: 179–187.
19. Emata AC, Ogata HY, Garibay ES, Furuita H (2003) Advanced broodstock diets for the mangrove red snapper and a potential importance of arachidonic acid in eggs and fry. Fish Physiol Biochem 28: 489–491.
20. Amaro HM, Guedes AC, Malcata FX (2011) Advances and perspectives in using microalgae to produce biodiesel. Appl Energy 88: 3402–3410.
21. Weyer KM, Bush DR, Darzins A (2010) Theoretical maximum algal oil production. Bioenerg Res 3: 204–213.
22. Dermirbas A, Dermirbas MF (2011) Importance of algae oil as a source of biodiesel. Energy Convers Manag 52: 163–170.
23. Mata TM, Martins AA, Caetano NS (2010) Microalgae for biodiesel production and other applications: A review. Renew Sustain Energ Rev 14: 217–232.
24. Knothe G, Gerpen JV, Krahl J (2005) The biodiesel handbook. Urbana, IL: AOCS Press.
25. Knothe G (2005) Dependence of biodiesel fuel properties on the structure of fatty acid alkyl esters. Fuel Process Technol 86: 1059–1070.
26. Abu-Rezq T, Al-Musallam L, Al-Shimmari J (1999) Optimum production conditions for different high-quality marine algae. Hydrobiologia 403: 91–107.
27. Chiu SY, Kao CY, Tsai MT, Ong SC, Chen CH, et al. (2009) Lipid accumulation and CO₂ utilization of Nannochloropsis oculata in response to CO₂ aeration. Bioresour Technol 100: 833–838.
28. Cho S, Ji SC, Hur S, Bae J, Park IS, et al. (2007) Optimum temperature and salinity conditions for growth of green algae Chlorella ellipsoidea and Nannochloris oculata. Fish Science 73: 1050–1056.
29. Li YQ, Horsman M, Wang B, Wu N, Lan CQ (2008) Effects of nitrogen sources on cell growth and lipid accumulation of green algae Neochloris oleoabundans. Appl Microbiol Biotechnol 81: 629–636.
30. Chen M, Tang H, Ma H, Holland TC, Ng KYS, et al. (2011) Effects of nutrient on growth and lipid accumulation in the green algae Dunaliella tertiolecta. Bioresour Technol 102: 1649–1655.
31. Huerlimann R, de Nys R, Heimann K (2010) Growth, lipid content, productivity, and fatty acid composition of tropical microalgae for scale-up production. Biotechnol Bioeng 107: 245–257.
32. Renaud SM, Thinh LV, Lambrinidis G, Parry DL (2002) Effect of temperature on growth, chemical composition and fatty acid composition of tropical Australian microalgae grown in batch cultures. Aquaculture 211: 195–214.
33. Araujo GS, Matos LJBL, Goncalves LRB, Fernandes FAN, Farias WRL (2011) Bioprospecting for oil producing microalgal strains: Evaluation of oil and biomass production for ten microalgal strains. Bioresour Technol 102: 5248–5250.
34. Lee S, Go S, Jeong G, Kim S (2011) Oil production from five marine microalgae for the production of biodiesel. Biotechnol Bioprocess Eng 16: 561–566.
35. Patil V, Kallqvist T, Olsen E, Vogt G, Gislerod HR (2007) Fatty acid composition of 12 microalgae for possible use in aquaculture feed. Aquaculture Internat 15: 1–9.
36. Converti A, Casazza AA, Ortiz EY, Perego P, Del Borghi M (2009) Effects of temperature and nitrogen concentration on the growth and lipid content of Nannochloropsis oculata and Chlorella vulgaris. Chem Eng Process 48: 1146–1151.
37. Illman AM, Scragg AH, Shales SW (2000) Increase in Chlorella strains calorific values when grown in low nitrogen medium. Enzyme Microb Technol 27: 631–635.
38. Borowitzka MA (1988) Fats, oils and hydrocarbons. Borowitzka, MA and LJ Borowitzka (Ed) Micro-Algal Biotechnology X+477p Cambridge University Press: New York, New York, USA; Cambridge, England, UK Illus: 257–287.
39. Roessler PG (1990) Environmental control of glycerolipid metabolism in microalgae – commercial implications and future-research directions. J Phycol 26: 393–399.
40. de la Pena MR, Villegas CT (2005) Cell growth, effect of filtrate and nutritive value of the tropical prasinophyte Tetraselmis tetrathele (Butcher) at different phases of culture. Aquaculture Res 36: 1500–1508.
41. Guillard RR, Ryther JH (1962) Studies of marine planktonic diatoms: I. Cyclotella nana (Hustedt) and Detonula confervacea (Cleve) Gran. Canad J Microbiol 8: 229–239.
42. Lorenz M, Friedl T, Day JG (2005) Perpetual maintenance of actively metabolizing microalgal cultures. In: Andersen RA, editor. Algal culturing techniques. Burlington, MA: Elsevier Academic Press. 145–156.
43. Chomczynski P, Sacchi N (1987) Single-step method of RNA isolation by acid guanidium thiocyanate phenol chloroform extraction. Analyt Biochem 162: 156–159.
44. Katoh K, Asimenos G, Toh H (2009) Multiple alignment of DNA sequences with MAFFT. Meth Mol Biol 537: 39–64.
45. Guidon S, Gascuel O (2003) A simple, fast and accurate algorithm to estimate large phylogenies by maximum likelihood. Systems Biol 52: 696–704.
46. Wood AM, Everroad RC, Wingard LM (2005) Chapter 18: Measuring growth rates in microalgal cluures. In: A AR, editor. Algal culturing techniques. Burlington, MA: Elsevier Academic Press. 269–285.
47. Pal D, Khozin-Goldberg I, Cohin Z, Boussiba S (2011) The effect of light, salinity and nitrogen availability on lipid production by Nannochloropsis sp. Appl Microbiol Biotechnol 90: 1429–1441.
48. Araujo SC, Garcia VMT (2005) Growth and biochemical composition of the diatom Chaetoceros cf. wighamii Brightwell under different temperature, salinity and carbon dioxide levels. I. Protein, carbohydrates and lipids. Aquaculture 246: 405–412.
49. Emdadi D, Berland B (1989) Variation in lipid class composition during batch growth of Nannochloropsis salina and Pavlova lutheri. Marine Chem 26: 215–225.
50. Rocha JMS, Garcia JEC, Henriques MHF (2003) Growth aspects of the marine microalga Nannochloropsis gaditana. Biomol Eng 20: 237–242.
51. Miyamoto K, Wable O, Benemann JR (1988) Vertical tubular reactor for microalgae cultivation. Biotechnol Lett 10: 703–708.
52. Alonzo F, Mayzaud P (1999) Spectrofluorometric quantification of neutral and polar lipids in zooplankton using Nile red. Marine Chem 67: 289–301.
53. Gong Y, Jiang M (2011) Biodiesel production with microalgae as feedstock: from strains to biodiesel. Biotechnol Lett 33: 1269–1284.
54. Hu H, Gao K (2003) Optimisation of growth and fatty acid composition of a unicellular marine picoplankton, Nannochloropsis sp., with enrichment carbon sources. Biotechnol Lett 25: 421–425.
55. Chen CH, Yeh K, Su H, Lo Y, Chen W, et al. (2010) Strategies to enhance cell growth and achieve high-level oil production of a Chlorella vulgaris isolate. Biotechnol Prog 26: 679–686.
56. Knothe G (2008) "Designer" biodiesel: optimising fatty ester composition to improve fuel properties. Energy Fuels 22: 1358–1364.
57. Reitan KI, Rainuzzo JR, Olsen Y (1994) Effect of nutrient limitation on fatty acid and lipid content of marine microalgae. J Phycol 30: 972–979.
58. Jeh EJ, Song SK, Seo JW, Hur BK (2007) Variation in the lipid class and fatty acid composition of Thraustochytrium aureum ATCC 34304. Korean J Biotechnol Bioeng 22: 37–42.
59. Dunstan GH, Volkman JK, Barret SM, Garland CD (1993) Changes in the lipid composition and maximization of the polyunsaturated fatty acid content of three microalgae grown in mass culture. J Appl Phycol 7: 71–83.
60. Shamsudin L (1992) Lipid and fatty acid composition of microalgae used in Malaysian aquaculture as live food for the early stage of penaeid larvae. J Appl Phycol 4: 371–378.
61. Okauchi M, Kawamura K (1997) Optimum medium for large-scale culture of Tetraselmis tetrathele. Hydrobiologia 358: 217–222.

4

Vertical Distribution and Estimated Doses from Artificial Radionuclides in Soil Samples around the Chernobyl Nuclear Power Plant and the Semipalatinsk Nuclear Testing Site

Yasuyuki Taira[1,8], Naomi Hayashida[1], Rimi Tsuchiya[4], Hitoshi Yamaguchi[3,7], Jumpei Takahashi[5], Alexander Kazlovsky[6], Marat Urazalin[7], Tolebay Rakhypbekov[7], Shunichi Yamashita[2], Noboru Takamura[1]*

1 Department of Global Health, Medical and Welfare, Nagasaki University Graduate School of Biomedical Sciences, Nagasaki, Japan, 2 Department of Radiation Medical Science, Nagasaki University Graduate School of Biomedical Sciences, Nagasaki, Japan, 3 Department of Ecomaterials Science, Nagasaki University Graduate School of Engineering, Nagasaki, Japan, 4 Nagasaki University School of Medicine, Nagasaki, Japan, 5 Center for International Collaborative Research, Nagasaki University, Nagasaki, Japan, 6 Department of Pediatrics, Gomel State Medical University, Gomel, the Republic of Belarus, 7 Department of Microbiology, Semey State Medical Academy, Semey, the Republic of Kazakhstan, 8 Nagasaki Prefectural Institute for Environmental Research and Public Health, Omura, Japan

Abstract

For the current on-site evaluation of the environmental contamination and contributory external exposure after the accident at the Chernobyl Nuclear Power Plant (CNPP) and the nuclear tests at the Semipalatinsk Nuclear Testing Site (SNTS), the concentrations of artificial radionuclides in soil samples from each area were analyzed by gamma spectrometry. Four artificial radionuclides (^{241}Am, ^{134}Cs, ^{137}Cs, and ^{60}Co) were detected in surface soil around CNPP, whereas seven artificial radionuclides (^{241}Am, ^{57}Co, ^{137}Cs, ^{95}Zr, ^{95}Nb, ^{58}Co, and ^{60}Co) were detected in surface soil around SNTS. Effective doses around CNPP were over the public dose limit of 1 mSv/y (International Commission on Radiological Protection, 1991). These levels in a contaminated area 12 km from Unit 4 were high, whereas levels in a decontaminated area 12 km from Unit 4 and another contaminated area 15 km from Unit 4 were comparatively low. On the other hand, the effective doses around SNTS were below the public dose limit. These findings suggest that the environmental contamination and effective doses on the ground definitely decrease with decontamination such as removing surface soil, although the effective doses of the sampling points around CNPP in the present study were all over the public dose limit. Thus, the remediation of soil as a countermeasure could be an extremely effective method not only for areas around CNPP and SNTS but also for areas around the Fukushima Dai-ichi Nuclear Power Plant (FNPP), and external exposure levels will be certainly reduced. Long-term follow-up of environmental monitoring around CNPP, SNTS, and FNPP, as well as evaluation of the health effects in the population residing around these areas, could contribute to radiation safety and reduce unnecessary exposure to the public.

Editor: Vishal Shah, Dowling College, United States of America

Funding: This work was supported by the Ministry of Education, Culture, Sports, Science and Technology of Japan through the Nagasaki University Global COE program. The funders had no role in the study design, data collection and analysis, decision to publish, or preparation of the manuscript.

Competing Interests: The authors have declared that no competing interests exist.

* E-mail: takamura@nagasaki-u.ac.jp

Introduction

On April 26, 1986, one of the most serious nuclear accidents involving radiation exposure occurred at Unit 4 of the Chernobyl Nuclear Power Plant (CNPP), located in Ukraine about 20 km south of the border with the Republic of Belarus. Significant releases of radioactive substances from Unit 4 of CNPP during the accident lasted 10 days and changes in the meteorological conditions during this period have resulted in a composite picture of contamination of vast territories [1,2]. Radioactive contamination from CNPP spread over 40% of Europe and wide territories in Asia, North Africa, and North America [3]. Nearly 400 million people resided in territories that were contaminated with radioactivity at a level higher than 4 kBq/m^2 (0.11 Ci/km^2) from April to July 1986 [3]. In 2000, the total inventories of the fuel component radionuclides in the upper 30 cm of the soil layer in the 30-km Chernobyl zone in Ukraine were estimated as 0.4–0.5% of the radionuclide amounts in the CNPP Unit 4 at the moment of the accident [2].

Since August 29, 1949, more than 450 nuclear explosions, including atmospheric, above-ground, and underground tests, have been conducted at the Semipalatinsk Nuclear Testing Site (SNTS). Since the site's closure in 1989, attention has been paid to clarifying the health effects in the population residing around SNTS [4–7]. According to some reports, fission products such as plutonium (Pu) and neutron-induced radioactivity were detected in the soil samples from SNTS.

The two main pathways leading to radiation exposure of the general public due to fallout are external exposure from radionuclides deposited on the ground and internal exposure through ingestion of contaminated foods produced in contaminated areas. It is extremely important to evaluate the environmental contamination and external and internal exposure risks due to nuclear disasters for radiation protection and public health.

On March 11, 2011, a 9.0-magnitude earthquake (The Great East Japan Earthquake) struck the east coast near Iwate, Miyagi, and Fukushima Prefectures, Japan. The earthquake in combination with the subsequent tsunami caused extensive damage to the Fukushima Dai-ichi Nuclear Power Plant (FNPP) and a radioactive plume derived from Units 1, 2, 3, and 4 of FNPP was dispersed in the atmosphere. The total amount of radioactive materials released into the atmosphere from FNPP corresponds to Level 7 on the International Nuclear and Radiological Event Scale (INES) by the International Atomic Energy Agency (IAEA). Although the effects of this accident are still being felt and will continue to affect the country, approximately 900 PBq of radioactive substances were emitted, a sixth of the amount of emissions from the Chernobyl accident when converted to ^{131}I (half-life: 8.0 d). There are now vast stretches of land, totaling 1,800 km^2, of Fukushima Prefecture with levels equaling a potential cumulative dose of 5 mSv/y or more (Available: http://naiic.go.jp/en/report/. Accessed 2012 Oct 25) [8]. Risks of internal exposure are extremely low because restrictions of food intake by the nation are strictly carried out after the FNPP accident (http://www.mhlw.go.jp/english/topics/2011eq/index.html. Accessed 2012 Oct 25). On the other hand, the risks of external exposure around the living space are attracting public attention due to considerable safety concerns. Although ongoing national efforts aimed at reducing the annual exposure dose closer to 1 mSv, which is the public dose limit specified by the International Commission on Radiological Protection (ICRP) in 1991, effective decontamination is not progressing smoothly around FNPP, primarily because of strong absorption of radiocesium by soil. In addition, secondary contamination of soil occurs after decontamination, creating another problem.

The evaluation of accumulated artificial radionuclides around CNPP and SNTS is extremely important for developing countermeasures such as those that will be required for future decontamination around FNPP. Therefore, to evaluate current environmental contamination and contributions from external exposure due to artificial radionuclides, concentrations of radionuclides and their vertical distribution in soil samples from areas around CNPP and SNTS were analyzed by gamma spectrometry (**Figure 1**). Furthermore, external effective doses were calculated from samples from these areas in order to estimate radiation exposure status.

Materials and Methods

Sample Sites

Soil samples around CNPP were collected around Masany (N51° 48′, E29° 96′) in the Republic of Belarus, a fixed-point observation site approximately 8 km from the Chernobyl reactor (N51° 39′, E30° 10′), around the 30-km zone in which the ^{137}Cs deposition exceeded 1,500 kBq/m^2 (**Figure 2**) [1]. Other samples around CNPP were collected at Minsk (N53° 91′, E27° 61′) and Gomel (N52° 42′, E30° 96′) in the Republic of Belarus approximately 340 km northwest and 135 km northeast from CNPP, respectively (**Figure 1**). At the same time, air dose rates in all sample sites were monitored in air 1 m above the ground by a portable detector for the management of radiation exposure (PDR-201®, Hitachi-Aloka Medical, Ltd., Tokyo, Japan).

Soil samples around SNTS were collected around the center of the explosion; the Experimental Field (N50° 20′, E77° 75′), an

atmospheric and surface nuclear testing site 70 km southwest of Kurchatov, that has very high radioactivity levels and Chagan (N49° 90′, E79° 05′), known as the Balapan Test Site for underground nuclear testing in the Republic of Kazakhstan (**Figure 3**).

Measurement of Radionuclides

For the evaluation of vertical distribution and external radiation exposure, core samples of soil (0–5 and 5–10 cm) were collected from CNPP areas between January 28 and February 3, 2012. Core samples of soil (0–5, 5–10 and 10–30 cm) were also collected from SNTS areas on August 29, 2011. Sampling of soil was carried out using soil coring at all sample sites. The size of the soil samples was 18.2 cm^2 (a diameter of 4.8 cm) and the density of the surface soil layer ranged from 0.98 to 1.8 g/cm^3-dry in CNPP and 1.2 to 1.6 g/cm^3-dry in SNTS.

The mass of soil samples collected in each area ranged from 57 to 127 g. After collection, soil samples were dried in a fixed temperature dryer (105°C, 24 h) before soil samples were sieved to remove pebbles and organic materials (>2 mm).

After preparation, samples were placed in polypropylene containers and analyzed with a high purity germanium detector (CANBERRA®, GC2520, Canberra Industries Inc., Meriden, CT, USA) coupled to a multi-channel analyzer (Lynx, Canberra Industries Inc., Meriden, CT, USA) for 80,000 s. We set the measuring time to detect objective radionuclide levels. Gamma-ray peaks used for measurements were 59.54 keV for ^{241}Am (half-life: 432.2 y), 122.06 keV for ^{57}Co (271.7 d), 604.66 keV for ^{134}Cs (2.1 y), 661.64 keV for ^{137}Cs (30.2 y), 724.18 and 756.72 keV for ^{95}Zr (64.0 d), 765.79 keV for ^{95}Nb (35.0 d), 810.76 keV for ^{58}Co (70.9 d), and 1173.21 and 1332.47 keV for ^{60}Co (5.3 y). Decay corrections were made based on sampling data. Detector efficiency calibration for different measurement geometries was performed using mixed activity standard volume sources (Japan Radioisotope Association, Tokyo, Japan). The relative detection efficiency of this instrument was 27.8%. In the present study, we analyzed each sample at least three times, considering with "Sum effect" and "Self-Absorption", and calculated standard errors by PASW statistics 18 software (SPSS Japan, Tokyo, Japan). Concentrations of artificial radionuclides were indicated as "counting values and ± standard errors". Sample collection, processing, and analysis were executed in accordance with standard methods of radioactivity measurement authorized by the Ministry of Education, Culture, Sports, Science, and Technology, Japan (MEXT) [9]. All measurements were performed at the Nagasaki Prefectural Institute for Environmental Research and Public Health, Nagasaki, Japan.

Effective Dose

After measurements, external effective doses (µSv/h and mSv/y) from soil samples were estimated from artificial radionuclide concentrations with the following formula:

$$H_{ext} = C \cdot D_{ext} \cdot f \cdot s \qquad (1)$$

in which C is the activity concentration of detected artificial radionuclides (^{241}Am, ^{134}Cs, ^{137}Cs, and ^{60}Co; half-life > 1y) [kBq/m^2; estimated from radionuclide concentration in Bq/kg and collected areas of surface soil (0–5 cm)]; D_{ext} is the dose conversion coefficient reported as the kerma-rate in air 1 m above the ground per unit activity per unit area [(µGy/h)/(kBq/m^2)], supposing that the kerma-rate in the air and the absorbed dose rate in the air are the same value, for radiocesium with the relaxation mass per unit

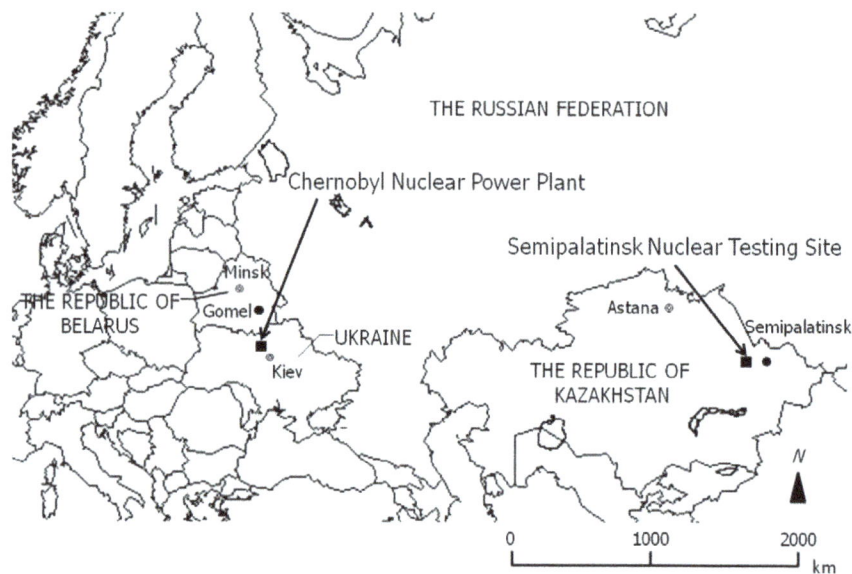

Figure 1. Locations around the Chernobyl Nuclear Power Plant (the Republic of Belarus, Ukraine, and the Russian Federation) and the Semipalatinsk Nuclear Testing Site (the Republic of Kazakhstan).

area (β: g/cm^2) set to 10 due to the passage of more than 20 years after the Chernobyl accident and nuclear tests of SNTS [1.7×10^{-5} (μGy/h)/(kBq/m^2) for ^{241}Am, 2.0×10^{-3} (μGy/h)/(kBq/m^2) for ^{134}Cs, 7.6×10^{-4} (μGy/h)/(kBq/m^2) for ^{137}Cs, and

Figure 2. The 30-km zone around the Chernobyl Nuclear Power Plant.

3.0×10^{-3} (μGy/h)/(kBq/m^2) for ^{60}Co, ICRU 1994] [10]; f is the unit conversion coefficient (0.7 Sv/Gy for effective dose rate in the body per unit absorbed dose rate in air) [11], and s is the decrease in the coefficient by a shielding factor against exposure with gamma rays from a deposit 1 m above the ground (0.7 under the condition of usual land) [12].

Results

The distribution of detected artificial radionuclides in soil samples from CNPP is shown in **Table 1**. The prevalent dose-forming artificial radionuclides were ^{241}Am, ^{134}Cs, ^{137}Cs, and ^{60}Co (these concentrations are shown in **Table 1**). Various radionuclides were especially detected near Unit 4 of CNPP. The concentrations of detected artificial radionuclides in surface soil samples around FNPP were higher than those of lower layers and the prevalent radionuclides were mainly accumulated in the surface layer.

On the other hand, the distribution of detected artificial radionuclides in soil samples from SNTS is shown in **Table 2**. The prevalent dose-forming artificial radionuclides were ^{241}Am, ^{57}Co, ^{137}Cs, ^{95}Zr, ^{95}Nb, ^{58}Co, and ^{60}Co (these concentrations are shown in **Table 2**). Various radionuclides were especially detected near the center of an explosion, as with CNPP. Also, the concentrations of detected artificial radionuclides other than ^{241}Am in surface soil samples around SNTS were higher than those of lower layers and those radionuclides were mainly accumulated in the surface layer.

For estimating the external effective doses, the activity concentrations in kBq/m^2 of detected artificial radionuclides in surface soil samples (0–5 cm) around CNPP and SNTS were calculated from these radionuclides concentrations in Bq/kg (these concentrations are shown in **Table 3** and **Table 4**).

The external effective doses from detected artificial radionuclides around CNPP and SNTS using Eq. (1) are summarized in **Table 5** and **Table 6**. Estimated external effective doses around CNPP were 1.3 μSv/h (12 mSv/y) in a contaminated area 12 km from Unit 4, 0.86 μSv/h (7.5 mSv/y) in a unknown area 12 km from Unit 4, 0.19 μSv/h (1.6 mSv/y) in a decontaminated area

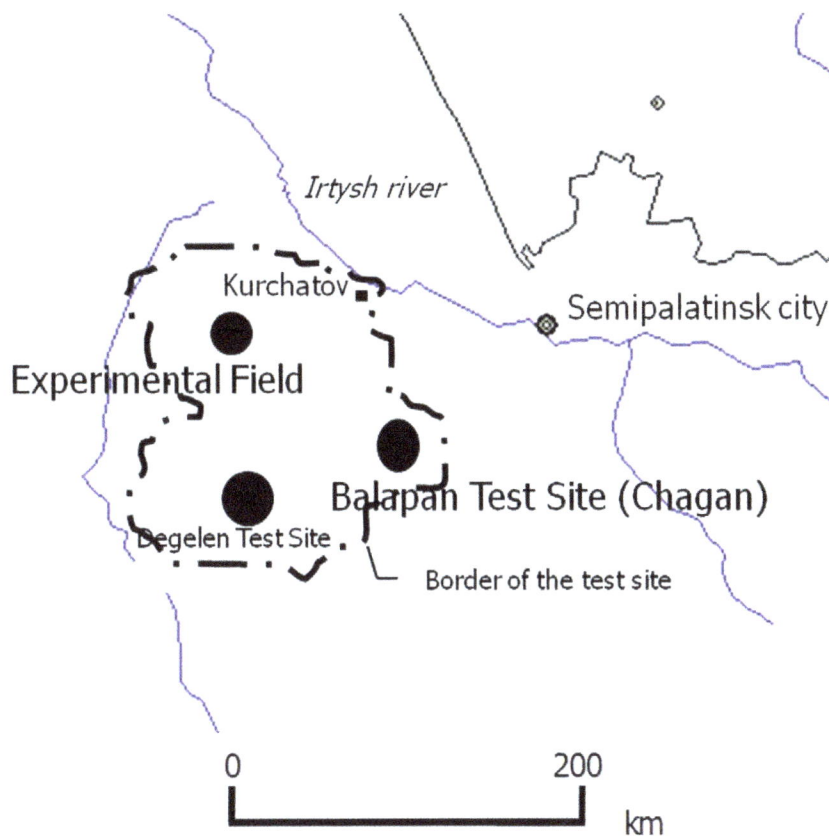

Figure 3. Test site around the Semipalatinsk Nuclear Testing Site.

12 km from Unit 4, and 0.17 µSv/h (1.5 mSv/y) in a contaminated area 15 km from Unit 4. Air dose rates were 0.80–4.2 µSv/h when soil samples were collected in areas around CNPP. Estimated external effective doses around CNPP were 4.2×10^{-5} µSv/h $(3.7 \times 10^{-4}$ mSv/y) in Minsk and 1.7×10^{-3} µSv/h $(1.5 \times 10^{-2}$ mSv/y) in Gomel. Air dose rates were 0.05–0.06 µSv/h when soil samples were collected in areas around CNPP.

Table 1. Distribution of detected artificial radionuclides in soil samples collected at the Chernobyl Nuclear Power Plant, Minsk and Gomel (Republic of Belarus).

Point	Distance[a] (km)	Depth (cm)	Artificial radionuclides in Bq/kg-dry			
			^{241}Am	^{134}Cs	^{137}Cs	^{60}Co
CNPP (Masany)	12	Contaminated 0–5	489±3.8[b]	n.d.	63341±23	2.1±0.2,2.5±0.2
		5–10	117±1.6	n.d.	9105±8.5	n.d.
		Unknown 0–5	531±3.4	8.3±1.4	47237±20	1.6±0.3,1.0±0.2
		5–10	8.5±0.5	n.d.	753±2.4	n.d.
		Decontaminated 0–5	137±1.9	n.d.	12458±11	n.d.
		5–10	56±1.2	n.d.	4209±6.1	n.d.
	15	Contaminated 0–5	97±2.5	n.d.	18729±17	n.d.
		5–10	14±0.7	n.d.	1763±4.2	n.d.
Minsk	340		n.d.	n.d.	2.8±0.2	n.d.
Gomel	135		n.d.	n.d.	83±0.9	n.d.

[a]distance from Unit 4 of the Chernobyl Nuclear Power Plant.
[b]error shows one sigma standard deviation from counting statistics.
Samples were collected at CNPP, Minsk and Gomel, Kazakhstan during January 28 and February 3, 2012. Radionuclides were analyzed with a germanium-detector (relative detection efficiency: 27.8% by Canberra) coupled to a multi-channel analyzer for 80,000 s at Nagasaki Prefectural Institute for Environmental Research and Public Health, Nagasaki, Japan.

Table 2. Distribution of detected artificial radionuclides in soil samples collected at the Semipalatinsk Nuclear Testing Site and Chagan (Kazakhstan).

Point	Distance[a] (km)	Depth (cm)	Artificial radionuclides in Bq/kg-dry						
			^{241}Am	^{57}Co	^{137}Cs	^{95}Zr	^{95}Nb	^{58}Co	^{60}Co
SNTS (Experimental Field)	Ground zero	0–5	900±6.4[b]	6079±4.8	42736±24	228±5.3, 133±5.6	15±2.9	97±2.6	347±3.5, 349±3.1
		5–10	1001±6.5	5694±4.5	39698±22	233±5.3, 124±5.6	15±2.8	83±2.3	319±3.3, 323±2.8
		10–30	590±3.7	3116±3.2	9816±11	106±3.8, 62±3.7	n.d.	48±1.9	132±2.2, 141±1.9
	1	0–5	552±2.3	25±0.3	499±2.1	n.d.	n.d.	n.d.	2.3±0.3, 2.2±0.3
		5–10	137±1.2	17±0.3	212±1.4	n.d.	n.d.	n.d.	1.9±0.4, 1.7±0.3
		10–30	318±2.0	12±0.3	138±1.3	n.d.	n.d.	n.d.	n.d.
	10	0–5	13±0.6	n.d.	26±0.6	n.d.	n.d.	n.d.	n.d.
		5–10	9.0±0.6	n.d.	24±0.6	n.d.	n.d.	n.d.	n.d.
		10–30	4.8±0.6	n.d.	12±0.5	n.d.	n.d.	n.d.	n.d.
Chagan (Balapan Test Site)		0–5	n.d.	n.d.	9.0±0.4	n.d.	n.d.	n.d.	n.d.
		5–10	n.d.	n.d.	5.9±0.3	n.d.	n.d.	n.d.	n.d.
		10–30	n.d.	n.d.	9.5±0.4	n.d.	n.d.	n.d.	n.d.

[a]distance from Unit 4 of the Chernobyl Nuclear Power Plant.
[b]error shows one sigma standard deviation from counting statistics.
Samples were collected at SNTS and Chagan, Kazakhstan on August 29, 2011. Radionuclides were analyzed with a germanium-detector (relative detection efficiency: 27.8% by Canberra) coupled to a multi-channel analyzer for 80,000 s at Nagasaki Prefectural Institute for Environmental Research and Public Health, Nagasaki, Japan.

On the other hand, estimated external effective doses around SNTS were 9.3×10^{-2} µSv/h (0.79 mSv/y) at Ground Zero (Experimental Field), 2.2×10^{-3} µSv/h (1.9×10^{-2} mSv/y) 1 km from the center of the explosion, 8.3×10^{-5} µSv/h (7.3×10^{-4} mSv/y) 10 km from the center of the explosion, and 3.7×10^{-5} µSv/h (3.2×10^{-4} mSv/y) in Chagan (Balapan Test Site).

Discussion

Deposition in the nearby contaminated zone (<100 km) around CNPP reflected the radionuclide composition of the fuel; volatile elements, including iodine and cesium, in the form of condensa-tion-generated particles were more widely dispersed in the far zone (from 100 km to approximately 2,000 km) [1]. The ^{137}Cs deposition was highest in a 30-km radius area surrounding the reactor, known as the 30-km zone, and deposition densities exceeded 1,500 kBq/m² in this zone and some areas (Gomel, Kiev, and Zhitomir regions) of the near zone to the west and northwest of the reactor [1]. According to the United Nations Scientific Committee on the Effects of Atomic Radiation (UNSCEAR), areas of ^{137}Cs deposition density greater than 555 kBq/m² (15 Ci/km²) are designated as areas of strict control following the CNPP accident on April 26, 1986 [1]. According to the 2006 IAEA report, the external doses around CNPP during 1986–2005 were about 1.2 times higher, and internal doses were

Table 3. Distribution of detected artificial radionuclides in soil samples collected at the Chernobyl Nuclear Power Plant, Minsk and Gomel (Republic of Belarus).

Point	Distance[a] (km)	Depth (cm)	Artificial radionuclides in kBq/m²			
			^{241}Am	^{134}Cs	^{137}Cs	^{60}Co
CNPP (Masany)	12	Contaminated 0–5	28±0.2[b]	n.d.	3592±1.3	0.1±0.01, 0.1±0.01
		5–10	6.5±0.1	n.d.	509±0.5	n.d.
		Unknown 0–5	26±0.2	0.4±0.07	2322±1.0	0.1±0.01, 0.05±0.01
		5–10	0.6±0.03	n.d.	51±0.2	n.d.
		Decontaminated 0–5	5.5±0.1	n.d.	501±0.5	n.d.
		5–10	3.0±0.1	n.d.	223±0.3	n.d.
	15	Contaminated 0–5	2.3±0.1	n.d.	451±0.4	n.d.
		5–10	0.5±0.03	n.d.	68±0.2	n.d.
Minsk	340		n.d.	n.d.	0.1±0.008	n.d.
Gomel	135		n.d.	n.d.	4.6±0.05	n.d.

[a]distance from Unit 4 of the Chernobyl Nuclear Power Plant.
[b]error shows one sigma standard deviation from counting statistics.

Table 4. Distribution of detected artificial radionuclides in soil samples collected at the Semipalatinsk Nuclear Testing Site and Chagan (Kazakhstan).

Point	Distance[a] (km)	Depth (cm)	Artificial radionuclides in kBq/m²						
			^{241}Am	^{57}Co	^{137}Cs	^{95}Zr	^{95}Nb	^{58}Co	^{60}Co
SNTS (Experimental Field)	Ground zero	0–5	5.0±0.04[b]	34±0.03	236±0.1	1.3±0.03, 0.7±0.03	0.1±0.02	0.5±0.01	1.9±0.02, 1.9±0.02
		5–10	6.4±0.04	37±0.03	256±0.1	1.5±0.03, 0.8±0.04	0.1±0.02	0.5±0.01	2.1±0.02, 2.1±0.02
		10–30	4.0±0.03	21±0.02	67±0.1	0.7±0.03, 0.4±0.03	n.d.	0.3±0.01	0.9±0.02, 1.0±0.01
	1	0–5	6.2±0.03	0.3±0.003	5.6±0.02	n.d.	n.d.	n.d.	0.03±0.003, 0.02±0.003
		5–10	1.3±0.01	0.2±0.003	2.1±0.01	n.d.	n.d.	n.d.	0.02±0.004, 0.02±0.003
		10–30	2.2±0.01	0.1±0.002	0.9±0.01	n.d.	n.d.	n.d.	n.d.
	10	0–5	0.1±0.01	n.d.	0.2±0.01	n.d.	n.d.	n.d.	n.d.
		5–10	0.1±0.004	n.d.	0.2±0.004	n.d.	n.d.	n.d.	n.d.
		10–30	0.03±0.004	n.d.	0.1±0.004	n.d.	n.d.	n.d.	n.d.
Chagan (Balapan Test Site)		0–5	n.d.	n.d.	0.1±0.004	n.d.	n.d.	n.d.	n.d.
		5–10	n.d.	n.d.	0.1±0.003	n.d.	n.d.	n.d.	n.d.
		10–30	n.d.	n.d.	0.1±0.004	n.d.	n.d.	n.d.	n.d.

[a]distance from Unit 4 of the Chernobyl Nuclear Power Plant.
[b]error shows one sigma standard deviation from counting statistics.

about 1.1–1.5 times higher, than those obtained during 1986–1995 (depending on soil properties and applied countermeasures) [13].

In the present study, four artificial radionuclides (^{241}Am, ^{134}Cs, ^{137}Cs, and ^{60}Co) were detected in surface soil samples from contaminated and unknown areas 12 km from Unit 4 of CNPP. Additionally, these were found in high concentrations compared with the data from a decontaminated area. The value of radioactive materials released into the environment by the Chernobyl accident corresponds to Level 7 of INES by IAEA. In the present study, parts of the CNPP area may be still contaminated with artificial radionuclides derived from the nuclear disaster because current levels around CNPP were over the public dose limit of 1 mSv/y (ICRP, 1991) [14]. In particular,

effective doses in a contaminated area 12 km from Unit 4, including an unknown area, were obviously high compared with an effective dose in a decontaminated area 12 km from Unit 4. In other words, these findings suggest that the environmental contamination and the effective dose on the ground are certainly decreased by decontamination such as removing surface soil, although the effective doses of sampling points around CNPP in the present study were all over the public dose limit. Thus, the remediation of soil as a countermeasure could be an extremely effective method and external exposure levels are certainly reduced. The existing remediation approaches and phytoextraction (phytoremediation) of radionuclides from contaminated soil have been examined [15,16]. However, the remediation of soil around FNPP contaminated by artificial radionuclides is attracting

Table 5. External effective doses from soil samples due to artificial radionuclides in the Chernobyl Nuclear Power Plant, Minsk and Gomel (Republic of Belarus).

Point	Distance (km)	Condition	External effective dose[a]		Air dose rate in µSv/h
			µSv/h	mSv/y	
CNPP (Manany)	12	Contaminated	1.3	12	4.2
		Unknown	0.86	7.5	3.2
		Decontaminated	0.19	1.6	0.80
	15	Contaminated	0.17	1.5	0.84
Minsk	340		4.2×10^{-5}	3.7×10^{-4}	0.06
Gomel	135		1.7×10^{-3}	1.5×10^{-2}	0.05

[a]External effective doses were calculated with the following formula: $H_{ext} = C \cdot D_{ext} \cdot f \cdot s$.
where C is the activity concentration of detected artificial radionuclides (^{241}Am, ^{134}Cs, ^{137}Cs and ^{60}Co; halh-life > 1y) (kBq/m²; calculated from radionuclide concentration in Bq/kg and collected areas of soils (0–5 cm)), D_{ext} is the dose conversion coefficient as kerma-rate in air at 1 m above ground per unit activity per unit area ((µGy/h)/(kBq/m²) for detected artificial radionuclides with the value of relaxation mass per unit area 10 g/cm² (ICRU 1994)), f is the unit conversion coefficient (0.7 Sv/Gy (UNSCEAR 2000)), s is the decrease in the coefficient by a shielding factor against exposure with gamma rays from a deposit at 1 m above ground (0.7 under the condition of usual land (IAEA-TECDOC-1162)).

Table 6. External effective doses from soil samples due to artificial radionuclides in the Semipalatinsk Nuclear Testing Site and Chagan (Kazakhstan).

Point	Distance (km)	External effective dose[a]	
		μSv/h	mSv/y
SNTS (Experimental Field)	Ground zero	9.3×10^{-2}	0.79
	1	2.2×10^{-3}	1.9×10^{-2}
	10	8.3×10^{-5}	7.3×10^{-4}
Chagan (Balapan Test Site)		3.7×10^{-5}	3.2×10^{-4}

[a]External effective doses were calculated with the following formula:
$H_{ext} = C \cdot D_{ext} \cdot f \cdot s$.
where C is the activity concentration of detected artificial radionuclides (^{241}Am, ^{137}Cs and ^{60}Co; halh-life > 1y) (kBq/m²; calculated from radionuclide concentration in Bq/kg and collected areas of soils (0–5 cm)), D_{ext} is the dose conversion coefficient as kerma-rate in air at 1 m above ground per unit activity per unit area ((μGy/h)/(kBq/m²) for detected artificial radionuclides with the value of relaxation mass per unit area 10 g/cm² (ICRU 1994)), f is the unit conversion coefficient (0.7 Sv/Gy (UNSCEAR 2000)), s is the decrease in the coefficient by a shielding factor against exposure with gamma rays from a deposit at 1 m above ground (0.7 under the condition of usual land (IAEA-TECDOC-1162)).

considerable public attention as the public seek confirmation that the areas are safe and that external exposure risks are reduced around the living space.

Some of the detected isotopes, namely europium-152 (^{152}Eu), europium-154 (^{154}Eu), ^{60}Co, and bismuth-217 (^{217}Bi), were reported to have been produced from the stable isotopes in the ground soil around SNTS as these isotopes were activated by the neutron-induced reactions from the bomb explosions [17].

In the present study, seven artificial radionuclides (^{241}Am, ^{57}Co, ^{137}Cs, ^{95}Zr, ^{95}Nb, ^{58}Co, and ^{60}Co) were detected in surface soil samples near the atmospheric testing site. Moreover, these levels were high compared with data from Chagan (Balapan Test Site). However, the current levels around SNTS were below the public dose limit of 1 mSv/y. These findings suggest that the remarkable accumulation of artificial radionuclides is not confirmed in surface soil samples around SNTS, although more than 450 nuclear explosions including atmospheric, above-ground, and under-ground tests were conducted at SNTS from 1949 to 1989 by the former Soviet Union. Also, the results suggest that artificial radionuclides derived from atmospheric tests were widely spread and transferred after nuclear explosions.

Although the amounts of artificial radionuclides released from nuclear reactors and diffusion scales remarkably differed between CNPP and SNTS, data on the environmental radioactivity levels around CNPP and SNTS are extremely important for taking countermeasures such as decontamination against future radiation exposure in Fukushima. In the present study, short-lived radionuclides, which have a half-life of less than 1 year such as ^{57}Co, ^{95}Zr, ^{95}Nb, and ^{58}Co, were detected in soil samples. It is suggested that the soil has incorporated large amounts of radionuclides due to the nuclear disaster.

There are several limitations in the present study. Radionuclides in soil samples may be unequally distributed around FNPP because the number of sampling points was relatively small. Moreover, several radionuclides could not be analyzed by an extraction procedure, including strontium-90 (^{90}Sr). However, the available depth profiles of radionuclides (fallout) in soil samples were mostly measured immediately after the nuclear disaster, although data of the initial distribution of deposited radionuclides in surface soil samples were essential for understanding their movement in the environment. In particular, radiocesium is selectively absorbed by fine soil particles, which have a greater specific surface area, and a positive relationship was found between the clay content of surface soil and the relaxation mass depth of ^{137}Cs [18,19]. Although 20 years or more have passed since the Chernobyl accident and atmospheric nuclear tests in SNTS, because the behavior of radionuclides depends on climate changes, radionuclide analysis of environmental samples by gamma spectrometry is extremely practical for the evaluation of current environmental contamination, vertical distribution, and estimated radiation dose rate. However, further investigation with detailed conditions about external and internal effective doses is needed.

In conclusion, we evaluated current environmental contamination and external radiation dose rates due to artificial radionuclides around CNPP and SNTS. Four artificial radionuclides (^{241}Am, ^{134}Cs, ^{137}Cs, and ^{60}Co) were detected in surface soil samples around CNPP and seven artificial radionuclides (^{241}Am, ^{57}Co, ^{137}Cs, ^{95}Zr, ^{95}Nb, ^{58}Co, and ^{60}Co) were detected in surface soil samples around SNTS. Current effective doses around CNPP were over the public dose limit of 1 mSv/y (ICRP, 1991). These levels in a contaminated area, including an unknown area, 12 km from Unit 4 were high, whereas levels in a decontaminated area 12 km from Unit 4 and a contaminated area 15 km from Unit 4 were comparatively low. On the other hand, the current effective doses around SNTS were below the public dose limit. These findings suggest that the environmental contamination and the effective dose on the ground were certainly decreased by decontamination such as removing surface soil, although the effective doses at sampling points around CNPP in the present study were all over the public dose limit. Thus, the remediation of soil as a countermeasure could be an extremely effective method not only for the areas around CNPP and SNTS but also for the areas around FNPP, and external exposure levels will be certainly reduced. Long-term follow-up of environmental monitoring around CNPP, SNTS, and FNPP, as well as evaluation of health effects in the population residing around these areas, could contribute to radiation safety and reduce unnecessary exposure to the public.

Author Contributions

Conceived and designed the experiments: YT NT . Performed the experiments: YT HY RT AK MU TR . Analyzed the data: YT NT. Contributed reagents/materials/analysis tools: RT JT. Wrote the paper: YT NH SY.

References

1. United Nations Scientific Committee on the Effects of Atomic Radiation (2000) Sources and effects of ionizing radiation. 2000 Report to General Assembly with scientific annexes. Annex J: Exposures and effects of the Chernobyl accident.

2. Kashparov VA, Lundin SM, Zvarych SI, Yoshchenko VI, Levchuk SE, et al. (2003) Territory contamination with the radionuclides representing the fuel component of Chernobyl fallout. Sci Total Environ 317: 105–119.

3. Yablokov AV, Nesterenko VB (2009) Chernobyl contamination through time and space. Ann N Y Acad Sci 1181: 5–30.

4. Yamamoto M, Tsukatani T, Katayama Y (1996) Residual radioactivity in the soil of Semipalatinsk nuclear test site in former USSR. Health Phys 71: 142–148.

5. Yamamoto M, Hoshi M, Takada J, Oikawa S, Yoshikawa I, et al. (2002) Some aspects of environmental radioactivity around the former Soviet Union's

Semipalatinsk nuclear test site: local fallout Pu in Ust'-Kamenogorsk district. J Radioanal Nucl Chem 252: 373–394.

6. Howard B J, Semioschkina N, Voigt G, Mukusheva M, Clifford J (2004) Radiostrontium contamination of soil and vegetation around the Semipalatinsk test site. Radiat Environ Biophys 43: 285–292.

7. Lehto J, Salminen S, Jaakkola T, Outola I, Pulli S, et al. (2006) Plutonium in the air in Kurchatov, Kazakhstan. Sci Total Environ 366: 206–217.

8. The official report of The Fukushima Nuclear Accident Independent Investigation Commission (2012) In:The National Diet of Japan. 37–41.

9. Ministry of Education, Culture, Sports, Science, and Technology, Japan. Environmental Radioactivity and Radiation in Japan (Available: http://www.kankyo-hoshano.go.jp/en/index.html. Accessed 2012 Oct 25)

10. International Commission on Radiation Units and Measurements (1994) Gamma-Ray Spectrometry in the Environment. ICRU Report 53.

11. United Nations Scientific Committee on the Effects of Atomic Radiation (2000) Sources and effects of ionizing radiation. 2000 Report to General Assembly with scientific annexes. Annex A: Dose assessment methodologies.

12. International Atomic Energy Agency (2000) Generic procedures for assessment and response during radiological emergency. IAEA-TECDOC-1162.

13. Balonov MI, Anspaugh LR, Bouville A, Likhtarev IA (2007) Contribution of internal exposures to the radiological consequences of the Chernobyl accident. Radiat Prot Dosim 127: 491–496.

14. International Commission on Radiological Protection (1991) The 1990 Recommendations of the International Commission on Radiological Protection. ICRP Publication 60; Annals of the ICRP, vol. 21, Nos. 1–3.

15. Zhu YG, Shaw G (2000) Soil contamination with radionuclides and potential remediation. Chemosphere 41: 121–128.

16. Jacob P, Fesenko S, Firsakova SK, Likhtarev IA, Schotola C, et al. (2001) Remediation strategies for rural territories contaminated by the Chernobyl accident. J Environ Radioact 56: 51–76.

17. Iwatani K, Sshizuma K, Hasai H, Nakamura S (1999) Radioactivity measurement of the soil samples collected at the former Semipalatinsk nuclear test site. Bulletin of Hiroshima Prefectural College of Health and Welfare 4: 1–5.

18. He Q, Walling DE (1996) Interpreting particle size effects in the adsorption of ^{137}Cs and unsupported ^{210}Pb by mineral soils and sediments. J Environ Radioact 30 (2): 117–137.

19. Kato H, Onda Y, Teramage M (2011) Depth distribution of ^{137}Cs, ^{134}Cs, and ^{131}I in soil profile after Fukushima Dai-ichi Nuclear Power Plant Accident. J Environ Radioact 2011 Oct 26 [Epub ahead of print].

5

Biogas Production by Co-Digestion of Goat Manure with Three Crop Residues

Tong Zhang[1], Linlin Liu[2], Zilin Song[1], Guangxin Ren[2], Yongzhong Feng[2], Xinhui Han[2], Gaihe Yang[2]*

1 College of Forestry and the Research Center of Recycle Agricultural Engineering and Technology of Shaanxi Province, Northwest A&F University, Yangling, Shaanxi, People's Republic of China, 2 College of Agronomy and the Research Center of Recycle Agricultural Engineering and Technology of Shaanxi Province, Northwest A&F University, Yangling, Shaanxi, People's Republic of China

Abstract

Goat manure (GM) is an excellent raw material for anaerobic digestion because of its high total nitrogen content and fermentation stability. Several comparative assays were conducted on the anaerobic co-digestion of GM with three crop residues (CRs), namely, wheat straw (WS), corn stalks (CS) and rice straw (RS), under different mixing ratios. All digesters were implemented simultaneously under mesophilic temperature at 35 ± 1 °C with a total solid concentration of 8%. Result showed that the combination of GM with CS or RS significantly improved biogas production at all carbon-to-nitrogen (C/N) ratios. GM/CS (30:70), GM/CS (70:30), GM/RS (30:70) and GM/RS (50:50) produced the highest biogas yields from different co-substrates (14840, 16023, 15608 and 15698 mL, respectively) after 55 d of fermentation. Biogas yields of GM/WS 30:70 (C/N 35.61), GM/CS 70:30 (C/N 21.19) and GM/RS 50:50 (C/N 26.23) were 1.62, 2.11 and 1.83 times higher than that of CRs, respectively. These values were determined to be the optimal C/N ratios for co-digestion. However, compared with treatments of GM/CS and GM/RS treatments, biogas generated from GM/WS was only slightly higher than the single digestion of GM or WS. This result was caused by the high total carbon content (35.83%) and lignin content (24.34%) in WS, which inhibited biodegradation.

Editor: Chenyu Du, University of Nottingham, United Kingdom

Funding: This work was supported by science and technology support projects "the biological technology integration and demonstration of high yield biogas digestion from the mix ingredients" (2011BAD15B03) from Ministry of Science and Technology Department of the People's Republic of China, and the Fundamental Research Funds for the Central Universities (QM2012002) from Ministry of Education of the People's Republic of China. The funders had no role in study design, data collection and analysis, decision to publish, or preparation of the manuscript.

Competing Interests: The authors have declared that no competing interests exist.

* E-mail: ygh@nwsuaf.edu.cn

Introduction

China is one of the largest agricultural countries in the world. The production of net available crop residues (CRs) in China is estimated to be over 800 million t/yr [1], which ranks first in the world. The use of agricultural waste as a major component of renewable energy is suitable for improving energy security and decreasing environmental disruption caused by carbon emissions [2,3]. Wheat straw (WS), rice straw (RS) and corn stalks (CS) are the top three agricultural wastes in China and account for 80.5% of the total output (15.7%, 24.2% and 40.6%, respectively) [1]. Thus, studying the energy generation potential of these three wastes is important.

Anaerobic digestion (AD) is a biological process that produces biogas from bio-degradable wastes by bacteria under poor or no oxygen conditions. In the past two decades, AD has been applied as an effective technology for solving the energy shortage and environmental pollution problems of biotechnology industries and residential activities caused by heating and electricity generation [4,5,6].

CRs and animal manure have recently been used together to produce biogas by AD. Compared with the single digestion of feedstock, the co-digestion of CRs and animal manures increases the rate of biogas production because of the greater balance between carbon and nitrogen [7] and improves AD efficiency [8].

Annual goat manure (GM) yield in China is approximately 3.21×10^8 t followed by dairy manure, swine manure and chicken manure [9]. The total nitrogen (TN) contents of fresh GM (1.01%) and chicken manure (1.03%) are significantly higher than those of dairy manure (0.35%) and swine manure (0.24%) [10]. High TN content is beneficial to co-digestion with CRs because it decreases the carbon-to-nitrogen (C/N) ratios of single CRs substrates. GM is also insensitive to acidification during anaerobic fermentation [11,12]. Hence, GM is an excellent raw material for AD. Although various raw materials, such as agricultural waste, animal manures, sewage sludge and food waste have been reported as potentially feasible for co-digestion [7,13,14,15,16,17,18,19,20], the suitable mixing ratios of multi-component substrates between GM and various CRs are largely unknown.

We investigated the biogas-producing efficiency of anaerobic co-digestion influenced by different GM and CR mixing ratios. The best ratio between these substrates was obtained by comparing the results. Furthermore, an optimum co-digestion condition for biogas production was proposed.

Materials and Methods

Feedstocks and inocula

GM was obtained from a local livestock farm near Northwest Agriculture and Forestry University (NWAFU), Yangling Shaanxi,

China. WS, CS and RS were collected from the experimental field of NWAFU. All of these straws were cut into sections at lengths of 2 cm to 3 cm by using a grinder. Inoculum was the anaerobic sludge of dairy manure, which was obtained from an anaerobic digester in a local village.

Experimental digester and design

The experiment was conducted according to Song et al. (2012) by using lab-scale anaerobic digesters fabricated from 1 L Erlenmeyer flasks. Batch reactors were used to determine the co-digestions of GM mixed with three CRs. The working volume of each digester was 700 mL, including 140 g inocula and an appropriate amount of digesting material. Deionized water was added to digesters to maintain a total solid (TS) content of 8% [5]. All reactors were gently mixed manually for approximately 1 min/d prior to biogas volume measurement.

To obtain the best mixing ratio of the co-digestion of GM supplemented with three CRs as external carbon sources, five different mixing mass ratios at 90:10, 70:30, 50:50, 30:70 and 10:90 were tested under mesophilic condition ($35\pm1^{\circ}C$) for 55 d. Unmixed GM (100:0) and CR (0:100) were anaerobically digested as controls. Each treatment was performed thrice with a control to investigate the effect of different mixed ratios on biogas production.

Analysis and statistics

The TS, volatile solids (VS), pH, volatile fatty acid (VFA), and TN content of the materials were determined in accordance with the *Standard Methods for the Examination of Water and Wastewater* of the American Public Health Association [21]. Total carbon (TC) and lignin contents were analyzed by using the method described by Cuetos et al. and Song et al. [5,22]. The amount of biogas produced from each digester was recorded every day by using the water displacement method during the digestion period. Each batch experiment was deemed complete when a clear downward trend in daily biogas volume produced was observed for 10 d.

ANOVA was performed to determine the significant differences among each treatment by using SAS version 8.12 (SAS Institute Inc., Cary, NC, USA).

Results and Discussion

Substrate characteristics

The C/N ratios of the different substrates and substrate mixtures in AD greatly influence biogas production [23,24]. A higher carbon content provides more carbon for CH_4 production, whereas a lower nitrogen content limits microbial activity because microbes need a considerable amount of nitrogen to maintain growth [8]. The ideal C/N ratios range from 9 to 30 for anaerobic digesters [25]. The chemical characteristics of substrates used in this study are shown in Table 1. The C/N ratio of GM was 17.97, which is too low for biogas production. However, the C/N ratios of WS, CS and RS were significantly higher (91.17, 88.13 and 92.91, respectively) than that of GM ($P<0.01$). This result suggested that CRs increased methane production when co-digested with GM under the optimal C/N ratio.

Biogas yields and production rates at different GM/CR ratios

The daily biogas production by the co-digestion of GM and CRs during 55 d of digestion was calculated under different mixing ratios (Fig. 1). Samples from the mixing ratios of GM/WS 30:70, GM/CS 30:70 and GM/RS 50:50 were measured, and their peak yield values were 570, 585 and 525 mL/d on the 17th,

Table 1. Chemical characterization of substrates used in the co-digestion experiments.

	GM	WS	CS	RS
pH	7.94±0.15	ND	ND	ND
TS (%)	33.65±3.23, b	81.08±7.62, a	81.74±7.43, a	77.92±6.97, a
VS (%)	82.21±8.93, a	90.29±9.25, a	91.42±9.33, a	94.23±9.42, a
TC (%)	18.22±1.14, c	35.83±3.17, a	28.82±2.03, b	31.96±2.92, ab
TN (%)	1.014±0.11, a	0.393±0.02, b	0.327±0.04, b	0.344±0.02, b
C/N	17.97±0.84, b	91.17±3.44, a	88.13±4.65, a	92.91±3.10, a
Lignin (%)	ND	24.34±1.89, a	15.38±1.21, b	9.49±0.33, c

TS, total solid; VS, volatile solids; TC, Total carbon; TN, total nitrogen.
The values are the mean ± standard deviation of the triplicate measurements.
ND = not detected.
The ANOVA test was conducted to determine the differences between each cultivar. Values with the same letters indicate no significant difference at $P<0.01$.

19th and 11th d, respectively (Fig. 1). The digestion of single GM substrate (100:0) produced biogas earlier than other combinations but had two relatively small peaks (402 and 500 mL/d) (Fig. 1). By contrast, the digestion of any single CR substrate (0:100) had only one peak (GM/WS 547, GM/CS 540 and GM/RS 477 mL/d) that occurred earlier than the other combinations (3rd d to 6th d) and decreased rapidly after the 16th d (Fig. 1). These results indicate that the co-digestion of GM and CRs could significantly delay the attainment of the highest gas production.

The final cumulative biogas productions by the co-digestion of GM and CRs at different mixing ratios are shown in Fig. 2. The cumulative biogas productions for GM/WS 10:90, 30:70, 50:50, 70:30 and 90:10 were 11890, 12765, 11253, 12685 and 9650 mL, respectively (Fig. 2A). These results showed an increase of 51.0%, 62.1%, 42.9%, 61.1% and 22.6% compared with single WS (7874 mL), and an increase of 51.0%, 62.1%, 42.9%, and 22.6% compared with single GM (10375 mL). However, the biogas production of GM/WS 90:10 (9650 mL) was lower than that of single GM (Fig. 2A). The same trends were observed for the GM/CS and GM/RS treatments, which had considerably higher increases (Fig. 2B and 2C). These data showed that the co-digestion of GM and CRs greatly improved biodegradability and biogas production at most mixing ratios compared with single substrate digestion. Our results supported those of Wu et al. [26], who found that co-digesting swine manure with CS, oat straw and WS significantly increase biogas production and net CH_4 volume at all C/N ratios.

To compare the effect of single substrate digestion and co-digestion with GM and CRs, the total biogas yield of each combination is shown in Fig. 3. The total biogas productions of most co-digestion systems were higher than the single digestion of either GM or CRs except those of GM/WS 90:10 and GM/RS 10:90. GM/CS 70:30 exhibited the highest total biogas yield of 16.02 L in all treatments, which was 83.02% and 54.44% higher than that of CS and GM alone, respectively. Among all the GM/RS treatments, the total biogas production of GM/RS 50:50 (15.70 L) was 111.28% and 51.31% higher than that of CS alone and GM alone, respectively. The co-digestion of GM/WS 30:70 was 62.12% and 23.04% higher than that of WS and GM alone, respectively. Compared with the TC contents of CS (28.82%) and RS (31.96%), the higher TC content of WS (35.83%) suppressed

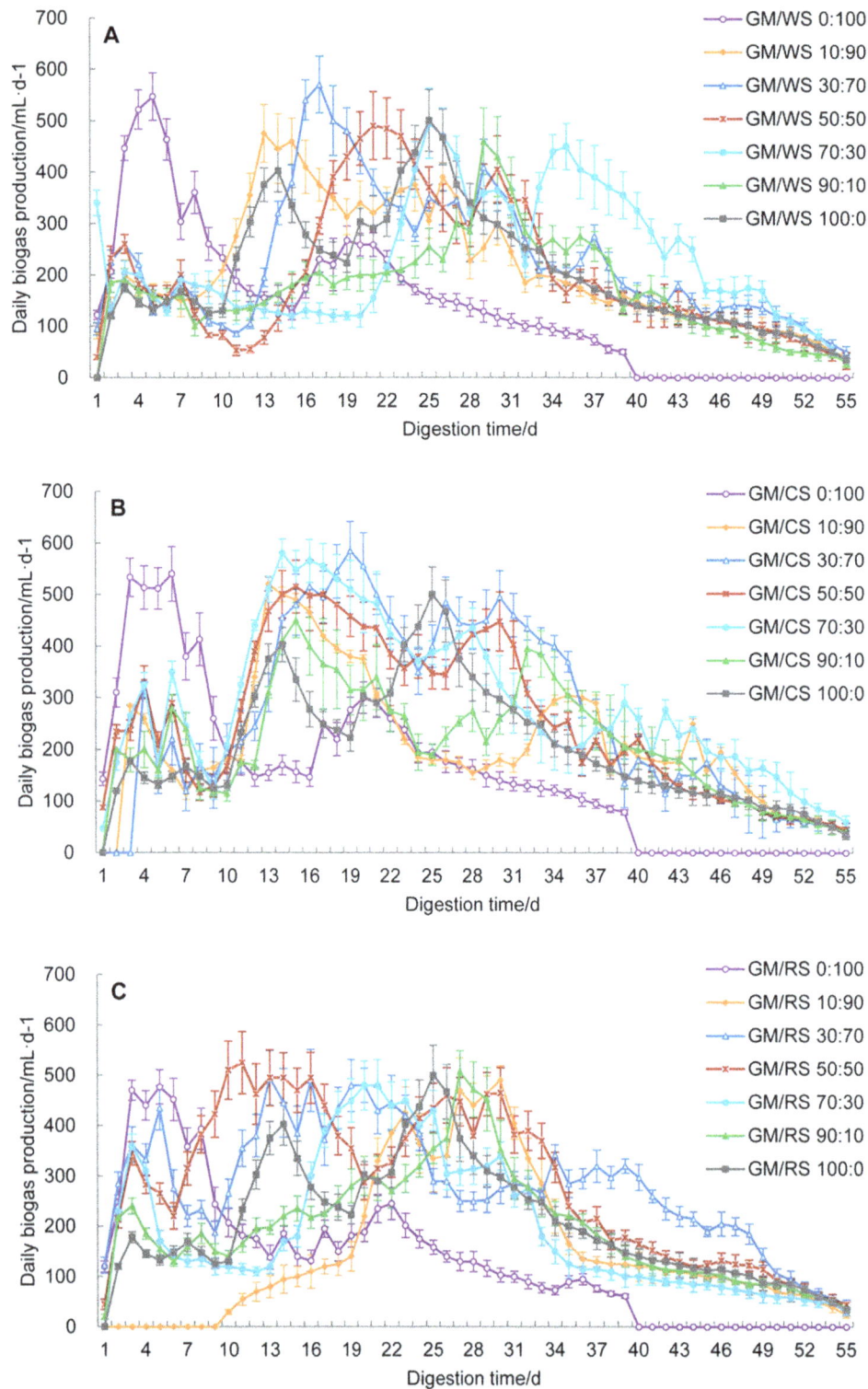

Figure 1. Daily biogas production from the co-digestion of GM and WS (A), CS (B) and RS (C) with different mixing ratios. Mean values originated from three independent replications. Vertical bars represent standard deviations.

microbial growth and methanogenesis because of the ammonium nitrogen deficiency and low pH [22,27,28].

These results indicated that co-digestion with suitable GM and CRs mixtures is an effective way to prolong the period of the highest gas production and improve biogas yield. The ANOVA indicated that the total biogas production of co-digestions were significantly higher ($P<0.01$) than the single digestion of GM or CRs (Fig. 3).

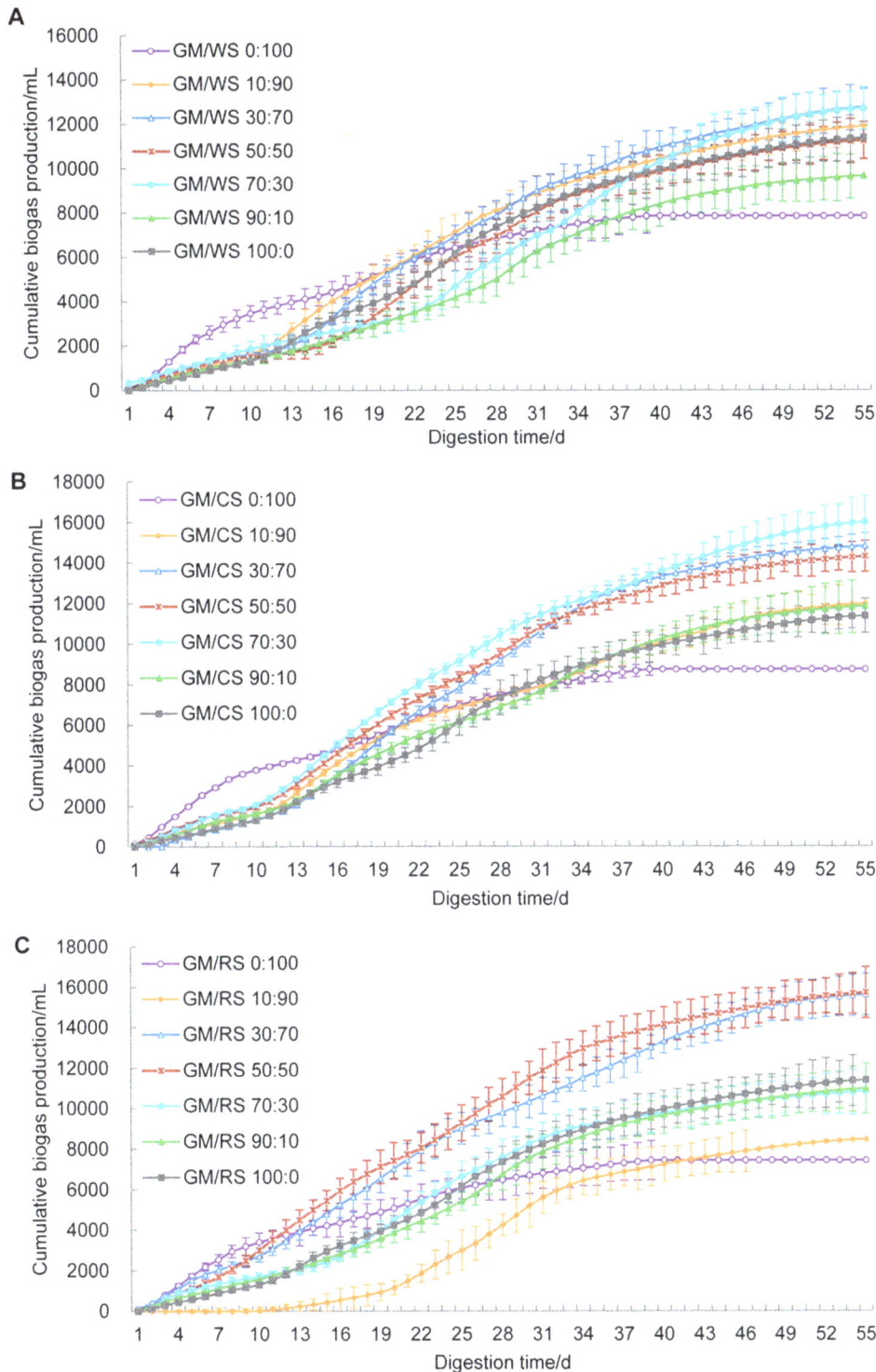

Figure 2. Cumulative biogas productions from co-digestion of GM and WS (A), CS (B) and RS (C) with different mixing ratios. Mean values originated from three independent replications. Vertical bars represent standard deviations.

Effect of C/N ratio on biogas production

The C/N ratio represents the relationship between the amount of carbon and nitrogen present in organic materials and is an important indicator for controlling biological treatment systems [23]. On one hand, a high C/N ratio indicates rapid nitrogen consumption by methanogens and leads to lower gas production.

On the other hand, a low C/N ratio results in ammonia accumulation and an increase in pH values, which is toxic to methanogenic bacteria [29]. The mean value of C/N ratios for each co-digestion combinations and single digestion ranged from 92.79 to 17.97 (Table 2). The C/N ratios of co-digestions were significantly lower than those of CR materials ($P<0.01$, Table 1),

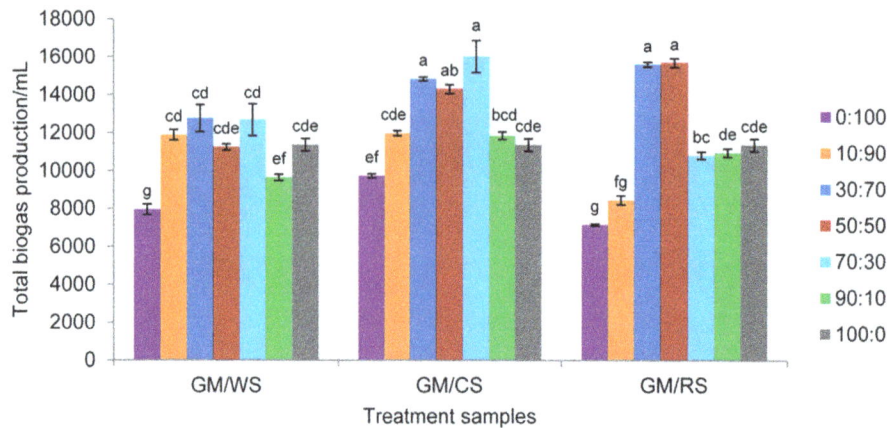

Figure 3. Total biogas productions from anaerobic co-digestion of GM with WS, CS and RS with different mixing ratios. Mean values originated from three independent replications. Vertical bars represent standard deviations. The ANOVA test was conducted to determine the differences between each cultivar. Values with the same letters indicate no significant difference at $P<0.01$.

thus indicating that co-digestion effectively reduced the C/N ratios of AD. Experimental data showed that the biogas yields of most co-digestions were higher than the corresponding single digestions. According to the cumulative biogas production (Fig. 3), the highest biogas yields (12765, 15698 and 16023 mL) at GM/WS 30:70 (C/N 35.64), GM/CS 70:30 (C/N 21.26) and GM/RS 50:50 (C/N 26.28) were 1.62, 2.11 and 1.83 times higher than that of CRs only, respectively. However, the total biogas yields of three GM/CR 10:90 treatments did not increase, and were even lower than that of single substrate. The reason for this result was that the C/N ratios of each GM/CR 10:90 treatment were less than 20 (Table 2). The results suggested that the ideal C/N ratio range is between 20 to 35 in the co-digestion of GM with CRs, which was consistent with the report of Verma [29], which revealed that the optimum C/N ratios in anaerobic digesters were between 20 to 30.

CRs typically contain high lignocellulosic contents. Problems such as low gas yield during the AD of these materials were usually associated with a high C/N ratio or high lignin content [30]. Although the C/N ratio was reduced by most co-digestions, no apparent increasing trend was observed in the biogas production of GM/WS, which even decreased slightly (GM/WS 90:10) compared with GM only. This phenomenon possibly resulted from the significantly higher lignin content (24.34%) of WS substrate than those of CS and RS (15.38% and 9.47%, respectively) ($P<0.01$, Table 1). To overcome the low degradability of lignin, reducing the particle size of CR substrate can increase the degradation rate of lignocelluloses and further improve biogas production [31].

Effects of pH and VFA

VFA and pH are the two key factors in AD [4]. The pH value and total VFA reflected the changing processes in the reactors (Fig. 4). The curves for the individual pH and total VFA of all mixtures and single substrates had similar trends. The growth of methanogens can be significantly influenced by the pH level [32]. The initial pH values of digesters gradually decreased from 6.5 to 6.0 with increasing CR percentage, and GM/RS 10:90 had the lowest pH value (5.5). The pH values increased from 6.5 as the percentage of GM increased in the 6th d, and then remained at approximately 6.8 until the 30th d. This stability confirmed that the daily biogas production of each mixture reached the methanogenesis stage, and that the pH value remained at approximately 6.8. Thereafter, the pH values dropped slightly to 6.0, thus indicating that the digestion changed in the later stages. However, the pH values of GM/CRs 0:100 decreased rapidly after the 18th d, thus showing the buffering capacity of GM. These results indicated that the best pH values for the co-digestion of GM and CRs ranged from 6.5 to 7.5.

VFAs are intermediate organic acid products, and the total VFA concentration is considered an important indicator of metabolic status in addition to the pH value during AD [33,34]. However, the VFA curves showed evidently contrasting trends with that of the pH values. VFA was initially approximately 7380 mg/L to 11767 mg/L for all treatments and then decreased to 4519 mg/L to 5484 mg/L at the 24th d. VFA increased again and finally decreased to 9812 mg/L to 11791 mg/L at the end of digestion (Fig. 3 and 4).

Table 2. Mean values for C/N ratios in the co-digestion of GM with three CRs.

Treatment	Co-digestion mixing ratios						
	0:100	10:90	30:70	50:50	70:30	90:10	100:0
GM/WS	91.05±3.44, a	58.24±0.48, b	35.64±0.58, c	29.71±1.22, d	22.06±0.82, e	19.12±0.83, f	17.97±0.84, f
GM/CS	88.51±4.65, a	53.43±2.50, b	32.64±1.46, c	25.13±1.13, d	21.26±0.97, e	18.90±0.87, e	17.97±0.84, e
GM/RS	92.79±3.10, a	57.46±0.30, b	34.82±0.61, c	26.28±0.77, d	21.80±0.82, e	19.05±0.83, f	17.97±0.84, f

The values are the mean ± standard deviation of triplicate measurements.
The ANOVA test was conducted to determine the differences between each cultivar. Values with the same letters indicate no significant difference at $P<0.01$.

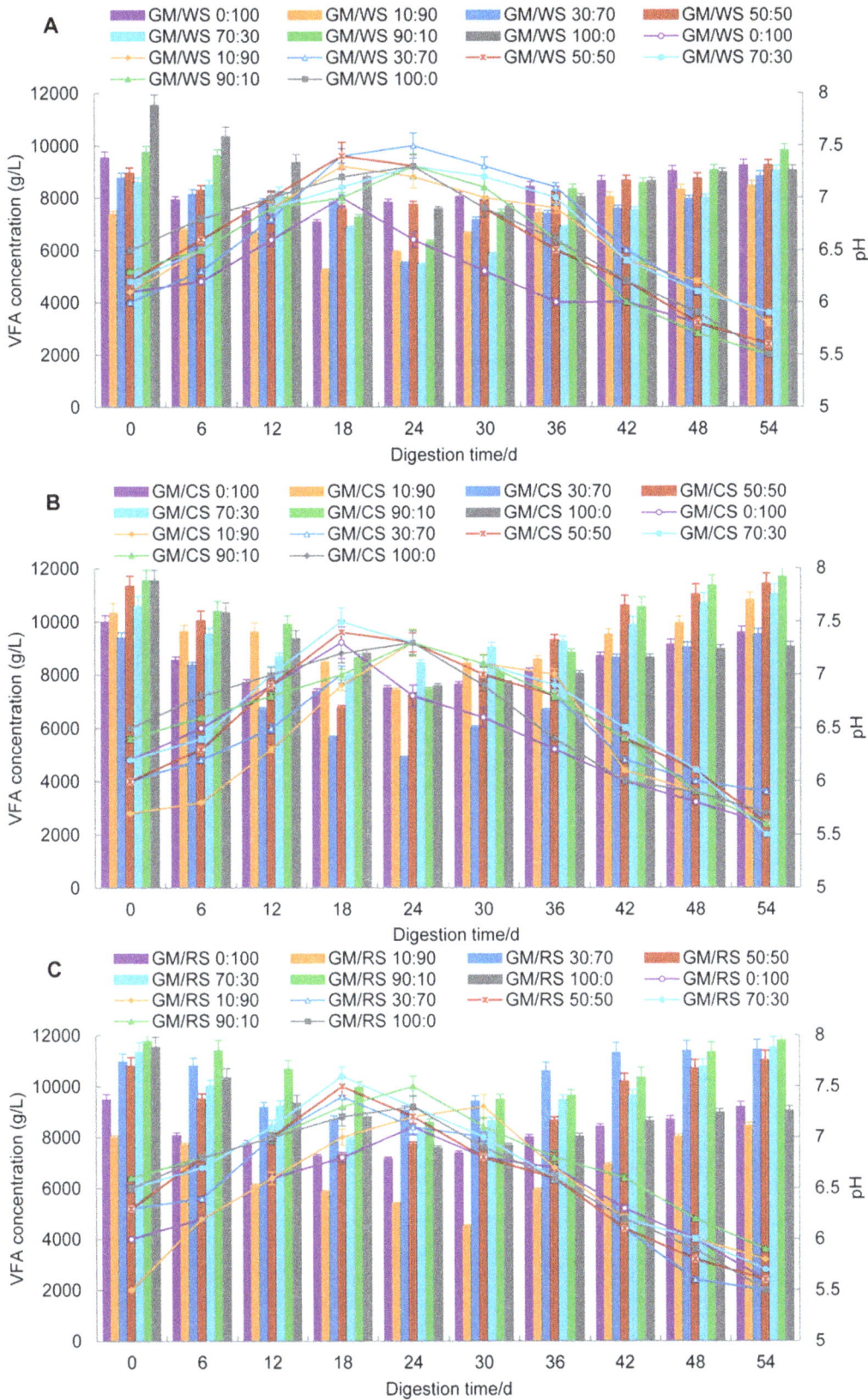

Figure 4. VFA concentrations and pH values from the co-digestion of GM and WS (A), CS (B) and RS (C) with different mixing ratios.
Mean values originated from three independent replications. Columns represent VFA, lines represent pH values, and vertical bars represent standard deviations. VFA, volatile fatty acid.

The ammonia produced by the biological degradation of proteins and urea often results in VFA accumulation. The accumulation of VFA leads to the decrease of pH value, thus affecting the growth of methanogens during the AD process [6,24,30]. Our results showed that pH and VFA were co-related with biogas yield in AD. Thus, the pH values were proportional to biogas yield, whereas VFAs were inversely proportional. These results further indicated that pH decreased with increasing VFA accumulation. High concentrations of VFA are toxic to methanogens and inhibits hydrolysis rates in reactors [35]. The interaction between pH and VFA may lead to an "inhibited steady state" with a lower methane yield [30,36,37]. The extended gas production peaks in each mixing treatment might be explained by the co-digestion of GM and CRs, which relieves the inhibited steady state caused by pH and VFA effectively. The co-digestion of GM and CRs improves the buffering capacity to VFA accumulation and inhibits the acidogenesis process, which is consistent with the previous study [38].

Conclusion

The anaerobic co-digestion of GM with CRs is a promising way for improving biogas production. This co-digestion not only resolves the environmental problems caused by straws burning, but also overcomes C/N ratio imbalances in single digestion substrates and enhances the AD process.

Our results showed that the anaerobic co-digestions of GM with CS and RS were efficient and produced more cumulative biogas by reducing the C/N ratios of substrates. The best ratios were GM/CS 30:70, GM/CS 70:30, GM/RS 30:70 and GM/RS 50:50. However, the co-digestion of GM with WS did not improve the biogas yield significantly, which is consistent with the result in previous research [26]. The higher TC content of WS suppressed microbial growth and methanogenesis because of the deficiency of ammonium nitrogen and low pH. For the pH and VFA ranges in this study, pH decreased with increasing VFA accumulation, thus leading to the inhibition of biowaste hydrolysis rates.

Acknowledgments

We thank Dr. Xiaodong Wang, Dr Furong Liu and Dr. Yuheng Yang for critical reading of this manuscript and editorial guidance.

Author Contributions

Conceived and designed the experiments: TZ YZF GHY. Performed the experiments: TZ LLL ZLS. Analyzed the data: TZ GXR XHH. Wrote the paper: TZ GHY.

References

1. Jiang D, Zhuang D, Fu J, Huang Y, Wen K (2012) Bioenergy potential from crop residues in China: Availability and distribution. Renew Sustain Energy Rev 16: 1377–1382.
2. Field CB, Campbell JE, Lobell DB (2008) Biomass energy: the scale of the potential resource. Trend Ecol Evolut 23: 65–72.
3. Yang Z, Zhang H (2008) Strategies for development of clean energy in China. Petrol Sci 5: 183–188.
4. Madsen M, Holm-Nielsen JB, Esbensen KH (2011) Monitoring of anaerobic digestion processes: A review perspective. Renew Sustain Energy Rev 15: 3141–3155.
5. Song Z, Yang G, Guo Y, Zhang T (2012) Comparison of two chemical pretreatments of rice straw for biogas production by anaerobic digestion. BioResources 7: 3223–3236.
6. Weiland P (2010) Biogas production: current state and perspectives. Appl Microbiol Biotechnol 85: 849–860.
7. El-Mashad HM, Zhang R (2010) Biogas production from co-digestion of dairy manure and food waste. Bioresour Technol 101: 4021–4028.
8. Zhu D (2010) Co-digestion of Different Wastes for Enhanced Methane Production: The Ohio State University.
9. Zhang P, Yang Y, Tian Y, Yang X, Zhang Y, et al. (2009) Bioenergy industries development in China: Dilemma and solution. Renew Sustain Energy Rev 13: 2571–2579.
10. Wang FH, Ma WQ, Dou ZX, Ma L, Liu XL, et al. (2006) The estimation of the production amount of animal manure and its environmental effect in China. China Environ Sci 26: 614–617.
11. Jain M, Singh R, Tauro P (1981) Anaerobic digestion of cattle and sheep wastes. Agr Wastes 3: 65–73.
12. Kanwar S, Kalia A (1993) Anaerobic fermentation of sheep droppings for biogas production. World J Microbiol Biotechnol 9: 174–175.
13. Dai X, Duan N, Dong B, Dai L (2012) High-solids anaerobic co-digestion of sewage sludge and food waste in comparison with mono digestions: Stability and performance. Waste Management.
14. Creamer K, Chen Y, Williams C, Cheng J (2010) Stable thermophilic anaerobic digestion of dissolved air flotation (DAF) sludge by co-digestion with swine manure. Bioresour Technol 101: 3020–3024.
15. Luostarinen S, Luste S, Sillanpää M (2009) Increased biogas production at wastewater treatment plants through co-digestion of sewage sludge with grease trap sludge from a meat processing plant. Bioresour Technol 100: 79–85.
16. Bouallagui H, Lahdheb H, Ben Romdan E, Rachdi B, Hamdi M (2009) Improvement of fruit and vegetable waste anaerobic digestion performance and stability with co-substrates addition. J Environ Manage 90: 1844–1849.
17. Álvarez J, Otero L, Lema J (2010) A methodology for optimising feed composition for anaerobic co-digestion of agro-industrial wastes. Bioresour Technol 101: 1153–1158.
18. Macias-Corral M, Samani Z, Hanson A, Smith G, Funk P, et al. (2008) Anaerobic digestion of municipal solid waste and agricultural waste and the effect of co-digestion with dairy cow manure. Bioresour Technol 99: 8288–8293.
19. Xie S, Lawlor P, Frost J, Hu Z, Zhan X (2011) Effect of pig manure to grass silage ratio on methane production in batch anaerobic co-digestion of concentrated pig manure and grass silage. Bioresour Technol 102: 5728–5733.
20. Nguyen VCN, Fricke K (2012) Energy recovery from anaerobic co-digestion with pig manure and spent mushroom compost in the Mekong Delta. J Vietnamese Environ 3: 4–9.
21. APHA (1995) Standard methods for the examination of water and wastewater: Washington. DC, American Public Health Association.
22. Cuetos MJ, Fernández C, Gómez X, Morán A (2011) Anaerobic co-digestion of swine manure with energy crop residues. Biotechnol Bioprocess Eng 16: 1044–1052.
23. Wang X, Yang G, Feng Y, Ren G, Han X (2012) Optimizing feeding composition and carbon-nitrogen ratios for improved methane yield during anaerobic co-digestion of dairy, chicken manure and wheat straw. Bioresour Technol 120: 78–83.
24. Kayhanian M (1999) Ammonia inhibition in high-solids biogasification: an overview and practical solutions. Environ Technol 20: 355–365.
25. Siddiqui Z, Horan N, Anaman K (2011) Optimisation of C: N ratio for co-digested processed industrial food waste and sewage sludge using the BMP test. Int J Chem React Eng 9.
26. Wu X, Yao W, Zhu J, Miller C (2010) Biogas and CH_4 productivity by co-digesting swine manure with three crop residues as an external carbon source. Bioresour Technol 101: 4042–4047.
27. Panichnumsin P, Nopharatana A, Ahring B, Chaiprasert P (2010) Production of methane by co-digestion of cassava pulp with various concentrations of pig manure. Biomass Bioenerg 34: 1117–1124.
28. Carucci G, Carrasco F, Trifoni K, Majone M, Beccari M (2005) Anaerobic digestion of food Industry Wastes: Effect of codigestion on methane yield. J Environ Eng 131: 1037–1045.
29. Verma S (2002) Anaerobic digestion of biodegradable organics in municipal solid wastes: Columbia University.
30. Chen Y, Cheng JJ, Creamer KS (2008) Inhibition of anaerobic digestion process: A review. Bioresour Technol 99: 4044–4064.
31. Palmowskl L, Müller J (2000) Influence of the size reduction of organic waste on their anaerobic digestion. Water Sci Technol: 155–162.
32. Duarte A, Anderson G (1982) Inhibition modelling in anaerobic digestion. Water Sci Technol 14: 749–763.
33. Fernández A, Sanchez A, Font X (2005) Anaerobic co-digestion of a simulated organic fraction of municipal solid wastes and fats of animal and vegetable origin. Biochem Eng J 26: 22–28.
34. Habiba L, Hassib B, Moktar H (2009) Improvement of activated sludge stabilisation and filterability during anaerobic digestion by fruit and vegetable waste addition. Bioresour Technol 100: 1555–1560.
35. Veeken A, Hamelers B (1999) Effect of temperature on hydrolysis rates of selected biowaste components. Bioresour Technol 69: 249–254.
36. Angelidaki I, Ahring B (1992) Effects of free long-chain fatty acids on thermophilic anaerobic digestion. Appl Microbiol Biotechnol 37: 808–812.
37. Angelidaki I, Ahring B (1993) Thermophilic anaerobic digestion of livestock waste: the effect of ammonia. Appl Microbiol Biotechnol 38: 560–564.
38. Angelidaki I (1997) Anaerobic digestion in Denmark. Past, present and future. Servicio de Publicaciones. pp. 335–342.

6

Comparative and Joint Analysis of Two Metagenomic Datasets from a Biogas Fermenter Obtained by 454-Pyrosequencing

Sebastian Jaenicke[1], Christina Ander[1], Thomas Bekel[1], Regina Bisdorf[1], Marcus Dröge[2], Karl-Heinz Gartemann[3], Sebastian Jünemann[1,4], Olaf Kaiser[2], Lutz Krause[5], Felix Tille[1], Martha Zakrzewski[1], Alfred Pühler[6], Andreas Schlüter[6,9], Alexander Goesmann[1*,9]

1 Computational Genomics, Center for Biotechnology (CeBiTec), Bielefeld University, Bielefeld, Germany, 2 Roche Diagnostics GmbH, Penzberg, Germany, 3 Department of Genetechnology/Microbiology, Bielefeld University, Bielefeld, Germany, 4 Department of Periodontology, University Hospital Münster, Münster, Germany, 5 Division of Genetics and Population Health, Queensland Institute of Medical Research, Herston, Australia, 6 Institute for Genome Research and Systems Biology, Center for Biotechnology (CeBiTec), Bielefeld University, Bielefeld, Germany

Abstract

Biogas production from renewable resources is attracting increased attention as an alternative energy source due to the limited availability of traditional fossil fuels. Many countries are promoting the use of alternative energy sources for sustainable energy production. In this study, a metagenome from a production-scale biogas fermenter was analysed employing Roche's GS FLX Titanium technology and compared to a previous dataset obtained from the same community DNA sample that was sequenced on the GS FLX platform. Taxonomic profiling based on 16S rRNA-specific sequences and an Environmental Gene Tag (EGT) analysis employing CARMA demonstrated that both approaches benefit from the longer read lengths obtained on the Titanium platform. Results confirmed *Clostridia* as the most prevalent taxonomic class, whereas species of the order *Methanomicrobiales* are dominant among methanogenic *Archaea*. However, the analyses also identified additional taxa that were missed by the previous study, including members of the genera *Streptococcus*, *Acetivibrio*, *Garciella*, *Tissierella*, and *Gelria*, which might also play a role in the fermentation process leading to the formation of methane. Taking advantage of the CARMA feature to correlate taxonomic information of sequences with their assigned functions, it appeared that *Firmicutes*, followed by *Bacteroidetes* and *Proteobacteria*, dominate within the functional context of polysaccharide degradation whereas *Methanomicrobiales* represent the most abundant taxonomic group responsible for methane production. *Clostridia* is the most important class involved in the reductive CoA pathway (Wood-Ljungdahl pathway) that is characteristic for acetogenesis. Based on binning of 16S rRNA-specific sequences allocated to the dominant genus *Methanoculleus*, it could be shown that this genus is represented by several different species. Phylogenetic analysis of these sequences placed them in close proximity to the hydrogenotrophic methanogen *Methanoculleus bourgensis*. While rarefaction analyses still indicate incomplete coverage, examination of the GS FLX Titanium dataset resulted in the identification of additional genera and functional elements, providing a far more complete coverage of the community involved in anaerobic fermentative pathways leading to methane formation.

Editor: Ramy K. Aziz, Cairo University, Egypt

Funding: T. Bekel acknowledges financial support from the Bundesministerium fur Bildung und Forschung (BMBF), SysMAP project (grant 0313704). S. Jaenicke acknowledges financial support by the BMBF grant 0313805A 'GenoMik-Plus'. S. Jünemann was supported by the BMBF PathoGeno-Mik Plus project 0313801N. The METAEXPLORE grant of the European Commission (KBBE-222625) is gratefully acknowledged. These funders had no role in study design, data collection and analysis, decision to publish, or preparation of the manuscript. M. Dröge and O. Kaiser are employees of Roche Diagnostics GmbH; sequencing of the obtained DNA sample was funded by Roche Diagnostics GmbH.

Competing Interests: Marcus Dro¨ge and Olaf Kaiser are employees of Roche Diagnostics GmbH, Germany.

* E-mail: agoesman@CeBiTec.Uni-Bielefeld.DE

9 These authors contributed equally to this work.

Introduction

The fraction of renewable energy forms for energy supply is constantly increasing since fossil fuels are running short and energy production from fossil fuels brings about emissions of the greenhouse gas carbon dioxide which has implications on the climate. In this context the production of biogas by means of fermentation of biomass becomes more and more important because biogas is regarded as a clean, renewable and environ- mentally compatible energy source [1,2]. Moreover, generation of energy from biogas relies on a balanced carbon dioxide cycle. In Germany biogas is mainly produced from energy crops such as maize and liquid manure in medium-sized agricultural biogas plants [1]. The microbiology of biogas formation from organic matter is complex and involves interaction of different microor- ganisms. In the first step of the digestion process, organic polymers of the substrate such as cellulose, other carbohydrates, proteins and lipids are hydrolysed to low-molecular weight compounds [3–5].

Cellulolytic *Clostridia* and *Bacilli* are among other bacteria important for this step. Subsequently, fermentative bacteria convert low-molecular weight metabolites into volatile fatty acids, alcohols, and other compounds which are then predominantly metabolised to acetate, carbon dioxide and hydrogen by syntrophic bacteria [6–11]. These latter compounds are in fact the substrates for methane synthesis which is accomplished by methanogenic *Archaea* [12,13]. Hydrogenotrophic *Archaea* are able to reduce carbon dioxide to methane using hydrogen as an electron donor, whereas aceticlastic *Archaea* convert acetate to methane [14–18]. The biochemistry and enzymology of methanogenesis is well known for model organisms, but the functioning of biogas-producing microbial communities on the whole is insufficiently explored. Community structures of biogas-producing microbial consortia were analysed for different systems and settings including a thermophilic municipal biogas plant [19], a thermophilic anaerobic municipal solid-waste digester [20], thermophilic upflow anaerobic filter reactors [21], a completely stirred tank reactor fed with fodder beet silage [22], a two-phase biogas reactor system operated with plant biomass [23], an anaerobic sludge digester [24], mesophilic anaerobic chemostats [25,26], a packed-bed reactor degrading organic solid waste [27] and many other habitats. Most of these studies were based on the construction of 16S rRNA clone libraries and subsequent sequencing of individual 16S rRNA clones. The resulting nucleotide sequences were then taxonomically and phylogenetically classified to deduce the structure of the underlying community. Also, *mcrA* clone libraries were used to elucidate methanogenic archaeal communities of different habitats [28–32]. The *mcrA* gene encodes the alpha subunit of methyl-coenzyme M reductase representing the final enzyme in the methanogenesis pathway. Since *mcrA* is present in all methanogenic *Archaea* analysed so far, it serves as a phylogenetic marker for this group of *Archaea*. Usually, analyses of *mcrA* and 16S rRNA clone libraries do not cover the whole complexity of the respective habitats since sequencing can only be done for limited numbers of clones. Moreover, results of clone library analyses are always biased by the choice of primers that are used for amplification of marker gene fragments and cloning efficiencies. In recent years, microbial communities have been studied on the basis of their metagenomes which became accessible by applying high-throughput sequencing technologies. Recently, the first metagenome sequencing approach for a biogas-producing community was described [33]. Community DNA isolated from a production-scale biogas plant fed with maize silage, green rye and low amounts of chicken manure was sequenced on the Genome Sequencer FLX platform which resulted in 142 million base pairs of sequence information. Bioinformatic methods were employed to deduce the taxonomic composition and functional characteristics of the intrinsic biogas community [34]. Analysis of the community revealed *Clostridia* as the most prevalent phylogenetic class, whereas species of the order *Methanomicrobiales* are dominant among methanogenic *Archaea*.

Similar results were obtained by parallel construction of 16S rRNA and *mcrA* amplicon libraries and subsequent sequencing of cloned fragments [35]. Moreover, bioinformatics results indicated that *Methanoculleus* species play a dominant role in methanogenesis and that *Clostridia* are important for hydrolysis of plant biomass in the analysed fermentation sample. Rarefaction analysis of the metagenome data showed that the sequencing approach was not carried out to saturation. Sufficient coverage of non-abundant microbial groups in the fermentation sample would require deeper sequencing. Therefore, the available total community DNA preparation from the biogas fermentation sample was additionally sequenced on the GS FLX Titanium platform, which provides longer read lengths and increased throughput compared to the GS

FLX platform. This paper describes an integrated analysis of the GS FLX and the GS FLX Titanium datasets with the objective to deepen the knowledge on the taxonomic structure and composition of a microbial community involved in biogas production within an agricultural, production-scale biogas plant. Moreover, the described analysis intends to elucidate the metabolic capacity of the community, functional roles of specific microorganisms and key organisms for the biogas production process.

Methods

Total community DNA preparation and sequencing

Total community DNA of a biogas fermentation sample obtained from an agricultural biogas plant was prepared by a CTAB-based DNA-isolation method [36] as described recently. More detailed information on the origin of the fermentation sample is given in a previous publication [33]. An aliquot of the DNA preparation that was recently used for whole genome shotgun sequencing on the Genome Sequencer (GS) FLX platform now served as template DNA for sequencing on the GS FLX Titanium platform. The sequencing library was constructed according to the protocol of the GS Titanium General Library Prep Kit (Roche Applied Science). After titration of the library using the GS Titanium SV emPCR Kit, a full sequencing run was carried out on the GS FLX Titanium platform.

Data normalization

To assess the overall comparability of the pyrosequencing datasets obtained from the GS FLX and Titanium platforms, the average GC content of all reads was determined for each dataset. For this, various Perl scripts were developed to determine the overall and individual GC content of the obtained reads. Results were visualized using the statistical computing software R [37].

To normalize the data with respect to the observed GC bias (see Results), an outlier detection approach was applied and the endpoints of the linear phases were determined; sequences longer than the computed thresholds were excluded. A linear regression was calculated for the GC plot beginning with 30 data points starting at the 100 bp position to exclude the portion of the dataset with high variance in GC content. In the next step, externally studentized residuals [38] were computed for the linear regression. Each data point was inspected whether it deviated from the linear trend using a Student-t-test. A data point was regarded as being an outlier if the $p-value$ of the studentized residuals test was below 0.05. If ten outliers in a row were found, the first one of these is representing the end of the linear phase and was taken as threshold value for filtering of the dataset.

To rule out sequencing errors as the reason for the observed decreasing GC content of the longer reads, the analysis was repeated for publicly available pyrosequencing datasets from both metagenome as well as genome sequencing projects which were obtained from NCBI's Short-Read Archive (SRA). The effect in question could be observed for all metagenome datasets, where the DNA fragments from a mixture of organisms have a broad distribution concerning their GC contents. Single genome sequencing projects, on the other hand, did not show this effect, thereby ruling out sequencing errors as an alternative explanation (see File S1). The comparably narrow GC distribution in the sequence data from single genome projects does not reveal this effect, and while backfolding might limit maximum obtainable read length, similar filtering steps would not be a prerequisite before e.g. subsequent assembly.

A recent study [39] described the occurence of artificially created duplicate reads in datasets generated using the pyrosequencing method. It is assumed that the duplication of individual DNA fragments occurs during the emulsion PCR reaction, which is a step of the library preparation. These nearly identical sequences might lead to inflated estimates of functional genetic elements or introduce an artificial shift in taxonomic profiles, unless they are filtered from the dataset. In this study, these duplicates were removed using the cdhit-454 [40] program, which filters almost identical reads beginning at the same position. Each dataset was processed separately using the accurate mode (option '-g 1') of the software.

Identification and taxonomic classification of 16S rRNA fragments

A BLAST [41] search versus the RDP database (Release 10.10) was conducted to identify reads carrying fragments of 16S rRNA genes. Since BLAST excludes regions with low sequence complexity by default, the sequence complexity filter was explicitly disabled (option '-F F'). Alignments with an E-value of 10^{-5} or better and a minimum length of 50 bp were extracted and processed using the RDP classifier [42], which employs a naive bayesian classifier to assign the sequences to taxonomic categories. Only 16S rRNA fragments with at least 80% assignment confidence were considered.

Taxonomic classification of Environmental Gene Tags (EGTs)

Metagenomic sequences were filtered using a BlastX search against the Pfam database [43]. Reads without a hit were additionally scanned for conserved protein domains by conducting a search for protein family members using Pfam HMMs. Community sequence reads that were predicted to contain fragments of genes (environmental gene tags, EGTs) were subsequently classified on different taxonomic ranks based on a phylogenetic tree of the metagenomic read itself and the matching Pfam protein family member sequences. A full description of the CARMA pipeline can be found in the original publication [44].

Here, an unpublished improved version of CARMA was applied to process both pyrosequencing datasets using the default settings. For reads containing more than one EGT and therefore more than one classification, the taxonomic classifications were merged into a single one that was determined as the lowest common ancestor of all classifications. While this approach decreases the number of classifications on the lower taxonomic ranks, it simultaneously eliminates contradicting classifications and reduces the influence of potential mis-classifications.

Community participation in substrate decomposition, fermentation and methane production

For further analysis both datasets were merged together, as no significant differences in the taxonomy as well as in the related enzyme composition could be observed. Orthologous groups of genes were retrieved by comparing the sequences to the eggNOG database [45]. For this, sequences were assigned to COGs and NOGs using BlastX with an E-value cutoff of 10^{-6} and annotated according to their best hit. Additionally, taxonomic information was added using the results obtained from the CARMA pipeline. Enzymes relevant for biochemical processes were identified with the help of MetaCyc [46] and the corresponding COGs (Clusters of Orthologous Groups) were selected and grouped. The conclusive tree construction based on the NCBI taxonomy [47], visualisation and data analysis were performed using unpublished software.

Identification of *Methanoculleus* variants

To confirm the presence of different *Methanoculleus* species in the biogas fermenter, 16S rRNA fragments that were identified by a

homology search and classified as belonging to the genus *Methanoculleus*, as described above, were assembled into longer contigs.

Here, the 16S rRNA nucleotide sequence from *Methanoculleus bourgensis* (GenBank accession AY196674) was used as an assembly template. The pyrosequencing reads were aligned to the reference sequence using 'align0' from the FASTA3 package [48] and then clustered into several groups based on their SNP (single nucleotide polymorphism) content. Afterwards, each of the resulting groups of reads was separately assembled employing Roche's GS de novo Assembler software. To put the resulting consensus sequences into context, a phylogenetic characterization was performed. Together with other 16S rRNA sequences obtained from GenBank, a multiple alignment of all sequences was computed using the MUSCLE tool [49] and a phylogenetic tree was generated with the fdnapars program [50], which implements the DNA parsimony algorithm.

Mapping of pyrosequencing reads to *Methanoculleus marisnigri* JR1

To identify the overall sequence similarity between the metagenomic reads and the genome of *Methanoculleus marisnigri* JR1, all reads were mapped to the published DNA sequence of this genome [51]. In a previous study [34], reads were aligned to reference sequences using BLAST with rather relaxed settings (E-Value $<10^{-4}$, aligned region >100 bp, sequence identity $>80\%$). Since this approach did not account for the relative fraction of a read taking part in an alignment, in this study the gsMapper software from Roche was employed. Coverage information for the genome and individual genes was extracted from the gsMapper output using various Perl scripts; the results for gene coverage were normalized and divided into two groups based on the prediction found in the HGT-DB [52], a database of procaryotic genomes that uses different statistical properties of coding sequences to predict whether they may have been acquired by horizontal gene transfer. While GC content is one of these parameters, several others such as codon and amino acid usage are also used. The R statistical software was employed to compute the kernel density estimates for both groups; for this, an Epanechnikov kernel with a bandwith of 0.4 was used.

Detection of hydrogenase gene fragments

To analyse the occurrence of metagenomic reads encoding hydrogenases or proteins involved in hydrogen uptake systems, corresponding genes annotated as hydrogenases or hydrogen uptake systems were extracted from the reference genome of *Methanoculleus marisnigri* JR1 [51]. Related gene fragments in the metagenome datasets were identified by a BLAST search using an e-value cutoff of 10^{-5} and a disabled sequence complexity filter (option '-F F'); for each sequence, only the best BLAST hit was considered.

Biodiversity and Rarefaction analysis

To gain an overview of the biodiversity of the studied microbial community, the Shannon index was computed based on 16S rRNA fragments classified on rank genus with at least 80% confidence as

$$H = -\sum_{i=1}^{n} p_i \cdot \ln p_i,$$

where p_i is the relative abundance of sequences assigned to genus i, and n denotes the total number of different genera.

Additionally, a rarefaction analysis was employed to assess the coverage of the microbial community by the datasets. The number of genera that would be observed for different sample sizes was estimated using Analytic Rarefaction (version 1.3). Rarefaction curves were obtained by plotting the sample sizes versus the estimated number of genera.

A separate rarefaction analysis was conducted to assess the coverage of the collective gene content of the microbial community. The number of Pfam protein families that would be identified in metagenomes of different sizes was estimated using Analytic Rarefaction. Rarefaction curves were generated by plotting the number of Pfam families versus the sample sizes.

Data availability

Sequence data from both the GS FLX and Titanium run has been deposited at the NCBI Short Read Archive (SRA) under the accessions SRR030746.1 for the GS FLX and SRR034130.1 for the Titanium dataset. Assembled 16S rRNA consensus sequences of the different *Methanoculleus* variants have been submitted to the GenBank database (accessions GU731070 to GU731076).

Results

Sequencing of the biogas microbial community on the GS FLX Titanium platform and data preprocessing

Metagenomic DNA from a biogas-producing microbial community residing in the fermenter of an agricultural biogas plant fed with maize silage, green rye and low amounts of chicken manure was recently sequenced on the Genome Sequencer (GS) FLX platform [33]. This approach resulted in 616,072 sequence reads with an average read length of 230 bases, accounting for approximately 142 million bases sequence information. Analysis of the obtained sequence data revealed that sequencing was not carried out to saturation [34]. To achieve a deeper coverage of the intrinsic biogas community and to elucidate its whole complexity, the same community DNA preparation that was used for sequencing on the Genome Sequencer FLX was now sequenced on the GS FLX Titanium platform. The single Titanium run yielded 1,347,644 reads with an average length of 368 bases resulting in 495.5 million bases sequence information, which represents a 3.5-fold increase in coverage of the sample compared to the previous sequence dataset. Statistical data of the Titanium run are summarized and compared to the sequencing approach on the GS FLX platform in Table 1. Although aliquots of the same DNA sample were used for sequencing on both platforms, the average GC content of the reads generated on the GS FLX platform was determined as 51.7%, while the GC content of the Titanium reads was determined as only 47.4%. Also, the average GC content for different read lengths was analysed. Results showed that GC content and obtained read length clearly correlate for both GS FLX and the Titanium platform. The applied sequencing methods did not only generate reads with different average levels of GC content, but also both pyrosequencing platforms show a significant decline in GC content once the read length exceeds a certain value (Figure 1).

While the differences in average GC content can be explained by variations in the sequencing chemistry and protocol used for sequencing on the GS FLX and Titanium platforms, the sharp decline of the GC content present in the longer reads is most probably caused by backfolding. It is assumed that the GC-rich parts of the synthesized ssDNA are forming stable secondary structures during the emulsion PCR process, which subsequently leads to a termination of the PCR reaction and only the non-folded part of the DNA fragment being amplified. On the other

hand, DNA fragments with lower GC content are not affected by this problem and can be amplified over their full length.

Previous studies have already identified GC biases in reads generated on the Illumina platform [53], but no such findings have yet been reported for Roche's pyrosequencing method. Such biases caused by the sequencing chemistry and protocol are likely to remain unnoticed in many cases, where research is focused on analysis of single datasets with no additional data available for comparison. In the context of metagenome studies, these effects can easily lead to a distortion of results, where e.g. taxonomic profiles would underestimate the relative abundance of GC-rich organisms.

In this study, both datasets were filtered to minimize the impact of the identified GC bias and exclude the artificial duplicate sequences. The endpoints of the linear phases of the GC plot were determined at 262 bp for the GS FLX and 535 bp for the Titanium dataset; subsequent removal of sequences longer than the computed thresholds resulted in the exclusion of 169,569 (27.52%) of the GS FLX reads and 26,795 (1.98%) sequences from the Titanium dataset, which is only marginally affected. This shows a significant improvement regarding GC bias for the GS FLX Titanium chemistry compared to the previous GS FLX chemistry.

After the removal of technical replicates, 407,558 reads from the GS FLX dataset and 1,019,333 of the Titanium reads passed all filtering steps and are suitable for further analysis. Throughout this study, these normalized datasets were used for the computation of relative abundances of either taxonomic groups or the functional characterization.

Taxonomic composition of the microbial community based on 16S rRNA analysis

The DNA sequence of the 16S rRNA gene has found wide application for taxonomic and phylogenetic studies. It is highly conserved between both *Archaea* and *Bacteria*, but also contains hypervariable regions that can be exploited for accurate taxonomic assignments. Using the previously filtered pyrosequencing reads to avoid a distortion of taxonomic profiles, a BLAST homology search using the RDP database was performed. In this step, 616 16S rRNA fragments with an average length of 159.5 bp were identified in the GS FLX dataset, while 2,709 fragments with an average length of 245.2 bp from the Titanium reads could be detected. The differing number of 16S rRNA sequences identified in the GS FLX dataset in comparison to the findings of a previous study [34] can be explained by the data normalization step as described above. Furthermore, the ARB database [54] was used instead of the RDP database. All identified 16S rRNA sequences were taxonomically classified by means of the RDP classifier [42]. For all taxonomic ranks except domain, the RDP classifier was able to assign a larger fraction of 16S rRNA fragments from the Titanium dataset with at least 80% confidence than from the FLX data. While 18.2% of the 16S rRNA fragments from the GS FLX reads could be classified on rank order, 25.3% of the Titanium 16S rRNA fragments were assigned on this rank, showing that the RDP classifier benefits from the longer reads generated on the Titanium platform. On rank genus, 6.0% of the GS FLX 16S rRNA fragments and 10.6% of the Titanium 16S rRNA fragments could be assigned. Nevertheless, it has to be noted that only a low fraction of pyrosequencing reads actually contains 16S rRNA fragments (0.15% of the filtered GS FLX reads; 0.26% of the filtered Titanium reads). While 16S rRNA genes are reliable phylogenetic markers, the low number that could be detected in the present metagenome datasets gives only a broad overview of the most abundant taxonomic groups and may be insufficient to obtain a detailed picture of the taxonomic composition of a metagenome.

Table 1. Sequencing results.

	number of reads	number of bases	avg. read length
GS FLX	616,072	141,685,079	230.0 bp
Titanium	1,347,644	495,506,659	367.7 bp
ratio Titanium/GS FLX	2.19	3.50	1.60

Overview of sequence data obtained from the studied biogas fermenter employing different pyrosequencing technologies.

Taxonomic classification of Environmental Gene Tags (EGTs)

One of the major limitations of the taxonomic classification of 16S rRNA fragments is the low fraction of reads actually containing 16S rRNA specific sequences. The CARMA software [44] overcomes this limitation as it is based on the identification of environmental gene tags (EGTs) in community sequences using profile hidden Markov models (pHMMs) and the phylogenetic deduction of the taxonomic origin of these fragments. The major strength of this approach is the high accuracy of pHMMs for the detection of short functional segments of Pfam protein family members. The search for conserved Pfam protein fragments resulted in a total of 100,546 identified EGTs (25% of all reads) for the GS FLX dataset and 329,550 (32% of all reads) for the Titanium dataset, respectively. At the taxonomic rank superkingdom, 88,528 of the GS FLX reads (22%) and 290,008 of the Titanium reads (28%) could be successfully classified. The fact that a higher fraction of Titanium reads could be assigned indicates

that CARMA as well benefits from the increased read lengths generated by the Titanium platform. The community composition as deduced from EGTs identified in both datasets is almost identical: The majority of EGTs was classified as belonging to the superkingdoms *Bacteria* (GS FLX: 67%; Titanium: 70%) and *Archaea* (11% and 8%, respectively). This conformity of taxonomic composition was also observed for almost all the prevalent and rare taxa at other taxonomic ranks (Figure 2). Despite this general accordance, there are some noteworthy differences between the GS FLX and Titanium datasets. While the summarized percentage of identified taxa at rank genus is roughly the same in both datasets, the actual abundance of specific genera differs between both datasets (Figure 3). A higher amount of variation exists especially for the most abundant genera between the two datasets: Both datasets support *Methanoculleus* as the most abundant genus, but a higher fraction of GS FLX reads was assigned to this genus than from the Titanium reads (4.36% of the GS FLX dataset; 3.46% of the Titanium data). For the genus *Clostridium*, this ratio is reversed: only 2.78% of the GS FLX reads, but 3.19% of the Titanium sequences were allocated to this genus. Similar differences can also be noted for *Bacteroides* and *Bacillus*. Even though differences exist between the two taxonomic profiles, results based on analysis of the Titanium dataset can be considered more reliable since they were deduced from a larger amount of sequence information. The discrepancy of the taxonomic profile for the GS FLX dataset as shown here in contrast to previously reported results [34] is due to the applied GC filtering step and adjustments within the CARMA software. Applying an unpublished, enhanced version of CARMA to the unfiltered GS FLX dataset, EGTs were identified for 167,134 (32%) of the reads, while only 133,337 (22%) EGTs were reported in the previous study.

GC content of pyrosequencing data

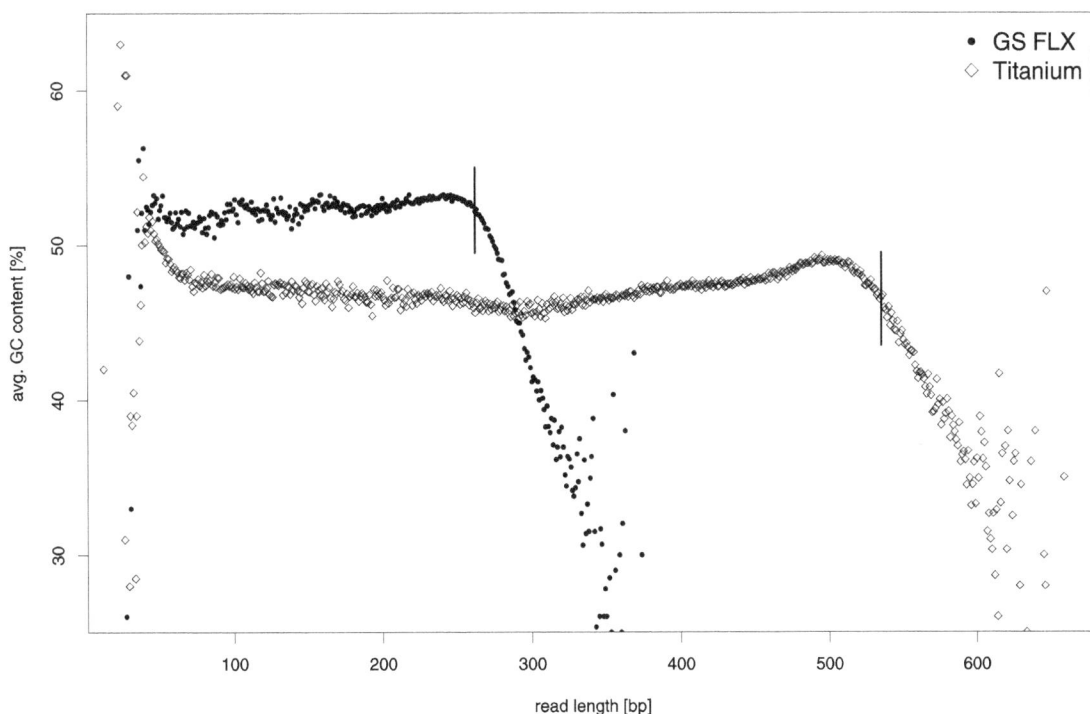

Figure 1. Read length and average GC content of pyrosequencing reads. A sharp decline in GC content can be seen once the read length exceeds a certain value. The vertical bars indicate the computed filtering thresholds for the GS FLX and the Titanium dataset, respectively.

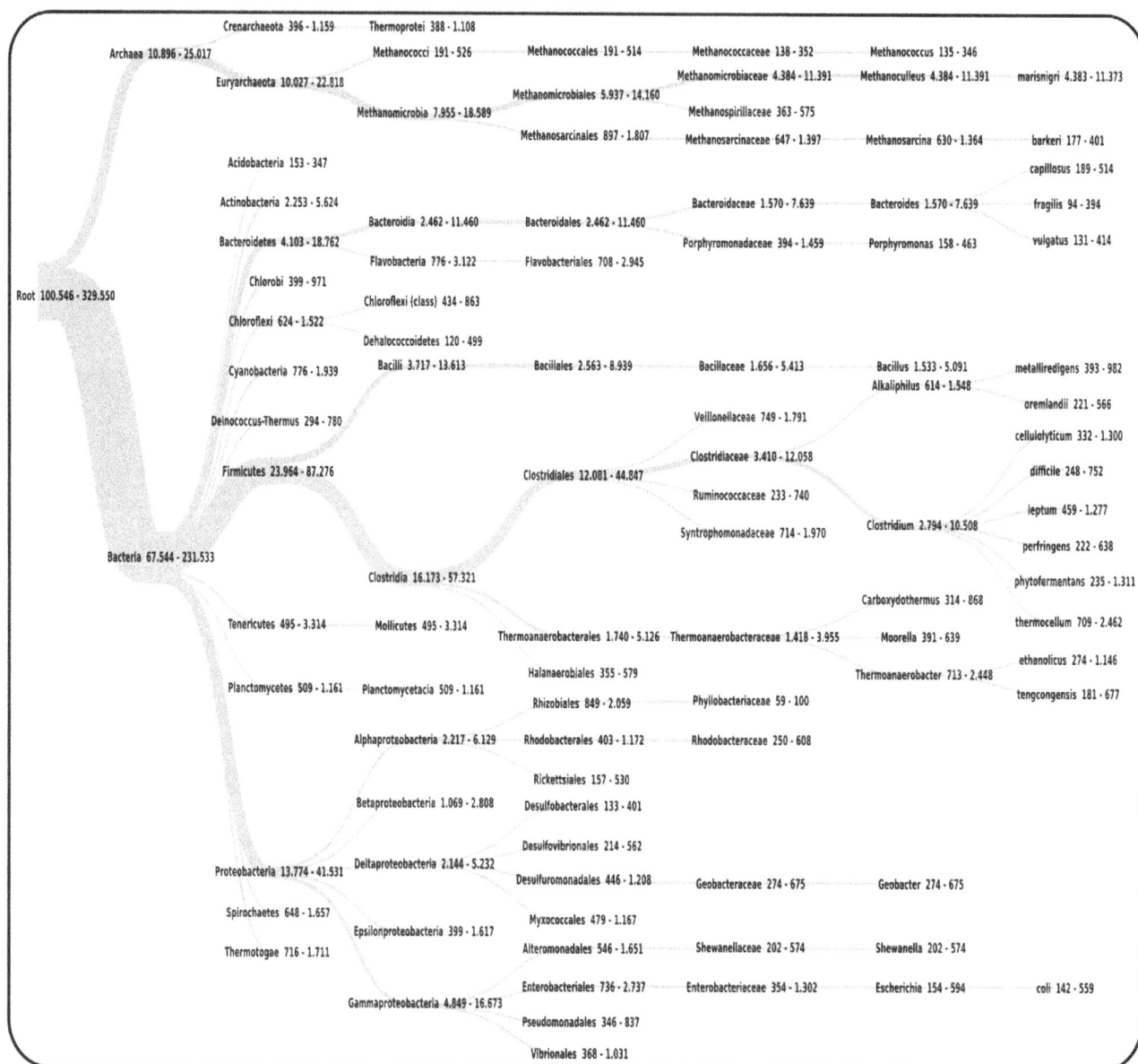

Figure 2. Characterization of the GS FLX and Titanium datasets based on the taxonomic classification of Environmental Gene Tags (EGTs). Displayed are only the most abundant taxa among *Bacteria* and *Archaea* lineages at various taxonomic ranks. For each group, the first number represents the number of assigned EGTs from the GS FLX dataset, the second number the EGTs from the Titanium dataset, respectively. Due to the different number of reads obtained from each sequencing platform, the amount of EGTs listed for the Titanium dataset is typically the fourfold of the number listed for the GS FLX dataset.

New genera identified in the GS FLX Titanium dataset

Due to the deeper coverage of the metagenome by the dataset obtained on the GS FLX Titanium system, new genera were identified in the corresponding taxonomic profile (see Table 4.1, File S1). These include *Streptococcus, Acetivibrio, Garciella, Tissierella, Gracilibacter, Gelria, Dysgonomonas* and *Arcobacter* to mention just a few. *Streptococcus* species were previously detected in different anaerobic habitats, especially in a mesophilic hydrogen-producing sludge and a glucose-fed methanogenic bioreactor [55–57]. In the latter habitat, an acetate- and propionate-based food chain was prevalent but the specific functions of the *Streptococcus* members dominating the bioreactor are not known [55]. Sequences related to the genus *Acetivibrio* (*Firmicutes*) were recently recovered from a community involved in methanogenesis utilizing cellulose under

mesophilic conditions [58]. *Acetivibrio* species most probably play a role in cellulose degradation [58,59]. Species of the genera *Garciella, Tissierella, Gracilibacter* and *Gelria* (all *Firmicutes*) are also adapted to anaerobic habitats where they are involved in different fermentative pathways [60–65]. Interestingly, a reference species of the genus *Gelria*, namely *G. glutamica*, was isolated from a propionate-oxidizing methanogenic enrichment culture and represents an obligately synthrophic, glutamate-degrading bacterium that is able to grow in co-culture with a hydrogenotrophic methanogen [65]. In this context it should be mentioned that hydrogenotrophic methanogens are dominant in the fermentation sample analysed in this study. The genus *Dysgonomonas* (see Table 4.1, File S1) belongs to the family *Porphyromonadaceae* of the order *Bacteroidales*. Members of the genus *Dysgonomonas* were *inter alia*

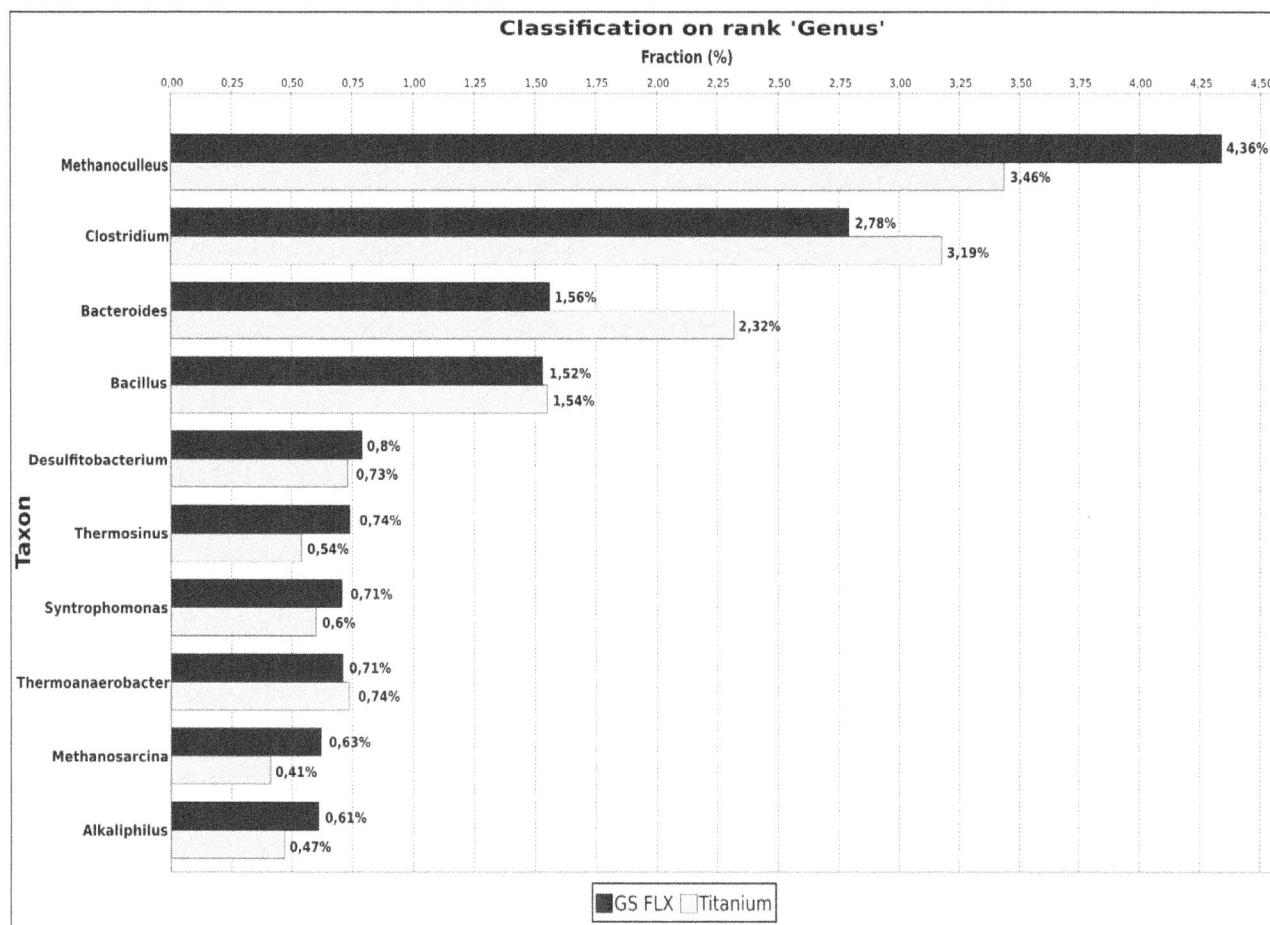

Figure 3. Comparison of taxonomic profiles on rank genus. The taxonomic profiles for the GS FLX (black bars) and the Titanium (lightgray bars) datasets were computed employing the CARMA pipeline. The percentage values correspond to the total amount of reads in the filtered datasets; included are the ten most abundant genera.

isolated from stool samples and are able to ferment glucose resulting in the production of acids [66]. Likewise, bacteria of the genus *Alkaliflexus* also cluster within the order *Bacteroidales* and represent anaerobic saccharolytic organisms [67]. Other genera such as *Arcobacter* were indeed found in an anaerobic community digesting a model substrate for maize but seem to play no dominant role in degradation of polysaccharides [68]. Some of the genera listed in Table 4.1 (File S1) have not been described for anaerobic, methanogenic consortia so far and hence their involvement in the biogas production process remains unknown. In summary, the more detailed taxonomic analysis presented in this study also revealed less-abundant genera that were missed in the previous taxonomic profile for the same community. Some of the newly identified genera presumably are of importance for the biogas production process.

Community participation in substrate decomposition, fermentation and methane production

As described in a previous study, *Firmicutes* and *Methanomicrobiales* play a crucial role in hydrolysis, acetogenesis and methanogenesis representing key steps in anaerobic degradation of plant biomass [34]. In this study corresponding pathways are investigated in more detail using the combined dataset consisting of the GS FLX and newly acquired Titanium reads, and an elaborated methodology to identify key organisms involved in the above mentioned

processes. For this purpose all reads were classified according to Cluster of Orthologous Groups (COG) categories to infer the functional potential of the underlying community. In a second step, obtained COG results were annotated with the taxonomic information generated by the CARMA software. This approach led to the identification of 292,782 reads (about 21% of the dataset) for which both functional as well as taxonomic information could be retrieved. Moreover, a subset of COG entries representing (a) the process of 'polysaccharide degradation', (b) 'acetogenesis' and (c) the 'methanogenesis' step within the fermentation process were chosen for a more detailed taxonomic analysis (see File S1). Even though not all reads classified into the selected COG categories may actually represent the pathways in the focus of this approach, this analysis provides insights into the relevance of different taxonomic groups for the hydrolysis, acetogenesis, and methanogenesis steps in fermentation of biomass. Misallocation of reads to these processes can be due to the fact that some COG entries include enzymes involved in different, but functionally related pathways.

One of the first steps in the fermentation process is the breakdown of polysaccharide components of plant cell material. Especially, the cell wall polymers cellulose, hemicellulose and pectin constitute a high amount of the carbon and energy resource that is available for bacteria in the fermentation sample. Accordingly, hydrolysis of plant biomass is considered to be the rate-limiting step in biogas production. Based on the assignment to COG entries (see File S1), a total of 4,762 reads were classified as potentially coding for enzymes involved in the degradation of complex polymers, namely cellulose, hemicellulose, and lignin (Figure 4a). Reads allocated to the cellulose degradation process account for 71% of the 4,762 reads. Hence, they constitute the most prevalent group followed by reads predicted to encode hemicellulose degrading enzymes (27%). As expected, reads related to lignin degradation (2%) are only rarely found. This is due to the fact that lignin degradation predominantly occurs under aerobic conditions, whereas fermentation of plant biomass for biogas production is an anaerobic process. Most of the reads (1,571) assigned to the 'polysaccharide degradation' context originate from members of the *Firmicutes* making this phylum the most important one. Additionally, *Bacteroidetes* (661) and *Proteobacteria* (319) are involved in this process, since significant numbers of reads were grouped into these taxa. The phylum *Actinobacteria* includes a high fraction of reads encoding fragments of lignin degrading enzymes. This is in accordance with literature, since *Actinobacteria* are known to express different ligninases [69].

The reductive acetyl-CoA pathway (also called Wood-Ljungdahl pathway) is important for acetogenic bacteria to autotrophically produce acetate from hydrogen and carbon dioxide or carbon monoxide, respectively [70], whereas aceticlastic methanogens reversely use this pathway to generate methane from acetate. Several COG classifications representing the Wood-Ljungdahl and the hydrogenotrophic methanogenesis pathway were taxonomically analysed to distinguish between aceticlastic [16] and hydrogenotrophic methanogens [71]. The fact that the Wood-Ljungdahl pathway is used by acetogenic bacteria like *Clostridia* as well as methanogenic *Archaea* is shown in Figure 4b. To address the total number of COG hits representing methanogenesis, only *Archaea* with 720 assignments were taken into account for subsequent analysis. The order *Methanomicrobiales* (520) constitutes 72% of all COG hits relevant for methanogenesis assigned to the superkingdom *Archaea*. Thus, the order *Methanomicrobiales* is by far the most abundant taxonomic group producing methane using CO_2 as a carbon source. Besides, it is noticeable that 25% of the reads assigned to *Methanomicrobiales* (131) could not be classified at lower taxonomic levels, indicating that either corresponding proteins originate from new taxa which are not represented in COG or highly conserved proteins shared by more than one taxon. Acetate as source for methane production seems to play a minor role, which is taxonomically indicated by the low fraction of *Methanosarcinales* (2,704 reads) within the group of known methanogens (27,693 reads). This observation correlates with the low fraction of hits (3.5%) indicative for acetate conversion to methane within the group of *Archaea* as well as the fact that reductive acetyl-CoA pathway enzymes are mainly found in the group of *Bacteria* (95%) with *Clostridia* as their most abundant class.

Identification of several different *Methanoculleus* species

Analysis of the community participation in methanogenesis revealed that members of the order *Methanomicrobiales* are dominant among the methanogenic *Archaea*. Recently, the complete genome sequence of *Methanoculleus marisnigri* JR1 of the order of *Methanomicrobiales* became available [51]. To gain an insight into the relatedness of dominant methanogens within the analysed fermentation sample to this reference species, several genes were analysed in more detail.

From both the (unfiltered) GS FLX and Titanium datasets, a total amount of 5,266 reads containing 16S rRNA sequence fragments was identified. Out of these, 44 of the GS FLX and 88 of the Titanium reads were taxonomically assigned to the genus

Methanoculleus by the RDP classifier, but could not be assembled into a single consensus sequence due to differences in sequence composition. Binning of the individual reads based on their SNP content (Figure 5A) and subsequent assembly resulted in seven consensus sequences, each of which comprised at least the partial sequence of a specific 16S rRNA gene. Subsequently, the consensus sequences were characterized in terms of their phylogeny together with several reference sequences obtained from GenBank (Figure 5B). All sequences except one were placed in close phylogenetic distance to *Methanoculleus bourgensis*; the one remaining sequence was placed in a remote branch formed by *Methanoculleus marisnigri*, *Methanoculleus palmolei*, *Methanoculleus chikugoensis* and *Methanoculleus thermophilus* (Figure 5B). These findings are consistent with the results of a previous study [35], where a phylogenetic characterization of the same biogas plant based on 16S rRNA clone library sequences was performed.

For verification, the same analysis was repeated for the 5S rRNA and the *mtrB* gene. While the *mcr* operon is partially duplicated in the genome of *Methanoculleus marisnigri* JR1, only one copy of the *mtrB* gene exists. Four different variants of the *mtrB* gene could be assembled, confirming the presence of several *Methanoculleus* species/strains (see File S1). Assembly of the 5S rRNA gene confirmed three variants (data not shown); downstream of the 5S rRNA gene, all three variants differ from *Methanoculleus marisnigri* JR1, indicating a different genomic context for the *rrn* cluster.

Mapping of metagenome reads to the *Methanoculleus marisnigri* JR1 reference genome

In an attempt to reconstruct the genome sequence of one of the dominant methanogenic species, both datasets were mapped onto the published genome of *Methanoculleus marisnigri* JR1, the only *Methanoculleus* strain for which a completely sequenced genome currently exists. Using only the sequence data from the GS FLX dataset, 39.8% of the reference genome was covered; mapping of the Titanium dataset resulted in 41.7% coverage. A joint mapping of both datasets produced a genomic coverage of 45.4%.

Upon further analysis, a correlation between poorly covered regions and areas with relatively low GC content was discovered (data not shown). Since variations in GC content often hint at horizontal gene transfer, the number of bases that could be mapped to the coding sequence of each gene in the genome of *Methanoculleus marisnigri* JR1 was determined and normalized with respect to the gene length. The results were divided into two different groups depending on the prediction found in the HGT-DB: one containing all genes that potentially have been acquired by horizontal gene transfer, and another one for all remaining genes. For both groups, the density function of all results was plotted (Figure 6). A comparison of the results for genes predicted as acquired by horizontal gene transfer in the HGT-DB and the remaining genes suggests that some mobile DNA segments are missing in one or several of the *Methanoculleus* species present in the studied biogas fermenter. The coverage of the *Methanoculleus marisnigri* JR1 genome as determined from the available metagenome sequence data does not suffice to determine the absence or presence of individual genes in the *Methanoculleus* species residing in the biogas fermenter. This is particularly relevant because several different *Methanoculleus* species/strains were identified in the studied biogas fermenter. Even with the currently available sequence data, the reconstruction of a complete genome of one of the dominant species still remains unfeasible due to the complexity of the microbial community and the close relationship of some of the dominant species.

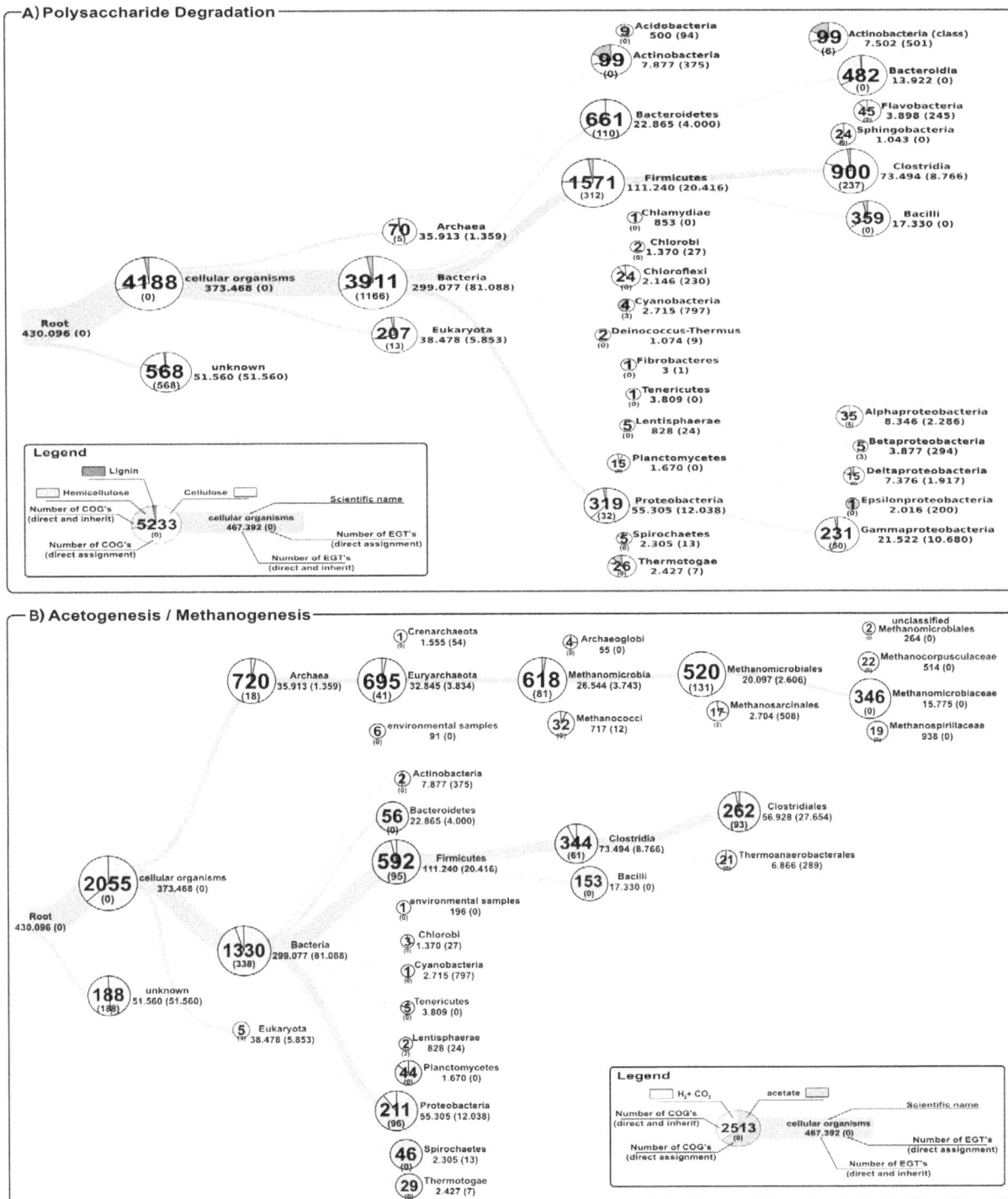

Figure 4. Taxonomic and physiological overview of relevant members in 'polysaccharide degradation' (a) and 'acetogenesis/ methanogenesis' (b). The trees are based on reads classified into the NCBI taxonomy by the CARMA software. For each taxonomic group the underlying number of reads is given. Numbers in brackets refer to the amount of reads which could not be classified at corresponding lower taxonomic ranks. Associated COG entries are depicted as pie charts, where the interpretation of the numbers is equivalent.

Hydrogenases of Methanomicrobiales

This study revealed that species of the genus *Methanoculleus* dominate among methanogenic *Archaea*. High abundance of *Methanoculleus* members has also been shown for other communities involved in fermentation of maize silage and related substrates [58,68,72,73] suggesting that in these habitats methane is mainly produced via the hydrogenotrophic pathway by reduction of carbon dioxide. Hydrogenotrophic methanogenesis usually is accompanied by syntrophic acetate oxidation which necessitates that released hydrogen from acetate oxidation is efficiently consumed by cooperating hydrogenotrophic methanogens. It is assumed that *Methanoculleus* species have a high hydrogen affinity

A

```
A C C C T A T A G G T T A C A G A T G C T G G A A T G C[T]C T G T A A C C C A A A G T T C C G[G]G C G C
                C T G G A A T G C[T]C T G T A A C C C A A A G T T C C G[G]G C G C   VAR1
                                                          C G[G]G C G C
        A T A G G T T A C A G A T G C T G G A A T G C[T]C T G T A A C C C A A A G T T C C G[G]G C G C

A C C C T A T A G[A]T T A C A G A T G C T G G A A T G C C C T G T A A[T]C C A A A G T T C C G A C G C   VAR2
A C C C T A T A G[A]T T A C A G A T G C T G G A A T G C C C T G T A A[T]C C A A A G T T C C G A C G C
A C C C T A T A G G T T A C A G A T G C T G G A A T G C C C T G T A A C C C A A A G T T C C G A C G C
```

B

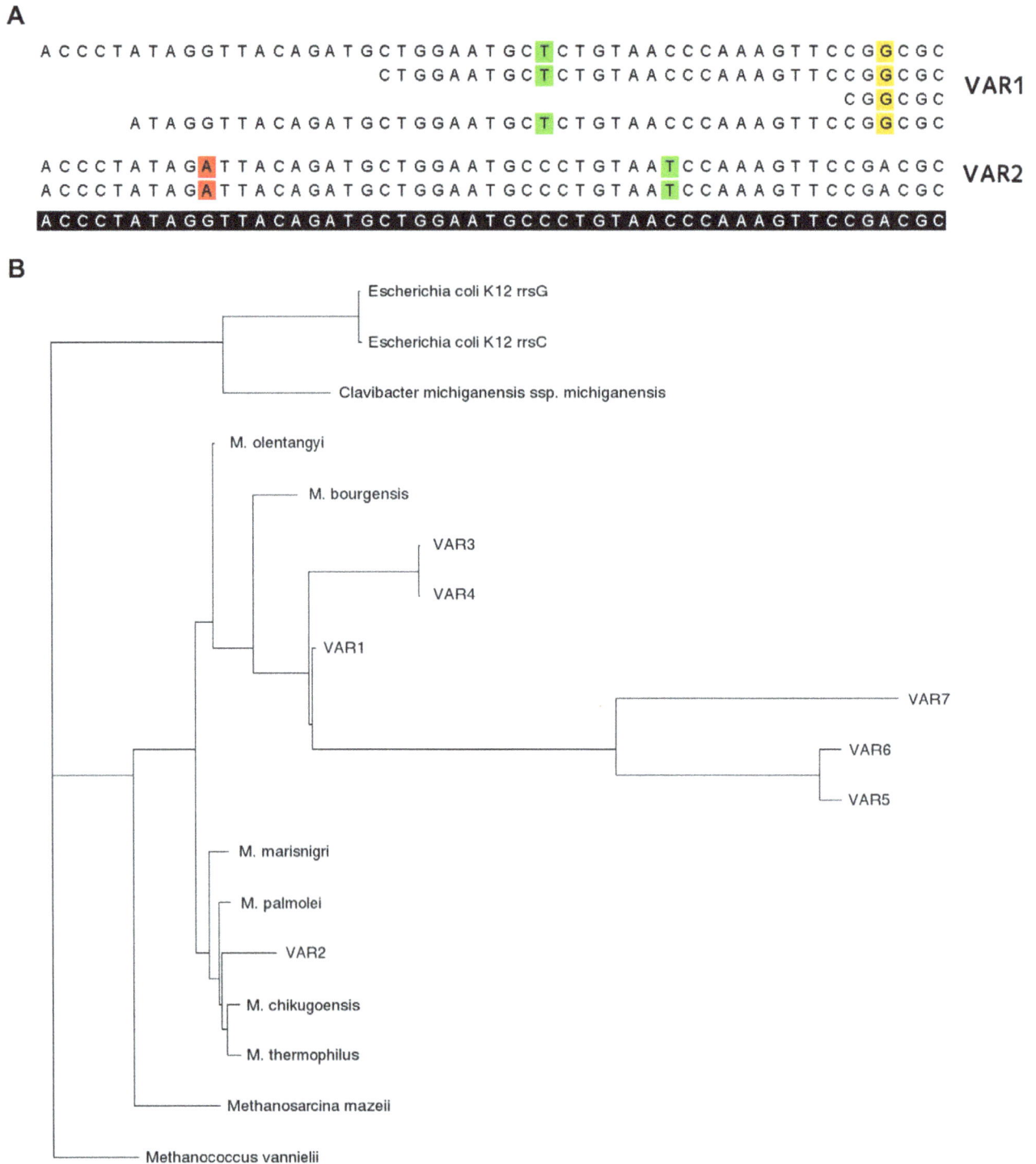

Figure 5. Identification of *Methanoculleus* variants. Partial view (A) of pyrosequencing reads aligned to the 16S rRNA sequence of *Methanoculleus bourgensis* (Genbank accession AY196674). Colored bases indicate differences between reads and the reference (shown in the bottom line). In the depicted part, two of the seven different variants are visible. To characterize the variants, a phylogenetic tree (B) was constructed together with various reference sequences. Most variants show close relationship to *M. bourgensis*; only variant VAR2 was placed in another branch formed by *M. marisnigri, M. palmolei, M. chikugoensis* and *M. thermophilus*. Several 16S rRNA sequences from the genus *Methanoculleus* were used: *M. olentangyi* (AF095270), *M. bourgensis* (AY196674), *M. palmaeoli* (Y16382), *M. thermophilus* (AB065297), *M. chikugoensis* MG62 (AB038795) and *M. marisnigri* JR1 (CP000562 (Memar_R0043)). Additional sequences in increasing taxonomic distance were included as outgroups: *Methanosarcina mazeii* (MMU20151), *Methanococcus vannielii* SB (CP000742 (Mevan_R0025)), *Clavibacter michiganensis ssp. michiganensis* NCPPB 382 (AM711867 (CMM_RNA_0001)) and two sequences from *Escherichia coli* K12 DH10B (NC_010473 (ECDH10B_3945 and ECDH10B_2759)).

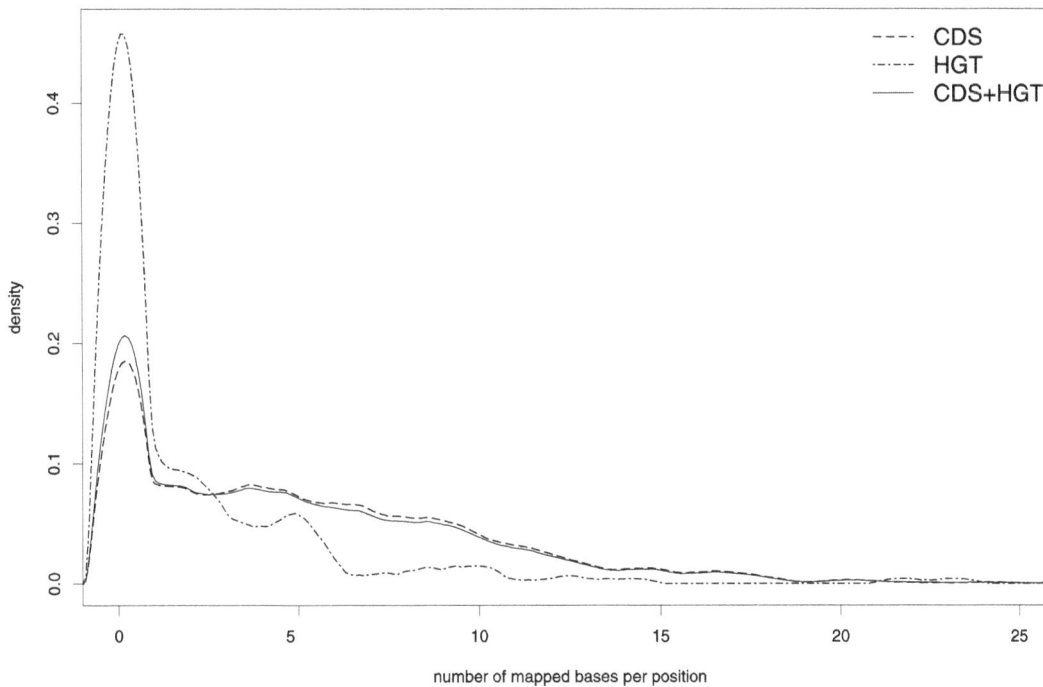

Figure 6. Comparison of kernel density estimates for metagenome reads mapped to coding regions of the *Methanoculleus marisnigri* **JR1 genome.** The figure shows the mapped bases per position in the genome separated into different groups: genes that have potentially been acquired by horizontal gene transfer (HGT, depicted as dot-dashed line) and all other genes without a prediction for horizontal gene transfer (CDS, shown as dashed line). For comparison, the density function for all genes combined is shown as well (solid line). Poor coverage of HGT genes hints at genomic features probably missing in the *Methanoculleus* species present in the studied biogas fermenter, further supporting their difference to *Methanoculleus marisnigri* JR1.

[73]. BLAST analyses revealed that several metagenome reads from the biogas community correspond to *M. marisnigri* JR1 hydrogenase genes (see Table 2). Among the identified genes are those possibly encoding the membrane-bound hydrogenases Eha (Memar_1172 - Memar_1185) and Ech (Memar_0359 - Memar_0364), which were predicted to participate in methanogenesis [51]. It is likely that Eha and Ech represent high-affinity hydrogenases contributing to efficient hydrogen oxidation in the course of methanogenesis. At least some *Methanoculleus* members within the community of the analysed fermentation sample possess membrane-bound hydrogenases related to Eha and Ech of *Methanoculleus marisnigri* JR1.

Coverage of the microbial community

To analyse whether the diversity of the biogas-producing microbial community is sufficiently covered by the sequence data, species richness, diversity and rarefaction calculations were conducted.

The Shannon index [74] was used to estimate the biological diversity of the underlying microbial community in the analyzed fermenter. It is widely applied in ecological studies as a measurement of biodiversity, accounting for both the number of different taxa as well as their relative abundances. For this purpose, 16S rRNA fragments were detected in the normalized pyrosequencing datasets and classified with the RDP classifier. Subsequently the Shannon index was computed considering only 16S rRNA sequences that could be classified on rank genus with at least 80% confidence.

This approach resulted in a Shannon index value of 1.90 for the GS FLX and 2.51 for the Titanium dataset, showing that sequence data obtained from the GS FLX platform clearly underestimates the biodiversity of the underlying community.

Since both datasets differ in size, the Shannon index was also computed for random subsets of 407,558 reads (i.e. the size of the normalized GS FLX dataset) extracted from the Titanium data; 5,000 iterations were calculated and the average of all results taken. The resulting index value of 2.58 shows that even for equally sized datasets, the Titanium dataset provides an higher estimate of biodiversity.

Rarefaction analyses of both unfiltered datasets revealed that the biogas-producing community is covered much deeper by the Titanium than by the GS FLX sequences (Figure 7). 16S rRNA fragments identified in the Titanium sequences were assigned to 38 different genera, but only 16 genera were observed for the GS FLX dataset. Three genera were specific for the GS FLX dataset, 13 were shared, and 25 were specific for the Titanium data. A full list of all identified genera can be found in the File S1. The rarefaction curve computed for the Titanium dataset was also far from reaching the plateau phase, indicating that considerably higher sequencing effort would be required to cover all phylogenetic groups of the underlying microbial community. Rarefaction analysis conducted at rank family is in accordance with this conclusion (see File S1).

The gene content of the studied metagenome samples was characterized by assigning reads to Pfam protein families. As expected, the collective gene content of the studied microbial community was covered much deeper by the Titanium than by the GS FLX sequence reads. In total, 4,759 different protein families were identified in the Titanium dataset, while 3,844 could be detected in the GS FLX reads. However, the rarefaction curves computed for the number of protein families also did not reach the plateau phase, suggesting that additional sampling would be required to capture the entire gene content of the underlying community (Figure 8). Protein families identified in the Titanium

Table 2. Metagenome reads comprising hydrogenase gene fragments.

Titanium[a]	GS FLX[a]	locus	description
462	134	Memar_0359[b]	4Fe-4S ferredoxin iron-sulfur binding domain-containing protein
66	44	Memar_0360[b]	NADH-ubiquinone oxidoreductase, chain 49kDa
14	16	Memar_0361[b]	ech hydrogenase, subunit EchD, putative
40	17	Memar_0362[b]	NADH ubiquinone oxidoreductase, 20 kDa subunit
221	111	Memar_0417	(NiFe) hydrogenase maturation protein HypF
46	32	Memar_0470	hydrogenase accessory protein HypB
111	46	Memar_0622	methyl-viologen-reducing hydrogenase, delta subunit
80	59	Memar_1007[c]	nickel-dependent hydrogenase, large subunit
55	28	Memar_1008[c]	NADH ubiquinone oxidoreductase, 20 kDa subunit
21	21	Memar_1014	hydrogenase maturation protease
146	51	Memar_1022	hydrogenase expression/formation protein HypE
11	10	Memar_1023	hydrogenase expression/synthesis, HypA
16	7	Memar_1024	hydrogenase assembly chaperone hypC/hupF
87	49	Memar_1044	coenzyme F420 hydrogenase/dehydrogenase beta subunit
72	45	Memar_1140	hydrogenase expression/formation protein HypD
21	24	Memar_1172[d]	hypothetical protein
12	8	Memar_1173[d]	hypothetical protein
14	10	Memar_1174[d]	hypothetical protein
18	5	Memar_1175[d]	hypothetical protein
37	18	Memar_1176[d]	hypothetical protein
18	15	Memar_1177[d]	uncharacterized membrane protein
3	6	Memar_1179[d]	hypothetical protein
3	5	Memar_1181[d]	hypothetical protein
28	13	Memar_1182[d]	hypothetical protein
24	14	Memar_1183[d]	NADH ubiquinone oxidoreductase, 20 kDa subunit
281	68	Memar_1185[d]	4Fe-4S ferredoxin iron-sulfur binding domain-containing protein
49	38	Memar_1378	coenzyme F420 hydrogenase/dehydrogenase beta subunit
17	6	Memar_1380	coenzyme F420 hydrogenase/dehydrogenase beta subunit
79	43	Memar_1623	coenzyme F420 hydrogenase/dehydrogenase beta subunit
93	45	Memar_2174	nickel-dependent hydrogenase, large subunit
31	21	Memar_2175	hydrogenase maturation protease
544	178	Memar_2176	coenzyme F420 hydrogenase
54	29	Memar_2177	coenzyme F420-reducing hydrogenase subunit beta

[a]number of reads assigned to a specific *M. marisnigri* JR1 locus (Memar);
[b]Ech operon encoding a membrane-bound hydrogenase [51];
[c]F_{420} non-reducing hydrogenase; d Eha operon encoding a membrane-bound hydrogenase [51].

reads were annotated with 1,623 different GO terms, whereas 1,338 GO terms were observed for the GS FLX dataset. Compared to the GS FLX data, the Titanium dataset provides a far more detailed overview of the gene content and metabolic potential of the studied microbial community.

Discussion

In a recent study, the metagenome of a biogas-producing microbial community was sequenced employing the GS FLX pyrosequencing platform. Since analysis results showed insufficient coverage of the community, the study was complemented by additional sequencing of the same DNA preparation on the GS FLX Titanium platform.

During the analysis, a previously unreported GC bias in pyrosequencing data was identified, which affected sequences from

both sequencing runs, thus indicating the importance of thorough data screening and filtering to avoid a contortion of results. However, sequence data generated using the GS FLX Titanium chemistry was only marginally affected by this issue and only a very small fraction of reads had to be excluded from further analysis. These differences result from improvements in the Titanium chemistry, leading to less bias compared to the previous GS FLX technology. Meanwhile, a new emPCR kit containing a specific additive has been launched by Roche Applied Science. Initial studies based on microbial genome sequencing revealed an almost bias free sequencing of even very high GC content regions (data not shown).

The composition of the microbial community was deduced from both the taxonomic classification of 16S rRNA fragments as well as the assignment of Environmental Gene Tags (EGTs) on different taxonomic ranks Obtained results essentially confirmed

Genus

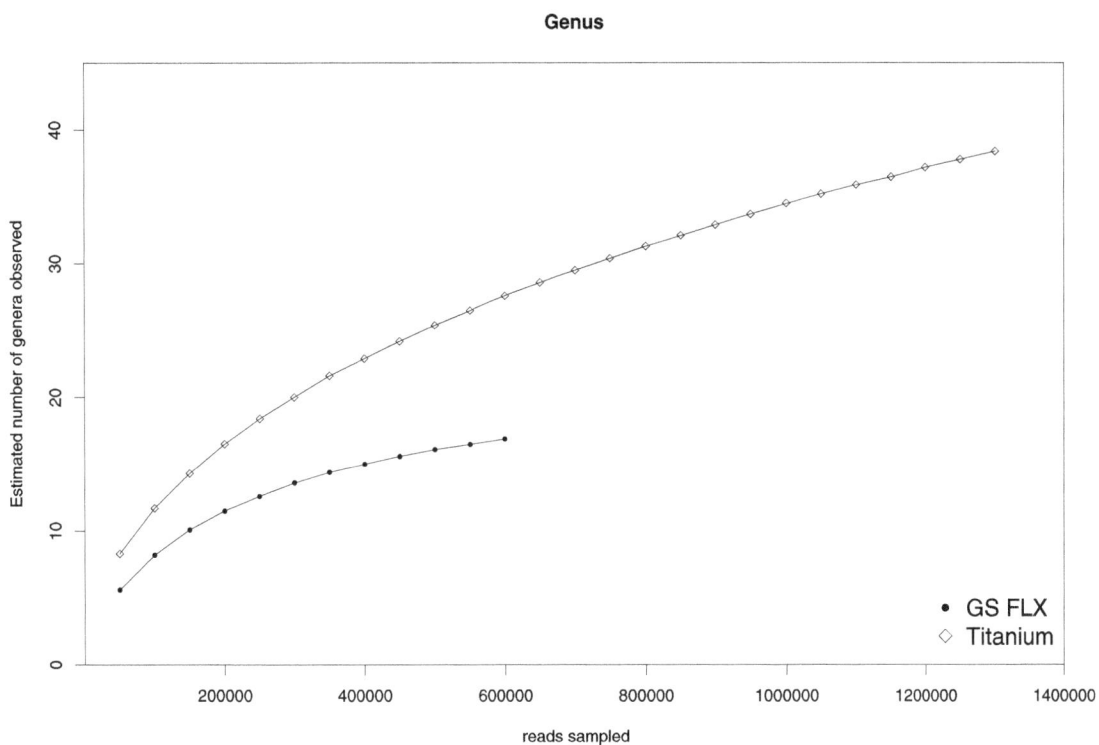

Figure 7. Rarefaction analysis of observed genera. The rarefaction curves represent the estimated number of genera that would be observed in biogas fermenter metagenomes of different sizes. The values were determined based on 16S rRNA fragments classified at rank genus identified in the entire Titanium and GS FLX datasets, respectively.

Protein families

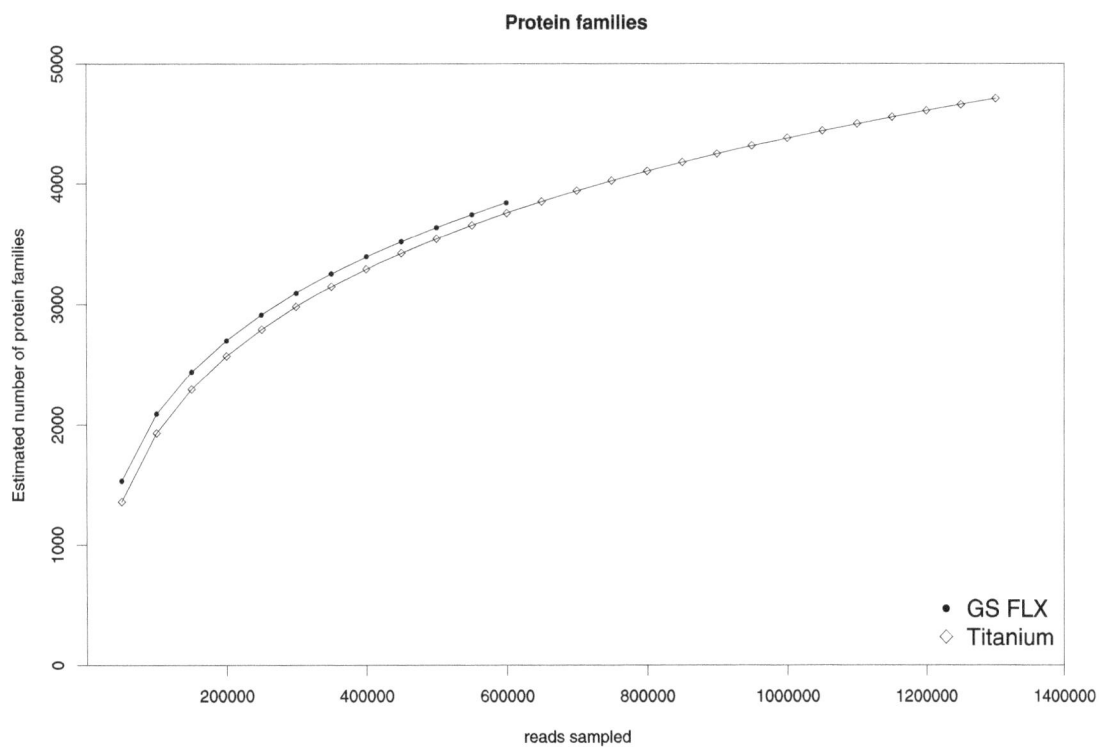

Figure 8. Rarefaction analysis of Pfam families. The estimated number of Pfam protein families that would be identified in biogas fermenter metagenomes of different sizes is shown. The values were computed based on the number of protein families identified in the entire Titanium and FLX metagenomes, respectively.

taxonomic profiles of the previous study. However, less abundant taxa could be identified by analyzing the GS FLX Titanium dataset thus justifying the additional sequencing effort. The fact that only a very small fraction of metagenomic reads actually contains fragments of the 16S rRNA gene emphasizes the advantages of using software such as the CARMA pipeline, which accurately classifies gene fragments detected in metagenome sequence data. A rarefaction analysis was performed to estimate the coverage of the microbial community in both sequencing datasets; as expected, sequencing data from the GS FLX Titanium platform provides a far more complete view of the underlying community, while the GS FLX sequencing run was not carried out to saturation.

During the functional characterization of the community, members of the phylum *Firmicutes* could be confirmed to represent the dominant organisms involved in the breakdown of polysaccharides together with *Bacteroidetes*. Beyond that, the novel analyses showed that *Proteobacteria* also play an important role in polysaccharide degradation. *Clostridia* were found to dominate within the functional context 'acetogenesis', as deduced by mapping of bacterial taxa to metagenome hits representing the Wood-Ljungdahl pathway which is also known as the reductive acetyl-CoA pathway. *Methanomicrobiales* are the most abundant order involved in methanogenesis using CO_2 as a carbon source, while acetate only seems to play a minor role, as indicated by a low fraction of *Methanosarcinales*.

Based on the identification of 16S rRNA fragments from the *Methanoculleus* genus and subsequent assembly, the presence of several *Methanoculleus* species closely related to *Methanoculleus bourgensis* in the studied biogas fermenter could be demonstrated. A rough characterization of the genomic content of these *Methanoculleus* species was conducted by mapping the metagenome sequence reads to the published genome of *Methanoculleus marisnigri* JR1. Comparison of the genomic content of dominant *Methano-*

culleus species within the analysed sample and the reference species *M. marisnigri* JR1 revealed that there are several differences mainly concerning genes that might have been acquired by horizontal gene transfer. Metagenome reads assigned to the genus *Methanoculleus* represent *inter alia* methanogenesis and membrane-bound hydrogenase genes predicted to be of importance for the pathway leading to the formation of methane within biogas. The close relationship of the *Methanoculleus* species in the studied biogas fermenter makes reconstruction of the genomic sequence of one of the dominant *Methanoculleus* species from the metagenomic sequence reads rather unlikely, since a reliable distinction between the most abundant strains can not be assured. In comparison, the GS FLX Titanium data offers a far more complete view of the analysed fermenter, even though analysis results give evidence that the available sequence data still does not fully cover the microbial community.

Acknowledgments

The authors thank the BRF system administrators for technical support and Jochen Blom for computing the linear regression. Aileen O'Connell and Dan Brooks are thanked for proof-reading the manuscript.

Author Contributions

Conceived and designed the experiments: S. Jaenicke AP AS AG. Performed the experiments: S. Jaenicke MD OK. Analyzed the data: S. Jaenicke CA TB RB KHG S. Jünemann LK FT MZ AS AG. Contributed reagents/materials/analysis tools: S. Jaenicke MD OK. Wrote the paper: S. Jaenicke S. Jünemann LK AS.

References

1. Weiland P (2003) Production and energetic use of biogas from energy crops and wastes in Germany. Applied Biochemistry and Biotechnology 109: 263–274.
2. Yadvika S, Sreekrishnan T, Kohli S, Rana V (2004) Enhancement of biogas production from solid substrates using different techniques - a review. Bioresource Technology 95: 1–10.
3. Bayer E, Belaich J, Shoham Y, Lamed R (2004) The cellulosomes: multienzyme machines for degradation of plant cell wall polysaccharides. Annual Review of Microbiology 58: 521–54.
4. Cirne D, Lehtomaki A, Bjornsson L, Blackall L (2007) Hydrolysis and microbial community analyses in two-stage anaerobic digestion of energy crops. Journal of Applied Microbiology 103: 516–527.
5. Lynd L, Weimer P, van Zyl W, Pretorius I (2002) Microbial cellulose utilization: fundamentals and biotechnology. Microbiology and molecular biology reviews 66: 506–577.
6. Drake H, Küsel K, Matthies C (2002) Ecological consequences of the phylogenetic and physiological diversities of acetogens. Antonie Van Leeuwenhoek 81: 203–213.
7. Drake H, Daniel S, Küsel K, Matthies C, Kuhner C, et al. (1997) Acetogenic bacteria: what are the in situ consequences of their diverse metabolic versatilities? Biofactors 6: 13–24.
8. Myint M, Nirmalakhandan N, Speece R (2007) Anaerobic fermentation of cattle manure: Modeling of hydrolysis and acidogenesis. Water Research 41: 323–332.
9. Schink B (1997) Energetics of syntrophic cooperation in methanogenic degradation. Microbiology and Molecular Biology Reviews 61: 262–280.
10. Shin H, Youn J (2005) Conversion of food waste into hydrogen by thermophilic acidogenesis. Biodegradation 16: 33–44.
11. Sousa D, Smidt H, Alves M, Stams A (2009) Ecophysiology of syntrophic communities that degrade saturated and unsaturated long-chain fatty acids. FEMS Microbiology Ecology 68: 257–272.
12. Deppenmeier U, Müller V (2008) Life close to the thermodynamic limit: how methanogenic archaea conserve energy. Results and problems in cell differentiation 45: 123–152.
13. Thauer R, Kaster A, Seedorf H, Buckel W, Hedderich R (2008) Methanogenic archaea: ecologically relevant differences in energy conservation. Nature Reviews Microbiology 6: 579–591.

14. Blaut M (1994) Metabolism of methanogens. Antonie van Leeuwenhoek 66: 187–208.
15. Deppenmeier U (2002) The unique biochemistry of methanogenesis. Progress in nucleic acid research and molecular biology 71: 223–283.
16. Ferry JG (1999) Enzymology of one-carbon metabolism in methanogenic pathways. FEMS Microbiol Rev 23: 13–38.
17. Liu Y, Whitman W (2008) Metabolic, phylogenetic, and ecological diversity of the methanogenic archaea. Annals of the New York Academy of Sciences 1125: 171–189.
18. Reeve J, Nölling J, Morgan R, Smith D (1997) Methanogenesis: genes, genomes, and who's on first? Journal of bacteriology 179: 5975–5986.
19. Weiss A, Jérôme V, Freitag R, Mayer H (2008) Diversity of the resident microbiota in a thermophilic municipal biogas plant. Applied Microbiology and Biotechnology 81: 163–173.
20. Tang Y, Shigematsu T, Morimura S, Kida K (2004) The effects of micro-aeration on the phylogenetic diversity of microorganisms in a thermophilic anaerobic municipal solid-waste digester. Water Research 38: 2537–2550.
21. Tang Y, Fujimura Y, Shigematsu T, Morimura S, Kida K (2007) Anaerobic treatment performance and microbial population of thermophilic upflow anaerobic filter reactor treating awamori distillery wastewater. Journal of Bioscience and Bioengineering 104: 281–287.
22. Klocke M, Mähnert P, Mundt K, Souidi K, Linke B (2007) Microbial community analysis of a biogas-producing completely stirred tank reactor fed continuously with fodder beet silage as mono-substrate. Systematic and Applied Microbiology 30: 139–151.
23. Klocke M, Nettmann E, Bergmann I, Mundt K, Souidi K, et al. (2008) Characterization of the methanogenic Archaea within two-phase biogas reactor systems operated with plant biomass. Systematic and Applied Microbiology 31: 190–205.
24. Chouari R, Le Paslier D, Daegelen P, Ginestet P, Weissenbach J, et al. (2005) Novel predominant archaeal and bacterial groups revealed by molecular analysis of an anaerobic sludge digester. Environmental Microbiology 7: 1104–1115.
25. Shigematsu T, Era S, Mizuno Y, Ninomiya K, Kamegawa Y, et al. (2006) Microbial community of a mesophilic propionate-degrading methanogenic consortium in chemostat cultivation analyzed based on 16S rRNA and acetate kinase genes. Applied Microbiology and Biotechnology 72: 401–415.

26. Tang Y, Shigematsu T, Morimura S, Kida K (2005) Microbial community analysis of mesophilic anaerobic protein degradation process using bovine serum albumin (BSA)-fed continuous cultivation. Journal of Bioscience and Bioengineering 99: 150–164.

27. Sasaki K, Haruta S, Ueno Y, Ishii M, Igarashi Y (2007) Microbial population in the biomass adhering to supporting material in a packed-bed reactor degrading organic solid waste. Applied Microbiology and Biotechnology 75: 941–952.

28. Friedrich M (2005) Methyl-coenzyme M reductase gene: unique functional markers for methanogenic and anaerobic methane-oxidizing Archaea. Methods in Enzymology 397: 428–442.

29. Juottonen H, Galand P, Yrjälä K (2006) Detection of methanogenic Archaea in peat: comparison of PCR primers targeting the mcrA gene. Research in Microbiology 157: 914–921.

30. Lueders T, Chin K, Conrad R, Friedrich M (2001) Molecular analyses of methyl-coenzyme M reductase alpha-subunit (mcrA) genes in rice field soil and enrichment cultures reveal the methanogenic phenotype of a novel archaeal lineage. Environmental Microbiology 3: 194–204.

31. Luton P, Wayne J, Sharp R, Riley P (2002) The mcrA gene as an alternative to 16S rRNA in the phylogenetic analysis of methanogen populations in landfill. Microbiology 148: 3521–3530.

32. Rastogi G, Ranade D, Yeole T, Patole M, Shouche Y (2008) Investigation of methanogen population structure in biogas reactor by molecular characterization of methyl-coenzyme M reductase A (mcrA) genes. Bioresource Technology 99: 5317–5326.

33. Schlüter A, Bekel T, Diaz N, Dondrup M, Eichenlaub R, et al. (2008) The metagenome of a biogas-producing microbial community of a production-scale biogas plant fermenter analysed by the 454-pyrosequencing technology. Journal of Biotechnology 136: 77–90.

34. Krause L, Diaz N, Edwards R, Gartemann K, Krömeke H, et al. (2008) Taxonomic composition and gene content of a methane-producing microbial community isolated from a biogas reactor. Journal of Biotechnology 136: 91–101.

35. Kröber M, Bekel T, Diaz N, Goesmann A, Jaenicke S, et al. (2009) Phylogenetic characterization of a biogas plant microbial community integrating clone library 16S-rDNA sequences and metagenome sequence data obtained by 454-pyrosequencing. Journal of Biotechnology 142: 38–49.

36. Zhou J, Bruns M, Tiedje J (1996) DNA recovery from soils of diverse composition. Applied and Environmental Microbiology 62: 316–322.

37. R Development Core Team (2008) R: A Language and Environment for Statistical Computing. R Foundation for Statistical Computing, Vienna, Austria. ISBN 3-900051-07-0.

38. Kutner M, Nachtsheim C, Neter J, Li W (2005) Applied linear statistical models. McGraw Hill.

39. Gomez-Alvarez V, Teal T, Schmidt T (2009) Systematic artifacts in metagenomes from complex microbial communities. The ISME Journal 3: 1314–1317.

40. Beifang N, Limin F, Shulei S, Weizhong L (2010) Artificial and natural duplicates in pyrosequencing reads of metagenomic data. BMC Bioinformatics 11: 187.

41. Altschul SF, Gish W, Miller W, Myers EW, Lipman DJ (1990) Basic local alignment search tool. J Mol Biol 215: 403–410.

42. Wang Q, Garrity G, Tiedje J, Cole J (2007) Naive bayesian classifier for rapid assignment of rRNA sequences into the new bacterial taxonomy. Appl Environ Microbiol 73: 5261–5267.

43. Finn R, Tate J, Mistry J, Coggill P, Sammut S, et al. (2008) The Pfam protein families database. Nucleic Acids Research 36: D281.

44. Krause L, Diaz N, Goesmann A, Kelley S, Nattkemper T, et al. (2008) Phylogenetic classification of short environmental DNA fragments. Nucleic Acids Research 36: 2230–2239.

45. Muller J, Szklarczyk D, Julien P, Letunic I, Roth A, et al. (2010) eggNOG v2. 0: extending the evolutionary genealogy of genes with enhanced non-supervised orthologous groups, species and functional annotations. Nucleic Acids Research 38: D190.

46. Caspi R, Foerster H, Fulcher C, Kaipa P, Krummenacker M, et al. (2008) The MetaCyc Database of metabolic pathways and enzymes and the BioCyc collection of Pathway/Genome Databases. Nucleic Acids Research 36: D623.

47. Sayers E, Barrett T, Benson D, Bolton E, Bryant S, et al. (2009) Database resources of the National Center for Biotechnology Information. Nucleic Acids Research 37: D5.

48. Pearson WR (2000) Flexible sequence similarity searching with the FASTA3 program package. Methods Mol Biol 132: 185–219.

49. Edgar R (2004) MUSCLE: multiple sequence alignment with high accuracy and high throughput. Nucl Acids Res 32: 1792–1797.

50. Felsenstein J (1989) PHYLIP - Phylogeny Inference Package (Version 3.2). Cladistics 5: 164–166.

51. Anderson I, Ulrich L, Lupa B, Susanti D, Porat I, et al. (2009) Genomic characterization of Methanomicrobiales reveals three classes of methanogens. PloS One 4: e5797.

52. Garcia-Vallve S, Guzman E, Montero M, Romeu A (2003) HGT-DB: a database of putative horizontally transferred genes in prokaryotic complete genomes. Nucleic Acids Research 31: 187–189.

53. Dohm J, Lottaz C, Borodina T, Himmelbauer H (2008) Substantial biases in ultra-short read data sets from high-throughput DNA sequencing. Nucleic Acids Research 36: e105.

54. Ludwig W, Strunk O, Westram R, Richter L, Meier H, et al. (2004) ARB: a software environment for sequence data. Nucleic Acids Research 32: 1363–1371.

55. Dollhopf S, Hashsham S, Dazzo F, Hickey R, Criddle C, et al. (2001) The impact of fermentative organisms on carbon flow in methanogenic systems under constant low-substrate conditions. Applied microbiology and biotechnology 56: 531–538.

56. Fang H, Zhang T, Liu H (2002) Microbial diversity of a mesophilic hydrogen-producing sludge. Applied microbiology and biotechnology 58: 112–118.

57. Fernandez A, Hashsham S, Dollhopf S, Raskin L, Glagoleva O, et al. (2000) Flexible community structure correlates with stable community function in methanogenic bioreactor communities perturbed by glucose. Applied and Environmental Microbiology 66: 4058.

58. Li T, Mazéas L, Sghir A, Leblon G, Bouchez T (2009) Insights into networks of functional microbes catalysing methanization of cellulose under mesophilic conditions. Environmental Microbiology 11: 889–904.

59. Noach I, Levy-Assaraf M, Lamed R, Shimon L, Frolow F, et al. (2010) Modular arrangement of a cellulosomal scaffoldin subunit revealed from the crystal structure of a cohesin dyad. Journal of Molecular Biology 2: 294–305.

60. Bae J, Park J, Chang Y, Rhee S, Kim B, et al. (2004) Clostridium hastiforme is a later synonym of Tissierella praeacuta. International journal of systematic and evolutionary microbiology 54: 947.

61. Kaster K, Voordouw G (2006) Effect of nitrite on a thermophilic, methanogenic consortium from an oil storage tank. Applied microbiology and biotechnology 72: 1308–1315.

62. Kim W, Hwang K, Shin S, Lee S, Hwang S (2010) Effect of high temperature on bacterial community dynamics in anaerobic acidogenesis using mesophilic sludge inoculum. Bioresource Technology 101: S17–S22.

63. Lee Y, Romanek C, Mills G, Davis R, Whitman W, et al. (2006) Gracilibacter thermotolerans gen. nov., sp. nov., an anaerobic, thermotolerant bacterium from a constructed wetland receiving acid sulfate water. International journal of systematic and evolutionary microbiology 56: 2089.

64. Miranda-Tello E, Fardeau M, Sepulveda J, Fernandez L, Cayol J, et al. (2003) Garciella nitratireducens gen. nov., sp. nov., an anaerobic, thermophilic, nitrate-and thiosulfate-reducing bacterium isolated from an oilfield separator in the Gulf of Mexico. International journal of systematic and evolutionary microbiology 53: 1509.

65. Plugge C, Balk M, Zoetendal E, Stams A (2002) Gelria glutamica gen. nov., sp. nov., a thermophilic, obligately syntrophic, glutamate-degrading anaerobe. International journal of systematic and evolutionary microbiology 52: 401.

66. Shah H, Olsen I, Bernard K, Finegold S, Gharbia S, et al. (2009) Approaches to the study of the systematics of anaerobic, Gram-negative, non-sporeforming rods: Current status and perspectives. Anaerobe 15: 179–194.

67. Zhilina T, Appel R, Probian C, Brossa E, Harder J, et al. (2004) Alkaliflexus imshenetskii gen. nov. sp. nov., a new alkaliphilic gliding carbohydrate-fermenting bacterium with propionate formation from a soda lake. Archives of microbiology 182: 244–253.

68. Pobeheim H, Munk B, Müller H, Berg G, Guebitz G (2010) Characterization of an anaerobic population digesting a model substrate for maize in the presence of trace metals. Chemosphere.

69. Kirby B, Le Roes M, Meyers P (2006) Kribbella karoonensis sp. nov. and Kribbella swartbergensis sp. nov., isolated from soil from the Western Cape, South Africa. Int J Syst Evol Microbiol 56: 1097–1101.

70. Pierce E, Xie G, Barabote R, Saunders E, Han C, et al. (2008) The complete genome sequence of Moorella thermoacetica (f. Clostridium thermoaceticum). Environ Microbiol 10: 2550–2573.

71. Ladapo J, Whitman WB (1990) Method for isolation of auxotrophs in the methanogenic archaebacteria: role of the acetyl-CoA pathway of autotrophic CO_2 fixation in Methanococcus maripaludis. PNAS 87: 5598–5602.

72. Hansen K, Ahring B, Raskin L (1999) Quantification of syntrophic fatty acid-beta-oxidizing bacteria in a mesophilic biogas reactor by oligonucleotide probe hybridization. Applied and environmental microbiology 65: 4767.

73. Schnürer A, Zellner G, Svensson B (1999) Mesophilic syntrophic acetate oxidation during methane formation in biogas reactors. FEMS microbiology ecology 29: 249–261.

74. Shannon CE (1948) A mathematical theory of communication. Bell System Technical Journal 27: 379–423 and 623–656.

7

Start-Up of an Anaerobic Dynamic Membrane Digester for Waste Activated Sludge Digestion: Temporal Variations in Microbial Communities

Hongguang Yu[1], Qiaoying Wang[1]*, Zhiwei Wang[1]*, Erkan Sahinkaya[2], Yongli Li[3], Jinxing Ma[1], Zhichao Wu[1]

1 State Key Laboratory of Pollution Control and Resource Reuse, School of Environmental Science and Engineering, Tongji University, Shanghai, PR China, **2** Istanbul Medeniyet University, Bioengineering Department, Kadıköy, Istanbul, Turkey, **3** Laboratory of Polymères, Biopolymères and Surfaces, UMR 6270, University of Rouen-CNRS-INSA, Boulevard Maurice de Broglie, Mont-Saint-Aignan, France

Abstract

An anaerobic dynamic membrane digester (ADMD) was developed to digest waste sludge, and pyrosequencing was used to analyze the variations of the bacterial and archaeal communities during the start-up. Results showed that bacterial community richness decreased and then increased over time, while bacterial diversity remained almost the same during the start-up. *Proteobacteria* and *Bacteroidetes* were the major phyla. At the class level, *Betaproteobacteria* was the most abundant at the end of start-up, followed by *Sphingobacteria*. In the archaeal community, richness and diversity peaked at the end of the start-up stage. Principle component and cluster analyses demonstrated that archaeal consortia experienced a distinct shift and became stable after day 38. *Methanomicrobiales* and *Methanosarcinales* were the two predominant orders. Further investigations indicated that *Methanolinea* and *Methanosaeta* were responsible for methane production in the ADMD system. Hydrogenotrophic pathways might prevail over acetoclastic means for methanogenesis during the start-up, supported by specific methanogenic activity tests.

Editor: Dwayne Elias, Oak Ridge National Laboratory, United States of America

Funding: This work is partially supported by the Key Special Program on the S&T for the Pollution Control and Treatment of Water Bodies (2011ZX07303-001) and by the Science and Technology Commission of Shanghai Municipality research project (12230707000). The funders had no role in study design, data collection and analysis, decision to publish, or preparation of the manuscript.

Competing Interests: The authors have declared that no competing interests exist.

* E-mail: qywang@tongji.edu.cn (QYW); zwwang@tongji.edu.cn (ZWW)

Introduction

With the widespread applications of biological processes for wastewater treatment, large quantities of waste activated sludge (WAS) are generated along with the degradation of organic pollutants. The disposal and utilization of WAS have drawn considerable attention worldwide [1]. Anaerobic digestion (AD) is one of the most widely used process because of its ability to produce biogas, reduce the amount of sludge, destroy pathogens, and limit the production of odor [2]. In traditional AD, WAS is usually thickened to reduce its volume before further processing. This requires thickening equipment. Hydraulic retention time (HRT) and solid retention time (SRT) are identical, which prolongs HRT and reduces the flexibility of the operation [1,2]. Anaerobic membrane digesters (AMDs), which integrate anaerobic digesters with anaerobic membrane bioreactors (AnMBRs), have several advantages over conventional AD. For example, AMD can provide a co-thickening potential, in which sludge thickening and digestion can be performed simultaneously [3]. AMD processes can also decouple HRT from SRT, which means that WAS can be treated with smaller footprint and higher organic loading rates.

In recent years, microfiltration (MF) and ultrafiltration (UF) membranes have been used in AMD processes [3,4]. However,

specific obstacles, such as the high cost of membrane modules and low membrane flux, still hinder the practical applications of AMDs [5]. In this respect, dynamic membrane technology might be a promising solution. Dynamic membranes can be formed and re-formed *in situ*, and the dynamic layer can be replaced by a newly deposited layer in case of membrane fouling. This cuts down the expense of purchasing and physically replacing new membranes [6]. High filtration flux can be achieved in anaerobic dynamic membrane reactors (AnDMBRs) [7]. To date, most related studies have focused on AnDMBR processes for municipal wastewater treatment [7,8]. However, the information on anaerobic dynamic membrane digesters (ADMDs) in the context of sludge digestion has been very limited. The use of ADMDs for WAS digestion is worth investigating due to the prominent advantages of the process over traditional AD processes.

In anaerobic WAS digesters, microorganisms play an important role in the process, and a variety of bacteria and archaea are involved in the metabolism. Recently, molecular biology approaches, such as polymerase chain reaction (PCR)-denaturing gradient gel electrophoresis (DGGE) and 16sRNA clone libraries, have been widely used in exploration of the microbial community in AD or AnMBR systems [9–11]. However, low sequencing depth of those methods, which represents a mere snapshot of the dominant members and hinders the comprehensive characterization of the microbial community structure [12]. Pyrosequencing developed by Roche 454 Life Science (Brandford, CT, U.S.) is a high-throughput analytical approach that can provide enough sequencing depth to cover the complex microbial communities [13,14]. It has been used in investigating microbial diversity in conventional AD and AnDMBR processes [15–18]. However, because of the use of dynamic membranes, the microbial composition and structure of ADMD might differ considerably from those of conventional AD and AnMBR within the context of wastewater treatment. For these reasons, it is essential to investigate the microbial composition and dynamics in a WAS-receiving ADMD. Pyrosequencing can provide comprehensive insights into the dynamics of the microbial communities during the start-up period of the ADMD system.

The overarching goal of this study is to characterize the temporal changes of bacterial and archaeal communities during the start-up of ADMD process for WAS digestion. By using 454 high-throughput pyrosequencing, the community structures and compositions of bacteria and archaea were investigated over time during the start-up stage in order to facilitate understanding the ADMD process.

Materials and Methods

Ethics statement

No specific permissions were required for these locations/activities of our field experiments. The field studies also did not involve endangered or protected species.

Experimental setup and operation

The experimental setup used in this study is shown in Figure S1. The ADMD system consisted of a completely mixed anaerobic digester (effective volume of 67 L) coupled with a submerged anaerobic dynamic membrane reactor (effective volume of 2 L). This facilitated convenient membrane cleaning and replacement in the membrane zone while maintaining the main digester strictly anaerobic at all times. The influent flow rate was about 17 L/d. The HRT and SRT of ADMD were 6 d and 20 d, respectively. A flat-sheet dynamic membrane module was mounted in the membrane zone, which was made of Dacron mesh (pore size = 39 μm). The surface area of dynamic membrane module was 0.038 m². A peristaltic pump was installed to recycle sludge from the anaerobic digester to the dynamic membrane zone at a recirculation ratio of 300%, and another peristaltic pump was used to withdraw permeate from the dynamic membrane module. The effluent flow rate and trans-membrane pressure (TMP) were monitored using a flowmeter and a pressure gauge, respectively. The membrane module was operated with an intermittent suction mode (10-min filtration and 2-min pause) with an instant flux of about 15 L/(m²·h). Biogas production was measured according to the volume of biogas collected in the wetted gas collector (LMF-1, Duoyuan Instrument Technology Co., Ltd., China), in which the gas pressure was maintained at a pressure of 1 atm. The liquor level in the system was controlled using an elevated influent tank. Electric heaters controlled by temperature sensors were used to maintain the temperature of the system at 35±2 °C. Biogas was recycled using a diaphragm gas pump (KNF, Germany) on an intermittent working mode (1-min on and 20-min off) to scour membrane surfaces for fouling control, and the specific gas demand (SGD) per unit projected area of the riser zone were controlled at 25.0 m³/(m²·h) during biogas scouring. Physical cleaning was conducted for the dynamic membrane when TMP increased to 30 kPa.

In the experiment, excess sludge from the Quyang wastewater treatment plant (WWTP) (Shanghai, China) (31°16′45″ N 121°29′08″ E) was used as the influent of the ADMD system after passing through a mesh (pore size = 0.9 mm). The characteristics of the influent WAS are shown in Figure 1. Inoculum was collected from an anaerobic digester of the Bailonggang WWTP in Shanghai, China (31°14′42″ N 121°43′51″ E). The main features of the inoculum are as follows: pH 7.18±0.04, total suspended solids (TSS) 41.69±0.08 g/L, volatile suspended solids (VSS) 21.41±0.30 g/L, total chemical oxygen demand (tCOD) 33111±363 mg/L, soluble chemical oxygen demand (SCOD) 1255±15 mg/L. The ratio of inoculum to substrate (v/v) was 1:1.

Microbial diversity analysis

DNA extraction and PCR amplification. For a full understanding of the microbial community dynamics during the start-up of ADMD system, sludge samples were collected on days 0, 7, 14, 28, 38, and 46, and stored at −20°C before further analysis.

Microbial DNA was extracted from sludge samples using an E.Z.N.A.® Soil DNA Kit (Omega Bio-tek, Norcross, GA, U.S.) according to manufacturer's protocols. The V1–V3 and V3–V5 regions of the bacteria and archaea 16S ribosomal RNA gene were amplified by polymerase chain reaction (95 °C for 2 min, followed by 25 cycles at 95 °C for 30 s, 55 °C for 30 s, and 72 °C for 30 s and a final extension at 72 °C for 5 min) using primers 27F (5′-AGAGTTTGATCCTGGCTCAG-3′)/533R (5′-TTACCGCG-GCTGCTGGCAC-3′) and Arch344F (5′-ACGGGGYGCAG-CAGGCGCGA-3′)/Arch915R (5′-GTGCTCCCCCGCCAAT-TCCT-3′), respectively. PCR reactions were performed in a 20 μL mixture containing 4 μL of 5× FastPfu Buffer, 2 μL of 2.5 mM dNTPs, 0.8 μL of each primer (5 μM), 0.4 μL of FastPfu Polymerase, and 10 ng of template DNA.

454 pyrosequencing. After purification using the AxyPrep DNA Gel Extraction Kit (Axygen Biosciences, Union City, CA, U.S.) and quantification using QuantiFluor™ -ST (Promega, U.S.), a mixture of amplicons was used for pyrosequencing on a Roche 454 GS FLX+ Titanium platform (Roche 454 Life Sciences, Branford, CT, U.S.) according to standard protocols [14]. The raw reads were deposited into the NCBI Sequence Read Archive (SRA) database (Accession Number: SRA082624).

Processing of pyrosequencing data. In total, 79085 and 75993 valid sequences were obtained from all 6 samples, with an average of 13181 and 12666 sequences per sample for bacteria and archaea, respectively. The resulting sequences were processed using Seqcln (http://sourceforge.net/projects/seqclean/) and Mothur (version: 1.28.0) [19]. After removing low quality sequences (quality score <25) and sequences shorter than 200 bp, with homopolymers longer than six nucleotides, and containing ambiguous base calls or incorrect primer sequences, a total of 56926 and 60634 high-quality sequences were produced with an average length of 474 and 508 bp per sequence for bacteria and archaea, respectively. Sequences were aligned against the silva database (version: SSU111 http://www.arb-silva.de/) using k-mer searching (http://www.mothur.org/wiki/Align.seqs). Potentially chimeric sequences were detected using UCHIME (http://drive5.com/uchime) and removed. The remaining reads were pre-clustered (http://www.mothur.org/wiki/Pre.cluster) and then clustered using uncorrected pairwise algorithm. In addition, operational taxonomic units (OTUs) were defined as sharing more than 97% sequence identity using Furthest neighbor method (http://www.mothur.org/wiki/Cluster). The total number of OTUs at 97% similarity level was 16663 and 5684 with an

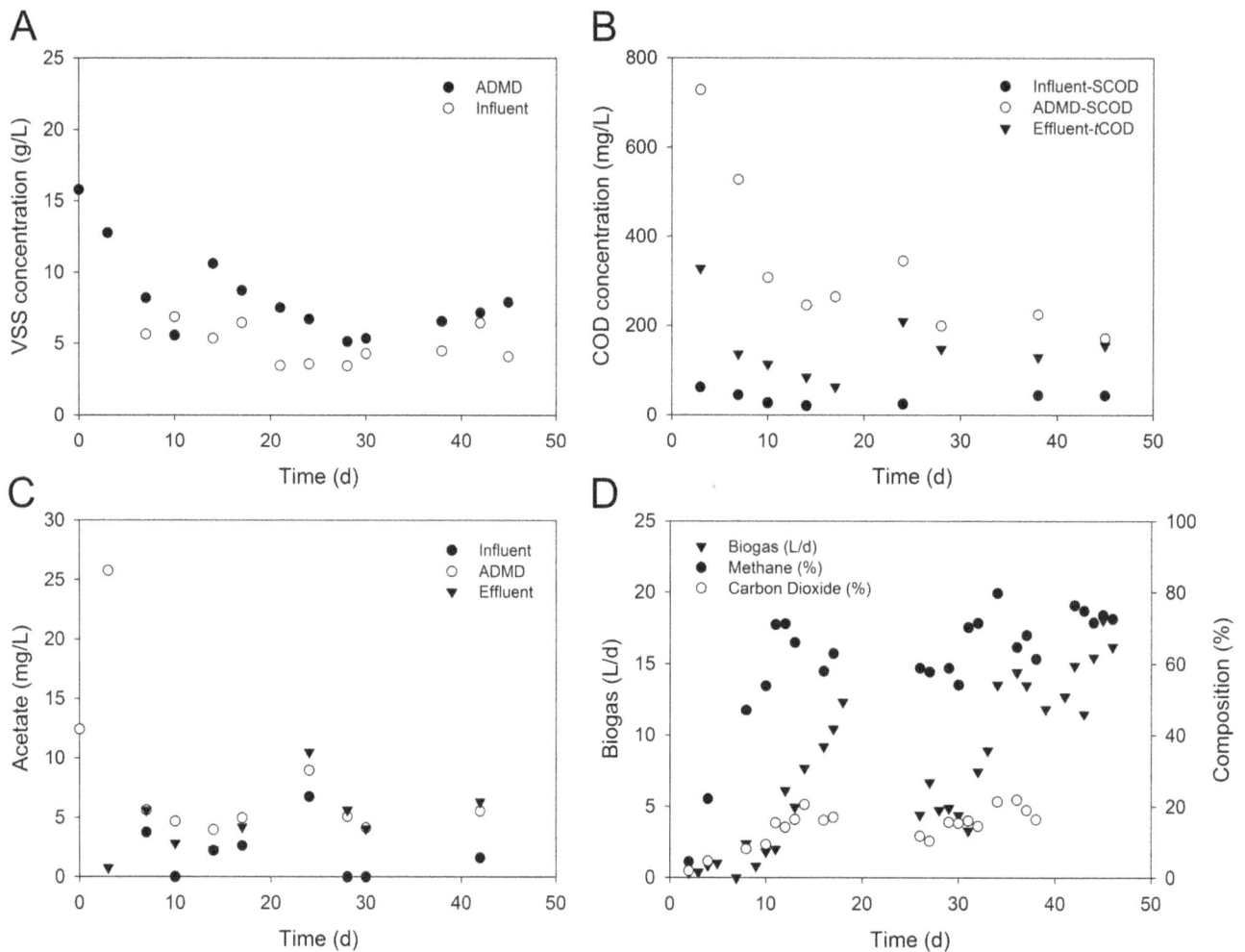

Figure 1. WAS digestion during the start-up stage. (A) VSS; (B) SCOD; (C) acetate; (D) biogas production and composition.

average of 2777 and 947 OTUs per sample for bacteria and archaea, respectively. Rarefaction curves, Chao1 richness estimator (http://www.mothur.org/wiki/Chao), the Shannon index (http://www.mothur.org/wiki/Shannon), and the Good's coverage (http://www.mothur.org/wiki/Coverage) were created using Mothur (version: 1.28.0) for each sample [19]. Taxonomic classification down to the phylum, class, order, family, and genus level was performed using Mothur (http://www.mothur.org/wiki/Classify.seqs) via the aforementioned silva database with a set confidence threshold of 80% [20]. Principle component analysis (PCA), heat map representation, and cluster analysis were performed via R Project (http://www.r-project.org/).

Specific methanogenic activity (SMA) tests

SMA tests were performed with 134 mL glass assay bottles sealed with rubber stoppers and aluminum crimps. Two different substrates, acetate and H_2/CO_2, were used for specific acetoclastic methanogenic activity (SAMA) and specific hydrogenotrophic methanogenic activity (SHMA) tests, respectively. For SAMA tests, the reaction volume (25 mL) in each bottle was composed of ADMD sludge (approximately 2 g VSS/L), 10 mL basal medium, and acetate (2.5 g COD/L) [21]. The pH was adjusted to 7.0–7.5. Bottles were flushed with N_2 gas for 5 min and then kept in a shaker (100 rpm) at 35 ± 1 °C. The gas in the bottle headspace was

analyzed periodically after overnight incubation [21]. All the analyses were performed in triplicate. This was followed by a blank control, which reproduced the test with deionized water instead of the substrate solution. For SHMA tests, the reaction volume (25 mL) in each bottle was composed of ADMD sludge (approximately 2 g VSS/L) and 10 mL basal medium [21], with pH adjusted to 7.0–7.5. Bottles were flushed with H_2/CO_2 (80%:20% v/v) for 5 min and pressurized to 1 atm, and then kept in a shaker (180 rpm) [22]. The temperature was maintained at 35 ± 1 °C. The gas in the bottle headspace was also analyzed periodically after overnight incubation [21]. All the analyses were performed in triplicate along with a blank control, which was treated with N_2 instead of H_2/CO_2. Both SAMA and SHMA were calculated according to the methods reported by Yukselen [23].

Other analytical methods

Sludge samples were centrifuged at 6000 rpm for 10 min and then filtered through normal quantitative papers. The filtrate was analyzed for volatile fatty acids (VFAs) and SCOD. The analyses of COD, TSS, and VSS were conducted in accordance with standard methods [24]. Gas composition (CH_4 and CO_2) was measured using a gas chromatography (6890N, Agilent, U.S.) equipped with a thermal conductivity detector (TCD). The determination of VFAs was described in a previous study [25].

Results and Discussion

Digestion performance

The ADMD system was operated for 46 days during the start-up stage (Figure 1). The influent VSS concentration was 4.90 ± 1.29 g/L (n = 11), while the average VSS concentration of ADMD sludge remained about 8.06 g/L after 7 days, which was about 1.6 times of that of the influent. This indicated that the ADMD system also had thickening functions along with the digestion of WAS. A VSS reduction rate of about 48.7% was achieved during the start-up stage. The SCOD concentration of influent was 35 ± 15 mg/L (n = 8), while SCOD of ADMD sludge and tCOD of effluent were 315 ± 183 mg/L (n = 10) and 152 ± 79 mg/L (n = 9), respectively. The SCOD of ADMD sludge was about 8 times of that of influent, indicating that soluble compounds had been released into the system. This is because sludge hydrolysis can be expressed by the increase of SCOD [26]. The tCOD of ADMD permeate was lower than the SCOD of ADMD sludge, indicating that dynamic membrane in ADMD was able to partially retain the dissolved organic matters. Acetate was the predominant VFA, accounting for 91–100% of total VFAs. Its concentration is shown in Figure 1C. The influent acetate concentration varied within a range of 0–2.1 mg/L. As for ADMD sludge, slight accumulation of acetate was observed on day 3, and the acetate concentration decreased afterwards and remained 5.4 ± 1.6 mg/L (n = 8). Meanwhile, the acetate concentration in effluents was 5.2 ± 2.6 mg/L (n = 8) from day 7 during the start-up, which was almost the same as that of ADMD sludge, indicating that acetate could permeate through dynamic membrane in the ADMD.

Figure 1D depicts biogas production in the ADMD system during start-up. Biogas production was converted to standard temperature and pressure (0 °C and 1 atm). Biogas production was observed on day 2 shortly after start-up, and it was found to increase in an almost linear fashion from day 2 to day 18. The deficiencies in biogas production from day 20 to day 31 were attributed to the malfunction of the gas collection system. From day 34 to day 46, biogas production remained 12.58 ± 4.59 L/d (n = 11). The methane content of the biogas increased from day 2 to day 11, and remained $67.8 \pm 7.6\%$ (n = 21) from day 11 to day 46. Biogas production became relatively steady at the end of the start-up stage, and the methane yield per gVSS-removed reached 0.32 L/(gVSS·d) from day 34 to day 46.

Richness and diversity of microbial communities

In the present study, 2447–3311 and 860–1055 OTUs were clustered at a dissimilarity level of 0.03 for analyzing the bacterial and archaeal communities, respectively. The number of effective sequence reads and of OTUs were both higher than that of conventional molecular biology methods [9–11]. This could indicate that 454 pyrosequencing could better characterize of microbial consortia [12]. As listed in Table 1, in the bacterial domain, the number of OTUs decreased from day 0 to day 28 and increased from day 28 to day 46. Changes in Chao1 were also used to estimate the total number of OTUs. It showed a trend similar to that of OTUs, but the lowest Chao1 estimators occurred on day 38. Both indicators, OTUs and Chao1, demonstrated that the richness of the bacterial community decreased and then increased during the start-up period. These findings were also confirmed by rarefaction curves (Figure S2A). The Shannon index, which estimates the diversity of microbial population, fluctuated within the range of 6.66–7.14 from day 0 to day 46, showing no significant changes in bacterial diversity during the start-up period. Unlike in the bacterial domain, in the archaeal

domain, the number of OTUs increased from day 0 to day 7, decreased from day 7 to day 14, remained relatively stable from day 14 to day 38, and increased again from day 38 to day 46. The changes in the Chao1 estimator were similar to those in OTUs. The Shannon index remained within the range of 3.85–3.96 from day 0 to day 14, and increased to 4.26–4.39 from day 28 to day 46. On day 46, OTUs, Chao1, and Shannon were the highest during the start-up period, indicating the richness and diversity of the archaeal community peaked at the end of the start-up stage. Good's coverage, which represents the probability that the next read will belong to an OTU that has already been observed, can be used to evaluate the level of information contained in microbial communities [11]. As listed in Table 1, the coverage values of bacterial and archaeal community were within the range of 0.69–0.73 and 0.93–0.95, respectively, indicating that the most common phylogenetic groups were detected in our study. Similar coverage values were also observed by other researchers, and the relatively low values of coverage for bacteria might be related to the large diversity encountered in anaerobic environment [11].

Bacterial consortia dynamics

Figure 2 illustrates the distribution of the bacterial community during the start-up of the ADMD system. As shown in Figure 2A, *Proteobacteria* and *Bacteroidetes*, whose relative abundance was within the range of 29.6–52.3% and 15.4–26.1%, respectively, were the two most predominant phyla. *Proteobacteria* and *Bacteroidetes* have been reported to be common and abundant in anaerobic sludge digesters [11,17,27], MBR and AnDMBR systems [18,20]. The relative abundance of *Proteobacteria* was 38.1% on day 0, decreased to 29.6% on day 7, and increased to 52.3% on day 46. The relative abundance of *Bacteroidetes* increased to 24.2% during the beginning 38 d and remained relatively stable within the range of 24.2–26.1% from day 38 to day 46. Results showed that at the end of the ADMD system's start-up stage, *Proteobacteria* and *Bacteroidetes* accumulated and overwhelmed other phyla. They were able to degrade a wide range of macromolecules and xenobiotic compounds [28].

As shown in Figure 2B, five classes of *Proteobacteria*, i.e., *alpha-*, *beta-*, *gamma-*, *delta-*, and *epsilon-*, were observed during the start-up of ADMD, but *beta-*, *gamma-*, and *deltaproteobacteria*, which accounted for 25.6–49.4% of total reads at the class level, were the major classes of *Proteobacteria*. *Betaproteobacteria* was the most predominant class, of which the relative abundance decreased from day 0 to day 7 and increased from day 7 to day 46. Previous studies also showed that *Betaproteobacteria* was the major class of *Proteobacteria* in conventional AD, MBR and AnDMBR systems [17,18,20]. *Betaproteobacteria* has been reported to be the most dominant group in propionate-, butyrate-, and acetate-utilizing microbial communities in anaerobic digesters [27]. Accordingly, the high abundance of *Betaproteobacteria* might lead to VFA consumption, and low VFA concentration was observed during the start-up of the ADMD system (Figure 1C).

In the phylum of *Bacteroidetes*, the relative abundance of the *Sphingobacteria* class was 8.4–15.5% during the experiment, and *Sphingobacteria* became the most abundant class at the end of the start-up (Figure 2B). In MBR systems for wastewater treatment, *Sphingobacteria* was also found to be most dominant class of *Bacteroidetes* [20]. With the relative abundance of 3.54–5.10% from day 14 to day 46, *Flavobacteria* was the second abundant class at the end of the start-up. The *Cytophaga-Flavobacteria* cluster, which belongs to the phylum of *Bacteroidetes*, was capable of consuming chitin, N-acetylglucosamine and protein, and degrading high molecular mass fraction of dissolved organic material [29,30]. In

Table 1. Diversity statistics of bacterial and archaeal community.

Time (d)	Bacterial Community				Archaeal Community			
	OTUs[a]	Chao1[a]	Shannon[a]	Coverage[a]	OTUs[a]	Chao1[a]	Shannon[a]	Coverage[a]
0	3311	10014	7.14	0.69	971	1757	3.96	0.95
7	2778	8185	6.66	0.73	1011	1969	3.94	0.95
14	2726	8069	6.98	0.73	860	1721	3.85	0.94
28	2447	7962	6.82	0.71	888	1887	4.29	0.93
38	2551	7823	6.85	0.73	899	1821	4.26	0.94
46	2850	8983	7.03	0.70	1055	2075	4.39	0.94

[a]Values were defined at a dissimilarity level of 0.03.

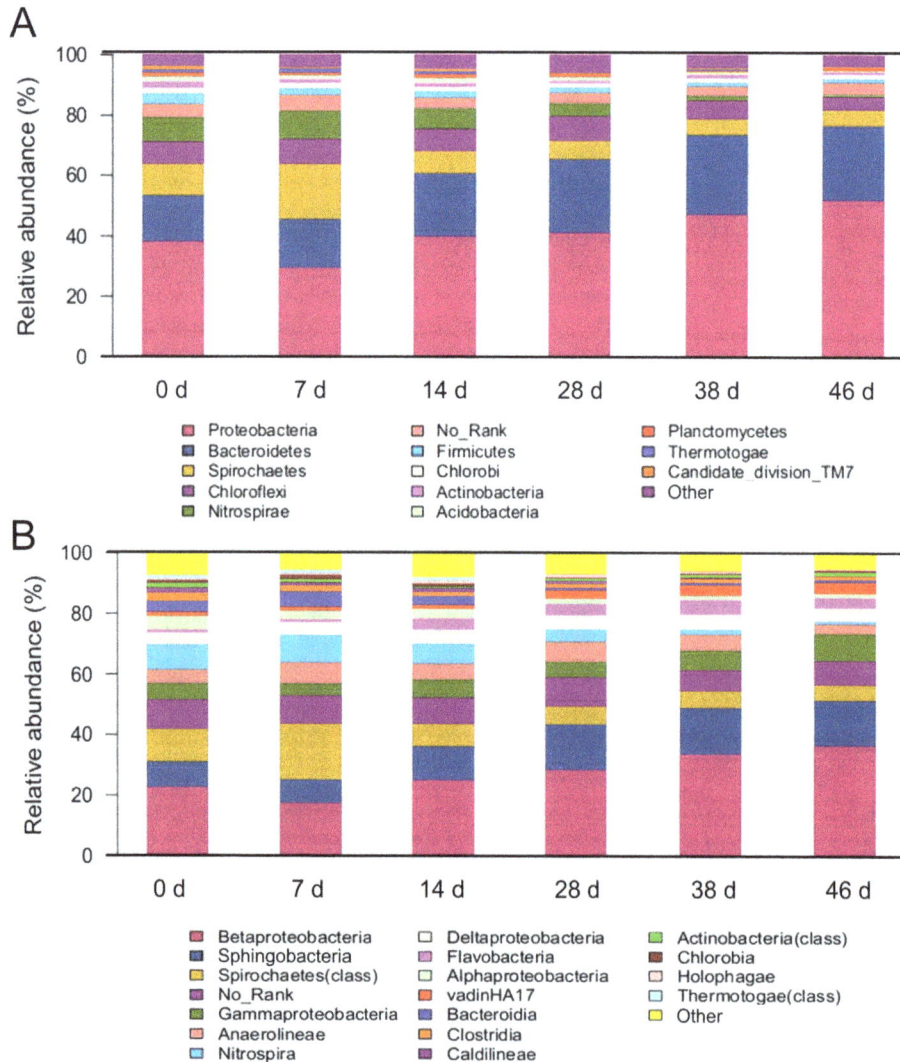

Figure 2. Changes in the bacterial community structure during the start-up of ADMD system. (A) phylum level; (B) class level. Relative abundance is defined as the number of sequences affiliated with that taxon divided by the total number of sequences per sample (%). Phyla or classes accounting for less than 1% of relative abundance are regarded as other.

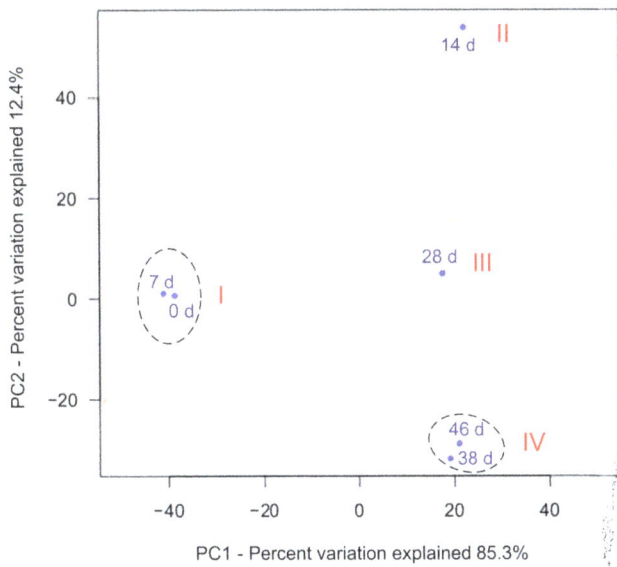

Figure 3. Principle component analysis (PCA) of ADMD samples taken at different times. PCA was conducted at a 3% cutoff OTU level.

this way, the presence of *Flavobacteria* might enhance the removal of WAS pollutants.

It is worth noting that some minor classes, such as *Alphaproteobacteria* and *Clostridia*, were observed during the start-up of ADMD system (Figure 2B). It has been reported that the genera of *Rhodobacter* and *Clostridium*, which belong to *Alphaproteobacteria* and *Clostridia*, respectively, were able to carry out the bio-hydrogen production [31]. In this study, the presence of these bacteria might produce hydrogen during the AD processes. The produced hydrogen could be utilized by methanogens via hydrogenotrophic pathway, which will be further elaborated in the following section.

Archaeal consortia dynamics

In order to analyze the similarity of archaeal consortia at different times, principle component analysis (PCA) was conducted using OTUs at a dissimilarity level of 0.03. As seen in Figure 3, ADMD samples were clustered into four groups: (1) Group I contains the samples on day 0 and day 7; (2) Group II contains the sample on day 14; (3) Group III is the ADMD sludge on day 28; (4) Group IV is the ADMD sludges on day 38 and day 46. Results showed that the samples on day 0 and day 7 were similar to each other. Considerable similarity was also observed between the

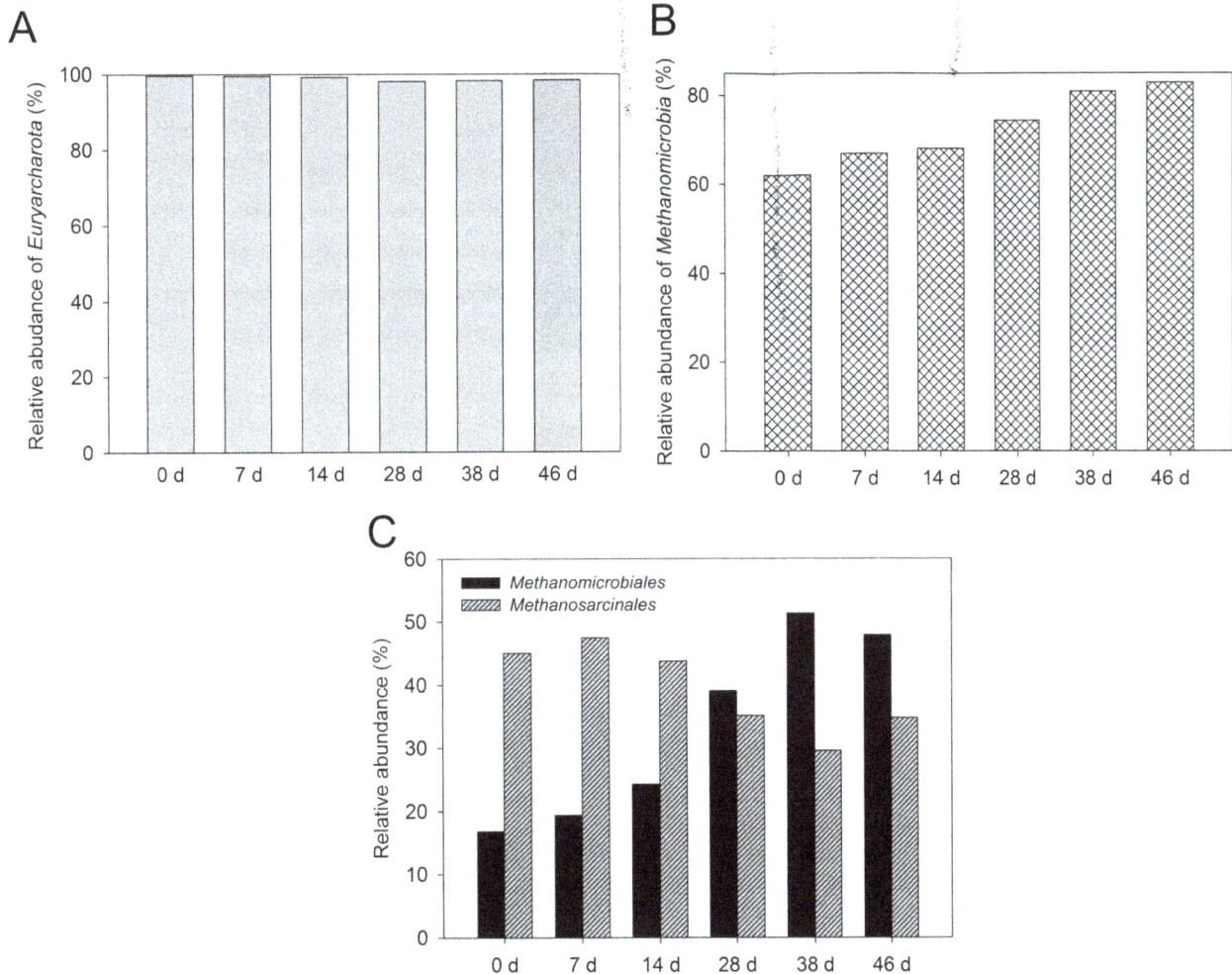

Figure 4. Changes in the archaeal community structure during the start-up of ADMD system. (A) phylum level; (B) class level; (C) order level. Relative abundance is defined as the number of sequences affiliated with that taxon divided by the total number of sequences per sample (%).

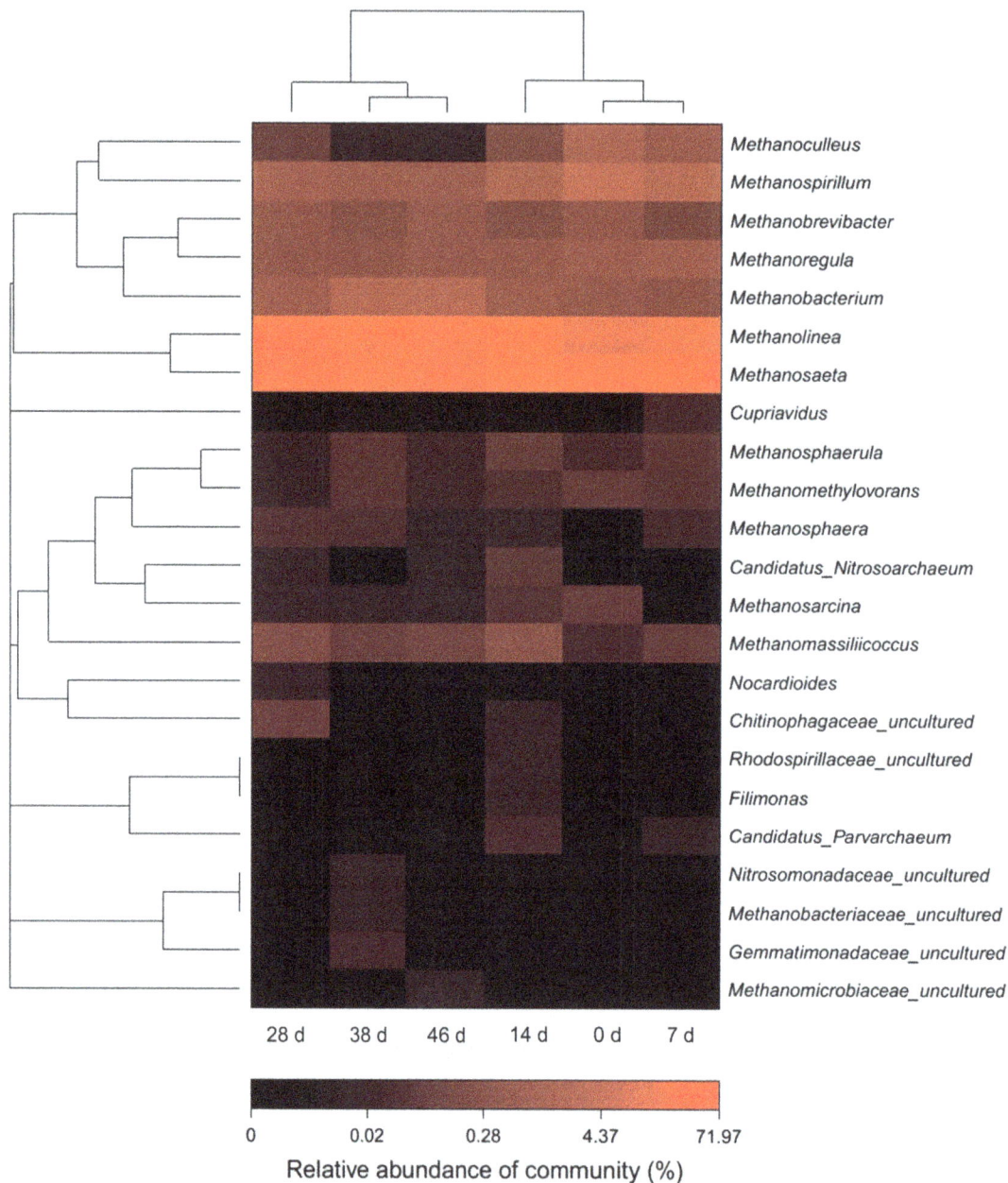

Figure 5. Heat map representation and cluster analysis of archaeal community at the genus level. Distance algorithm: Bray-Curtis; clustering method: complete. The color intensity of scale demonstrates the relative abundance of each genus. Relative abundance is defined as the number of sequences affiliated with that taxon divided by the total number of sequences per sample (%).

samples taken on day 38 and those taken on day 46. In a horizontal view, the difference of Groups II, III, and IV was small, but each of them differed from Group I considerably. The above-mentioned observations indicate that a distinct shift of archaeal populations of ADMD system occurred after day 7 because PC1 was the major principle component, which accounted for 85.3% of the total variation. Archaeal consortia became relatively stable after 38 d. This was in accordance with biogas production (Figure 1D) and archaeal community structure at the order level (Figure 4C). The results indicate that the ADMD could be well started up within 38–46 d.

The changes in the archaeal consortia can be demonstrated with the variations of archaeal community structure. As shown in

Figure 4A, although a small decrease in relative abundance was observed during the start-up stage, the relative abundance of *Euryarcharota* was within the range of 98.4–99.6% at the phylum level, indicating that *Euryarcharota* was the predominant phylum in the ADMD system. The relative abundance of *Methanomicrobia* increased from 62.0% to 82.9% during the start-up (Figure 4B). It was the major class in the ADMD system. At the order level (Figure 4C), *Methanomicrobiales* and *Methanosarcinales*, which together accounted for 61.9–82.6% of total reads on the order level, were the two predominant orders during the start-up stage. *Methanomicrobiales* and *Methanosarcinales* have also been found to be abundant in AnMBR system for swine manure treatment [32]. The sum of relative abundance of both orders increased over time. However,

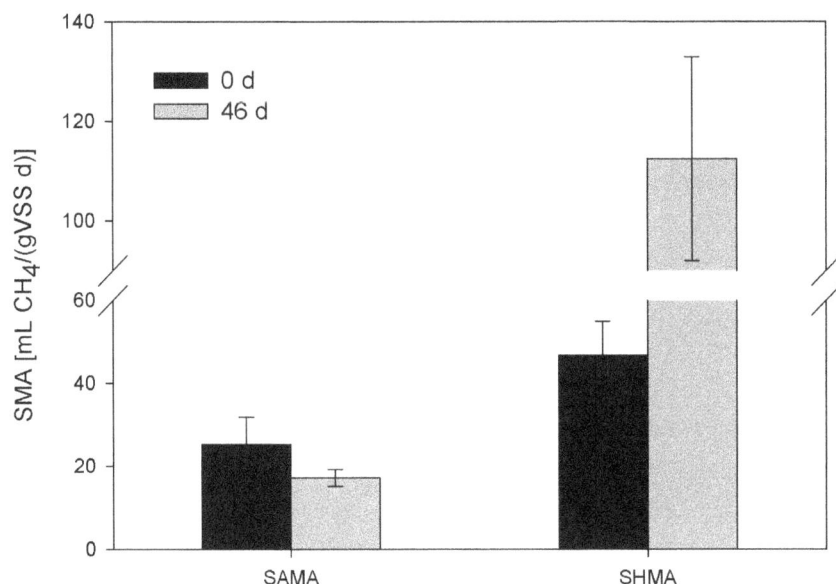

Figure 6. SAMA and SHMA tests of ADMD samples on day 0 and day 46. Error bars represent standard deviations of triplicate tests.

changes in individual orders were different. The relative abundance of *Methanomicrobiales* increased from day 0 to day 38 and remained relatively stable within the range of 47.9–51.3% from day 38 to day 46. However, the relative abundance of *Methanosarcinales* decreased from day 0 to day 28 and remained relatively stable within the range of 29.5–35.1% afterward. During the start-up stage, the population of *Methanomicrobiales* gradually exceeded that of *Methanosarcinales*. It became the most abundant order on day 46. It has been reported that *Methanomicrobiales* comprises hydrogenotrophic methanogens and *Methanosarcinales* contains acetoclastic methanogens [33]. The results indicated that methanogenic pathways might be altered during the start-up stage of ADMD system.

To further investigate the changes of archaeal community and functional groups, cluster analysis was conducted at the genus level. As shown in Figure 5, four clusters were identified from the six archaeal consortia using hierarchical cluster analysis: (1) Cluster I contains the ADMD sludges on day 0 and day 7; (2) Cluster II contains the ADMD sludge on day 14; (3) Cluster III is the ADMD sludge on day 28; (4) Cluster IV is the ADMD sludges on days 38 and 46. The archaeal consortia on the genus level in Cluster I (0 d and 7 d) showed considerable homology, which was also observed in Cluster IV (38 d and 46 d). These above-mentioned results are consistent with PCA at a 3% cutoff OTU level (Figure 3) and cluster analysis of bacterial community at the genus level (see Figure S3). *Methanolinea* and *Methanosaeta* exhibited considerable abundance during the start-up of ADMD system. On day 0, the relative abundance of *Methanosaeta* and *Methanolinea* were 50.0% and 10.3%, respectively. However, with the increase in abundance of the former and the decrease in abundance of the latter during the start-up, *Methanolinea* surpassed *Methanosaeta* to become the most predominant genus from day 28 to day 46. Previous studies have been done on the archaeal community in AD systems. They have reported that the acetoclastic *Methanosaeta* was the most abundant archaeal genus and the hydrogenotrophic *Methanolinea* only accounted for a small fraction of genus-assigned sequences [17]. The results of the present study were quite different, indicating that archaeal community in the ADMD

system was distinct from conventional AD systems. *Methanosaeta* were found to be the dominant acetoclastic methanogens in a variety of anaerobic bioreactors when acetate concentration was low [34]. In the present study, the high abundance of *Methanosaeta* was in accordance with low acetate concentration during the start-up (Figure 1C). However, carbon dioxide contained 8–22% of biogas after 7 days (Figure 1D), which was below the range of 30–35% in AD processes [2]. These results indicated that some of the carbon dioxide might be consumed by hydrogenotrophic methanogens and used as a carbon source. This is consistent with the predominant abundance of *Methanolinea*. Based on these results, it might be speculated that *Methanolinea* and *Methanosaeta* were responsible for most of the methane production in the ADMD system. Because *Methanolinea* and *Methanosaeta* are hydrogenotrophic and acetoclastic, respectively, methanogenesis might be carried out via both hydrogenotrophic and acetoclastic pathways, but the hydrogenotrophic pathway might prevail.

To validate the available methanogenic pathways during the start-up of the ADMD system, SMA tests were performed on day 0 and day 46 using acetate and H_2/CO_2 as substrates, respectively. As shown in Figure 6, SAMA on day 0 and day 46 were 25.3±6.5 mL CH_4/(g VSS·d) and 17.1±2.0 mL CH_4/(g VSS·d), respectively, indicating that acetoclastic methanogens became less dominant at the end of start-up. However, SHMA was 46.6±8.4 mL CH_4/(g VSS·d) and 112.4±20.5 mL CH_4/(g VSS·d) on day 0 and day 46, respectively, indicating that hydrogenotrophic methanogens became more dominant at the end of start-up. The results of SAMA and SHMA could not be compared directly because the concentrations of substrates were different. However, the effects of substrate concentrations could be eliminated via the comparison of SHMA/SAMA ratio. The average SHMA/SAMA on day 46 was about 3.6 times of that on day 0. The results indicated that both hydrogenotrophic and acetoclastic pathways were observed on day 0 and day 46 via SMA tests, but hydrogenotrophic pathways were more predominant at the end of the start-up of ADMD system. This is in accordance with the analysis of archaeal community structure.

Conclusions

During the start-up, methane production improved gradually and became relatively steady in the end. Pyrosequencing showed that *Proteobacteria* and *Bacteroidetes* were the predominant phyla in bacterial consortia. In archaeal consortia, principle component and cluster analyses demonstrated that archaeal consortia experienced a distinct shift and became relatively stable after day 38, indicating that the reactor can be well started-up within 38 days. *Methanomicrobiales* gradually exceeded *Methanosarcinales*, and became the most abundant at the class level. Further analysis indicated that *Methanolinea* and *Methanosaeta* were the functional genera for methane production. Hydrogenotrophic pathways might prevail over acetoclastic pathways for methanogenesis in the process.

Supporting Information

Figure S1   Schematic of ADMD system.

Figure S2   Rarefaction curves based on pyrosequencing of microbial communities. (A) bacteria; (B) archaea. The OTUs were defined by clustering sequences at a dissimilarity level of 0.03.

Figure S3   Heat map representation and cluster analysis of bacterial community at the genus level. The y-axis is the clustering of the 100 most abundant genera in reads. Distance algorithm: Bray-Curtis; clustering method: complete. The color intensity of scale demonstrates the relative abundance of each genus. Relative abundance is defined as the number of sequences affiliated with that taxon divided by the total number of sequences per sample (%).

Acknowledgments

The instruments used in this research are supported by the Collaborative Innovation Center of Advanced Technology and Equipment for Water Pollution Control and the Collaborative Innovation Center for Regional Environmental Quality.

Author Contributions

Conceived and designed the experiments: HGY ZWW. Performed the experiments: HGY JXM. Analyzed the data: HGY ZWW JXM. Contributed reagents/materials/analysis tools: ZWW QYW YLL ZCW. Wrote the paper: HGY. Critical revision of the manuscript for important intellectual content: ZWW ES ZCW.

References

1. Wang Z, Yu H, Ma J, Zheng X, Wu Z (2013) Recent advances in membrane bio-technologies for sludge reduction and treatment. Biotechnol Adv 31: 1187–1199.
2. Appels L, Baeyens J, Degreve J, Dewil R (2008) Principles and potential of the anaerobic digestion of waste-activated sludge. Prog Energy Combust Sci 34: 755–781.
3. Dagnew M, Parker WJ, Seto P (2010) A pilot study of anaerobic membrane digesters for concurrent thickening and digestion of waste activated sludge (WAS). Water Sci Technol 61: 1451–1458.
4. Xu M, Wen X, Yu Z, Li Y, Huang X (2011) A hybrid anaerobic membrane bioreactor coupled with online ultrasonic equipment for digestion of waste activated sludge. Bioresource Technol 102: 5617–5625.
5. Liao BQ, Kraemer JT, Bagley DM (2006) Anaerobic membrane bioreactors: Applications and research directions. Crit Rev Environ Sci Technol 36: 489–530.
6. Fan B, Huang X (2002) Characteristics of a self-forming dynamic membrane coupled with a bioreactor for municipal wastewater treatment. Environ Sci Technol 36: 5245–5251.
7. Zhang XY, Wang ZW, Wu ZC, Lu FH, Tong J, et al. (2010) Formation of dynamic membrane in an anaerobic membrane bioreactor for municipal wastewater treatment. Chem Eng J 165: 175–183.
8. An Y, Wang Z, Wu Z, Yang D, Zhou Q (2009) Characterization of membrane foulants in an anaerobic non-woven fabric membrane bioreactor for municipal wastewater treatment. Chem Eng J 155: 709–715.
9. Connaughton S, Collins G, O'Flaherty V (2006) Development of microbial community structure and actvity in a high-rate anaerobic bioreactor at 18 degrees C. Water Res 40: 1009–1017.
10. Gao WJ, Qu X, Leung KT, Liao BQ (2012) Influence of temperature and temperature shock on sludge properties, cake layer structure, and membrane fouling in a submerged anaerobic membrane bioreactor. J Membr Sci 421–422: 131–144.
11. Riviere D, Desvignes V, Pelletier E, Chaussonnerie S, Guermazi S, et al. (2009) Towards the definition of a core of microorganisms involved in anaerobic digestion of sludge. Isme J 3: 700–714.
12. Zhang T, Shao MF, Ye L (2012) 454 Pyrosequencing reveals bacterial diversity of activated sludge from 14 sewage treatment plants. Isme J 6: 1137–1147.
13. Shendure J, Ji HL (2008) Next-generation DNA sequencing. Nat Biotechnol 26: 1135–1145.
14. Margulies M, Egholm M, Altman WE, Attiya S, Bader JS, et al. (2005) Genome sequencing in microfabricated high-density picolitre reactors. Nature 437: 376–380.
15. Krause L, Diaz NN, Edwards RA, Gartemann K-H, Krömeke H, et al. (2008) Taxonomic composition and gene content of a methane-producing microbial community isolated from a biogas reactor. J Biotechnol 136: 91–101.
16. Kröber M, Bekel T, Diaz NN, Goesmann A, Jaenicke S, et al. (2009) Phylogenetic characterization of a biogas plant microbial community integrating clone library 16S-rDNA sequences and metagenome sequence data obtained by 454-pyrosequencing. J Biotechnol 142: 38–49.
17. Nelson MC, Morrison M, Yu Z (2011) A meta-analysis of the microbial diversity observed in anaerobic digesters. Bioresource Technol 102: 3730–3739.
18. Ma J, Wang Z, Zou X, Feng J, Wu Z (2013) Microbial communities in an anaerobic dynamic membrane bioreactor (AnDMBR) for municipal wastewater treatment: Comparison of bulk sludge and cake layer. Process Biochem 48: 510–516.
19. Schloss PD, Westcott SL, Ryabin T, Hall JR, Hartmann M, et al. (2009) Introducing mothur: Open-Source, Platform-Independent, Community-Supported Software for Describing and Comparing Microbial Communities. Appl Environ Microbiol 75: 7537–7541.
20. Ma J, Wang Z, Yang Y, Mei X, Wu Z (2013) Correlating microbial community structure and composition with aeration intensity in submerged membrane bioreactors by 454 high-throughput pyrosequencing. Water Res 47: 859–869.
21. Donlon BA, Razoflores E, Field JA, Lettinga G (1995) Toxicity of N-substituted aromatics to acetoclastic methanogenic activity in granular sludge. Appl Environ Microbiol 61: 3889–3893.
22. Coates JD, Coughlan MF, Colleran E (1996) Simple method for the measurement of the hydrogenotrophic methanogenic activity of anaerobic sludges. J Microbiol Methods 26: 237–246.
23. Yukselen MA (1997) Preservation characteristics of UASB sludges. J Environ Sci Health 32: 2069–2076.
24. APHA (2012) Standard Methods for the Examination of Water and Wastewater. 22nd ed. Washington, DC, USA: American Public Health Association/ American Water Works Association/Water Environment Federation.
25. Yu H, Wang Z, Wang Q, Wu Z, Ma J (2013) Disintegration and acidification of MBR sludge under alkaline conditions. Chem Eng J 231: 206–213.
26. Hatziconstantinou GJ, Yannakopoulos P, Andreadakis A (1996) Primary sludge hydrolysis for biological nutrient removal. Water Sci Technol 34: 417–423.
27. Ariesyady HD, Ito T, Okabe S (2007) Functional bacterial and archaeal community structures of major trophic groups in a full-scale anaerobic sludge digester. Water Res 41: 1554–1568.
28. Chouari R, Le Paslier D, Daegelen P, Ginestet P, Weissenbach J, et al. (2005) Novel predominant archaeal and bacterial groups revealed by molecular analysis of an anaerobic sludge digester. Environ Microbiol 7: 1104–1115.
29. Cottrell MT, Kirchman DL (2000) Natural assemblages of marine proteobacteria and members of the Cytophaga-Flavobacter cluster consuming low- and high-molecular-weight dissolved organic matter. Appl Environ Microbiol 66: 1692–1697.
30. Kirchman DL (2002) The ecology of Cytophaga-Flavobacteria in aquatic environments. Fems Microbiol Ecol 39: 91–100.
31. Kapdan IK, Kargi F (2006) Bio-hydrogen production from waste materials. Enzyme Microb Technol 38: 569–582.
32. Padmasiri SI, Zhang J, Fitch M, Norddahl B, Morgenroth E, et al. (2007) Methanogenic population dynamics and performance of an anaerobic membrane bioreactor (AnMBR) treating swine manure under high shear conditions. Water Res 41: 134–144.
33. Garcia J-L, Patel BKC, Ollivier B (2000) Taxonomic, Phylogenetic, and Ecological Diversity of Methanogenic Archaea. Anaerobe 6: 205–226.
34. Zheng D, Raskin L (2000) Quantification of Methanosaeta species in anaerobic bioreactors using genus- and species-specific hybridization probes. Microbiol Ecol 39: 246–262.

8

Euphorbia tirucalli L.–Comprehensive Characterization of a Drought Tolerant Plant with a Potential as Biofuel Source

Bernadetta Rina Hastilestari[1], Marina Mudersbach[2], Filip Tomala[2], Hartmut Vogt[2], Bettina Biskupek-Korell[2], Patrick Van Damme[3,4], Sebastian Guretzki[1], Jutta Papenbrock[1]*

1 Institute of Botany, Gottfried Wilhelm Leibniz University Hannover, Hannover, Germany, **2** Technology of Renewable Resources, University of Applied Sciences Hannover, Hannover, Germany, **3** Department of Plant Production, Laboratory for Tropical and Subtropical Agriculture and Ethnobotany, Ghent University, Ghent, Belgium, **4** Institute of Tropics and Subtropics, Czech University of Life Sciences Prague, Prague, Czech Republic

Abstract

Of late, decrease in mineral oil supplies has stimulated research on use of biomass as an alternative energy source. Climate change has brought problems such as increased drought and erratic rains. This, together with a rise in land degeneration problems with concomitant loss in soil fertility has inspired the scientific world to look for alternative bio-energy species. *Euphorbia tirucalli* L., a tree with C_3/CAM metabolism in leaves/stem, can be cultivated on marginal, arid land and could be a good alternative source of biofuel. We analyzed a broad variety of *E. tirucalli* plants collected from different countries for their genetic diversity using AFLP. Physiological responses to induced drought stress were determined in a number of genotypes by monitoring growth parameters and influence on photosynthesis. For future breeding of economically interesting genotypes, rubber content and biogas production were quantified. Cluster analysis shows that the studied genotypes are divided into two groups, African and mostly non-African genotypes. Different genotypes respond significantly different to various levels of water. Malate measurement indicates that there is induction of CAM in leaves following drought stress. Rubber content varies strongly between genotypes. An investigation of the biogas production capacities of six *E. tirucalli* genotypes reveals biogas yields higher than from rapeseed but lower than maize silage.

Editor: Haibing Yang, Purdue University, United States of America

Funding: B.R. Hastilestari was supported by Katholischer Akademischer Ausländer-Dienst (KAAD). The funders had no role in study design, data collection and analysis, decision to publish, or preparation of the manuscript.

Competing Interests: The authors have declared that no competing interests exist.

* E-mail: Jutta.Papenbrock@botanik.uni-hannover.de

Introduction

Agriculture faces a range of serious environmental problems such as soil salinisation and depletion of water resources. Additionally, agricultural production and unsustainable human intervention often leave the land under stress, leading to an increase in non-arable land area [1]. The supply of fossil fuel in future will also soon start decreasing. Therefore, efforts are made to find substitute sources of energy. One such source is solar energy, which is unlimited. Plants capture this energy through photosynthesis. Faced with a decrease in arable land and crude oil supply, it is important to find species for growing in marginal, non-arable land. These plants should have high drought and salinity tolerance as well as contain compounds that could be used in phytochemical, pharmaceutical or nutraceutical applications.

Euphorbia tirucalli L. belongs to the dicotyledonous order Euphorbiales, family Euphorbiaceae, subsection *tirucalli* [2]. The natural distribution of *E. tirucalli* comprises the Paleotropical region of Madagascar, the Cape region (South Africa), East Africa, and Indochina [3]. This plant is also grown as garden plant in numerous tropical countries, also in America. *E. tirucalli* seems to have high salinity and drought tolerance [4] and survives in a wide range of habitats even under conditions in which most crops c.q.

plants cannot grow. These include tropical arid areas with low rainfall, poor eroded or saline soils and high altitudes but *E. tirucalli* cannot survive frost [3]. Its high stress tolerance can be explained at least in part by its photosynthetic system. The family of *E. tirucalli*, the Euphorbiaceae, consists of five subfamilies [5] and its species have C_3, C_4, intermediate C_3–C_4 and/or Crassulacean Acid Metabolism (CAM) photosynthetic systems dependent on the ecological conditions [6]. Batanouny et al. [6] reported that *Euphorbia* species having the C_3 photosynthetic pathway grow under conditions of better water resources and lower temperature, whereas CAM and C_4 plants grow under high temperature. The photosynthetic system of *E. tirucalli* stems has been identified to follow CAM [7]. It has been classified based on the C-isotope ratio. The range of values -8 to -18 are characteristic of plants with C_4 or CAM [8], while "Kranz" anatomy provides strong evidence of C_4 system. Meanwhile Ting et al. [9] described values in the range of -15.4 to -16.2 were classified as CAM plants, whereas -12.6 and -11.3 as C_4. Bender [7] showed $^{13}C/^{12}C$ ratios of *E. tirucalli* was -15.3. This value indicated that *E. tirucalli* did not follow C_4; this was also supported that there was no Kranz syndrome in *E. tirucalli* stem [10]. Its photosynthetic system followed C_3 in non-succulent leaves and CAM pathway in succulent stems based on gas exchange observations [3]. In

CAM plants one can observe an opening/closure of stomata during night/day allowing nightly CO_2 uptake accompanied with malate oscillation that follows stomatal opening and closure [11,12]. Hence, malate presence confirms CAM photosynthetic pathway in *E. tirucalli*. Under unfavorable conditions, its non-succulent C_3 leaves soon die and the plant will then continue its metabolism via the CAM photosynthetic pathway in the stem. The combination of C_3 leaves and CAM stems can explain *E. tirucalli*'s fast accumulation of biomass since C_3 maximizes growth during favorable conditions and CAM during drought to reduce water loss and maintain photosynthetic integrity [13]. C_3 photosynthetic pathway takes place when leaves are present and in combination with CAM stem, whereas CAM stem takes up CO_2 when conditions deteriorate. However, to date there is no evidence that there is a change from C_3 and CAM at leaf level following drought events, a mechanism that has been evidenced in *Mesembryanthemum crystallinum* L. [14] and the genus *Sedum* [15].

E. tirucalli has been reported to present numerous pharmacological activities. The species has been patented for modern drugs such as prostate cancer medicine [16] and has a very high ethnomedicinal value [17–20]. *E. tirucalli* produces and stores abundant amounts of latex in so-called laticifers [21]. *E. tirucalli* latex contains high amounts of sterols and triterpenes [22] and might be used for rubber fractionation and has been investigated for its diesel oil properties [17,23–27]. Through the hydrocarbons of its latex, the species was documented in 1978 to produce the equivalent to 10–50 barrels oil L ha^{-1} [24], whereas its biomass can yield 8,250 m^3 ha^{-1} biogas (in the tropical, subhumid conditions of Colombia [28]). Furthermore, *E. tirucalli* latex has pesticidal properties against such pests as mosquitoes (*Aedes aegypti* and *Culex quinquefasciatus*) [29], bacteria (*Staphylococcus aureus*) [30], molluscs (*Lymnaea natalensis*) and nematodes such as *Haplolaimus indicus*, *Helicotylenchus indicus* and *Tylenchus filiformis* [31]. *E. tirucalli* latex can also be used as glue and adhesive [32].

The morphological characteristics of different *E. tirucalli* accessions do not allow differentiating them amongst themselves, except for one US accession that has yellow tips and has been promoted for ornamental uses. Hence, classification of *E. tirucalli* based on its genetic characteristic will be more precise than using morphological descriptors. Until now, genetic diversity between *E. tirucalli* genotypes from different areas has not been investigated. Analysis of genetic diversity among genotypes is also a prerequisite if one wants to start selecting and/or breeding for increased drought tolerance, gain in biomass, rubber content and biogas production. Our final aim is to recommend the best genotypes first for field research experiments and then for initiating commercial *E. tirucalli* plantations in arid areas for the respective applications.

Materials and Methods

2.1 Plant material, propagation and growth conditions

Mother plants of genotypes Morocco, Senegal, Burundi, Rwanda, Kenya and USA were collected by Van Damme over the last 20 years from wild individuals and grown in greenhouses at Ghent University, Department of Plant Production, Laboratory for Tropical and Subtropical Agriculture and Ethnobotany, Belgium. Genotype India was collected in Ajmer and Jaipur from naturalized plants but genotype Jaipur could not be propagated as it died after delivery. Genotype Indonesia was collected in Yogyakarta from a wild-grown individual, genotype Italy was collected in Calabria from a cultivated ornamental, genotype Togo was collected in Togo from wild plants by Torsten Schmidt (Hannover, Germany), whereas genotype Hannover was an ornamental specimen of unknown origin. No specific permissions

were required for collecting on these locations because the plants grow like weed on locations that are not privately-owned or protected in any way and the *E. tirucalli* species does not belong to endangered or protected species.

Propagation for our experiments was done vegetatively by cuttings taken on no predefined part of the respective mother plants. The 10–15 cm cuttings obtained from healthy plants and planted in pots with volume of 436 cm^3 according to the formula of truncated cones that contained a mixture of clay-loam:sand (2:1). These cuttings were cultivated in the greenhouse of Institute of Botany, Leibniz University Hannover, for six months at 14 h/ 24°C (day) and 10 h/22°C (night) with a light intensity of 350 μmol m^{-2} s^{-1}; and watered once every two days. In control conditions fertilizer Wuxal Top N (Aglukon, Düsseldorf, Germany) consisting of 0.6% NPK and 99.4% water was applied once every two days (about 8.6 ml per pot). For the water stress conditions the same concentration of fertilizer was added in a smaller volume of water.

2.2 Molecular analysis through genetic marker

2.2.1 DNA extraction and quantification. DNA was extracted from twelve genotypes of the *E. tirucalli* collection. DNA isolation procedure using NucleoSpin® Plant II Kit (Macherey & Nagel GmbH & Co. KG, Düren, Germany) was used to extract genomic DNA from 60 mg of young leaf samples. Freshly extracted DNA was quantified photometrically using an Uvikon xs photometer (Biotek Germany, Bad Friedrichshall, Germany). Quantification was done by measuring 2 μl of non-diluted DNA sample at 260 nm wavelength. Extracted DNA was stored at −20°C until use.

2.2.2 Amplified Fragment Length Polymorphism (AFLP). AFLP analysis was performed essentially as described by Vos et al. [33]. Restriction fragments were produced by digestion of 250 ng genomic DNA for 1 h at 37°C with 0.5 μl *Eco*RI (10 U/μl) and 0.3 μl *Mse*I (10 U/μl) in a total volume of 25 μl containing 2.5 μl 10×RL Buffer, 100 mM Tris HCl, 100 mM MgAc, 500 mM KAc, 50 mM DTT, pH 7.5, and H_2O. The digestion was followed by ligation of specific *Mse*I (50 pmol) and *Eco*RI (5 pmol) adapters (MWG Biotech Eurofins, Ebersberg, Germany) with 5 μL reaction mix (0.5 μl of *Eco*RI adapter, 0.5 μl of *Mse*I adapter, 0.6 μl of 10 mM ATP, 0.5 μl 10× RL-Buffer, 0.05 μl of T4-DNA-Ligase (1 U μl^{-1}), and 2.85 μl H_2O) which was added to the restricted DNA and incubated for 3.5 h at 37°C.

For the pre-amplification a reaction mix (5 μl of digested and ligated DNA, 1.5 μl *Eco*RI+0 (5′ GACTGCGTACAA TTC 3′) and *Mse*I+0 (5′ GATGAGTCCTGAGTAA 3′) or *Eco*RI+A/ *Mse*I+A primer combinations (50 ng μl^{-1}), 5 μl dNTPs (2 mM), 5 μl 10×Williams Buffer (100 mM Tris/HCl, pH 8.3; 500 mM KCl; 20 mM MgCl$_2$; 0.01% gelatine; H_2O), 1 μl *Taq* polymerase (5 U μl^{-1}) and 31 μl H_2O) was amplified in a thermocycler with 94°C/5 min, then 20 cycles of 94°C/30 s, 60°C/30 s, 72°C/60 s and finally 72°C/10 min. Selective amplifications were performed using primer pairs containing three selective nucleotides. For selective amplification, 2.5 μl of a 20-fold diluted pre-amplification mixture with reaction mix (2.5 μl *Eco*RI-IRD primer (2 ng μl^{-1}), 0.3 *Mse*I primer (50 ng μl^{-1}), 1 μl dNTPs (2 mM), 0.05 μl *Taq* polymerase (5 U μl^{-1}), 1 μl 10×Williams Buffer and 2.65 μl H_2O) was amplified consisting of 94°C/5 min, one cycle of 94°C/30 s, 65°C/30 s and 72°C/60 s, then lowering the annealing temperature to about 0.7°C reduction per cycle for next 11 cycles, thereafter 24 cycles of 94°C/30 s, 56°C/30 s, 72°C/60 s and lastly 72°C/10 min. IRD 700 labelled *Eco*RI primers and *Mse*I primers with three selective nucleotides at their 5′ end was used

Table 1. Primer combinations for selective amplification.

Primer combination	*Eco*RI 700	*Mse*I
1	GACTGCGTACAA TTC ACA	GATGAGTCCTGAG TAA ACT
2	GACTGCGTACAA TTC ACA	GATGAGTCCTGAG TAA ACT
3	GACTGCGTACAA TTC ACA	GATGAGTCCTGAG TAA ACA
4	GACTGCGTACAA TTC ACC	GATGAGTCCTGAG TAA ATTA
5	GACTGCGTACAA TTC ACC	GATGAGTCCTGAG TAA ATGG
6	GACTGCGTACAA TTC ACA	GATGAGTCCTGAG TAA ATGG
7	GACTGCGTACAA TTC ACA	GATGAGTCCTGAG TAA ACAT

(Table 1). After PCR, an equal volume of sequencing loading buffer (98% formamide, 10 mM EDTA, pararosaniline 0.05%) was added. The mixture was heated to 90°C for 3 min and then cooled on ice.

Marked fragments were separated over 6% polyacrylamide gel from Sequa gel X® (16 ml of monomer solution, 4 ml of complete buffer and 160 µl of 10% APS) with 1×TBE buffer. A sizing standard was labeled with IRD 700 at their 5′ end (MWG Biotech Eurofins). Samples were analyzed on a LICOR Gene Reader 4300 automated sequencer (LI-COR Biosciences, Lincoln, USA), at condition 1500 V, 35 A, 40 W, 45°C, slow scan speed and 30 min pre-run.

2.2.3 PCR product detection and phylogenetic analysis. Detection of AFLP products and phylogenetic analysis of DNA AFLP fingerprints was conducted based on the number, frequency and distribution of amplified DNA fragments. AFLP product diversity was determined from the difference in gel migration of PCR products from each individual sample. Based on the presence or absence of AFLP bands, band profiles were translated into binary data. Data were analyzed using fingerprint analysis with missing data 1.0 (FAMD) (program available from http://homepage.univie.ac.at/philipp.maria.schlueter/famd.html) [34]. The tree was generated using Unweighted Pair Group Method with Arithmetic Mean (UPGMA). The tree was visualized using the TreeView program version 1.6.6 [35].

2.3 Investigation of drought tolerance

Investigation of drought effects was conducted based on Jefferies [36] with some modifications. Six month old *E. tirucalli* plants from Morocco and Senegal with a height of 27–29 cm were selected. This experiment was conducted in a climatic chamber for 8 weeks with condition 24/20°C day (14 h)/night (10 h), at light intensity 155 µmol m^{-2} s^{-1} and 60% humidity. Twenty plants from each genotype were grown in clay-loam and sand substrate with four different volumetric water contents (VWC) 25%, 15%, 10% and 5% monitored using Fieldscout® based on time domain reflectometry (TDR) (Spectrum Technologies, Plainfield, USA). Dry set value was 1% below and wet value was 1% above the respective VWCs. According to the manual of this instrument, sandy-clay-loam substrate has water holding capacity of 25% VWC, and a wilting point at 15% VWC. Soil moisture was measured based on water deficit (D) values which indicate the amount of irrigation water necessary to raise the soil water content to the target point. Water was added based on calculation of D values times 8.66 ml for a pot with 7 cm height.

As *E. tirucalli* grows in semi-arid and arid areas, two VWC points below 15% were investigated for their effect on the species'

physiology. Selected VWC points were 10% and 5%. Growth parameters such as plant height, root length, dry matter production, and water content were measured. Plant height and tap root length were measured with a scale. For fresh and dry biomass determination shoots and root of plants were harvested separately and measured after 8 weeks of treatment. Shoots and roots were dried in an incubator at 90°C for 36 h. Investigation on whether there was an effect of drought on photosynthesis during drought application, chlorophyll fluorescence measurements were conducted every week during 8 weeks during drought treatment using the non-invasive method of Imaging PAM (M series, Heinz Walz GmbH, Effeltrich, Germany). Hence, quantum efficiency (Fv/Fm) was measured at leaves having C_3 photosynthetic pathway and stems having CAM photosynthetic pathway.

2.4 Investigation of the C3 and CAM photosynthetic pathways: malate determination

Stems and leaves of genotypes Morocco and Senegal were harvested at the end of the dark period (5 am) and the end of the light period (7 pm). The end of the dark period is the phase where malate concentration is highest, whereas the end of the light period is the phase where this value is lowest [37]. Harvested material with 3 replications was put in liquid nitrogen and stored in the freezer at −80°C before malate extraction.

Malate was extracted by putting 60 mg of leaves and stems of each genotype separately in 1.4 ml H_2O and vortexing the mixture for 1 min; the mixture as then kept at room temperature for 10 min and mixed again for 1 min. A centrifugation by 13,000 rpm at 4°C for 10 min followed whereupon the supernatant was pipetted into new tubes and centrifuged again at 13,000 rpm for 10 min at 4°C. The supernatant was then pipetted into new tubes and kept at −20°C until measurement by capillary electrophoresis (CE). A P/ACE^{TM} MDQ capillary electrophoresis system (Beckman Coulter, Krefeld, Germany) was used for CE analyses. Separations were performed in a eCAP^{TM} CE-MS capillary (fused silica, 75 µm i.d., 57 cm total length, 50 cm effective length, Beckman Coulter). Before starting the analyses the capillary was equilibrated with the background electrolyte Basic Anion Buffer for HPCE (Agilent Technologies, Waldbronn, Germany) at 14.5 psi for 4 min. Injection was done by applying 0.7 psi for 3.5 s. Separation of the samples was performed by applying 14 kV for 10 min at 22°C. After each run, the capillary was washed with the background electrolyte for 4 min. Buffer was changed after 8 to 10 runs. Samples were detected at 235 nm with a bandwidth of 10 nm. Calibration graphs were generated with 0.313 to 10 mM malic acid. Elaboration of the electropherograms was done using Karat 32 7.0 software (Beckman Coulter).

2.5 Latex analysis

E. tirucalli latex consists of 2.8% to 8.3% rubber and 50.4% to 82.1% resin [38]. Latex of *E. tirucalli* has attracted a lot of attention because it has an economical potential as source of rubber. Therefore, rubber content was investigated in different genotypes. Rubber content analysis was conducted by LipoFit Analytic GmbH (Regensburg, Germany) using nuclear magnetic resonance (NMR, 600 MHz Bruker Avance^{+} spectrometer, Bruker Daltonic GmbH, Bremen, Germany). Samples were taken from Burundi, Hannover, Kenya, Morocco, Rwanda, Senegal, Togo and USA genotypes. The input material was 100 to 500 mg fresh weight of stems.

To fresh plant material, 1.5 ml water p.a. (0.03% NaN$_3$) and a sharp aglet were added. By shaking 10 min the material was mechanically milled. The aglet was extracted from the suspension by a magnet. The suspension was centrifuged (20 min;

14,500 rpm; 20°C) to separate cell debris. Sodium phosphate buffer pH 6.8 (final concentration 100 mM), D20 (5%) and sodium trimethyl silyl propionate (0.1 mM) were added to the supernatant. The suspension was then transferred to 5 mm-NMR-tubes.

Relative rubber concentrations refer to the average of the spectra measured in the *E. tirucalli* samples. The average is calculated out of the integral from all the spectra which are expected to contain rubber signals. The reference for the absolute concentrations was 1,4-polyisoprene with a molar mass of 47,300 g mol^{-1}. The reference was also measured by NMR. In reference to polyisoprene, only the spectra with the same pattern as the reference were calculated.

2.6 Biogas production

Plant material of genotypes Kenya, Morocco, Rwanda, Senegal, Togo, and USA was harvested from the greenhouse (Hannover, Germany), dried, and chopped into 0.5 to 4 cm pieces before being used in biogas batch tests. Biogas yields of the selected genotypes were determined through anaerobic batch digestion tests according to the German Standard Procedure VDI 4630 [39]. The inoculum was biogas slurry from an agricultural biogas plant mainly fed with maize silage. Organic dry matter (ODM), density and chemical oxygen demand (COD) were determined for all samples and the inoculum according to standard methods. Based on results, the weighted samples of the substrates and the inoculum were balanced to obtain a Slurry Loading Rate (SLR; $ODM_{substrate}$ to $ODM_{inoculum}$) of 0.3 as recommended by VDI 4630. Each substrate and one control without the addition of substrate, was incubated in triplicate in gas-tight 1,250 ml dark DURAN glass bottles. Experiments were conducted for 28 days at 38°C in a warming cupboard. Biogas yields (L kg^{-1} ODM) were calculated based on the pressure in the bottles following biogas production. Rise in pressure was recorded with LabView software connected to the batch plant. After tests were finished, the concentration of CH_4 in the biogas produced were analyzed as follows: In each bottle, 20 ml of a 10 molar NaOH solution were injected through the septum with the help of a syringe. The NaOH solution fixes the CO_2 in the biogas by reacting to sodium carbonate which precipitates in the liquid phase. As a result, in the bottles a decrease in pressure occurs and on the basis of this data, the methane ratio in the produced biogas can be calculated. H_2S in biogas samples of genotypes Morocco, Kenya and USA were quantified using gas chromatography.

2.7 Statistical analysis

All statistical analysis was conducted with Statistix 8 version 2 (Analytical software, Tallahassee, USA). Interaction between means was calculated by the least significant different (LSD) at *p<0.05*. Graphs were drawn using SigmaPlot Version 12.2 (Systat Software Inc., San Jose, USA).

Results

3.1 Genetic marker analysis

AFLP technique was used as a tool for assessing species relationships within the *E. tirucalli* collection. Seven primer combinations were selected for AFLP analysis (Table 1). Total number of polymorphic bands was 243 with a mean of 34.7. We were able to derive two main groups from the phylogenetic analysis of the 12 accessions of *E. tirucalli* cluster analysis using UPGMA with 1000 bootstrap replicates (Fig. 1). Nevertheless, the genotypes tested share a lot of similarities as evidenced from the low bootstrap values. The first group consists of two clades and

comprises mainly genotypes from Africa: Burundi, Morocco, Senegal and Togo accessions that are clustered with a bootstrap value of 63. The second group consists of four clades with mainly non-African genotypes (except Kenya and Rwanda): Ajmer (India), Hannover (Germany), Indonesia, Italy, Jaipur (India), Kenya, Rwanda and USA with a bootstrap value of 72. A dendrogram derived from NJ calculation showed the same pattern (data not shown). All genotypes have been propagated by cuttings and cultivated in the greenhouse since a long time or at least for a couple of years. Therefore they should have the same amount of endophytes, if any. In our AFLP analysis the genotypes differ in several hundred bands. In case there are some bands originating from endophytes they would not influence the results significantly.

3.2 Stress tolerance

We were interested to analyze physiological differences among members of the genetically quite homogeneous African group. Therefore the response to different soil water contents of *E. tirucalli* genotypes Morocco and Senegal that were grown on clay-loam:sandy soil type after eight weeks of treatment was evidenced through the measurement of growth parameters.

Plant height was significantly reduced by applying drought stress in the experiment (Fig. 2A). It decreased in line with the decrease in VWC (%). Average plant height before treatment was 29.06 cm for Morocco and 27.93 cm for Senegal. After eight weeks the highest height of genotype Morocco was with plants grown in VWC 25% (54.30±1.48 cm) whereas lowest values were obtained in VWC 5% (40.30±1.89 cm). Genotype Senegal had the highest (54.91±3.45 cm) and the lowest (32.80±0.86 cm)

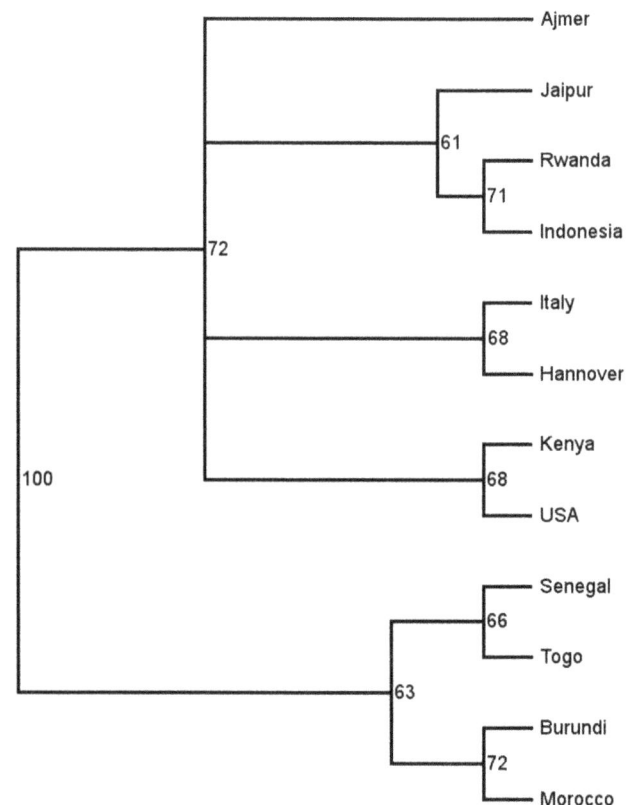

Figure 1. Dendrogram of twelve *E. tirucalli* genotypes calculated with UPGMA showing the phenetic relationships within the colletion. Bootstrap values≥50% are above the branches.

heights in the same respective VWCs. Plant height decreased linearly with decrease in water content. Thus, genotype Morocco grew by 86.85% at normal water content and 38.67% at high water limitation. Meanwhile, growth in genotype Senegal was 96.59% at VWC 25% and 17.43% at VWC 5%. Growth percentage showed that genotype Senegal grew faster than genotype Morocco when water was well available, but that drought highly decreased the growth rate.

Increased water limitation caused reduction of dry weight (Fig. 2C) and water content (Fig. 2D) in both genotypes. Genotype Senegal had higher biomass accumulation at VWC 25% (12.74±0.51) than genotype Morocco (10.53±0.54). The first genotype also had higher yield at the lowest VWC (6.71±0.39 g) than genotype Morocco (5.74±0.22 g). Decrease in water content percentage was small due to water limitation: genotype Senegal was 88% and Morocco 84% at VWC 25%, and 84% and 79% at VWC 5%, respectively.

Drought stress increased tap root length (Fig. 2B) and root/shoot ratio (Fig. 2E) in both genotypes. Genotype Senegal showed a ratio of 0.09±0.01 at VWC 25% and 0.19±0.03 at VWC 5%, genotype Morocco 0.09±0.01–0.17±0.02 in VWC (%) 25 to 5, respectively. The result implies that both genotypes partitioned photosynthetic products more in root biomass following drought stress. Plant height, dry weight, water content percentage and root/shoot ratio of genotypes Morocco and Senegal showed a significant reduction when plants were subjected to a drought stress of eight weeks. The stress responses of both genotypes differed indicating differences in phenotypic plasticity.

3.3 Chlorophyll fluorescence

Quantum efficiency of genotypes Morocco and Senegal in the photosystems of leaves and stems over eight weeks decreased linearly with water limitation (Fig. 3). Stems (Fig. 3B, 3D) of both genotypes showed higher quantum efficiency than leaves (Fig. 3A, 3C). Quantum efficiency of Morocco leaves for all VWCs (%) was in a range of 0.757–0.605. These values were higher than those for genotype Senegal (0.758–0.579) at similar VWCs. Genotype Morocco also had higher values at stem level (0.780–0.643) than genotype Senegal (0.780–0.616). In the leaves of both genotypes, there was no significant difference between different VWCs in the first three weeks, but there was a significant difference from week four onwards. When considering stems, however, genotypes performed differently. In genotype Morocco, significant differences between VWCs started to develop in week five, while in genotype Senegal (Fig. 3D) changes started in week four. This shows that genotype Morocco had higher drought tolerance than genotype Senegal.

3.4 Malate content

Differences in photosynthetic pathways were ascertained by comparing malate content of leaves and stems before drought stress and after exposure to drought stress. Our results show that before drought exposure, there was malate content oscillation between day and night in both genotypes' stems (Fig. 4). In genotype Morocco, malate content of stems at the end of light period was 58.9% lower than that at the end of dark period. Meanwhile, decrease in genotype Senegal was only 17.4%.

With increasing drought stress, malate content increased in stems of both genotypes (Fig. 5). We noted a significant difference in malate content in stems and leaves of the plants, but there was no significance difference between genotypes. The highest malate oscillation between day and night at stem level for genotype Morocco was 68.75% in VWC 15% whereas for genotype Senegal it was 69.55% at VWC 10%.

In leaves, there were significant differences between day and night malate content at VWCs 10% and 5%. In VWC 10%, malate content was 48.22% and 33.16% lower during the day than during the day for genotypes Morocco and Senegal, respectively. In VWC 5%, we only evidenced a significant different in genotype Senegal. At this VWC, day-time malate content was 50% lower than that at night. These values would indicate that there is CAM induction in leaves following drought stress which strength might be genotype-dependent.

3.5 Rubber content

E. tirucalli can be a source of rubber. The rubber content analysis was done by NMR for eight genotypes in our collection, including Morocco and Senegal. The analysis showed strong differences in the concentration of rubber between the genotypes (Fig. 6). Senegal, with 10.74 mg g^{-1} fresh weight, had the highest amount of rubber among genotypes tested, followed by USA 8.80 mg g^{-1} fresh weight. The lowest rubber concentration was found in genotype Togo which had 1.42 mg g^{-1} fresh weight. There is no correlation of rubber content and genotype classification (Fig. 1 and Fig. 6), at least in greenhouse conditions.

3.6 Biogas production

The results of the mesophilic anaerobic digestion of dried samples of six different genotypes of *E. tirucalli* indicate a promising potential with regard to the use of dried biomass of this species as a feedstock for biogas production. Specific biogas production (L biogas kg^{-1} ODM) was in the range of 114 for genotype Togo and 637 for genotype Kenya. Both genotypes which has been investigated in more detail in the drought stress experiments show values around 440 L biogas kg^{-1} ODM, about 70% of the highest value. The methane concentrations lie between 43% and 69%, depending on the genotype. These are preliminary results based on two independent experiments. Not for all genotypes data for all the three replicates in each experiment could be obtained due to initial technical problems with our bench-scale biogas plant. Therefore, we are currently not able to calculate any reliable standard deviations. The experiment will be repeated shortly for all genotypes with optimised equipment. Remarkable are the high amounts of H_2S which reached up to 1,750 ppm (Table 2).

Discussion

4.1 Molecular analysis through genetic markers

The division in two groups as presented in Figure 1 is congruent with the geographic division in an African group and a mostly non-African group (except for Kenya and Rwanda). More samples have to be collected for example from Pakistan, Egypt, and Somalia to analyze whether they belong to the non-African group. Analysis of the genotypes from Brazil might help to estimate the phylogenetic position of the USA genotype, if this is domestic species. Genotypes of *E. tirucalli* are propagated vegetatively since many years in the greenhouse. Therefore the genotype originally collected is not changed since the cultivation due to pollination of flowers. Therefore the genetic drift between generations is low. This vegetative propagation also occurs naturally and/or is conducted by man because this plant seldom produces viable seeds [3]. The dendrogram shows that there is no correlation between morphological characters, as genotype Kenya and USA that have different stem color are clustered as a monophyletic group. Genotype USA has the most distinctive morphological character, i.e. yellow tips. This morphological character is useful for marketing purposes as this accession is sold as an ornamental. The division in two groups within the collection may indicate the

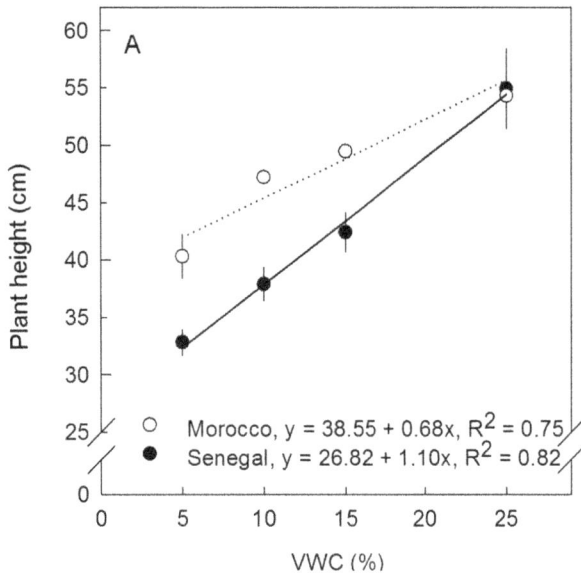

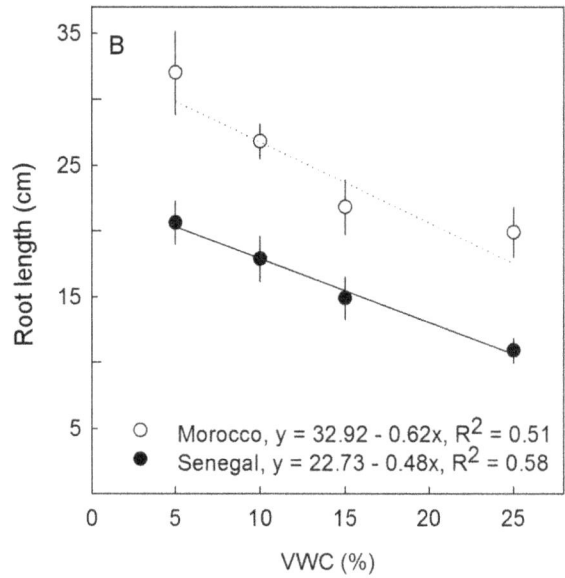

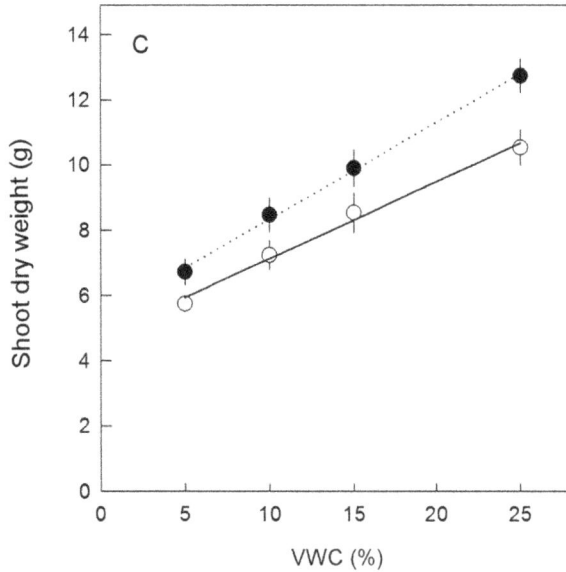

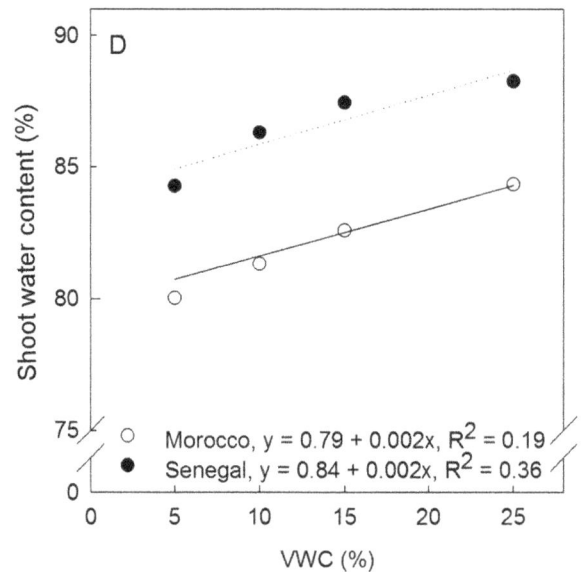

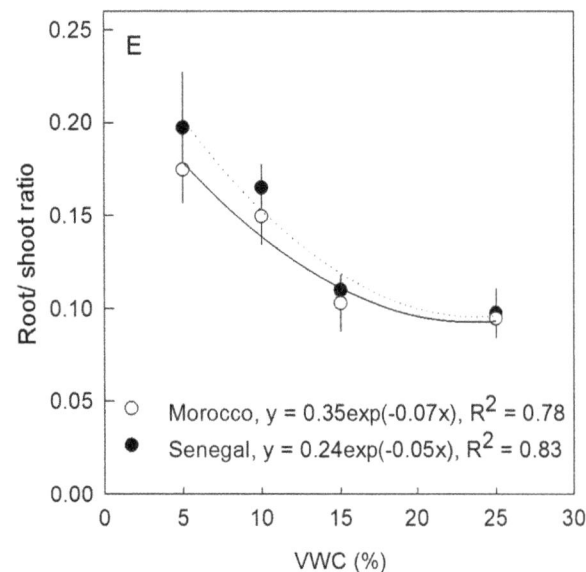

Figure 2. Effect of water limitation on (A) plant height,(B) root length, (C) shoot dry weight, (D) shoot water content and (E) root/ shoot ratio of *E. tirucalli* genotypes Morocco and Senegal after 8 weeks drought stress treatment. Vertical error bars denote standard error of mean (SEM), n = 5.

breeding potential for different utilizations that can be explored. Generally, genotypes in the African group grow faster and produce more biomass than those in the non-African group (data not shown). It indicates that genotypes in the African group may be suitable as source of biomass and therefore bioenergy, while genotypes in the other group may be suitable for other purposes such as ornamental plant.

4.2 Response of plants to different drought treatments

Variation in drought tolerance within a genotype collection is important for subsequent selection work. Analysis of physiological parameters shows that plant height, dry weight and water content decreased with higher drought stress. Research on other plant

species, such as *Amaranthus* and wheat, showed also that there is reduction in plant height and biomass with increase in drought stress in the soil [40,41]. In general, decrease in biomass production rate due to stress exposure has been found to be associated with cessation of photosynthesis, metabolic dysfunction and damage of cellular structure [42]. Further, in response to drought stress, *E. tirucalli* genotypes Morocco and Senegal altered their root dry mass ratio and root length as one of the mechanisms to adapt to drought stress. Root dry mass in drought conditions is higher than in normal condition; this is in accordance with early studies [40,43] and in line with the theory of functional balance which indicates that plants will respond to low water contents with a relative increase in the flow of assimilates to roots and increased root dry mass [44]. The root grows longer which enables the plant

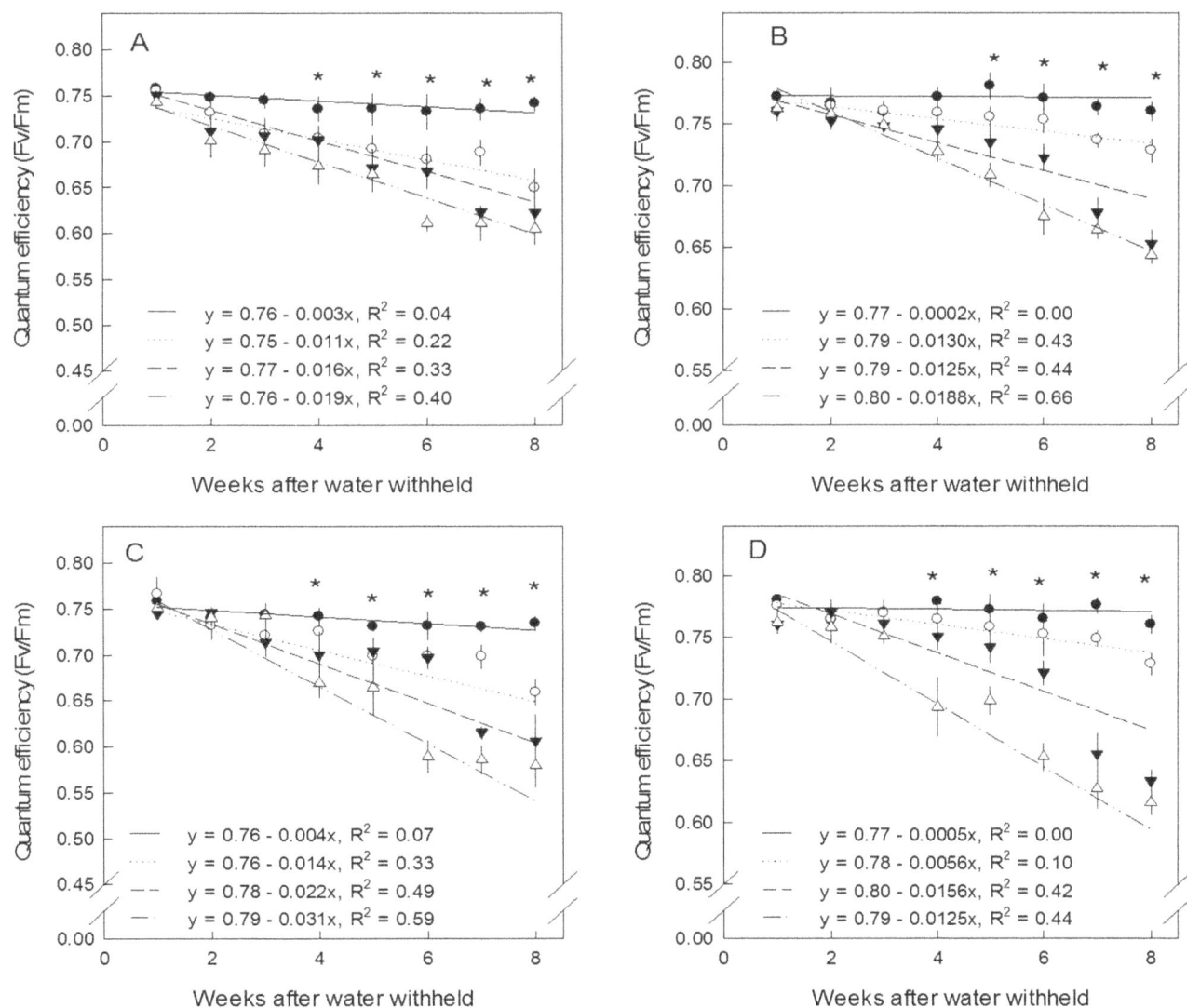

Figure 3. Effect of water limitation on quantum effciency during 8 weeks drought stress treatment. n = 5 (A) Morocco leaves, (B) Morocco stem, (C) Senegal leaves (D) Senegal stem, (●) VWC 25%, (○) VWC 15%, (▼) VWC 10% and (Δ) VWC 5%, n = 5. Vertical error bars denote the standard error of mean (SEM). Stars above the point denote significant difference between VWC in each week treatment following the Tukey procedure ($p<0.05$).

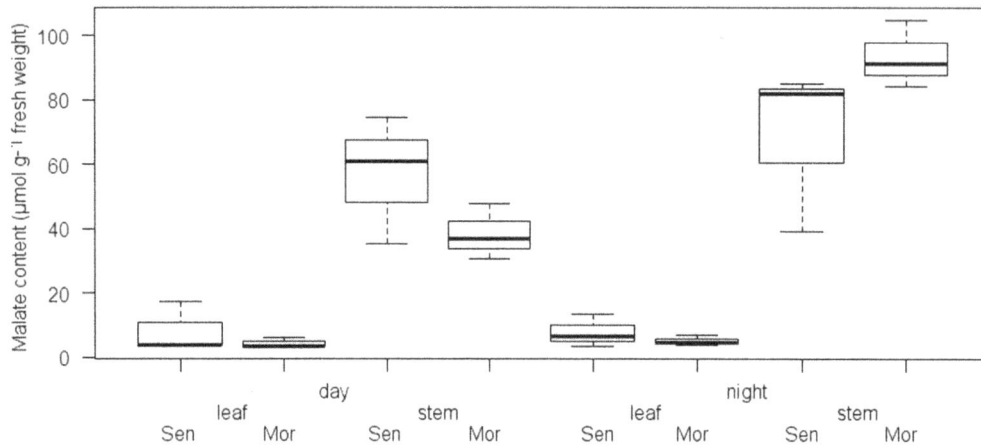

Figure 4. Box plot (n = 3) of malate contents of stems and leaves (μmol g^{-1} fresh weight) of *E. tirucalli* genotypes Morroco (Mor) and Senegal (Sen).

getting to deeper water layers thus escaping from water deficits near the surface [45]. Root elongation reduces shoot dry weight as photosynthesis yield is used for root development at the expense of shoots. Our results of responses to different water content RWC showed that 15% VWC was a critical threshold, below which plants partitioned assimilates to roots which might reduce stem yield.

C_3 leaves wither and die quickly after the onset of stress, and also *E. tirucalli* becomes leafless. CAM stems can proceed with photosynthesis with closed stomata during the day. This provides an ecological advantage of CAM as it allows supplying CO_2 [46] through decarboxylation of malate; hence it can prevent photorespiration damage during stress [47]. However, during prolonged drought stress, CO_2 release from decarboxylation may be insufficient to protect chloroplast membranes from oxidative stress. This oxidative stress derives from partially reduced forms of atmospheric O_2 and influences the repair of PSII during stress [48]. Cessation of photosynthesis is supported by a decline in Fv/Fm along with prolonged drought in both genotypes. The decline

of Fv/Fm becomes higher at lower VWCs, whereby VWC 5% shows the highest decline. The decrease of Fv/Fm at high water limitation has been related to a decline in functioning of primary photochemical reactions, primarily involving inhibition of PSII that is located in the thylakoid membrane system [49]. The values between leaves and stems are not significantly different in the three first weeks of the experiments, during which stress symptoms such as leaf senescence did not appear yet. After prolonged stress, values at stems of both genotypes are higher than at leaves. Quantum efficiency values for all VWC values of genotype Morocco at leaf (0.757–0.605) and stem (0.780–0.643) levels were higher than in genotype Senegal for both leaf (0.758–0.579) and stem (0.780–0.616) levels, respectively. This indicates that quantum efficiency difference is also determined genetically. Drought significantly decreases quantum efficiency at week five for stems of genotype Morocco and at week four for stems of genotype Senegal. Lower photosynthetic efficiency under stress is associated with a damaged photosystem due to stress and reflects a certain degree of environmental stress [50]. The CAM photosynthetic pathway in

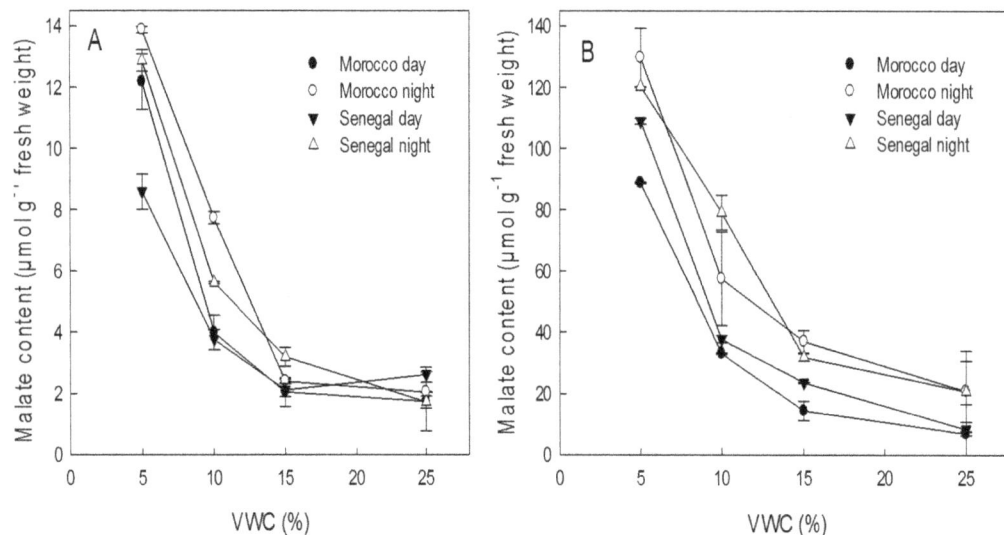

Figure 5. Malate content of (A) leaves (B) stem of genotypes Morocco and Senegal at day and night on different VWC after eight weeks of drought stress treatment. Vertical error bars denote standard error of mean (SEM), n = 3.

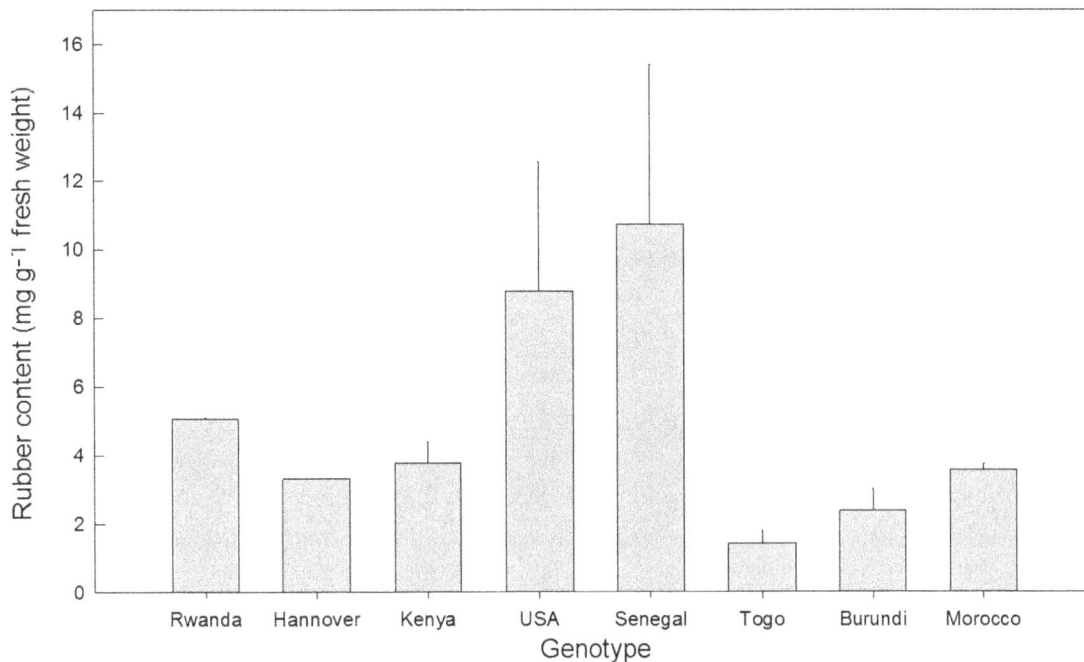

Figure 6. Rubber content of eight *E. tirucalli* genotypes. Each bar illustrates the mean (n = 3). Vertical error bars denote standard error of mean (SEM).

the stem provides an ecological advantage by supplying CO_2 through decarboxylation of malate [51]; hence, it can prevent formation of reactive oxygen species (ROS) and limit photorespiration during stress [47]. However, during prolonged drought stress or higher water limitation, the release of CO_2 from decarboxylation may be insufficient to protect chloroplast membranes from oxidative stress, which affects the repair of PSII during stress [48].

Stomatal conductance and infrared thermography measurements are suitable for genotype screening towards their drought tolerance. However, due to the cylindrical morphology of the *E. tirucalli* stem it is impossible to use a regular porometer. We obtained some results using a thermography camera T360 (FLIR Systems, Wilsonville, USA). In several parameters determined we observed differences in drought tolerance among the two genotypes supporting the data shown in Figure 2 to 5. However, due to the *E. tirucalli* morphology the results could not be exactly

Table 2. Specific biogas production (L biogas kg^{-1} ODM) and gas composition in the biogas produced.

Genotype	Biogas production	CH₄(%)	H₂S (ppm)
Togo	114	69	n.a.
USA	367	44	~1350
Morocco	435	43	~1630
Senegal	440	54	n.a.
Rwanda	522	41	n.a.
Kenya	637	50	~1750

n.a., not analyzed. In case standard deviations could be calculated, they were always less than 10%.

calculated and compared. In summary, the genotype Morocco is more tolerant to drought than genotype Senegal.

Water use efficiency, and assimilation rate to transpiration rate ratio increase in CAM is higher than in C_3 and C_4 [51]. However, biomass accumulation in CAM plants is usually very low, so that growth rate of plants that only rely on CAM is often limited [52]. However, in some species such as *M. crystallinum*, a plant with facultative CAM, photosynthetic rate is higher than that C_3 species due to a high CO_2 fixation rate at night which contributes for a great part to biomass production [53].

E. tirucalli genotypes Morocco and Senegal were both shown to tolerate severe drought stress (VWC 5%) without causing any plant death. Thus, our result confirms that the species has very good potential to be grown in arid area. Genotype Morocco had 84% water content and 16% dry weight in VWC 25%; those values decreased down to 79% and 21% in severe drought stress. Meanwhile, genotype Senegal had 88% water content and 12% dry weight, those values decreased down to 84% and 16% at the same VWCs. *E. tirucalli* water content and dry weight differs between studies: 76.6% water content and 23.4% dry weight [28], 88.33% water content and 11.67% dry weight [54], or 90% water content and 10% dry weight [3]. Different percentages of water content and dry weight might be due to differences in genotypes and growth environment.

4.3 CAM and C_3 photosynthetic pathways in *E. tirucalli*

The analysis of malate content in two genotypes of *E. tirucalli* shows that there are significant differences in leaves and stem. This clearly indicates that there is a difference in photosynthetic pathways between both parts. This result confirms the findings of Van Damme [55] evidenced by gas exchange experiments that there are two photosynthetic pathways allowing to distinguish C_3 leaves from CAM. Malate content before exposure to water limitation shows that the highest content is in nocturnal stems which confirms dark nocturnal CO_2 uptake [56]. More gas exchange experiments are needed to quantify the CO_2 uptake. We observed open stomata at night and closed stomata during the day. Wax patches appear as a dotted white line along the stem axis in a magnified view and surround the stomata (data not shown). These epicuticular wax patches do not melt in greenhouse conditions to seal or block the stomata. Therefore CO_2 influx at night is not hindered by melted wax. Malate content under higher water limitation increases both in stems and leaves, maybe as an indication of CAM induction in the latter. In stems, the highest percentage of malate day–night oscillation of genotype Morocco is at VWC 15% whereas for genotype Senegal we evidenced it at VWC 10%. Malate might be transported from the stem into the leaves. However, so far it was not reported that malate or other water-soluble compounds are transported via the non-articulate laticifers from organ to organ. Phosphoenolpyruvate (PEP) carboxylase enzyme activity and its gene expression could be investigated in stems and leaves to prove our hypothesis that there might be CAM induction in leaves under drought stress.

Photosynthesis in non-succulent leaves of *E. tirucalli* is reported as C_3 and CAM in succulent stems [3]. Having two photosynthetic pathways in two very distinct plant parts is reasonable as it is supported by different anatomy. In genotype Morocco, we evidenced a significant difference in malate content (in μmol g^{-1} fresh weight) at VWC 10% between 3.9 (day) and 7.7 (night) and at VWC 5% between 13.9 (day) and 12.2 (night) while genotype Senegal shows differences at VWC 10% of 3.7 (day) and 5.5 (night) and at VWC 5% of 8.0 (day) and 12.9 (night). This result, however, reveals that there may be an induction of CAM in leaves due to drought stress as there is oscillation in nocturnal and diurnal malate content. This result which may seem at odds with previous results needs further investigation because anatomically leaves of *E. tirucalli* are non-succulent, in contrast to the stems. It is thereby tempting to question whether the leaves are really non-succulent. Indeed, CAM is a syndrome that impliesa certain degree of succulence based on the presence of large vacuoles for malate storage [11]. We therefore recommend *E. tirucalli* leaves would be anatomically investigated for large vacuoles for supporting malate storage. Species such as *Tillandsia usneoides* L. that perform CAM with non-succulent anatomy still have large vacuoles [57,58].

Environmental conditions can influence the plasticity of photosynthetic pathways. Strong stress leads to conversion of C_3 to CAM photosynthetic pathway, for example in the genus *Clusia* [59]. Change of C_3 to CAM has been documented in other, succulent, species such as *M. crystallinum* [14], genus *Sedum* [15], and some species of *Peperomia* and *Clusia* [60,61]. CAM induction during stress positively influences the activities of enzymes involved in malate metabolism [14,62,63]. These enzymes are nicotinamine adenine dinucleotide-dependent malic enzyme (NAD-ME) [64], nicotinamide adenine dinucleotide phosphate dependent malic enzyme (NADP-ME) [14], and PEP carboxylase [65].

With two photosynthetic pathways present at leaf and stem levels, and certain plasticity in switching between C_3/CAM metabolism in *E. tirucalli*, it is not surprising that this plant is recommended as source of biomass for biofuel production that can be grown in marginal conditions. Loke et al. [28] mentioned the prospect of planting *E. tirucalli*; they are already monitoring plantations in Colombia, and are planning to have more in Somalia and other dry African countries. The species can yield 22–25 t dry weight biomass ha^{-1} y^{-1} under optimal conditions whereby optimal planting density is estimated at 14,000 plants ha^{-1}. However, the data presented by the latter authors are not complemented by detailed information on cropping conditions such as irrigation, planting density, and genotypes used. In

addition, Van Damme (unpublished data) was able to show that a 3 years' old plantation in Kenya was able to fetch around 500 t ha^{-1} of fresh material.

4.4 Potential use as source of rubber and biogas

Our results indicate that rubber content varies between genotypes, independently of the affiliation to one AFLP group. This result is supported by a study with several other genotypes: rubber content was different in each genotype depending on soil, climate and year [66], whereas it is not clear whether this is due only to genetic determinants or whether there are also some environmental influences that intervene. Akpan et al. [67], who analysed latex yield of *Hevea brasiliensis* L. found that rubber yield was influenced by clone and soil type. The authors revealed that when soil fertility was better, rubber (latex) yield was also higher. We evidenced the highest rubber content in genotype Senegal. This result supports Van Damme [55] who mentioned that the Senegal genotype was promising as a source of rubber.

Latex of *E. tirucalli* has drawn a lot of attention because it contains high levels of rubber. It has been used as such since the early 20th century [68]. The type of rubber of *E. tirucalli* is a mixture of long chain ketones and cis-1,4 polyisoprene, and is slightly soluble in hot alcohol [66,69]. Beside rubber, the latex of this plant also consists of a resin which prevents long-term stability of latex [54]. Although the rubber has lower quality than that of *H. brasiliensis*, its properties should be further explored in order to fully exploit its potential as a naturally occurring polymer. The detailed composition of sterols and triterpenoids in greenhouse-grown plants and field-grown plants has to be analyzed by GC-MS in the future. Also the expression of the rate limiting enzyme of the mevalonate pathway, 3-hydroxy-3-methylglutaryl-CoA reductase, should be analyzed for its expression in different *E. tirucalli* genotypes to analyze the genetic dependency of the biosynthesis of latex components.

The use of *E. tirucalli* as a source of energy is promising because it grows fast whilst having at the same time low water requirements and a low demand for nutrients [3]. It was stated that this species could be used for biofuel production due to its high latex content [24]. Our results indicate that the biogas production in our batch tests varies among genotypes (Table 2). The results also show that *E. tirucalli* definitely has potential to serve as a feedstock for the production of biogas.

To date only a few experimental results concerning the biogas production potential of *E. tirucalli* have been published. Sow et al. [70] reported a potential annual methane production of around 3,000 m^3 ha^{-1} per year based on research carried out in Kenya with a stand density of 80,000 plants per hectare and a biomass yield of 20 t ha^{-1} y^{-1} (DM). In field experiments in Colombia, 30 t ha^{-1} y^{-1} (DM) of *E. tirucalli* biomass brought about 8,250 m^3 ha^{-1} biogas [28]. Assuming a methane content of approx. 50% (Table 2), the methane yield of *E. tirucalli* seems to be smaller compared to the yields of maize silage (5,800 m^3 ha^{-1} y^{-1}) and forage beet plus leaves (5,800 m^3 ha^{-1}y^{-1}); however, its yield exceeds that of wheat (2,960 m^3 ha^{-1} y^{-1}) and rapeseed (1,190 m^3 ha^{-1} y^{-1}) [71].

In the results presented here it is remarkable that the H_2S concentrations are the comparatively high in the *E. tirucalli*-derived biogas. H_2S contents are indeed lower than those from the fermentation of manure, biowaste and food waste which are in the range of 2,000–6,000 ppm due to a high content of sulfur-containing proteins [72], but higher than those of maize silage-derived biogas with approx. 500 ppm. H_2S can impair the utilization of biogas, as it has the ability to corrode the metal parts of the fermenting installation and can cause health problems in high doses and long exposures [73]. To decrease H$_2$S content during processing, different techniques are available, such as biofilters consisting of phototrophic (*Cholorobium limicola*) or chemotrophic bacteria (*Thiobacillus* spp.) [74]. In order to improve the reliability of the method, further biogas batch tests with *E. tirucalli* should comprise a systematical variation of the following parameters: age of plant material (because the older the plant, the higher the lignin content), particle size of the substrate in order to investigate its influence on biodegradability of feedstock, optimization of choice and pre-treatment of the inoculum [75], and last but not least genotype-dependent differences.

The presented data are based on lab-scale experiments. Further field experiments will be necessary before a specific *E. tirucalli* genotype can be proposed for practical application. Among the genotypes tested, Kenya has the highest yield in biogas per organic dry matter and should be further analyzed for its biomass gain during drought stress conditions in the greenhouse and in the field. Senegal is promising as a source of biomass and biogas as well. When water availability is limited, using genotype Morocco with higher drought tolerance as a source of bio-energy is recommended, because biogas production using genotype Morocco is as high as with genotype Senegal. Genotype USA is promising as an ornamental plant and source of biogas, but its drought tolerance is not yet known. Combining these valuable characteristics through breeding may bring more benefit. Stocked genotypes could be distributed to interested farmers and researchers in arid areas for performing field experiments and challenge the greenhouse results by natural conditions.

Conclusion

E. tirucalli has a high potential as drought-tolerant crop plant because of its unique combination of photosynthetic pathways and

as source of biofuel, rubber and maybe even phytochemicals. The genetic relationship within the collection was analyzed by AFLP. There may be induction of CAM in leaves due to stress. Despite these substantial results, several questions remain to be addressed. The confirmation of *E. tirucalli* photosynthetic pathways' plasticity at leaf level, that may play an important role to survive during drought stress, needs to be investigated in more detail. Thus, it will be interesting to analyze how enzymes influence metabolic adjustment to stress conditions in leaves and stem. To explore the use of *E. tirucalli*, determination of rubber composition in different genotypes, and quality and technical optimization of fermentation processes for the production of biogas need to be performed. The characterized genotypes from our greenhouse should be used in field experiments in tropical regions to verify and extend the data obtained in greenhouse conditions.

Acknowledgments

Samples from India were kindly provided by Dr. Vijendra Shekhawat, University of Mumbai, India. We would like to thank the gardeners for growing plants and Pamela von Trzebiatowski for malate analysis. We acknowledge support by Deutsche Forschungsgemeinschaft and Open Access Publishing Fund of Leibniz Universität Hannover.

Author Contributions

Conceived and designed the experiments: JP PVD BBK. Performed the experiments: BH FT MM HV SG. Analyzed the data: BH FT MM HV SG. Contributed reagents/materials/analysis tools: JP PVD BBK. Wrote the paper: JP PVD BBK SB BH.

References

1. Dai A (2012) Increasing drought under global warming in observations and models. Nature Clim Change doi:10.1038/nclimate1633.
2. Bruyns PV, Mapaya RJ, Hedderson T (2006) A new subgeneric classification for *Euphorbia* (Euphorbiaceae) in southern Africa based on ITS and *psbA-trnH* sequence data. Taxon 55: 397–420.
3. Van Damme PLJ (2001) *Euphorbia tirucalli* for high biomass production. In: Schlissel A, Pasternak D, editors. Combating desertification with plants, Kluwer Academic Pub. pp. 169–187.
4. Janssens MJ, Keutgen N, Pohlan J (2009) The role of bio-productivity on bio-energy yield. J Agr Rural Dev Trop 110: 39–47.
5. Webster GL (1975) Conspectus of a new classification of the Euphorbiaceae. Taxon 24: 593–601.
6. Batanouny KH, Stichler W, Ziegler H (1991) Photosynthetic pathways and ecological distribution of *Euphorbia* species in Egypt. Oecologia 87: 565–569.
7. Bender MM (1971) Variation in the 13C/12C ratios of plants in relation to the pathway of photosynthetic carbon dioxide fixation. Phytochemistry 10: 1239–1244.
8. Pearcy RW (1975) C4 photosynthesis in form *Euphorbia* species from Hawaiian rainforest sites. Plant Physiol 55: 1054–1056.
9. Ting IP, Bates L, Sternberg LO, Denior MJ (1985) Physiological and isotopic aspects of photosynthesis in peperomia. Plant Physiol 78: 246–249.
10. Smith BN (1982) General characteristics of terrestrial plants (agronomic and forests)-C$_3$, C$_4$ and Crassulacean Acid Metabolism plants. CRC Handbook of biosolar resources 1 (2), 99–113.
11. Nuernbergk EL (1961) Endogener Rhythmus und CO$_2$ Stoffwechsel bei Pflanzen mit diurnalem Säurerhythmus. Planta 56: 28–70.
12. Osmond CB (1978) Crassulacean acid metabolism: a curiosity in context. Annu Rev Plant Biol 29: 379–414.
13. Cushman JC, Borland AM (2002) Induction of Crassulacean acid metabolism by water limitation. Plant Cell Environ 25: 295–310.
14. Holtum JAM, Winter K (1982) Activity of enzymes of carbon metabolism during the induction of Crassulacean acid metabolism in *Mesembryanthemum crystallinum* L. Planta 155: 8–16.
15. Gravatt DA, Martin CE (1992) Comparative ecophysiology of five species of *Sedum* (Crassulaceae) under well watered and drought stressed conditions. Oecologia 92: 532–541.
16. Aylward JH, Parsons PG (2008) Treatment of prostate cancer. Peplin Research May, 27 2008: US 7378445 Available: http://appft1.uspto.gov/. Accessed 2010 Dec 28.
17. Duke J (1983) *Euphorbia tirucalli* L., handbook of energy crops. Purdue University centre for new crops and plant products. www.hort.purdue.edu. Accessed 5 December 2010.
18. Kumar A (1999) Some potential plants for medicine from India, Ayurvedic medicines, University of Rajasthan, Rajasthan. pp. 1–12.
19. Schmelzer GH, Gurib-Fakim A (2008) Medicinal plants. Plant Resources of Tropical Africa. pp. 412–415.
20. Van Damme PLJ (1989) Het traditioneel gebruik van *Euphorbia tirucalli*. African Focus 5: 176–193.
21. Uchida H, Yamashita H, Kajikawa M, Ohyama K, Nakayachi O et al. (2009) Cloning and characterization of a squalene synthase gene from a pretroleum plant, *Euphorbia tirucalli* L. Planta 229: 1243–1252.
22. Nielsen PE, Nishimura H, Liang Y, Calvin M (1979) Steroids from *Euphorbia* and other latex-bearing plants. Phytochemistry 18: 103–104.
23. Furstenberger G, Hecker E (1977) New highly irritant euphorbia factors from latex of *Euphorbia tirucalli* L. Experentia 33: 986–988.
24. Calvin M (1978) Chemistry, population, resources. Pure Appl Chem 50: 407–425.
25. Calvin M (1980) Hydrocarbons from plants: Analytical methods and observations. Naturwissenschaften 67: 525–533.
26. Kalita D (2008) Hydrocarbon plant - New source of energy for future. Renew Sust Energ Rev 12: 455–471.
27. Mwine J, Van Damme P (2011) *Euphorbia tirucalli* L. (Euphorbiaceae) – The miracle tree: Current status of available knowledge. Sci Res Essay 6: 4905–4914.
28. Loke J, Mesa LA, Franken JY (2011) *Euphorbia tirucalli* biology manual: Feedstock production, bioenergy conversion, application, economics Version 2. FACT.
29. Rahuman AA, Gopalakrishnan G, Venkatesan P, Geetha K (2008) Larvicidal activity of some Euphorbiaceae plant extracts against *Aedes aegypti* and *Culex quinquefasciatus* (Diptera: Culicidae). Parasitol Res 102: 867–873.
30. Lirio LG, Hermano ML, Fontanilla MQ (1998) Antibacterial activity of medicinal plants from the Philippines. Pharm Biol 36: 357–359.
31. Vassiliades G (1984) Note on the molluscidal properties of two Euphorbiaceae plants – *Euphorbia tirucalli* and *Jatropha curcas*. Rev Elev Med Vet Pays Trop 37: 32–34.

32. Murali R, Mwangi JG (1998) *Euphorbia tirucalli* resin: potential adhesive for wood-based industries, in: F. d. FAO corporate document repository (Ed.), International conference on domestication and commercialization of non-timber forest products in Agrosystems. FAO. Rome.

33. Vos P, Hogers R, Bleeker M, Reijans M, Van de Lee T et al. (1995) AFLP: a new technique for DNA fingerprinting. Nucleic Acids Res 23: 4407–4414.

34. Schlüter PM, Harris SA (2006) Analysis of multilocus fingerprinting data sets containing missing data. Mol Ecol Notes 6: 569–572.

35. Page RDM (1996) TREEVIEW: An application to display phylogenetic trees on personal computers. Comput Appl Biosci 12: 357–358. Available: http://taxonomy.zoology.gla.ac.uk/rod/treeview.html. Accessed 2011 Mar 25.

36. Jefferies RA (1994) Drought and chlorophyll fluorescence in field-grown potato (*Solanum tuberosum*). Physiol Plant 90: 93–97.

37. Kluge M (1971) Veränderliche Markierungsmuster bei $^{14}CO_2$-Fütterung von *Bryophyllum tubiflorum* zu verschiedenen Zeitpunkten der Hell-Dunkelperiode II. Beziehungen zwischen dem Malatgehalt des Gewebes und dem Markierungsmuster nach $^{14}CO_2$-Lichtfixierung. Planta 98: 20–30.

38. Duke J (1983) *Euphorbia tirucalli* L., handbook of energy crops. Purdue University centre for new crops and plant products. www.hort.purdue.educ. Accessed on 5 December 2010.

39. VDI 4630 (2006) Fermentation of organic materials, Characterisation of the substrate, sampling, collection of material data, fermentation tests. Beuth Verlag. Berlin, Germany. 92 p.

40. Liu F, Stützel H (2004) Biomass partitioning, specific leaf area, and water use efficiency of vegetable amaranth (*Amaranthus* spp.) in response to drought stress. Sci Hortic 15: 15–27.

41. Zhang J, Hao C, Ren Q, Chang X, Liu G et al. (2011) Association mapping of dynamic developmental plant height in common wheat. Planta 234: 891–902.

42. Krasensky J, Jonak C (2012) Drought, salt, and temperature stress-induced metabolic rearrangements and regulatory networks. J Exp Bot 64: 1593–1608.

43. Dias PC, Araujo WL, Moraes GABK, Barros RS, DaMatta FM (2007) Morphological and physiological responses of two coffee progenies to soil water availability. J Plant Physiol 164: 1639–1647.

44. Brouwer R (1963) Some aspects of the equilibrium between overground and underground plant parts. In: Jaarboek IBS, Wageningen. pp. 31–39.

45. Schenk HJ, Jackson RB (2002) Rooting depths, lateral root spreads and below-ground/above-ground allometries of plants in water-limited ecosystems. J Ecol 90: 80–494.

46. Martin CE, Jackson JL (1986) Photosynthetic pathways in a midwestern rock outcrop succulent, *Sedum nuttallianum* Raf. (Crassulaceae). Photosyn Res 8: 17–29.

47. Borland A, Elliot S, Patterson S, Taybi T, Cushman J et al. (2006) Are the metabolic components of Crassulacean acid metabolism up-regulated in response to an increase in oxidative burden? J Exp Bot 57: 319–328.

48. Nishiyama Y, Allakhverdiev SI, Murata N (2006) A new paradigm for the action of reactive oxygen species in the photoinhibition of photosystem II. Biochim Biophys Acta 1757: 742–749.

49. Souza RP, Machado EC, Silva JAB, Lagôa AMMA, Silveira JAG (2003) Photosynthetic gas exchange, chlorophyll fluorescence and some associated metabolic changes in cowpea (*Vigna unguiculata*) during water stress and recovery. Environ Exp Bot 51: 45–56.

50. Maxwell K, Johnson GN (2000) Chlorophyll fluorescence-a practical guide. J Exp Bot 51: 659–668.

51. Herrera A (2008) Crassulacean acid metabolism and fitness under water deficit stress: if not for carbon gain, what is facultative CAM good for? Ann Bot 103: 645–653.

52. Heldt HW, Piechulla B (2011) Plant Biochemistry. 4th edition, Elsevier, London, UK. 656 p.

53. Bloom AJ, Troughton JH (1979) High productivity and photosynthetic flexibility in a CAM plant. Oecologia 38: 35–43.

54. Orwa C, Mutua A, Kindt R, Jamnadass R, Simons A (2009) Agroforestree Database: A Tree Reference and Selection Guide Version 4.0. Available: http://www.worldagroforestry.org/af/treedb/. Accessed 2010 Dec 20.

55. Van Damme PLJ (1990) Gebruik van *Euphorbia tirucalli* als rubberleverancier en energiewas. African Focus 6: 19–44.

56. Kluge M, Heininger B (1973) Untersuchungen über den Efflux von Malat aus den Vacuolen der assimilierenden Zellen von *Bryophyllum* und mögliche Einflüsse dieses Vorganges auf den CAM. Planta 113: 333–343.

57. Kluge M, Lange OL, Eichmann V, Schmid R (1973) Diurnaler Säurerhythmus bei *Tillandsia usneoides*: Untersuchungen über den Weg des Kohlenstoffs sowie die Abhängigkeit des CO_2-Gaswechsels von Lichtintensität, Temperatur und Wassergehalt der Pflanze. Planta 112: 357–372.

58. Loeschen VS, Martin CE, Smith M, Eder SL (1993) Leaf anatomy and CO_2 recycling during Crassulacean acid metabolism in twelve epiphytic species of *Tillandsia* (Bromeliaceae). Int J Plant Sci 154: 100–106.

59. Taybi T, Nimmo HG, Borland AM (2004) Expression of phosphoenolpyruvate carboxylase and phosphoenolpyruvate carboxylase kinase genes. Implications for genotypic capacity and phenotypic plasticity in the expression of Crassulacean acid metabolism. Plant Physiol 135: 587–598.

60. Ting IP, Hann J, Sipes DL, Patel A, Walling LL (1993) Expression of p-enolpyruvate carboxylase and other aspects of CAM during the development of *Peperomia camptotricha* leaves. Bot Acta 106: 313–319.

61. Borland AM, Tecsi LI, Leegood RC, Walker RP (1998) Inducibility of crassulacean acid metabolism (CAM) in *Clusia* species: physiological/biochemical characterisation and intercellular localization of carboxylation and decarboxylation processes in three species which exhibit degress of CAM. Planta 205: 342–351.

62. Ostrem JA, Vernon DM, Bohnert HJ (1990) Increased expression of a gene coding for NAD-glyceraldehyde-3-phosphate dehydrogenase during the transition from C3 photosynthesis to Crassulacean acid metabolism in *Mesembryanthemum crystallinum*. J Biol Chem 256: 3497–3502.

63. Cushman JC (1992) Characterization and expression of a NADP-malic enzyme cDNA induced by salt stress from the facultative crassulacean acid metabolism plant, *Mesembryanthemum crystallinum*. Eur J Biochem 208: 259–266.

64. Dittrich P, Campbell WH, Black CC Jr (1973) Phosphoenolpyruvate carboxykinase in plants exhibiting crassulacean acid metabolism. Plant Physiol 52: 357–361.

65. Ting IP (1968) CO Metabolism in Corn Roots. III. Inhibition of p-enolpyruvate carboxylase by L-malate. Plant Physiol 43: 1919–1924.

66. Uzabakiliho B, Largeau C, Casadevall E (1987) Latex constituents of *Euphorbia candelabrum, E. grantii, E. tirucalli* and *Synadenium grantii*. Phytochemistry 26: 3041–3045.

67. Akpan AU, Edem SO, Ndaeyo NU (2007) Latex yield of rubber (*Hevea brasiliensis* Muell Argo) as influenced by clone planted and locations with varying fertility status. J Agricul Soc Sci 3: 1813–2235.

68. Scasselati-Sforzolini G (1916) L'Euphorbia tirucalli. Istituto Agricolo Coloniale Italiano S. 25: 40.

69. Blaschek W, Hänsel R, Keller K, Reichling J, Rimpler H, et al (1998) Hagers Handbuch der Pharmazeutischen Praxis, Drogen A-K. Berlin, Heidelberg, Springer Verlag. 909 p.

70. Sow D, Ollivier B, Viaud P, Garcia JL (1989) Mesophillic and thermophilic methane fermentation of *Euphorbia tirucalli*. Mircen J Appl Microb 5: 547–550.

71. Weiland P (2003) Production and energetic use of biogas from energy crops and wastes in Germany. Appl Biochem Biotechnol 109: 263–274.

72. Schieder D, Quicker P, Schneider R, Winter H, Prechtl S, et al. (2003) Microbiological removal of hydrogen sulfide from biogas by means of a separate biofilter system: experience with technical operation. Water Sci Technol 48: 209–212.

73. Binder R, Deninger A, Grous-Göldner A, Huter E, Jungwirth M et al. (2009) Gefahrenpotential von Schwefelwasserstoff beim Betrieb von Biogasanlagen. Available: http://www.lea.at/download/Biogas/H2S_Leitfaden%20Biogasanlagen_2009.pdf. Accessed 2012 Sep 8.

74. Syed M, Soreanu G, Falletta P, Béland M (2006) Removal of hydrogen sulfide from gas streams using biological processes–A review. Can Biosyst Eng 48: 1–14.

75. Tomala F (2012) Entwicklung einer Methodik zur Ermittlung der Biogas und Methanausbeuten verschiedener Herkünfte von *Euphorbia tirucalli* als vielversprechende Energiepflanze. Bachelor thesis. Hannover, University of Applied Science Hannover.

9

Comparison of Seven Chemical Pretreatments of Corn Straw for Improving Methane Yield by Anaerobic Digestion

Zilin Song[1], GaiheYang[2], Xiaofeng Liu[1]*, Zhiying Yan[2]*, Yuexiang Yuan[1], Yinzhang Liao[1]

1 Chengdu Institute of Biology, Chinese Academy of Science, Chengdu, Sichuan, PR China, 2 Research Center of Recycle Agricultural Engineering Technology of Shaanxi Province, Northwest A&F University, Yangling, Shaanxi, PR China

Abstract

Agriculture straw is considered a renewable resource that has the potential to contribute greatly to bioenergy supplies. Chemical pretreatment prior to anaerobic digestion can increase the anaerobic digestibility of agriculture straw. The present study investigated the effects of seven chemical pretreatments on the composition and methane yield of corn straw to assess their effectiveness of digestibility. Four acid reagents (H_2SO_4, HCl, H_2O_2, and CH_3COOH) at concentrations of 1%, 2%, 3%, and 4% (w/w) and three alkaline reagents ($NaOH$, $Ca(OH)_2$, and $NH_3 \cdot H_2O$) at concentrations of 4%, 6%, 8%, and 10% (w/w) were used for the pretreatments. All pretreatments were effective in the biodegradation of the lignocellulosic straw structure. The straw, pretreated with 3% H_2O_2 and 8% $Ca(OH)_2$, acquired the highest methane yield of 216.7 and 206.6 mL CH_4 g VS^{-1} in the acid and alkaline pretreatments, which are 115.4% and 105.3% greater than the untreated straw. H_2O_2 and $Ca(OH)_2$ can be considered as the most favorable pretreatment methods for improving the methane yield of straw because of their effectiveness and low cost.

Editor: Dwayne Elias, Oak Ridge National Laboratory, United States of America

Funding: This study was supported by the National 973 of China (No. 2013CB733502), Applied Basic Research Program of Sichuan Province, China (No. 2013JY0050) and Deployment Project of Chinese Academy of Sciences (No. KGZD-EW-304-1). The funders had no role in study design, data collection and analysis, decision to publish, or preparation of the manuscript.

Competing Interests: The authors have declared that no competing interests exist.

* E-mail: lxf3636@163.com (XL); zhiyingyan2010@yeah.net (ZY)

Introduction

Biomass is considered as a valuable alternative energy source to fossil fuels worldwide because it can be converted into various available forms of energy, such as heat, electricity, steam, biogas, hydrogen, and liquid transportation biofuels [1,2]. As the largest agricultural country in the world, China has an abundance of biomass resources. Approximately 800 million tons of various crop residues are produced in China per year, of which corn and wheat straw account for 216 and 135 million tons, respectively [3]. Crop straws have not been widely used for bioenergy production because of the undeveloped conversion technology. Instead, many crop straws are burnt or directly dumped into the fields, causing serious environmental pollution and degraded soil conditions [4]. Therefore, the development of inexpensive and effective technologies for corn straw utilization is necessary.

Anaerobic digestion (AD) of agricultural straw for bioenergy production is widely used as a promising and alternative energy source to fossil fuels [5]. This technology has been considered as the main commercially viable option for the both treatment and recycling of biomass wastes, and thus is of great interest from an environmental and bioenergy source perspective [6]. However, the efficiency of this technology in treating agricultural straws is limited because the components of straw (lignin, cellulose, and hemicellulose) are difficult to degrade; thus, soluble compounds with low molecular weights are less available for anaerobic microorganisms [7]. Straw pretreatments prior to AD is a simple

and effective method of improving the biodegradability of lignocellulosic materials because it can decompose cellulose and hemicellulose into relatively readily biodegradable components while breaking down the linkage between polysaccharide and lignin to make cellulose and hemicellulose more accessible to bacteria [8,9].

Pretreatment methods mainly include physical methods [2,10], chemical methods [11–14], biological methods [1,15], and a combination of the abovementioned methods [16,17]. Compared with physical and biological treatment methods, chemical pretreatment methods are predominantly used because they are inexpensive and are effective for enhancing the biodegradation of complex materials [18]. In chemical pretreatment methods, sulphuric acid (H_2SO_4), hydrochloric acid (HCl), hydrogen peroxide (H_2O_2), acetic acid (CH_3COOH), sodium hydroxide ($NaOH$), lime ($Ca(OH)_2$), and aqueous ammonia ($NH_3.H_2O$) are the common chemicals to improve AD performance of agricultural residues [19–26]. For instance, Fernández-Cegrí et al. [2] reported that the methane yield of sunflower oil cake with $Ca(OH)_2$ is 130 CH_4 g^{-1} COD, which is 25% higher that of the untreated sample. Zhu et al. [12] found that NaOH-pretreated corn stover yields 37.0% to 72.9% higher biogas productions than the untreated sample. Kang et al. [23] showed that the optimal conditions for the ethanol production of rapeseed straw is through immersion in aqueous ammonia containing 19.8% ammonia water at 69.0°C for 14.2 h. In addition, H_2SO_4, HCl, and

Table 1. Effect of acid pretreatment on the chemical composition of corn straw.

Pretreatment	Concentration	Cellulose %	Hemicellulose %	Lignin %	TC %	C/N
H_2SO_4	1%	47.1±2.5 a	26.9±1.2 a	7.5±0.5 a	37.3±2.0 b	45.5±2.0 b
	2%	41.3±1.8 b	22.5±1.8 b	7.3±0.4 a	30.6±1.9 c	38.7±1.6 c
	3%	38.0±1.6 bc	16.2±1.2 c	7.3±0.4 a	30.3±2.1 c	37.9±1.1 c
	4%	36.1±1.6 c	13.0±0.9 d	6.7±0.5 a	25.3±1.5 d	30.9±3.2 d
Untreated		49.3±1.8 a	28.8±1.4 a	7.5±0.4 a	42.3±2.8 a	51.4±3.6 a
HCl	1%	46.7±2.2 a	26.2±1.9 a	7.9±0.5 a	37.1±2.0 b	44.2±1.8 b
	2%	40.4±2.0 b	22.2±2.0 b	7.2±0.7 a	32.4±2.0 c	39.5±0.9 c
	3%	38.2±1.6 b	17.3±1.0 c	6.4±0.6 a	29.2±1.9 c	38.4±2.1 c
	4%	35.4±0.8 c	14.5±1.3 d	6.9±1.0 a	26.1±1.2 d	32.6±0.7 d
Untreated		49.3±1.8 a	28.8±1.4 a	7.5±0.4 a	42.3±2.8 a	51.4±3.6 a
CH_3COOH	1%	43.8±1.9 b	26.8±2.6 a	7.1±0.9 a	38.6±2.9 a	47.7±1.6 ab
	2%	37.4±2.4 c	21.7±1.1 b	6.7±0.5 a	34.8±0.9 b	46.4±1.9 b
	3%	34.2±0.9 d	18.1±1.4 c	6.8±0.5 a	29.5±0.9 c	36.9±2.6 c
	4%	30.4±1.5 e	15.1±0.5 d	6.7±0.7 a	26.4±1.6 d	32.2±0.7 d
Untreated		49.3±1.8 a	28.8±1.4 a	7.5±0.4 a	42.3±2.8 a	51.4±3.6 a
H_2O_2	1%	40.5±1.5 b	25.0±1.4 b	7.0±0.2 a	34.4±2.6 b	44.7±2.3 b
	2%	34.6±2.1 c	20.8±2.3 c	6.5±0.3 b	28.7±0.8 c	37.3±1.1 c
	3%	30.8±0.8 d	14.3±1.2 d	5.7±0.4 c	25.1±1.2 d	30.6±2.4 d
	4%	22.5±0.6 e	9.5±0.7 e	5.1±0.2 d	20.4±1.3 e	25.2±2.1 e
Untreated		49.3±1.8 a	28.8±1.4 a	7.5±0.4 a	42.3±2.8 a	51.4±3.6 a

Data are expressed as mean ± deviation of triplicate measurements. TC: Total carbon.
The ANOVA test was conducted to determine the differences between each pretreatment. Values with the same letters in each pretreatment indicate no significant difference at $P<0.05$.

CH_3COOH pretreatments have been used to improve the AD of lignocellulosic materials [24,25]. However, the most economically and effectively favorable treatments, among these, have yet to be identified. Additionally, the optimal concentration for the favorable pretreatment has been scarcely reported. Such information is important for the reasonable and efficient utilization of agricultural residues. The present study compared the effects of four acid and three alkaline pretreatments on the lignocellulosic compositions and methane yield of corn straws by AD. Our objective was to determine the most cost-effective pretreatment methods for enhancing the methane yield of straws.

Materials and Methods

Raw Material

Corn straw was obtained from a local villager near the Northwest A&F University (Yangling, Shaanxi, China). Prior to use, the straws were air dried, cut into lengths of 20 mm to 30 mm using a grinder, and then individually homogenized for further use. The full composition and main features of the corn straw were as follows (mean values of three determinations ± standard deviations): total solids (TS), 93.6%±2.8%; volatile solids (VS), 86.7%±1.9%; total carbon (TC), 42.3%±2.8%; total nitrogen (TN), 0.82%±0.05%; hemicellulose, 28.8%±1.4%; cellulose, 49.3%±1.8%; and lignin, 7.5%±0.4%.

Pretreatment Process

Seven pretreatment methods were used in this study, including four acid treatments (H_2SO_4, HCl, CH_3COOH, and H_2O_2) and three alkaline treatments (NaOH, $Ca(OH)_2$, and $NH_3 \cdot H_2O$). The

reagents were purchased from Sinophram Chemical Reagent Co. Ltd, Beijing, China. The chosen pretreatment conditions were based on previous studies [9,20] and carried out using different concentrations of reagents. Acid reagents (H_2SO_4, HCl, H_2O_2, and CH_3COOH) at concentrations of 1%, 2%, 3%, and 4% (w/w) and alkaline reagents (NaOH, $Ca(OH)_2$, and $NH_3 \cdot H_2O$) at concentrations of 4%, 6%, 8%, and 10% (w/w) were used for the pretreatments. The corn straw not pretreated with any chemicals was used as the control. Each pretreatment was conducted in triplicate.

Dried corn straw (500 g) was soaked in the prepared 1.5 L solutions contained in beakers, yielding straw samples with 75% moisture. All prepared beakers were covered with plastic films, secured with a plastic ring, and then stored in a chamber at an ambient temperature of 25±2°C for 7 days. After the pretreatment, the straws were removed from the beakers, dried in an electronic oven (HengFeng SFG-02.600, Huangshi, China) at 80°C for 48 h, and then kept in a refrigerator for composition determination and AD experiments to investigate the effect of different chemical treatments on methane yield.

Anaerobic Digestion

The digestion experiment was conducted according to methods described by Song et al. [22] using laboratory-scale simulated anaerobic digesters in 1 L Erlenmeyer flasks. The batch reactors were used to determine the digestion levels of the straws with different pretreatments. Each pretreated straw was used as the digestion material, with the untreated straw as the control. The digestion inoculum was collected from an anaerobic digester in a model village powered by household biogas (Yangling, Shaanxi,

Table 2. Effect of alkaline pretreatment on the chemical composition of corn straw.

Pretreatment	Concentration	Cellulose %	Hemicellulose %	Lignin %	TC %	C/N
NaOH	4%	48.0±3.9 a	23.8±1.4 b	6.7±0.5 a	39.3±0.8 a	49.1±1.0 a
	6%	46.1±3.0 a	20.6±0.9 c	5.5±0.5 b	35.4±2.3 b	46.0±0.8 b
	8%	46.7±2.2 a	16.2±0.9 d	4.6±0.3 c	33.7±1.6 b	42.1±2.1 c
	10%	47.4±2.6 a	11.3±1.2 e	4.0±0.2 d	28.1±1.2 c	34.7±1.3 d
Untreated		49.3±1.8 a	28.8±1.4 a	7.5±0.4 a	42.3±2.8 a	51.4±3.6 a
Ca(OH)₂	4%	47.5±1.8 a	24.6±2.2 b	6.8±0.2 a	37.8±1.5 a	45.0±1.8 b
	6%	46.1±2.4 a	21.2±1.4 c	6.0±0.3 b	32.8±3.1 b	40.0±2.0 c
	8%	46.3±1.9 a	16.4±1.1 d	5.4±0.2 c	29.4±2.0 b	38.7±0.7 c
	10%	48.0±1.1 a	12.3±1.2 e	4.6±0.3 d	22.6±1.8 c	28.3±1.9 d
Untreated		49.3±1.8 a	28.8±1.4 a	7.5±0.4 a	42.3±2.8 a	51.4±3.6 a
NH₃•H₂O	4%	48.1±1.2 a	25.7±1.9 a	7.0±0.6 ab	39.2±1.9 a	48.4±2.1 a
	6%	45.4±3.3 a	22.4±0.8 b	6.6±0.3 b	36.6±2.5 b	48.8±0.4 a
	8%	45.9±3.0 a	18.6±1.8 c	6.2±0.2 c	33.2±0.9 b	41.5±1.8 b
	10%	45.1±2.9 a	17.8±1.1 c	5.5±0.2 d	30.7±1.6 c	37.4±1.8 b
Untreated		49.3±1.8 a	28.8±1.4 a	7.5±0.4 a	42.3±2.8 a	51.4±3.6 a

Data are expressed as mean ± deviation of triplicate measurements. TC: Total carbon.
The ANOVA test was conducted to determine the differences between each pretreatment. Values with the same letters in each pretreatment indicate no significant difference at $P<0.05$.

China). This particular inoculum was selected because of its high methanogenic activity. The characteristics and features of the anaerobic inoculum used were as follows: pH, 7.6±0.1; TS, 86.6%; and VS, 47.5%. The digestion material (500 g) and inoculums (200 g) were added to each digester, followed by deionized water to obtain an 8% TS content. They were stirred and placed in a thermostatic water bath at the mesophilic condition of 37±1°C for 35 d of AD. All reactors were tightly sealed with rubber septa and screw caps. All reactors were gently mixed manually at approximately 1 min d^{-1} prior to biogas

volume measurement to ensure mixing of the reactor contents. Moreover, 200 g of the inoculums was digested to serve as the blank in determining the normalized methane yield of the inoculum by itself. The digestion of each pretreatment was performed in triplicate.

Analysis and Calculations

The volume of biogas was measured by water displacement. The methane content in the produced biogas was analyzed with a fast methane analyzer (Model DLGA-1000, Infrared Analyzer,

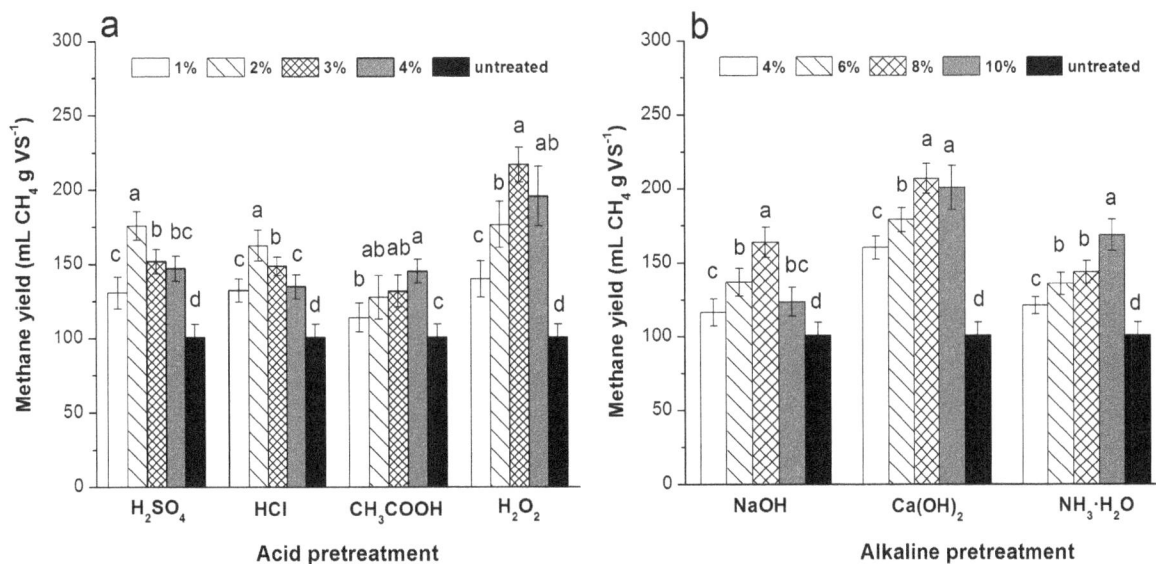

Figure 1. Effect of pretreatments on the methane yield of corn straw. (a) Acid pretreatment; (b) Alkaline pretreatment. Data was expressed at mean ± deviation of triplicate measurements. The ANOVA test was conducted to determine the differences between each pretreatment. Values with the same letters in each pretreatment indicate no significant difference at $P<0.05$.

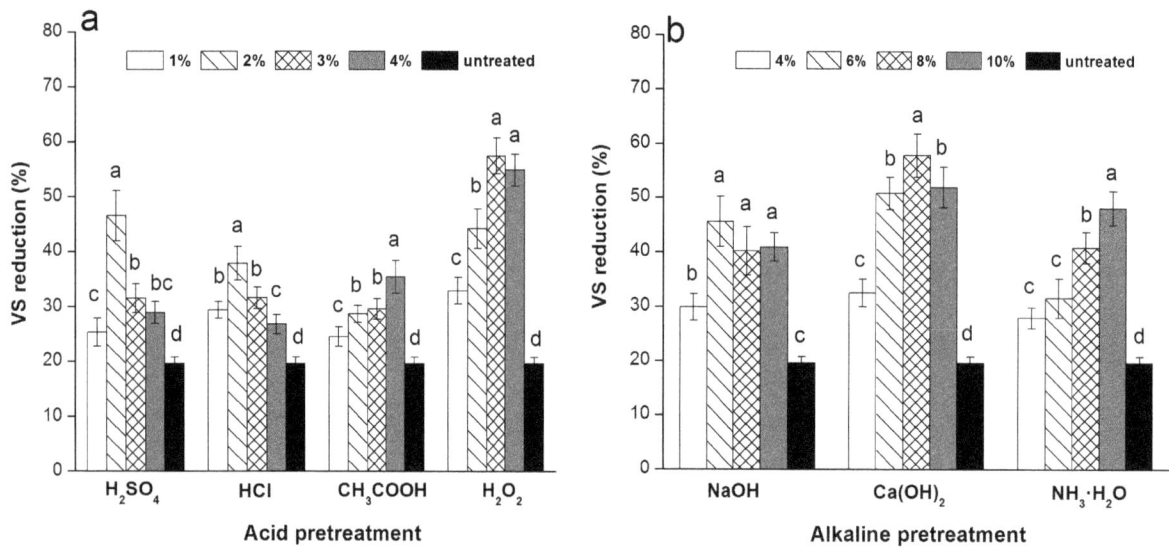

Figure 2. Effect of pretreatments on the VS consumption of corn straw. (a) Acid pretreatment; (b) Alkaline pretreatment. Data was expressed at mean ± deviation of triplicate measurements. The ANOVA test was conducted to determine the differences between each pretreatment. Values with the same letters in each pretreatment indicate no significant difference at $P<0.05$.

Dafang, Beijing, China). The TS, VS, TN, and pH of the materials were measured according to the *Standard Methods for the Examination of Water and Wastewater* of the American Public Health Association [27]. The pH was tested once every 5 d. TC content was analyzed using the method described by Cuetos et al. [28]. The C/N ratio was determined by dividing the total organic carbon content to the TN content. The volatile fatty acid (VFA) was analyzed using a colorimetric method [29], and the result was expressed in terms of acetic acid content. The cellulose, hemicellulose, and lignin contents were analyzed based on the methods previously described by Wang and Xu [30].

Data Analysis

Data is expressed as mean ± standard deviation (SD) of the triplicate measurements. Differences between mean values were examined by ANOVA. Comparisons among means were made using the Duncan multiple range test, and significance was set at $P<0.05$. All statistical analyses were performed using the software program SPSS 15.0. (SPSS Inc., Chicago, USA).

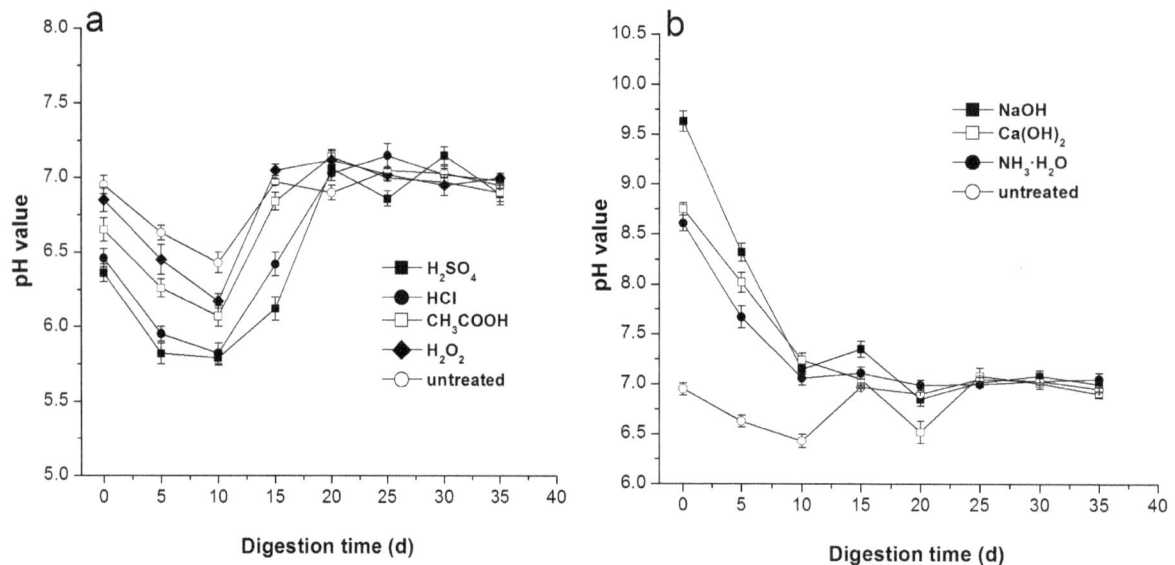

Figure 3. Change in the pH of pretreated corn straw during digestion. (a) Acid pretreatment; (b) Alkaline pretreatment. Data was expressed at mean ± deviation of triplicate measurements.

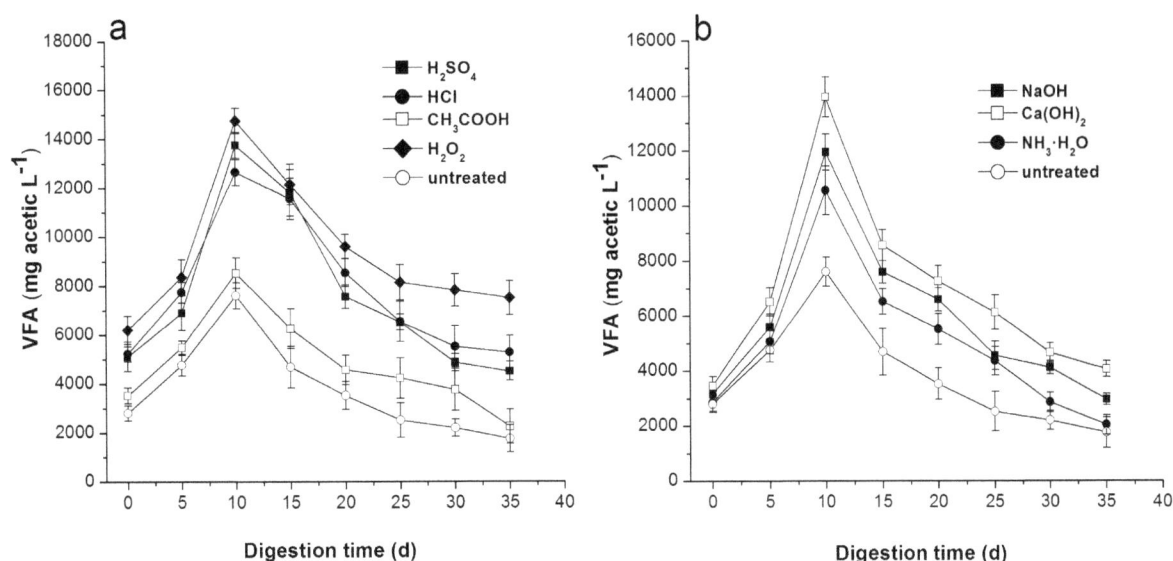

Figure 4. Change in the VFA of pretreated corn straw during digestion. (a) Acid pretreatment; (b) Alkaline pretreatment. Data was expressed at mean ± deviation of triplicate measurements.

Results and Discussion

Effects of Pretreatments on the Chemical Composition of Corn Straw

The aim of the pretreatments was to change the raw material properties, remove or dissolve lignin and hemicellulose, and reduce the crystallinity of cellulose [31]. In the present study, both acid and alkaline pretreatments changed the lignocellulosic composition of corn straw (Tables 1 and 2). Compared with the untreated straw, the hemicellulose and cellulose contents of the acid-treated straw significantly decreased by 6.6% to 66.0%, and 4.4% to 54.3% ($P<0.05$), and the hemicellulose and lignin contents of alkaline-treated corn straw decreased by 10.7% to 46.7%, and 10.8% to 60.7%. These results indicated that pretreatments are more effective in breaking down the lignocellulose matrix and in changing the chemical components of straw. Considerable amounts of lignocellulose appeared to be decomposed and converted into other soluble components that are available to anaerobic microorganisms [32].

Guo et al. [20] reported that corn stalk mainly lost its hemicellulose and cellulose fractions after the acid treatment and lost its lignin fraction after the alkaline treatment. Fernández-Cegrí et al. [2] observed that H_2SO_4 cannot dissolve the lignin of sunflower oil cake, maintaining the same proportion as that of the untreated case. They also found that alkali pretreatments give higher removal levels of lignin compared with other reagents regardless of the temperature effect. The present study revealed a similar phenomenon that acid and alkaline pretreatments had different effects on the lignocellulose composition. In the case of acid reagents, hemicellulose and cellulose contents significantly decreased while the lignin content remained constant in the treated and untreated samples, except when the H_2O_2 was used that the lignin content decreased by 6.7% to 32.0%. The alkaline treatment was mainly effective in removing the lignin fraction. The effectiveness of degrading the lignocellulosic structure usually depends on the type of pretreatment method used, because of the attack on the different parts of the substrate by different chemicals. Acid pretreatment results in disruption of covalent bonds, hydrogen bonds, and Van der Waals forces that hold together the biomass components, which consequently causes the solubilization of hemicellulose and the reduction of cellulose [33]. In contrast, alkali treatment breaks the links between lignin

Table 3. Economic performance of the different pretreatments.

	Chemicals	Concentration	Price [a](CNY)	Cost [b](CNY)	Methane yield (mL CH_4 gVS^{-1})
Acid	H_2SO_4	2%	21	2.57	175.6
	HCl	2%	15	4.92	163.4
	CH_3COOH	4%	12.5	9.34	145.1
	H_2O_2	3%	6	3.6	216.7
Alkaline	NaOH	8%	9	4.2	163.5
	$Ca(OH)_2$	8%	9.5	4.58	206.6
	$NH_3 \cdot H_2O$	10%	9	19.28	168.3

[a]The price was collected from the Sinophram Chemical Reagent Co. Ltd, Beijing China, and the unit of H_2SO_4, HCl, CH_3COOH, H_2O_2, and $NH_3.H_2O$ price was per 500 mL, NaOH and $Ca(OH)_2$ was per 500 g. CNY is the abbreviation for Chinese Yuan, and a dollar is equivalent to 6.12 CNY on Oct 1, 2012; Bank of China. [b] The cost was calculated based on the pretreatment of 1 kg corn straw.

monomers or between lignin and polysaccharides that makes the lignocelluloses swell through saponification reactions [34]. Among the pretreatments, H_2O_2 and NaOH showed the highest solubilization of hemicellulose cellulose, and lignin contents. This trend can be attributed to the strong oxidation ability of H_2O_2 [35] and the high alkalinity of NaOH that allow them to break down the lignocellulose matrix to change the chemical components of the straw. The increased degradation of lignocellulosic materials by H_2O_2 and NaOH suggests that these two chemicals are the most effective in degrading the lignocellulosic structure of corn straw.

The C/N ratio of anaerobic feedstock is significant for AD performance [36]. Analysis of the C/N ratio showed that the percentage of C in the pretreated straw significantly decreased with increasing chemical concentration ($P<0.05$, Tables 1 and 2). The decrease in TC content also affirmed this result. Although the C/N ratio in the pretreated straw was lower than that of the untreated sample, it was still higher than the optimum C/N ratio of feedstock materials (between 20 and 30) [36]. Therefore, the pretreated straw still represents a good co-digestion biomass because it provides a higher carbon fraction for digestion.

Effects of Pretreatments on the Methane Yield of Corn Straw

The methane yield, defined as CH_4 production per unit volatile solids (in mL CH_4 g VS^{-1}), was determined to compare the energy conversion efficiency and the improvement in biodegradability (Fig. 1). As shown in Fig. 1, the straws pretreated by acid and alkaline had significantly increased methane yields ($P<0.05$), i.e., an approximate 10.3% to 115.4% higher yield than for the untreated samples. These results are consistent with previous studies [11,17] which verified the effectiveness of chemical pretreatment in improving biodegradability and enhancing bioenergy production. This phenomenon can be explained by the fact that alkaline and acid pretreatments promote organic solubilization and increase the surface area available for enzymatic action [31]. Chemical pretreatments have different effects on the anaerobic digestibility of corn straw. The methane yield was not improved as the chemical concentration increased. The highest methane yield was achieved at different concentrations for the seven pretreatments. For instance, the highest methane yield was achieved by H_2SO_4 and HCl at 2% concentration, CH_3COOOH at 4%, H_2O_2 at 3%, $Ca(OH)_2$ and NaOH at 8%, and $NH_3 \cdot H_2O$ at 10%. The reason may due to the fact that successful biogasification is not only affected by the sufficient soluble component available but also by anaerobic bacteria. More soluble components from the biodegradation of the lignocellulosic composition need more bacterial to assimilate them. In the present study, the same amount of inoculums (200g) was applied in each digestion experiment, thus, the relative shortage of inoculums could be responsible for the lower methane yield of the chemical pretreatment with high concentration. Among the acid and alkaline treatments, H_2O_2 and $Ca(OH)_2$ respectively produced the highest methane yield in the straw. This result suggests that H_2O_2 and $Ca(OH)_2$ are best for improving the methane yield of corn straws compared with the other pretreatments. The methane yield was significantly heightened as the H_2O_2 concentration increased from 1% to 3% and 4%. However, the methane yield did not increase with further dose increases, showing no significant difference between 3% and 4%. The same trend was also observed for the $Ca(OH)_2$ pretreatment at concentrations between 8% and 10%. The presence of excessive H^+ in 4% H_2O_2 and OH^- in the 10% $Ca(OH)_2$ pretreatment can cause toxicity to the methanogens thereby inhibiting their activity and interfering with their metabolism [37]. Therefore, 3% and 8% are the most suitable concentrations for the H_2O_2 and $Ca(OH)_2$ pretreatments of corn straw, respectively.

Effects of Pretreatments on VS Reduction of Corn Straw

Methane is generated from the conversion of substrates; thus, the methane yield can be determined by reductions in the amount of dry matter of the substrate, as represented by VS. The VS reductions in the straw are shown in Fig. 2. Consistent with previous studies [22], the chemically-treated corn straw obtained higher VS reductions than untreated samples and exhibited reduction of 57.3% to 70.0% for the acid pretreatment and 57.5% to 70.8% for the alkaline pretreatment. 3% H_2O_2 and 8% $Ca(OH)_2$ yielded the greatest reduction in the amount of dry matter of the substrate. The pretreatment triggers the conversion of VS into soluble compounds, including sugar, starch, pectin, tannin, cyclitol, and some inorganics, which become available to anaerobic microorganisms. Generally, this treatment contributes to a substantial improvement in the biodegradability of corn straw. High methane production requires more substrates for digestion; thus, increased VS reductions could explain why the methane yield of the treated straw was highly improved.

Effects of Pretreatments on pH during AD

To investigate the effect of pretreatment on the VFA and pH during the AD of corn straw, the optimal concentration of each pretreatment for methane production was selected as follows: 2% H_2SO_4, 2% HCl, 4% CH_3COOH, 3% H_2O_2, 8% NaOH, 8% $Ca(OH)_2$, and 10% $NH_3 \cdot H_2O$.

Fermentative microorganisms can function in a wider pH range of between 4.0 and 8.5 [38]. In the present study over the first 10 d, the pH of the fermentation broth of the acid-pretreated corn straws was below 7.0 (Fig. 3), whereas that of the three alkaline-pretreated corn straws was over 7.0. The pH curves of all pretreatments were similar, showing a decreasing trend in the initial 10 d and an increasing trend thereafter, slight fluctuations between days 10 to 20. At the end of the fermentation, all pretreatments maintained a pH of approximately 7.0. This trend can be attributed to the variation in VFA concentration because the production of VFA during AD decreases pH. The highly concentrated substrate at the initial phase of AD supplies sufficient organic acid from the degradation of hemicellulose, cellulose, lignin, and VS for the methanogens [20], which decreases pH and accelerates methanogen growth. As digestion proceeded, the content of organic acid gradually decreased with the consumption by the methanogens, which increased the pH. The shortage in organic acid limited the activities of the methanogens but stimulated the acidogens, which increased the amount of organic acids and the dropped the pH. The activity of the methanogens increased again when the organic acid accumulated to an extent, which increased the pH. However, compared with the dramatic fluctuation in the initial phase of AD, the change in the pH in the middle–late phase was slightly heightened because the concentration of the organic acid in the substrate was not as high as the initial concentration. The lack of significant differences in the pH for all pretreatments at the end of AD indicates that these pretreatments can recover the pH. As shown in Fig. 3, the pH of the fermentation broth of the pretreated corn straw markedly declined compared with that of the untreated corn straw. This result can be ascribed to the various acids in the soluble substance of the pretreated straw being significantly higher than that of the untreated straw.

Effects of Pretreatments on VFA during AD

The VFA concentration of each pretreatment initially increased (Fig. 4) and then decreased, which is contrary to the trend of the pH curve. The VFA content of the fermentation broth from the pretreated straw increased more sharply than that of the untreated corn straw. This result can be attributed to the significantly higher soluble substance content of the pretreated corn straw compared with the untreated samples. Among the seven pretreatments, the average VFA concentrations (mg acetic L^{-1}) of the pretreatments during the AD were as follows: 7629 (H_2SO_4), 7879 (HCl), 4821 (CH_3COOH), 9321 (H_2O_2), 5810(NaOH), 6818 ($Ca(OH)_2$), and 4964 ($NH_3 \cdot H_2O$). The highest VFA values were observed for H_2O_2 in the acid treatment, whereas the lowest was observed for CH_3COOH. This result is consistent with the results of the hemicellulose, cellulose, and lignin decomposition and methane yield (Table 1), which further confirmed the effectiveness of H_2O_2 in biodegrading the lignocellulosic structure of straws. Large amounts of hemicellulose and cellulose are converted into simple sugars, lipids (fats) into fatty acids, amino acids, and short-chain organic acids (butyric acid, propionic acid, acetate, and acetic acid), all of which are utilized by methanogens for methane production [15]. In the alkaline pretreatments, the highest VFA content was observed after using $Ca(OH)_2$. This result was consistent with the observations from the methane yield experiments, but contradicted the lignocellulosic composition results where degradation of the lignin fraction was highest after NaOH pretreatments. This disparity can be explained by the fact that successful biogasification is not only affected by the sufficient soluble component available for the anaerobic bacteria but also by the balance between methanogens and acidogens [39]. The excessively high concentration of OH^- in NaOH likely inhibited acetogenesis and disturbed this balance. However, this hypothesis warrants further investigation.

Economic Performance of the Pretreatment Methods

The effectiveness of a pretreatment is not only based on the effectiveness of AD but also on the economic performance. Table 3 compares the economic performance of the pretreatments at the optimal concentrations for methane yield. H_2O_2 and H_2SO_4 showed the lowest costs among the acid pretreatments. However, H_2O_2 was more favorable because it produced higher methane yields than H_2SO_4. In the alkaline pretreatments, although no great difference in the expenses was observed between the $Ca(OH)_2$ and NaOH pretreatments, $Ca(OH)_2$ produced is slightly advantageous over NaOH as it generates a higher methane yield. Therefore, with respect to economic performance and effectiveness, H_2O_2 and $Ca(OH)_2$ can be considered as the most suitable pretreatments for corn straw.

Recently, some researchers combined chemical and physical treatments to improve the biodegradability of lignocellulose composition. High temperature (120–250°C) is often used in combination with dilute acids or base in a pressure cell for much shorter durations. For instance, Saha et al. [40] found the 74% higher saccharification yield wheat straw was subjected to 0.75%

v/v of H_2SO_4 at 121°C for 1 h. Cara et al. [41] shown that olive tree biomass pretreated with 1.4% H_2SO_4 at 210°C resulted in 76.5% of hydrolysis yields. Rocha et al. [42] reported that ethanol yield as high as 0.47 g/g glucose was achieved in fermentation tests with cashew apple bagasse pretreated with diluted H_2SO_4 at 121°C for 15 min. These studies showed the advantage of combination treatment on solubilizing the lignocellulosic composition and shortening the pretreatment time. Nevertheless, depending on the process temperature, some sugar degradation compounds such as furfural and aromatic lignin degradation compounds are detected, and affect the microorganism metabolism in the fermentation step [40]. Furthermore, the pretreatment of high temperature combined with chemicals consumes a substantial amount of energy, and need high facility investment and high treatment cost.

In the present study, although pretreatment time (7 day) was longer than that of chemical treatment with the addition of heat and pressure, the contents of hemicelluloses, cellulose, and lignin fractions of corn straw was greatly reduced, which was contribute to the enhancement of methane production. Furthermore, using single chemicals have no excessive energy consumption and less operation cost. Since cost reduction and low energy consumption are required for an effective pretreatment, chemical pretreatment without the addition of heat and pressure would be desirable to optimize the effectiveness on the process. As for the longer incubation time of the chemical pretreatment, more efforts should be made to investigate the combination of chemicals and low temperature (Below 100°C) pretreatment to shorten the incubation time and improve the anaerobic digestion efficiency.

Conclusions

Four acid pretreatments (H_2SO_4, HCl, CH_3COOH, and H_2O_2) and three alkaline pretreatments (NaOH, $Ca(OH)_2$, and $NH_3 \cdot H_2O$) for improving the methane yield of corn straw were compared. All pretreatments were effective in the biodegradation of the lignocellulosic structure. Straw pretreated with 3% H_2O_2 and 8% $Ca(OH)_2$ elicited the highest methane yields of 216.7 and 206.6 mL CH_4 g VS^{-1}, which are 115.4% and 105.3% higher than that of the untreated straw, respectively. H_2O_2 and $Ca(OH)_2$ are economically and effectively superior to the other pretreatments. Therefore, H_2O_2 and $Ca(OH)_2$ are both recommended as the pretreatments for improving the methane yield of straw.

Acknowledgments

We thank Dr. Chao Zhang for assistance of improving the paper and advice on statistical treatments.

Author Contributions

Conceived and designed the experiments: ZLS GHY XFL. Performed the experiments: ZLS XFL. Analyzed the data: ZLS ZYY. Contributed reagents/materials/analysis tools: YXY YZL. Wrote the paper: ZLS.

References

1. Zhong WZ, Zhang ZZ, Luo YJ, Sun SS, Qiao W, et al. (2011) Effect of biological pretreatments in enhancing corn straw biogas production. Bioresour Technol 102: 11177–11182.

2. Fernández-Cegrí F, Raposo F, de la Rubia MA, Borja R (2013) Effects of chemical and thermochemical pretreatments on sunflower oil cake in biochemical methane potential assays. J Chem Technol Biotechnol 88: 924–929.

3. National Bureau of Statistics of China (2010) China Statistical Yearbook of 2009. China Statistics Press, Beijing, China.

4. Pang YZ, Liu YP, Li XJ (2008) Improving biodegradability and biogas production of corn stover through sodium hydroxide solid state pretreatment. Energ Fuel 22: 2761–2766.

5. Murphy JD, Power NM (2009) An argument for using biomethane generated from grass as a biofuel in Ireland. Biomass Bioenerg 33: 504–512.

6. Amon T, Amon B, Kryvoruchko V, Machmüller A, Hopfner SK, et al. (2007) Methane production through anaerobic digestion of various energy crops grown in sustainable crop rotations. Bioresour Technol 98: 3204–3212.

7. Taherzadeh MJ, Karimi K (2008) Pretreatment of lignocellulosic waster to improve ethanol and biogas production: a review. Int J Mol Sci 9: 1621–1651.

8. Teghammar A, Yngvesson J, Lundin M, Taherzadeh MJ, Horváth IS (2010) Pretreatment of paper tube residuals for improved biogas production. Bioresour Technol 101: 1206–1212.

9. Ferreira LC, Donoso-Bravo A, Nilsen PJ, Fdz-Polanco F, Pérez-Elvira SI (2013) Influence of thermal pretreatment on the biochemical methane potential of wheat straw. Bioresour Technol 143: 251–257.

10. Sapci Z (2013) The effect of microwave pretreatment on biogas production from agricultural straws. Bioresour Technol 128: 487–494.

11. Zheng MX, Li XJ, Li LQ, Yang XJ, He YF (2009) Enhancing anaerobic biogasification of corn stover through wet state NaOH pretreatment. Bioresour Technol 100: 5140–5145.

12. Zhu J, Wan C, Li Y (2010) Enhanced solid-state anaerobic digestion of corn stover by alkaline pretreatment. Bioresour Technol 101: 7523–752.

13. Cao WX, Sun C, Liu RH, Yin RZ, Wu XW (2012) Comparison of the effects of five pretreatment methods on enhancing the enzymatic digestibility and ethanol production from sweet sorghum bagasse. Bioresour Technol 111: 215–221.

14. Michalska K, Miazek K, Krzystek L, Ledakowicz S (2012) Influence of pretreatment with Fenton's reagent on biogas production and methane yield from lignocellulosic biomass. Bioresour Technol 119: 72–78.

15. Gomez-Tovar F, Celis LB, Razo-Flores E, Alatriste-Mondragón F (2012) Chemical and enzymatic sequential pretreatment of oat straw for methane production. Bioresour Technol 116: 372–378.

16. Zhang QH, Tang L, Zhang JH, Mao ZG, Jiang L (2011) Optimization of thermal–dilute sulfuric acid pretreatment for enhancement of methane production from cassava residues. Bioresour Technol 102: 3958–3965.

17. Chandra R, Takeuchi H, Hasegawa T, Kumar R (2012) Improving biodegradability and biogas production of wheat straw substrates using sodium hydroxide and hydrothermal pretreatments. Energy 43: 273–282.

18. Zhou SX, Zhang YL, Dong YP (2012) Pretreatment for biogas production by anaerobic fermentation of mixed corn stover and cow dung. Energy 46: 644–648.

19. González G, Urrutia H, Roeckel M, Aspé E (2005) Protein hydrolysis under anaerobic, saline conditions in presence of acetic acid. J Chem Technol Biotechnol 80: 151–157.

20. Guo P, Mochidzuki K, Cheng W, Zhou M, Gao H, et al. (2011) Effects of different pretreatment strategies on corn stalk acidogenic fermentation using a microbial consortium. Bioresour Technol 102: 7526–7531.

21. Li Q, Gao Y, Wang HS, Li B, Liu C, et al. (2012) Comparison of different alkali-based pretreatments of corn stover for improving enzymatic saccharification. Bioresour Technol 125: 193–199.

22. Song ZL, Yang GH, Guo Y, Ren GX, Feng YZ (2012) Comparison of two chemical pretreatments of rice straw for biogas production by anaerobic digestion. BioResources 7: 3223–3236.

23. Kang KE, Jeong GT, Swoo CS, Park DH (2012) Pretreatment of rapeseed straw by soaking in aqueous ammonia. Bioprocess Biosyst Eng 35: 77–84.

24. Pakarinen OM, Kaparaju PLN, Rintala JA (2011) Hydrogen and methane yields of untreated, water-extracted and acid (HCl) treated maize in one-and two-stage batch assays. Int J Hydrogen Energy 36: 14401–14407.

25. Monlau F, Latrille E, Da Costa AC, Steyer JP, Carrère H (2013) Enhancement of methane production from sunflower oil cakes by dilute acid pretreatment. Appl Energ 102: 1105–1113.

26. Us E, Perendeci NA (2012) Improvement of methane production from greenhouse residues: optimization of thermal and H_2SO_4 pretreatment process by experimental design. Chem. Eng J 182: 120–131.

27. APHA (1998) Standard methods for the examination of water and wastewater, 20th ed, American Public Health Association, Washington, DC, USA.

28. Cuetos MJ, Fernandez C, Gomez X, Moran A (2011) Anaerobic co-digestion of swine manure with energy crop residues. Biotechnol Bioprocess Eng 16: 1044–1052.

29. Chengdu Institute of Biology, Chinese Academy of Sciences (1984) Routine Analysis of Biogas Fermentation, Science Technology Press, Beijing, China.

30. Wang YW, Xu WY (1987) The method for measurement of the content of cellulose, hemicellulose and lignin in solid materials, Microbiology China 2: 81–85.

31. Silverstein RA, Chen Y, Sharma-Shivapp RR, Boyette MD, Osborne J (2007) A comparison of chemical pretreatment methods for improving saccharification of cotton stalks. Bioresour Technol 98: 3000–3011.

32. Lin YQ, Wang DH, Wu SQ, Wang CM (2009) Alkali pretreatment enhances biogas production in the anaerobic digestion of pulp and paper sludge. J Hazard Mater 170: 366–373.

33. Li C, Knierim B, Manisseri C, Arora R, Scheller HV, et al. (2010) Comparison of dilute acid and ionic liquid pretreatment of switchgrass: biomass recalcitrance, delignification and enzymatic saccharification. Bioresource Technol 101: 4900–4906.

34. Xiao B, Sun XF, Sun R (2001) Chemical, structural, and thermal characterizations of alkali-soluble lignins and hemicelluloses, and cellulose from maize stems, rye straw, and rice straw. Polymer Degrad Stability 74: 307–319.

35. Li ZL, Chen CH, Hegg EL, Hodge DB (2013) Rapid and effective oxidative pretreatment of woody biomass at mild reaction conditions and low oxidant loadings. Biotechnol Biofuels 6: 119. doi:10.1186/1754-6834-6-119.

36. Estevez MM, Linjordet R, Morken J (2012) Effects of steam explosion and codigestion in the methane production from Salix by mesophilic batch assays. Bioresour Technol 104: 749–756.

37. Chen Y, Cheng JJ, Creamer KS (2008). Inhibition of anaerobic digestion process: A review. Bioresour Technol 99: 4044–4064.

38. Hwang MH, Jang NJ, Hyum SH, Kim IS (2004) Anaerobic bio-hydrogen production from ethanol fermentation: the role of pH. J Biotechnol 111: 297–309.

39. Zhang T, Liu L, Song Z, Ren G, Feng Y (2013) Biogas production by co-digestion of goat manure with three crop residues. PLoS ONE 8(6): e66845.

40. Saha BC, Iten LB, Cotta MA, Wu YV (2005) Dilute acid pretreatment, enzymatic accharification and fermentation of wheat straw to ethanol. Process Biochem 40: 3693–3700.

41. Rocha MV, Rodrigues TH, de Macedo GR, Gonçalves LR (2009) Enzymatic hydrolysis and fermentation of pretreated cashew apple bagasse with alkali and diluted sulfuric acid for bioethanol production. Appl Biochem Biotechnol 155: 407–417.

42. Cara C, Ruiz E, Oliva JM, Sáez F, Castro E (2008) Conversion of olive tree biomass into fermentable sugars by dilute acid pretreatment and enzymatic saccharification. Bioresour Technol 99: 1869–1876.

10

Conversion of Solid Organic Wastes into Oil via *Boettcherisca peregrine* (Diptera: Sarcophagidae) Larvae and Optimization of Parameters for Biodiesel Production

Sen Yang[1], Qing Li[1,2], Qinglan Zeng[3], Jibin Zhang[1], Ziniu Yu[1], Ziduo Liu[1]*

1 State Key Laboratory of Agricultural Microbiology, National Engineering Research Centre of Microbial Pesticides, College of Life Science and Technology, Huazhong Agricultural University, Wuhan, Hubei, People's Republic of China, 2 College of Science, Huazhong Agricultural University, Wuhan, Hubei, People's Republic of China, 3 Department of Biological Engineering, Xianning Vocational Technical College, Xianning, China

Abstract

The feedstocks for biodiesel production are predominantly from edible oils and the high cost of the feedstocks prevents its large scale application. In this study, we evaluated the oil extracted from *Boettcherisca peregrine* larvae (BPL) grown on solid organic wastes for biodiesel production. The oil contents detected in the BPL converted from swine manure, fermentation residue and the degreased food waste, were 21.7%, 19.5% and 31.1%, respectively. The acid value of the oil is 19.02 mg KOH/g requiring a two-step transesterification process. The optimized process of 12:1 methanol/oil (mol/mol) with 1.5% H_2SO_4 reacted at 70°C for 120 min resulted in a 90.8% conversion rate of free fatty acid (FFA) by esterification, and a 92.3% conversion rate of triglycerides into esters by alkaline transesterification. Properties of the BPL oil-based biodiesel are within the specifications of ASTM D6751, suggesting that the solid organic waste-grown BPL could be a feasible non-food feedstock for biodiesel production.

Editor: Chenyu Du, University of Nottingham, United Kingdom

Funding: This study was supported by grants from the open project of State Key Laboratory of Huazhong Agricultural University (ALM0810), the China National Natural Sciences Foundation (u1170303), and the Genetically Modified Organisms Breeding Major Projects of China (2011ZX08001-001). The funders had no role in study design, data collection and analysis, decision to publish, or preparation of the manuscript.

Competing Interests: The authors have declared that no competing interests exist.

* E-mail: lzd@mail.hzau.edu.cn

Introduction

With economic development and population growth, large quantities of solid organic wastes such as animal waste (e.g., manure), food waste and fermentation industry waste (e.g., filter cake, etc.) are generated all over the world, especially in developing countries. These wastes, if not properly managed, will not only cause environmental pollution but also resource wasting. On the other hand, the reserves of conventional energies are non-renewable, and the major energy resources for almost every country are progressively decreasing and predicted to exhaust in the near future [1]. Biodiesel, one of the recyclable energies, has been considered an ideal substitute for fossil fuels. Currently, about 84% of the world's biodiesel is produced from rapeseed oil, followed by sunflower oil (13%), palm oil (1%), and soybean oil and others (2%) [2]. However, a large-scale biodiesel production from edible oils can bring a global imbalance to the food supply and demand market, and lead to deforestation and destruction of ecosystem on the planet [3,4]. Therefore, a promising solution to environmental and energy crises is to use non-edible oils, such as insect grease/oil derived from solid organic wastes [5].

Insects, one of the largest biomass in the world, could be found nearly in every corner of the earth, and insect fat/oil is igniting particular interest among researchers [6,7,8]. Many saprophagous insect larvae fed on solid organic wastes (e.g., animal manure, food waste etc.) can be converted into insect oil and nutrition [5]. For example, black soldier fly (*Hermetia illucens*) larvae converted from

organic wastes (cattle manure, pig manure, and chicken manure) can be used as a resource for biodiesel production [9]. Besides, black soldier fly can also help reduce the solid residual fraction (hereafter SRF) accumulation of restaurant waste after typical grease extraction, and increase the overall biodiesel yield [10]. Li et al. reported that 5-days *Chrysomya megacephala* (Fabricius) larvae fed on restaurant garbage could be used as a raw material for biodiesel production [11]. However, little information is currently available on the use of *Boettcherisca peregrine* (flesh fly) larvae as an alternative biodiesel feedstock.

B. peregrine belongs to the Boettcherisca genus, Sarcophagidae family, and Diptera order, and is widely used as a death investigator in forensic entomology [12], with a wide geographic distribution from semitropical to tropical regions. Its life cycle comprises of three stages: larva (2–3 days), pupa (8–10 days) and adult at a temperature between 25–30°C [13,14]. Solid organic wastes such as animal manure and carrion are the main breeding grounds for *B. peregrine* larvae (hereafter BPL) and cattle liver can be used to establish a laboratory colony [15]. Occasionally, BPL are found to grow thriftily on the decayed gibberellin A3 fermentation residue (hereafter GFR), a by-product derived from *Fusarium fujikuroi* fermentation liquor in gibberellin A3 production [16], but *C. megacephala* larvae can not develop normally on GFR.

In this paper, we report the use of GFR, swine manure and SRF as media for rearing BPL and the extraction of oil from 4-days BPL for biodiesel production. Additionally, we optimized four variables affecting the yield of acid-catalyzed production of methyl esters: the molar ratio of methanol to oil, catalyst amount, reaction time and temperature. The major fatty acid components of the BPL oil-based biodiesel were also compared with those of three known biodiesels. Finally, the properties of the produced fatty acid methyl esters (hereafter FAME) were evaluated against ASTM D6751, aiming to explore the feasibility of biodiesel production from BPL fed on solid organic wastes.

Materials and Methods

Collection of Fly Source and Solid Organic Wastes

The initial *B. peregrine* larvae were gathered from the swine manure at the National Engineering Research Centre of Microbial Pesticides, Huazhong Agricultural University (at 29–31°N, 113–115°E). The larvae were reared with a medium made of wheat bran and fish meal (7:3, w/w) before the mature larvae were collected from the digested medium in a cage for eclosion. The adults were kept at a constant temperature, humidity and photoperiod (25±5°C, 70±5% RH, 12L:12D) in experimental cages (50×50×50 cm) and provided ad libitum with water and food (brown sugar: milk powder = 1:1). Decayed fish meal (65% moisture) was used as an oviposition substrate to induce the adults to produce larvae. This laboratory colony was established during the spring of 2011, and has been maintained for more than 10 generations.

GFR was a gift from Jiang Xi New Reyphon Biochemical Co., LTD, which contained 60% water, 25% crude protein, 14.9% crude fat, 7.0% crude fiber, 5.6% crude ash, 0.9% phosphorus, 0.5% calcium and 0.3% potassium (dry matter base, w/w). Swine manure was collected from the Pig Breeding Farm of Huazhong Agricultural University, which contained 76% water, 17.5% crude protein, 13.8% crude fat, 16.0% crude fiber, 12.7% crude ash, 1.5% phosphorus, 1.4% calcium and 0.5% potassium (dry matter base, w/w). Raw restaurant waste was collected from restaurants in Wuhan City, China, by Hubei Tianji Bioengineer Co. Ltd., a government authorized biodiesel processing plant located in Huazhong Agricultural University. Waste grease was extracted from the raw restaurant waste in the processing plant and used for biodiesel production. On average, about 3% (w/w) of the raw restaurant waste was separated as waste grease, which was further processed into biodiesel with an overall biodiesel yield of 2.7% (w/w) (Unpublished data from Tianji Bioengineer Co. Ltd.) [10]. The SRF of the restaurant waste after typical grease extraction was used to rear BPL, which contained 80% water, 27.8% crude protein, 11.6% crude fat, 7.4% crude fiber, 13.6% crude ash, 0.7% phosphorus, 0.8% calcium, and 1.3% potassium (dry matter base, w/w).

Conversion of Solid Organic Wastes into Larvae

Based on our previous studies, 40% rice straw powder (w/w, dry matter) was added into GFR to adjust its permeability for BPL feeding. Approximately 7 kg of solid organic wastes (GFR containing rice straw powder, swine manure and SRF, 65% moisture) was placed, respectively, in a plastic tank (0.65×0.43×0.14 m) to a thickness of 5 cm and exposed for biodegradation by 21,000 BPL (Fig. 1). After 4d of cultivation, the mature larvae were separated from residue by sieving, dried at 105°C and ground for oil extraction.

Oil Extraction

BPL oil was extracted using a modified method described by Li et al [9]. The ground BPL powder was placed in a filter bag and soaked in petroleum ether (2000 ml) for 48 h at room tempera-

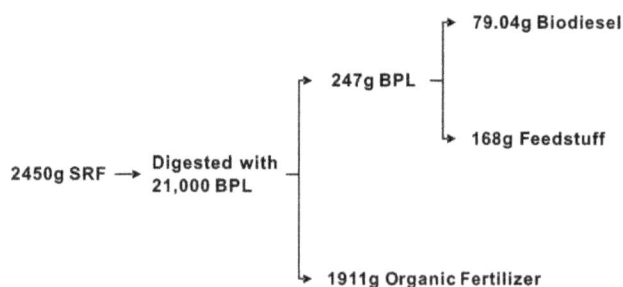

Figure 1. Conversion process from SRF into biodiesel via BPL.

ture. Crude BPL oil was obtained by evaporating petroleum ether with a rotary evaporator.

Optimization of Parameters Affecting Biodiesel Production

The acid value of the BPL oil is 19.02 mg KOH/g, which was determined by titration with the standard method (ISO 660, 1996).To improve the FFA conversion efficiency by acid-catalyzed esterification, four principal variables were investigated: reaction time, reaction temperature, catalyst amount, and the molar ratio of methanol to oil [17]. The optimization of the four parameters was performed by pretreating twenty groups with methanol using H_2SO_4 as the catalyst under following conditions: five groups were pretreated (1.0% catalyst, 65°C, 60 min, 200 rpm) with five different methanol to oil molar ratios of 6:1, 8:1, 10:1, 12:1 and 14:1; five groups were pretreated (12:1 methanol to oil molar ratio, 65°C, 60 min, 200 rpm) with five different catalyst amounts of 0.5%, 1.0%, 1.5%, 2.0% and 2.5%; five groups were pretreated (1.5% catalyst, 12:1 methanol to oil molar ratio, 60 min, 200 rpm) at five different temperatures of 60°C, 65°C, 70°C, 75°C and 80°C; five groups were pretreated (1.5% catalyst, 12:1 methanol to oil molar ratio, 70°C, 200 rpm) for five different time periods of 30 min, 60 min, 90 min, 120 min and 150 min. Each group had 3 replicates of 10 g of crude BPL oil. The esterification apparatus consists of a water bath with a mechanical stirrer, a digital temperature controller and a 50 ml three-neck round bottom flask.

After esterification, the resulting mixture was poured into a 50 ml extraction bottle, and was separated by gravity. Subsequently, the upper layer was transferred to a reactor, and dried in an oven at 105°C for 2 hours before being mixed with methanol (8:1 methanol to oil molar ratio) and the catalyst NaOH (1%, w/w). After a 60°C water bath for 30 min with stirring, the mixture was separated by gravity. Finally, the upper biodiesel layer was separated from the lower layer with a funnel and distilled at 80°C to remove the residual methanol and water.

Analysis

During esterification, 2 ml samples were withdrawn periodically to determine the acid value by titration. The acid value (AV) of the crude oil was determined by titration with potassium hydroxide. FFA conversion rate (%) was calculated using the following formula:

$$FFA\ conversion\ (\%) = (AV_{1i} - AV_{1t})/AV_{1i} \times 100$$

where AV_i represents the initial acid value, AV_t represents the acid value at reaction time t.

The fatty acid compositions were determined by a GC/MS (Thermo-Finnigan, USA) equipped with a polyethylene glycol phase capillary column (Agilent, USA) according to the method from EN 14103 [27]. Other biodiesel characteristics such as density (EN ISO3675), viscosity (EN3105), flash point (EN ISO3679) and cetane index (EN ISO5165) were also measured.

Results and Discussion

BPL Yield

As shown in Fig. 1, about 21,000 BPL reduced 2450 g of SRF to 1911 g of dry digested SRF within 4 days and 247 g of dried larvae in the end. Fig. 2 presents the BPL yields and digested residue derived from four different diets: swine manure, GFR (mixed with 40% rice straw powder, w/w), SRF and artificial diet made of wheat bran and fish meal mixture (7:3, w/w). The GFR achieved the highest yield (12.6%), followed by the swine manure (10.7%), the artificial diet (10.0%) and the SRF (9.9%), indicating that GFR could be the most efficient resource for culturing BPL. Nevertheless, the GFR influence on the growth of BPL needs further studying. The SRF achieved the lowest BPL yield, because SRF medium varied in composition [18]. Yields of digested residue ranged from 65.7% to 79.0% for those solid organic wastes.

In this study, the conversion of the solid organic wastes into BPL for oil extraction was accomplished in only 4 days, much shorter than the time for black soldier fly larvae (15 days) and *C. megacephala* (Fabricius) larvae (5 days). The possible reason is that BPL are ovoviviparous and the larva stage only lasts about 3 days under the optimal condition [13], which suggests a shorter production cycle and a lower cost in treating organic wastes with BPL on a large scale.

Crude Oil Extraction

Fig. 3 shows the crude oil content in the BPL grown on solid organic wastes (dry matter base), indicating a connection of the oil content with the larvae media. The BPL derived from SRF has the highest oil content (31.1%), slightly lower than the black soldier fly

larvae fed on SRF (39.2%), but higher than the 5-days *C. megacephala* (Fabricius) larvae (24.40% to 26.29%) fed on restaurant garbage [10,11]. The BPL converted from swine manure contains 21.7% of oil, lower than rapeseed (37%), but higher than soybean (20%) [1]. The BPL converted from GFR contains the least oil (19.5%), still higher than the oil content of okra seed (12%), the main feedstock for biodiesel production in tropical and subtropical regions [19].

Oil Properties

Table 1 shows the properties of BPL oil. The BPL acid value is 19.02 mg KOH/g, which is close to that of kusum (*Schleichera triguga*) oil (21.30 mg KOH/g), an ideal feedstock for biodiesel production. Acid value indicates the amount of free fatty acid in oil [17]. Saponification value is used for measuring the average molecular weight of oil and expressed in milligrams of potassium hydroxide (mg KOH/g oil). BPL oil has a saponification value of 220.5 mg KOH/g oil, higher than that of *H. illucens* larvae oil (146.6 mg KOH/g oil) and *C. Megacephala* larvae oil (202.1 mg KOH/g oil), respectively [10,11]. The iodine value of BPL oil (98.5 cg I/g oil) is higher than that of *C. Megacephala* larvae oil (73 cg I/g oil), but lower than that of rapeseed oil (108.1 cg I/g oil). Iodine value indicates the degree of unsaturation of an oil [20]. Additionally, BPL oil has a lower melting point (2.0°C), but a higher peroxide value than *H. illucens* larvae oil (4.0°C). The melting point refers to the freezing point of oil, and the peroxide value indicates the level of rancidity during storage. All the data suggest that BPL oil is suitable for biodiesel production.

Table 2 shows the fatty acid compositions of BPL oil compared with three different oils: *H. illucens* larvae oil, *C. Megacephala* larvae oil and rapeseed oil. 11 different fatty acids were detected and identified, and the main fatty acids identified in BPL oil were oleic acid, palmitic acid, and palmitoletic acid. In most organisms, the major acid is palmitic, followed by oleic and other acids typical of the species [21]. The concentration of oleic acid (44.5%) in BPL oil is nearly double that in *H. illucens* larvae oil (23.6%) and *C. Megacephala* larvae oil (24.4%), but lower than that in rapeseed oil (64.4%). In BPL oil, the concentration of palmitic acid (18.9%) is close to that of palmitoletic acid (18.2%). Interestingly, the concentration of total odd-carbon fatty acids (C15:0, C17:0,

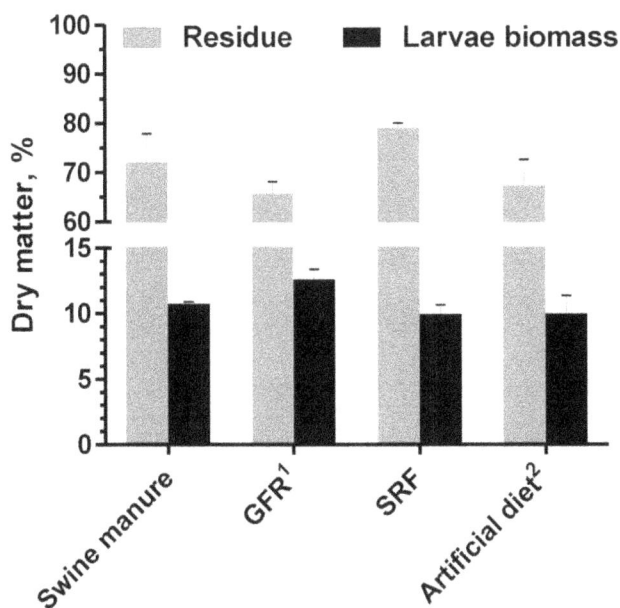

Figure 2. Yields of BPL biomass and residues derived from solid organic wastes. [1]GFR: gibberellin A3 fermentation residue and rice straw powder (6:4, w/w); [2]artificial diet: wheat bran and fish meal (7:3, w/w).

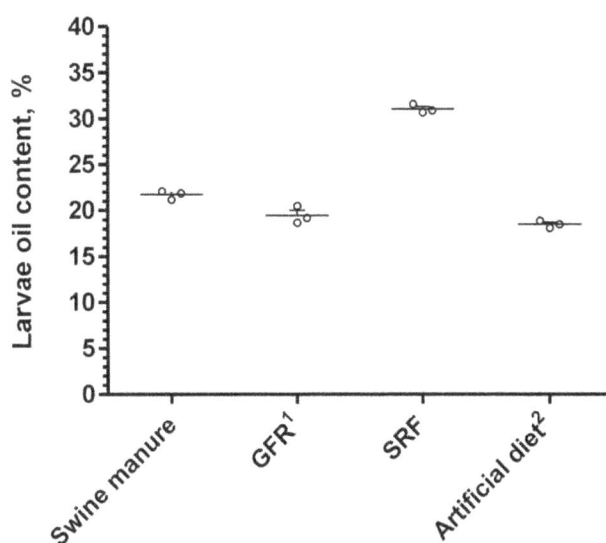

Figure 3. Crude oil content in BPL biomass grown on solid organic wastes.

Table 1. Properties of *B. peregrine* larvae oil and three known oils.

Properties	Units	B. peregrine	H. illucens[a]	C. Megacephala[b]	Rapeseed[c]
Acid value	mg KOH/g	19.02	7.1	1.10	0.20
Density	kg/m^3	944	899	n/a	912
Saponification value	mg KOH/g	220.5	146.6	202.1	197.1
Iodine value	cg I/g oil	98.5	89.7	73.0	108.1
Melting point	°C	2.0	4	n/a	n/a
Peroxide value	meq/kg	0.2	0.04	n/a	n/a

n/a stands for not reported;
[a]data from literature [10];
[b]data from literature [11];
[c]data from literature [20].

C19:0 and C17:1) is 2.4% in BPL oil, but only 1.4% and 0.6% in *H. illucens* larvae oil and *C. Megacephala* larvae oil, respectively. The odd-carbon fatty acids are major fatty acids in microorganisms like thraustochytrids which have the potential as a feedstock for biofuel production [22], but are typically minor (<1%) components in lower plants or animals [21]. It is reported that the fatty acid methyl esters derived from odd-carbon fatty acids have a better low temperature performance than those from even-carbon fatty acids [23]. Furthermore, BPL oil not only contains more total saturated fatty acids (23. 2%) than peanut oil (17.8%), rapeseed oil (4.4%), sunflower oil (9.3%), and soybean oil (16%), respectively [20], but also contains more total unsaturated fatty acids (75.1%)

than *H. illucens* larvae oil (40.8%) and *C. Megacephala* larvae oil (54.6%), respectively [10,11].

Optimization of Parameters Affecting Biodiesel Production

Optimized in this study for FFA acid-esterification are four parameters: methanol to oil molar ratio, catalyst amount, reaction temperature and reaction time.

Fig. 4A shows the effect of the methanol to oil molar ratio on FFA conversion. The results indicate that the higher the ratio, the higher the conversion rate. A molar ratio less than 12:1 resulted in a lower conversion rate, while a ratio more than 12:1 could not

Table 2. Fatty acid compositions of *B. peregrine* larvae oil and three known oils.

Fatty acid (%)	Structure	B. peregrine	H. illucens[a]	C. Megacephala[b]	Rapeseed[c]
Capric acid	C10:0	n/d	3.1	n/a	n/a
Lauric acid	C12:0	n/d	35.6	n/a	n/a
Myristic acid	C14:0	1.4	n/a	3.9	n/a
Pentadecanoic acid	C15:0	1.6	n/a	0.3	n/a
Palmitic acid	C16:0	18.9	14.8	35.5	3.5
Margaric acid	C17:0	0.4	n/a	0.3	n/a
Stearic acid	C18:0	0.2	3.6	2.8	0.8
Noadecanic acid	C19:0	0.5	n/a	n/a	n/a
Arachidic acid	C20:0	n/d	n/a	0.4	n/a
Myristic acid	C14:1	n/d	7.6	n/a	n/a
Palmitoletic acid	C16:1	18.3	3.8	13.0	n/a
Heptadecenoic acid	C17:1	0.4	n/a	n/a	n/a
Oleic acid	C18:1	44.5	23.6	24.4	64.4
Linoleic acid	C18:2	4.1	5.8	15.3	22.3
Linolenic acid	C18:3	7.6	n/a	1.3	n/a
Nonadecanoic acid	C19:1	n/d	1.4	n/a	n/a
Saturated fatty acid		23.2	57.1	43.2	4.3
Unsaturated fatty acid		75.1	40.8	54.6	94.9
Odd-carbon fatty acid		2.4	1.4	0.6	n/a
Total		98.3	97.9	97.8	99.2

n/d stands for not detected; n/a stands for not reported;
[a]data from literature [10];
[b]data from literature [11];
[c]data from literature [20].

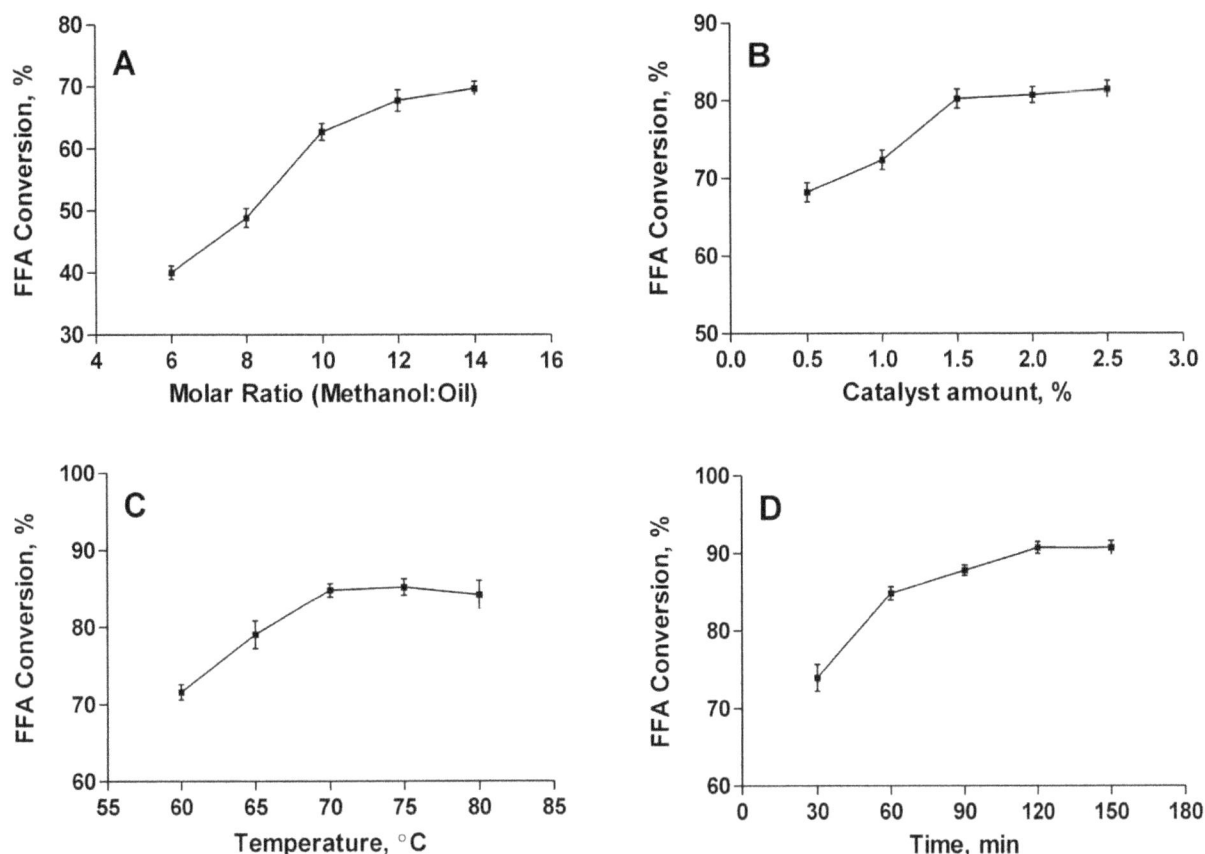

Figure 4. Optimization of four esterification parameters. A) methanol to oil molar ratio (1.0% catalyst, 65°C, 60 min, 200 rpm); B) catalyst amount (12:1 methanol to oil molar ratio, 65°C, 60 min, 200 rpm); C) temperature (1.5% catalyst, 12:1methanol to oil molar ratio, 60 min, 200 rpm); D) reaction time (1.5% catalyst, 12:1methanol to oil molar ratio, 70°C, 200 rpm). Each experiment was repeated for three times.

improve the conversion significantly. Hence, the methanol to oil molar ratio 12:1 was selected for esterification. A higher methanol to fatty acid ratio could result in increased ester formation, but could also interfere with the separation of glycerin in later stages [9].

Fig. 4B shows the effect of catalyst concentration on the FFA acid-esterification. The FFA conversion rate was very low (72.38%) at a catalyst concentration of 1.0%, but dramatically increased to 80.26% at 1.5%. At a concentration of more than 1.5%, however, the conversion rate did not increase significantly. Therefore, 1.5% was determined as the "optimal" catalyst concentration, because a lower amount of acid catalyst is not able to reduce the acid value of the reactants to the desired limit,

Table 3. Properties of *B. peregrine* larvae oil-based biodiesel and three known biodiesels.

Properties	Units	ASTM	B. peregrine	H. illucens[a]	C. megacephala[b]	Rapeseed [c]
Density 15°C	kg/m^3	n/a	884.2	885	874.3	880
Viscosity at 40°C	mm^2/s	1.9–6.0	5.6	5.8	4.0	6.35
Ester content	%	n/a	98.6	97.2	n/a	n/a
Water	%	0.05	<0.03	0.03	<0.03	0.03
Flash point	°C	130	146	123	170	n/a
Cetane number		47	52	53	54.8	45
Acid value	mg KOH/g	0.8	0.62	1.1	0.35	0.3
Distillation temperature	°C	360	323	360	337.0	352

n/a stands for not reported;
[a]data from literature [10];
[b]data from literature [11];
[c]data from literature [20].

whereas a higher amount will result in the darkening of the product [17].

Fig. 4C depicts the effect of temperature on FFA conversion. It was found that the temperature has a positive influence on FFA conversion, which could be attributed to the mass transfer efficiency, a high solubility of crude fat required for efficient mass transfer [24]. However, when the temperature was higher than 70°C, the conversion improvement was not significant. Therefore, 70°C was selected as the optimal esterification temperature.

The effect of reaction time on the FFA conversion rate is presented in Fig. 4D. Among the five sets of reaction time: 30 min, 60 min, 90 min, 120 min and 150 min, 120 min was selected for the optimal esterification time, because FFA conversion rate increased dramatically from 84.8% to 90.8% while the time was increased from 60 min to 120 min at a 30 min interval. However, a reaction time longer than 120 min did not increase the conversion significantly.

The above optimized parameters (the reaction temperature 70°C, the methanol to fat molar ratio 12:1, the catalyst amount 1.5% and the reaction time 120 min) were applied for pretreatment of the crude oils extracted from the BPL grown on three types of solid organic wastes. After the pretreatment, the acid value of BPL oil was reduced to 1.75 mg KOH/g, and the usual acid value of feedstock for alkaline transesterification should be lower than 5 mg KOH/g [24]. The conversion rate of triglycerides (crude oil) to esters reached 92.3% by alkaline trans-esterification using 1% sodium hydroxide (NaOH) as catalyst [9].

Fuel Properties

Table 3 shows the properties of BPL oil-based biodiesel in comparison to two other insect oil-based and one plant oil-based biodiesels. Most of the properties of the biodiesel from BPL fed on solid organic wastes have met the specifications of ASTM D6751, including density (884.2 kg/m^3), viscosity (5.6 mm^2/s), flash point (146°C), cetane number (52), ester content (98.6%) and acid value (0.62 mg KOH/g). Biodiesel properties are strongly influenced by the characteristics of individual fatty esters. The viscosity of BPL biodiesel is 5.6 mm^2/s, higher than that of *C. megacephala* (4.0 mm^2/s) biodiesel, but lower than that of *H. illucens* (5.8 mm^2/s) and rapeseed (6.35 mm^2/s) biodiesels. Viscosity affects the atomization of a fuel upon injection into the combustion chamber and ultimately the formation of engine deposits [25]. The flash point of BPL biodiesel (146°C) agrees well with the minimum specifications in ASTM D6751 (130°C), higher than that in *H. illucens* biodiesel (123°C). A higher flash point value means a higher

handling and storing safety index. The cetane number is a dimensionless indicator related to the ignition quality of a fuel in a diesel engine. It is generally believed that the higher the cetane number, the better the ignition quality of a fuel, and vice versa [26]. The BPL biodiesel has a higher cetane number than ASTM D6751 (47) and rapeseed biodiesel (45), but lower than the other two insect oil-based biodiesels (53 and 54.8 in *H. illucens* and *C. megacephala*,), respectively. The cetane number increases with the increase of the chain length and saturation of a feedstock oil. It is worth noting that the BPL oil contains 44.5% oleic acid, nearly double that in *H. illucens* (23.6%) and *C. megacephala* (24.4%) oils. Knothe pointed that methyl oleate has been proposed as a suitable major component in biodiesel fuels for improving their properties [26].

Conclusions

This study indicates that *Boettcherisca peregrine* holds a high promise for converting solid organic wastes into an alternative feedstock for biodiesel production due to its high oil content (19.5–31.1%) and short production cycle (4 days). The yields of BPL converted from swine manure, GFR, and SRF reach to 10.7%, 12.6% and 9.9% respectively. In addition, four parameters for acid-esterification of FFA in BPL oil were optimized: 12:1 methanol to oil molar ratio, 1.5% catalyst at 70°C for 120 min, under which conditions, the conversion rate of FFA and triglycerides into biodiesel reached 90.8% and 92.3%, respectively. Compared with the other two insects (*H. illucens* and *C. Megacephala*) BPL oil contains more unsaturated fatty acid (75.1%) and Odd-carbon fatty acid (2.4%). The properties of the biodiesel from BPL oil met the standard ASTM D6751 including density (884.2 kg/m^3), viscosity (5.6 mm^2/s), flash point (146°C), cetane number (52) and ester contents (98.6%). The results of this study demonstrated that BPL can recycle different solid organic wastes into clean energy, and reduce environmental pollution of wastes.

Acknowledgments

The authors would like to acknowledge the Pig Breeding Farm of Huazhong Agricultural University for supplying the swine manure.

Author Contributions

Conceived and designed the experiments: ZL. Performed the experiments: SY. Analyzed the data: QL ZY. Contributed reagents/materials/analysis tools: QZ JZ. Wrote the paper: SY.

References

1. Gui MM, Lee KT, Bhatia S (2008) Feasibility of edible oil vs. non-edible oil vs. waste edible oil as biodiesel feedstock. Energy 33: 1646–1653.
2. Thoenes P (2007) Biofuels and commodity markets-palm oil focus. FAO, Commodities and Trade Division. http://www.oil palm world.com/bio-fuels.pdfS.
3. Mongobay, Butler RA (2006) Why is oil palm replacing tropical rainforests? Why are biofuels fueling deforestation? http://news.mongabay.com/2006/0425-oil_palm.htmlS.
4. Martindale W, Trewavas A (2008) Fuelling the 9 billion. Nat Biotechnol 26: 1068–1070.
5. Li Q, Zheng L, Hou Y, Yang S, Yu Z (2011) Insect fat, a promising resource for biodiesel. J Pet Environ Biotechnol S2.
6. Angela ED (2007) Symbiotic microorganisms: untapped resources for insect pest control. Trends Biotechnol 25: 338–342.
7. Liu WC, Bonsall MB, Godfray HC (2007) The form of host density-dependence and the likelihood of host–pathogen cycles in forest-insect systems. Theor Popul Biol 72: 86–95.
8. John EB, Riccardo B, Barbara E (2008) Population response to resource separation in conservation biological control. Biological Control 47: 141–146.
9. Li Q, Zheng L, Cai H, Garza E, Yu Z, Zhou S (2011) From organic waste to biodiesel: black soldier fly, *Hermetia illucens*, makes it feasible. Fuel 90: 1545–1548.
10. Zheng L, Li Q, Zhang J, Yu Z (2011) Double the biodiesel yield: Rearing black soldier fly larvae, *Hermetia illucens*, on solid residual fraction of restaurant waste after grease extraction for biodiesel production. Renew Energy 41: 1–5.
11. Li Z, Yang D, Huang M, Hu X, Shen J, et al. (2012) *Chrysomya megacephala* (Fabricius) larvae: A new biodiesel resource. Appl Energy 94: 349–354.
12. Goff ML, Brown WA, Hewadikaram KA, Omori AI (1991) Effect of heroin in decomposing tissues on the development rate of *Boettcherisca peregrina* (Diptera, Sarcophagidae) and implications of this effect on estimation of postmortem intervals using arthropod development patterns. J of Forensic sci 36: 537–542.
13. Wang H, Shi Y, Liu X, Zhang R (2010) Growth and development of *Boettcherisca peregrine* under different temperature conditions and its significance in forensic entomology. J of Enviro Entom 32: 166–172.
14. Schroeder H, Klotzbach H, Puschel K (2003) Insects colonization of human corpses in warm and cold season. Int J Legal Med 5: 372–374.
15. Wu G, Li K, Ye G (2006). Isolation, purification and characterization of metallothioein from *Boettcherisca peregrina* (Diptera: Sarcophagidae). Acta Entomologica Sinica 49: 22–28.
16. Rodrigues C, Vandenberghe L, Oliveira J, Soccol C (2011). New perspectives of gibberellic acid production: a review. Crit Rev Biotechnol 1–11.
17. Sharma YC, Singh B (2010) An ideal feedstock, kusum (*Schleichera triguga*) for preparation of biodiesel: Optimization of parameters. Fuel 89: 1470–1474.

18. Neves L, Goncalo E, Oliveira R, Alves MM (2008) Influence of composition on the biomethanation potential of restaurant waste at mesophilic temperatures. Waste Manage 28: 965–972.
19. Kafuku G, Mbarawa M (2010) Biodiesel production from *Croton megalocarpus* oil and its process optimization. Fuel 89: 2556–2560.
20. Demirbas A (2003) Biodiesel fuels from vegetable oils via catalytic and non-catalytic supercritical alcohol transesterifications and other methods: a survey. Energ Convers Manage 44: 2093–2109.
21. Řezanka T, Sigler K (2009) Odd-numbered very-long-chain fatty acids from the microbial, animal and plant kingdoms. Prog Lipid Res 48: 3206–3238.
22. Chang K, Mansour M, Dunstana G, Susan I, Koutoulisc A, et al. (2011). Odd-chain polyunsaturated fatty acids in thraustochytrids. Phytochemistry 72: 1460–1465.
23. Xu M (1999) Study on relationship of low temperature performance of pentaerythritol ester base oil and acid composition in feedstock. Lubricating oil 14: 56–58.
24. Suyin G, Hoon K, Chun W, Nafisa O, Mohd A (2010) Ferric sulphate catalysed esterification of free fatty acids in waste cooking oil. Bioresource Technol 101: 7338–7343.
25. Knothe G (2005) Dependence of biodiesel fuel properties on the structure of fatty acid alkyl esters. Fuel Process Technol, 86: 1059–1070.
26. Knothe G (2008) "Designer" biodiesel: Optimizing fatty ester composition to improve fuel properties. Energ Fuel 22: 1358–1364.
27. EN 14103 (2003) Fat and oil derivatives-Fatty Acid Methyl Esters (FAME) Determination of ester and linolenic acid methyl ester contents. Brussels, Belgium: European Committee for Standardization.

11

Abatement Cost of GHG Emissions for Wood-Based Electricity and Ethanol at Production and Consumption Levels

Puneet Dwivedi[1]*, Madhu Khanna[2]

1 Warnell School of Forestry and Natural Resources, University of Georgia, Athens, Georgia, United States of America, **2** Energy Biosciences Institute, University of Illinois at Urbana-Champaign, Urbana, Illinois, United States of America

Abstract

Woody feedstocks will play a critical role in meeting the demand for biomass-based energy products in the US. We developed an integrated model using comparable system boundaries and common set of assumptions to ascertain unit cost and greenhouse gas (GHG) intensity of electricity and ethanol derived from slash pine (*Pinus elliottii*) at the production and consumption levels by considering existing automobile technologies. We also calculated abatement cost of greenhouse gas (GHG) emissions with respect to comparable energy products derived from fossil fuels. The production cost of electricity derived using wood chips was at least cheaper by 1 ¢MJ^{-1} over electricity derived from wood pellets. The production cost of ethanol without any income from cogenerated electricity was costlier by about 0.7 ¢MJ^{-1} than ethanol with income from cogenerated electricity. The production cost of electricity derived from wood chips was cheaper by at least 0.7 ¢MJ^{-1} than the energy equivalent cost of ethanol produced in presence of cogenerated electricity. The cost of using ethanol as a fuel in a flex-fuel vehicle was at least higher by 6 ¢km^{-1} than a comparable electric vehicle. The GHG intensity of per km distance traveled in a flex-fuel vehicle was greater or lower than an electric vehicle running on electricity derived from wood chips depending on presence and absence of GHG credits related with co-generated electricity. A carbon tax of at least \$7 Mg CO_2e^{-1} and \$30 Mg CO_2e^{-1} is needed to promote wood-based electricity and ethanol production in the US, respectively. The range of abatement cost of GHG emissions is significantly dependent on the harvest age and selected baseline especially for electricity generation.

Editor: Jorge Aburto, INSTITUTO MEXICANO DEL PETRÓLEO, Mexico

Funding: The authors are thankful to the funding support provided by the Energy Biosciences Institute at the University of Illinois at Urbana-Champaign/ University of California, Berkeley. The funders had no role in study design, data collection and analysis, decision to publish, or preparation of the manuscript.

Competing Interests: The authors have declared that no competing interests exist.

* Email: puneetdwivedi@gmail.com

Introduction

The electricity and transportation sectors of the US economy emitted 57% of total GHG emissions (6753 million Mg CO_2e) in 2011 [1]. Therefore, policy makers have announced several incentives to promote electricity generation from various renewable sources including biomass to reduce GHG emissions from the electricity sector [2]. It is projected that these incentives will increase biomass-based electricity generation at the national level from 11.5 to 49.3 billion kWh between 2010 and 2035 [3]. There is an emphasis on reducing GHG emissions from the transportation sector as well. The Energy Independence and Security Act of 2007 has set a target of producing 60.5 billion liters of cellulosic biofuels by 2022 nationwide [4].

Biomass obtained from the nation's forestlands would play a critical role in supplying required biomass for renewable electricity generation and production of cellulosic ethanol [5]. A few studies have analyzed economic and environmental potential of utilizing forest biomass for generating electricity [6–12] and producing ethanol [13–19]. Typically, these studies indicate that wood-based energy products could save significant amounts of GHG emissions (about 80% or more) but are costlier (at least 15% or more) than equivalent energy products derived from fossil fuels. These studies use different species, energy pathways, system boundaries, and modeling assumptions; therefore, it is practically very difficult to compare these studies with each other to get an insight about the cost-effectiveness of various woody feedstocks in reducing GHG emissions. No study has done a side-by-side comparison of the economic and environmental performance of wood-based electricity and ethanol at the production and consumption levels for existing automobile technologies using similar assumptions under realistic system boundaries. Comparable existing studies only focus on agriculture feedstocks and typically consider environmental [20–22] and economic performances [23,24] of energy products disjointedly. A consideration of both economic and environmental performances of different bioenergy products in a single framework is critical to compare cost-effectiveness of various GHG mitigation options to minimize total cost related with the reduction of GHG emissions at the national and regional levels [25]. Additionally, these information will help in determining the minimum carbon tax that would be needed to promote production and consumption of wood-based energy products in the US. Furthermore, existing studies [6–19] measure economic and environmental performances of biomass-based energy products either at production or consumption levels but not at both levels

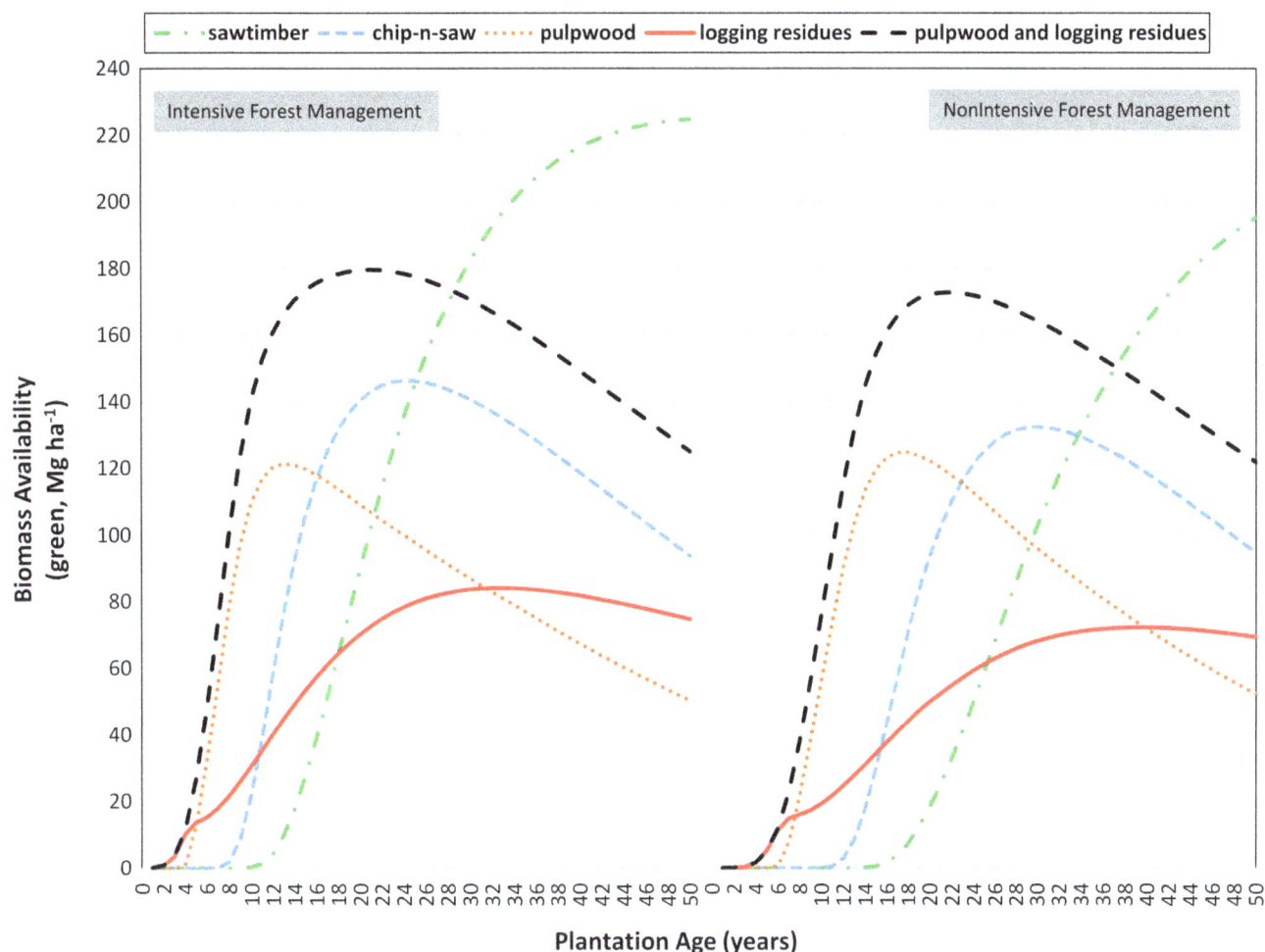

Figure 1. Availability of timber products at different plantation ages. Site index is 21.4 meters at 25th year of plantation. Initial plantation density is 1236 seedlings ha^{-1}.

simultaneously. This gives an incomplete picture as it is our assertion that the performance of biomass-based energy products could vary significantly at the selected level of analysis as fuel economy of automobiles operating on ethanol and electricity differ from each other [26].

We analyzed four energy pathways in this study. Focus of first two energy pathways was on electricity generation while the last two energy pathways focused on ethanol production. Under the first energy pathway, wood was converted to wood pellets and then manufactured wood pellets were burned at a nearby power plant to generate electricity. This pathway was based on the fact that the US has become a major exporter of wood pellets to power plants located in European countries [27]. We wanted to test the economic and environmental feasibility of utilizing manufactured wood pellets within the US only assuming that power plant owners in the country will follow a similar trend in the future as well. Under the second energy pathway, wood was chipped at the forest site and then wood chips were directly burned at a nearby power plant to generate electricity. For the third energy pathway, feedstock was chipped at the forest site and then sent to an ethanol mill for ethanol production [5]. The co-generated electricity was supplied to the grid for additional income and GHG credits. Under the fourth energy pathway, feedstock was chipped at the forest site and then sent to an ethanol mill for ethanol production

[5]. However, co-generated electricity was not supplied to the grid and therefore, no additional income and GHG-credits were accrued.

For each energy pathway, we analyzed 186 scenarios (three feedstocks – logging residues only, pulpwood only, both logging residues and pulpwood; two forest management choices – intensive and non-intensive; 31 harvest ages – age 10 to age 40 in steps of 1 year). We selected pulpwood as a potential feedstock as evidence suggests that it is increasingly being used to manufacture wood pellets [27]. Under intensive forest management, herbicides were applied at the establishment year followed by fertilizers at plantation ages 2 and 12. No herbicides and fertilizers were applied under non-intensive forest management choice. Intensive forest management represents industrial plantations whereas non-intensive forest management represents plantation owned by non-industrial private forestland owners. The geographical focus of this study is US South as this region contributed about 62% of total roundwood removals in 2006 nationwide [28]. We selected slash pine as a representative species as this species is a popular commercial forest species of the region [28]. Additionally, pine plantations contribute maximum to the overall roundwood harvest in southern forestry landscape and therefore, focusing on a popular pine species will define the role of existing forest resources in the region in mitigating GHG emissions. This becomes even more important as the majority of existing studies focus on short rotation woody crops like willow [8,11,16], eucalyptus [18], and poplar [19].

Methods

Feedstock Availability

We used a growth and yield model of slash pine [29] to estimate availability of three timber products: sawtimber, chip-n-saw, and pulpwood under intensive and non-intensive forest management choices at different plantation years. The availability of logging residues at a plantation year was calculated as the difference between total biomass present in logs and total biomass present in merchantable portion of logs (sawtimber, chip-n-saw, and pulpwood) plus 20% of biomass present in sawtimber, chip-n-saw, and pulpwood at the same plantation year [30]. Additional 20% biomass was added as a proxy for biomass available in branches and tree tops [30].

GHG Intensity of First Energy Pathway

We calculated total wood pellets produced (WP in Mg ha^{-1}) at a harvest age using Equation (1).

$$WP_{h,f,i} = B_{h,f,i}^{green} \times MC_{wood} \times BU \times \left(\frac{100}{100 - MC_{WP}} \right) \quad (1)$$

where, B^{green} is the biomass available at a given harvest age (h), feedstock type (f), and forest management intensity (i); MC_{wood} is the moisture content of the green wood (50%); BU is the ratio of biomass used for wood pellet production (80%) [6]; and MC_{WP} is the moisture content of wood pellets (5%) [6]. We calculated total electricity generated (EC^{WP} in MJ ha^{-1}) from wood pellets using Equation (2).

$$EC_{h,f,i}^{WP} = WP_{h,f,i} \times CV_{WP} \times CE \times (100 - TRAN) \times 1000 \quad (2)$$

where, CV_{WP} is the calorific value of wood pellets (18.5 MJ kg^{-1}), CE is the conversion efficiency of a 100 MW power plant (31.70%) [31], and TRAN is the electricity transmission losses (7%) [32]. A 100 MW power plant is considered based on the fact that several large-scale facilities have recently been established in the US and Europe which will utilize wood pellets/wood chips to generate electricity [33,34]. We calculated GHG intensity ($GHGI^{Elec-WC}$ in g CO$_2$e MJ^{-1}) of generated electricity from wood pellets using Equation (3).

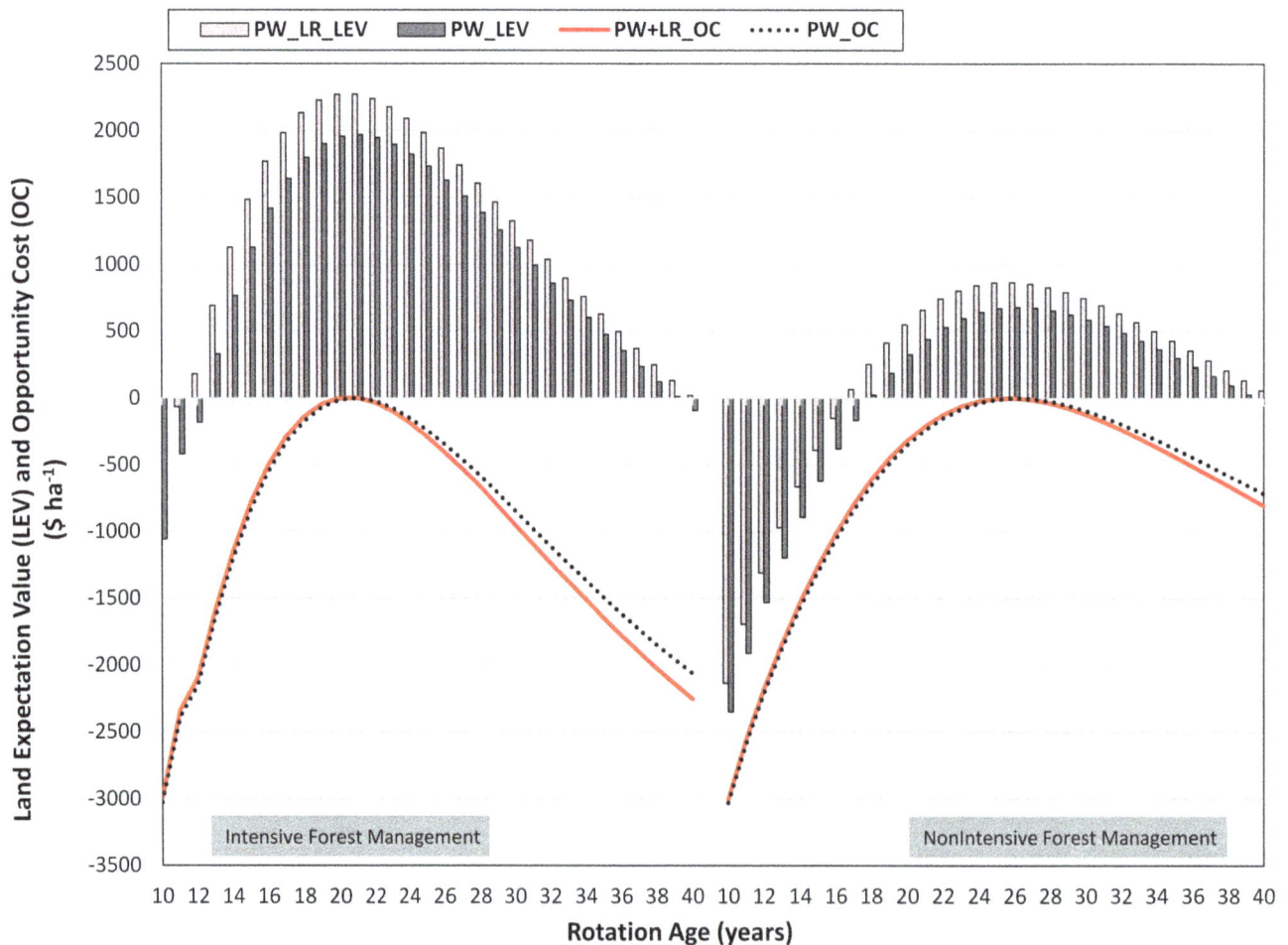

Figure 2. Distribution of land expectation values (LEVs) and opportunity costs (OCs). Opportunity cost is calculated by subtracting land expectation value at a given harvest age from the land expectation value at the optimal rotation age. The land expectation value is highest at the optimal rotation age.

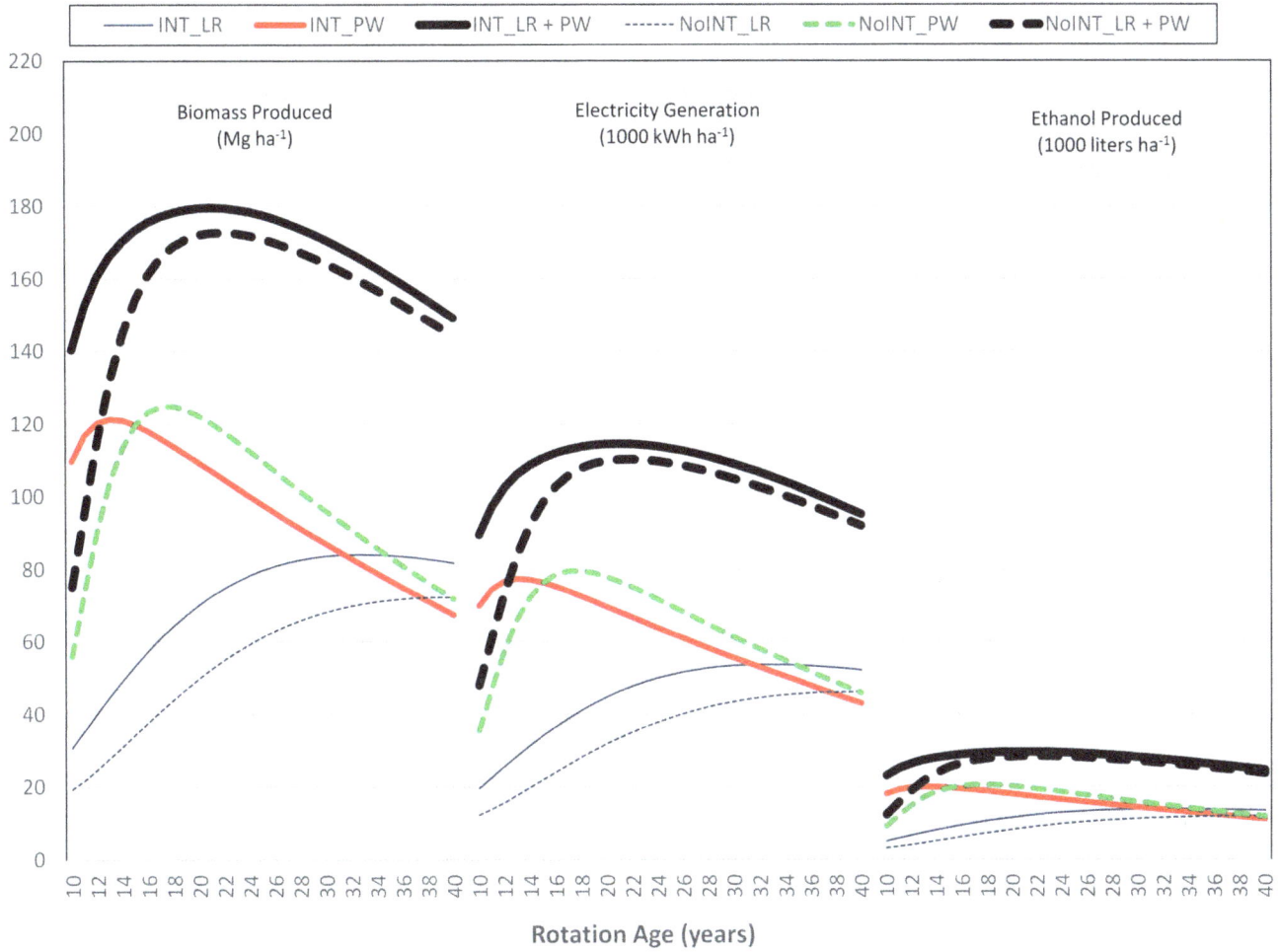

Figure 3. Availability of feedstocks, electricity generated, and ethanol produced. LR: logging residues; PW: pulpwood; INT: intensive forest management; NoINT: non-intensive forest management.

$$GHGI_{h,f,i}^{Elec-WP} =$$

$$\frac{\left(E_{h,f,i}^{Bio\,Pr\,o} + E_{h,f,i}^{Bio-Tran} + E_{h,f,i}^{Bark} + E_{h,f,i}^{WP} + E_{h,f,i}^{WP-Tran} + E_{h,f,i}^{WP-Burn}\right)}{EC_{h,f,i}^{WP}} \quad (3)$$

where, $E^{Bio\,Pro}$ (Mg CO_2e ha^{-1}) represents GHG emissions related to wood production. The total GHG emission under intensive forest management was 4803 kg CO_2e ha^{-1} when the harvest age was equal or greater than 12 years whereas it was 2431 kg CO_2e ha^{-1} when the harvest age was 10 and 11 years [6]. For non-intensive forest management choice, total GHG emission was 2200 kg CO_2e ha^{-1} for the selected range of harvest ages [6]. We updated the value of nitrous oxide emission based on the GREET model [35]. These GHG emissions were allocated to feedstocks based on the percentage of mass occupied by feedstocks out of total timber products available at a given harvest age. The parameter $E^{Bio-Tran}$ reflects GHG emissions related to transportation of biomass from a harvest site to a nearby wood pellet plant. It was a product of GHG emission factor (0.133 kg CO_2e Mg^{-1} km^{-1}) [36], total green biomass transported, and average distance traveled (100 km one way). The parameters E^{Bark} reflects non-biogenic GHG emissions related with bark burning in a boiler (34.4 g CO_2e kg^{-1} of burned material) [37]. Percentage of bark

was 20% of incoming biomass [6]. The parameter E^{WP} reflects GHG emissions related with manufacturing of wood pellets (155.7 g CO_2e kg^{-1}) [6]. The parameter $E^{WP-Tran}$ reflects GHG emissions related to transportation of wood pellets from wood pellet mill to a nearby power plant. It was a product of GHG emission factor (0.133 kg CO_2e Mg^{-1} km^{-1}) [36], total wood pellets transported, and average distance traveled (50 km one way). We followed steps for estimating parameter E^{Bark} for quantifying non-biogenic GHG emissions related with the burning of wood pellets ($E^{WP-Burn}$) at a power plant.

GHG Intensity of Second Energy Pathway

We calculated total wood chips produced (WC in Mg ha^{-1}) at a harvest age using Equation (4).

$$WC_{h,f,i} = B_{h,f,i}^{green} \quad (4)$$

We calculated total electricity generated (EC^{WC} in MJ ha^{-1}) from wood chips using Equation (5).

$$EC_{h,f,i}^{WC} = WC_{h,f,i} \times CV_{WC} \times CE \times (100 - TRAN) \times 1000 \quad (5)$$

where, CV_{WC} is the calorific value of wood chips (10 MJ kg^{-1}).

Table 1. Range of production costs and GHG intensities for selected energy pathways.

Pathway		Unit Cost												GHG Intensity											
		Production Level (¢ MJ⁻¹)						Consumption Level (¢ km⁻¹)						Production Level (g CO_2e MJ⁻¹)						Consumption Level (g CO_2e km⁻¹)					
		Intensive			Non-Intensive			Intensive			Non-Intensive			Intensive			Non-Intensive			Intensive			Non-Intensive		
		LR	PW	LR+PW	LR	PW	LR+PW	LR	PW	LR+PW	LR	PW	LR+PW	LR	PW	LR+PW	LR	PW	LR+PW	LR	PW	LR+PW	LR	PW	LR+PW
Electricity from Wood Pellets	Min	4.0	4.3	4.2	4.0	4.3	4.2	2.9	3.1	3.0	2.9	3.1	3.0	49.0	49.0	47.1	47.1	47.1	20.8	35.1	35.7	35.1	33.7	34.0	33.7
	Max	8.3	5.6	5.1	10.8	6.6	5.9	5.9	4.0	3.7	7.7	4.7	4.3	54.2	56.2	54.2	57.6	62.0	57.6	38.8	40.3	38.8	41.2	44.4	41.2
Electricity from Wood Chips	Min	3.0	3.2	3.1	3.0	3.2	3.1	2.2	2.3	2.2	2.2	2.3	2.2	20.8	21.5	20.8	19.3	19.6	19.3	14.9	15.4	14.9	13.8	14.0	13.8
	Max	6.3	4.2	3.9	8.3	5.0	4.5	4.5	3.0	2.8	5.9	3.6	3.2	24.8	26.4	24.8	27.4	30.9	27.4	17.8	18.9	17.8	19.6	22.1	19.6
Ethanol with co-generated electricity	Min	3.7	3.9	3.8	3.7	3.9	3.9	8.4	8.9	8.7	8.4	8.9	8.7	3.5	4.3	3.5	1.8	2.2	1.8	8.0	9.7	8.0	4.2	5.1	4.2
	Max	7.4	5.1	4.7	9.5	6.0	5.4	16.7	11.5	10.6	21.6	13.5	12.2	7.9	9.7	7.9	10.9	14.6	10.9	18.0	22.1	18.0	24.6	33.2	24.6
Ethanol without co-generated electricity	Min	4.5	4.7	4.6	4.5	4.7	4.6	10.1	10.6	10.4	10.1	10.6	10.4	21.9	22.7	21.9	20.2	20.6	20.2	49.7	51.4	49.7	45.9	46.8	45.9
	Max	8.1	5.8	5.4	10.3	6.7	6.1	18.4	13.2	12.3	23.3	15.2	13.9	26.3	28.1	26.3	29.2	33.0	29.2	59.8	63.8	59.8	66.4	75.0	66.4

LR: logging residues; PW: pulpwood; Intensive: intensive forest management; Non-intensive: non-intensive forest management.

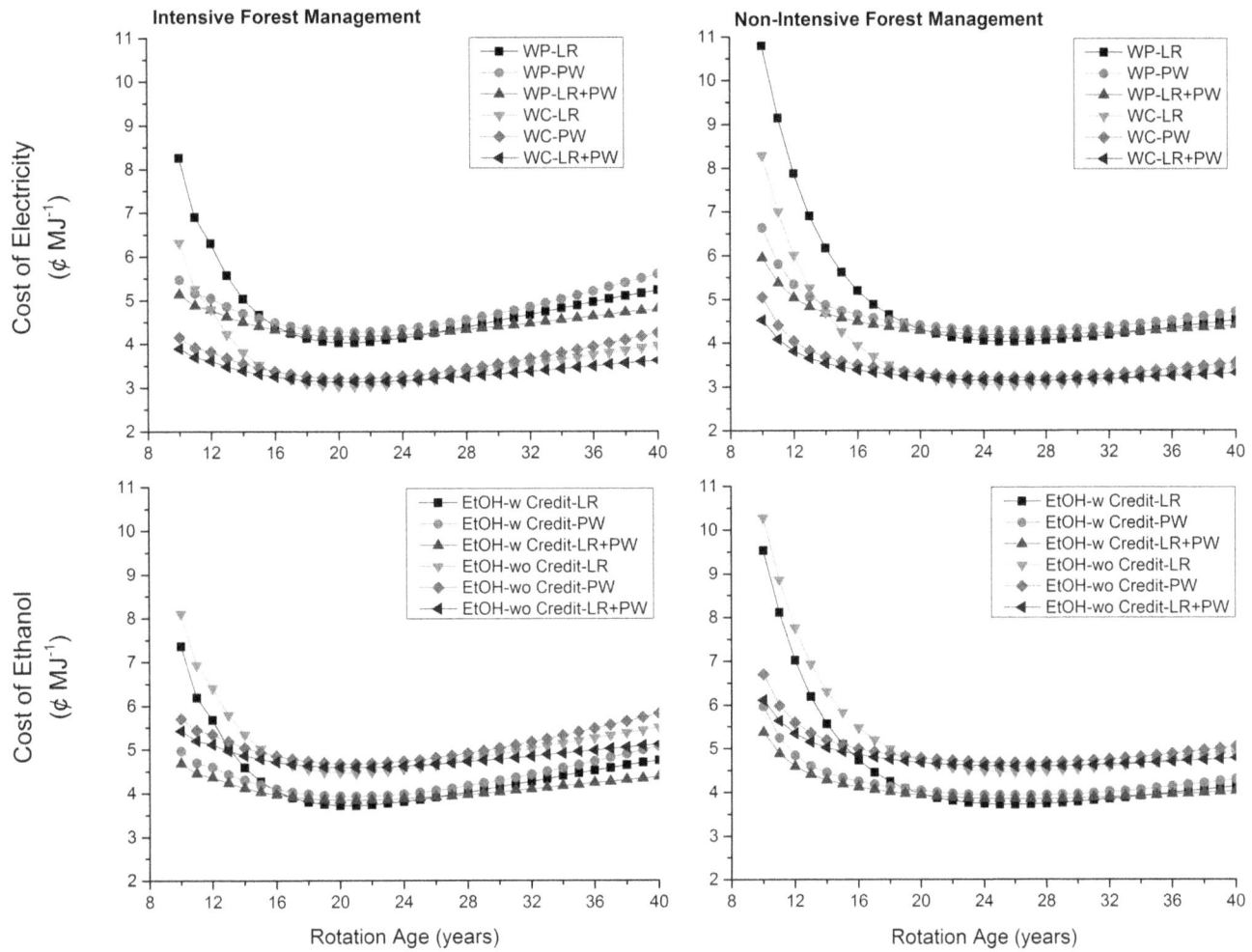

Figure 4. Cost of energy products at the production level. LR: logging residues; PW: pulpwood; WP: wood pellets; WC: wood chips; w: with income from cogenerated electricity; wo: without income from cogenerated electricity.

We calculated GHG intensity ($GHGI^{Elec-WC}$ (in g CO_2e MJ^{-1}) of generated electricity from wood chips using Equation (6).

$$GHGI_{h,f,i}^{Elec-WC} = \frac{\left(E_{h,f,i}^{Bio\,Pro} + E_{h,f,i}^{Chipping} + E_{h,f,i}^{WC-Tran} + E_{h,f,i}^{WC-Burn}\right)}{EC_{h,f,i}^{WC}} \quad (6)$$

where, the parameters $E^{chipping}$ refers to GHG emissions related to chipping of feedstocks ($E^{Chipping}$, 4 kg CO_2e Mg^{-1}) on the forest site [32] and $E^{WC-Burn}$ reflects non-biogenic GHG emissions related with burning of wood chips in a boiler (34.4 g CO_2e kg^{-1} of burned material) [37]. The parameter $E^{WC-Tran}$ reflects GHG emissions related to transportation of wood chips from a harvest site to a nearby power plant. It was a product of GHG emission factor (0.133 kg CO_2e Mg^{-1} km^{-1}) [36], total wood chips transported, and average distance traveled (100 km one way). The parameters $E^{WC-Burn}$ reflects non-biogenic GHG emissions related with burning of wood chips in a boiler (34.4 g CO_2e kg^{-1} of burned material) [37].

GHG Intensity of Third Energy Pathway

Ethanol yield from a metric ton of bone dry feedstock was 329.6 l [38]. The conversion technology was assumed as dilute acid-

pretreatment of feedstock followed by enzymatic hydrolysis [38]. The value of co-generated electricity at the time of ethanol production was 0.48 kWh l^{-1} of ethanol [38]. We multiplied total available biomass (B^{green}) with the half of ethanol yield to estimate total ethanol availability (EE in l ha^{-1}). We used Equation (7) to estimate the GHG intensity ($GHGI^{EtOH-NoCredits}$ in g CO_2e MJ^{-1}) of ethanol.

$$GHGI_{h,f,i}^{EtOH-Credits} = \frac{\left(E_{h,f,i}^{Bio\,Pro} + E_{h,f,i}^{Chipping} + E_{h,f,i}^{WC-Tran} + E_{h,f,i}^{EtOH-Credits}\right)}{EE_{h,f,i} \times 21.3} \quad (7)$$

where, $E^{EtOH-Credits}$ refers to GHG emissions related to conversion of biomass into ethanol and transporting it to a nearby pump station (50 km one side). We obtained this value (–106.5 g CO_2e l^{-1} or –5.0 g CO_2e MJ^{-1} of ethanol produced) from the GREET model after updating default values of ethanol yield and co-generated electricity with values used in this study [35]. Calorific value of ethanol was 21.3 MJ l^{-1} [35].

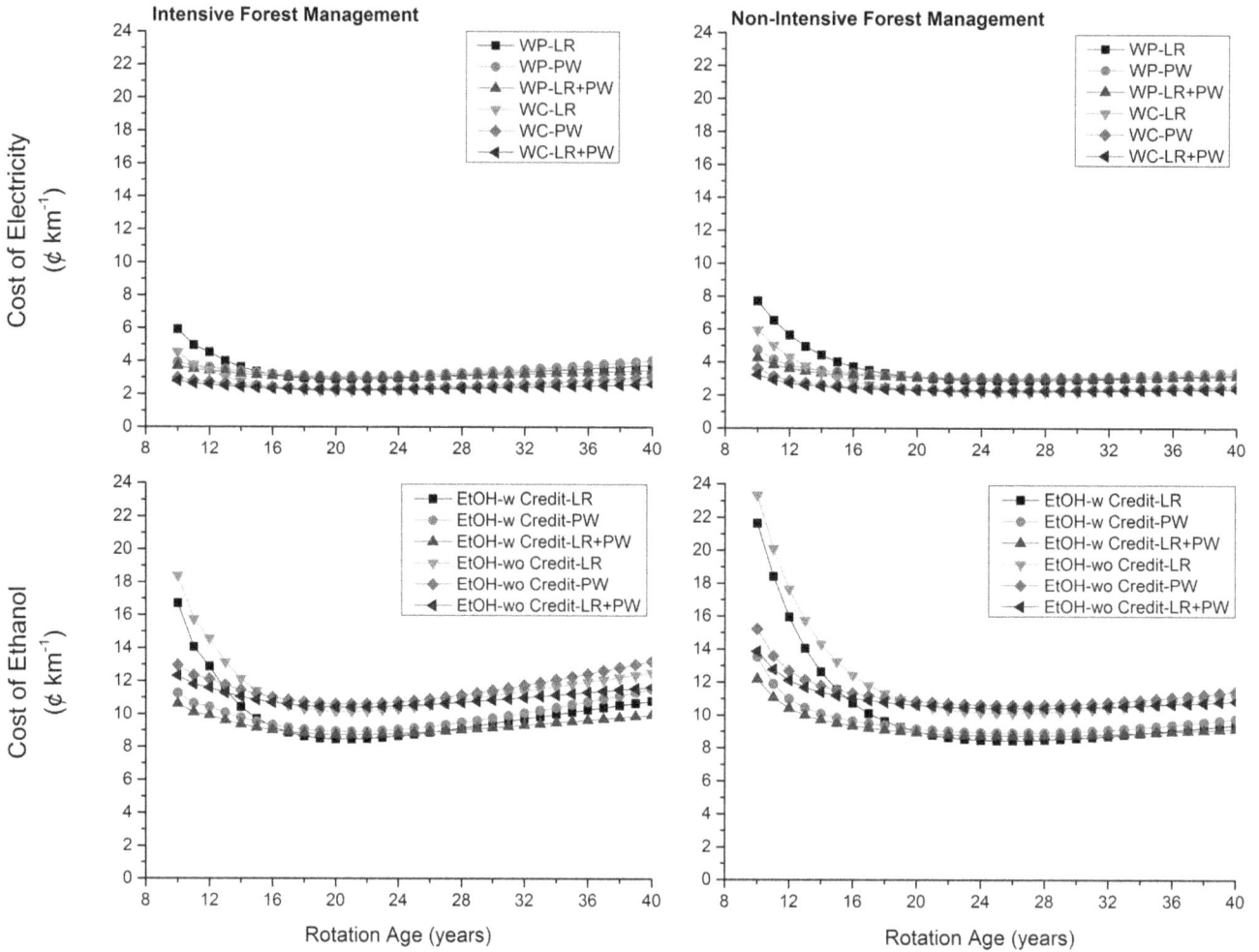

Figure 5. Cost of energy products at the consumption level. LR: logging residues; PW: pulpwood; WP: wood pellets; WC: wood chips; w: with income from cogenerated electricity; wo: without income from cogenerated electricity.

GHG Intensity of Fourth Energy Pathway

We used Equation (8) to estimate the GHG intensity ($GHGI^{EtOH-NoCredits}$ in g CO_2e MJ^{-1}) of ethanol.

$$GHGI^{EtOH-NoCredits}_{h,f,i} = \frac{\left(E^{Bio\,Pro}_{h,f,i} + E^{Chipping}_{h,f,i} + E^{WC-Tran}_{h,f,i} + E^{EtOH-NoCredits}_{h,f,i}\right)}{EE_{h,f,i} \times 21.3} \quad (8)$$

where, $E^{EtOH-NoCredits}$ refers to GHG emissions related to conversion of biomass into ethanol and transporting it to a nearby pump station (50 km one way). We obtained this value (191.7 g CO_2e l^{-1} or 9.0 g CO_2e MJ^{-1} of ethanol produced) from the GREET after updating default values of ethanol yield and co-generated electricity with values used in this study [35].

Unit Cost Estimation

We calculated land expectation value (LEV in \$ ha^{-1}) at different harvest ages under intensive forest management using Equation (9). The LEV is defined as the net present value of bare forestland over infinite forest rotations [39]. We used parameters given in Table S1 in File S1for calculating LEVs.

$$LEV_h =$$

$$\frac{(p^{st} \times Q^{st} + p^{cns} \times Q^{cns} + p^{pw} \times Q^{pw} + p^{lr} \times Q^{lr}) \times e^{-r \times h} - (T+M) \times \left(\frac{1-e^{-r \times h}}{r}\right)}{1-e^{-r \times h}}$$

$$\frac{- F^{@2year}_{h \geq 2years} \times e^{-2 \times r} - F^{@12year}_{h \geq 12year} \times e^{-12 \times r} - C}{1-e^{-r \times h}} \quad (9)$$

where, p^{st}, p^{cs}, p^{pw}, and p^{lr} represent prices of sawtimber, chip-n-saw, pulpwood, and logging residues, respectively. Parameters Q^{st}, Q^{cs}, Q^{pw}, and Q^{lr} represent quantities of sawtimber, chip-n-saw, pulpwood, and logging residues available at a given harvest age, respectively. Parameters C, T, and M represent site preparation cost, annual taxes, and annual cost of plantation management, respectively. Parameter F represents cost of fertilizers applied at the 2nd and 12th year of plantation. Parameter r stands for the real discount rate (4%). We selected the highest LEV out of all LEVs and declared the corresponding harvest age as the optimal rotation age. Then, we subtracted LEVs for different harvest ages from the LEV at the optimal rotation age to determine the opportunity cost of changing harvest age. We made suitable changes in Equation (9) to ascertain LEVs at different harvest ages for non-intensive forest management. We have not considered the income obtained from logging residues while calculating LEVs for intensive and non-intensive forest management choices when they were not used as a

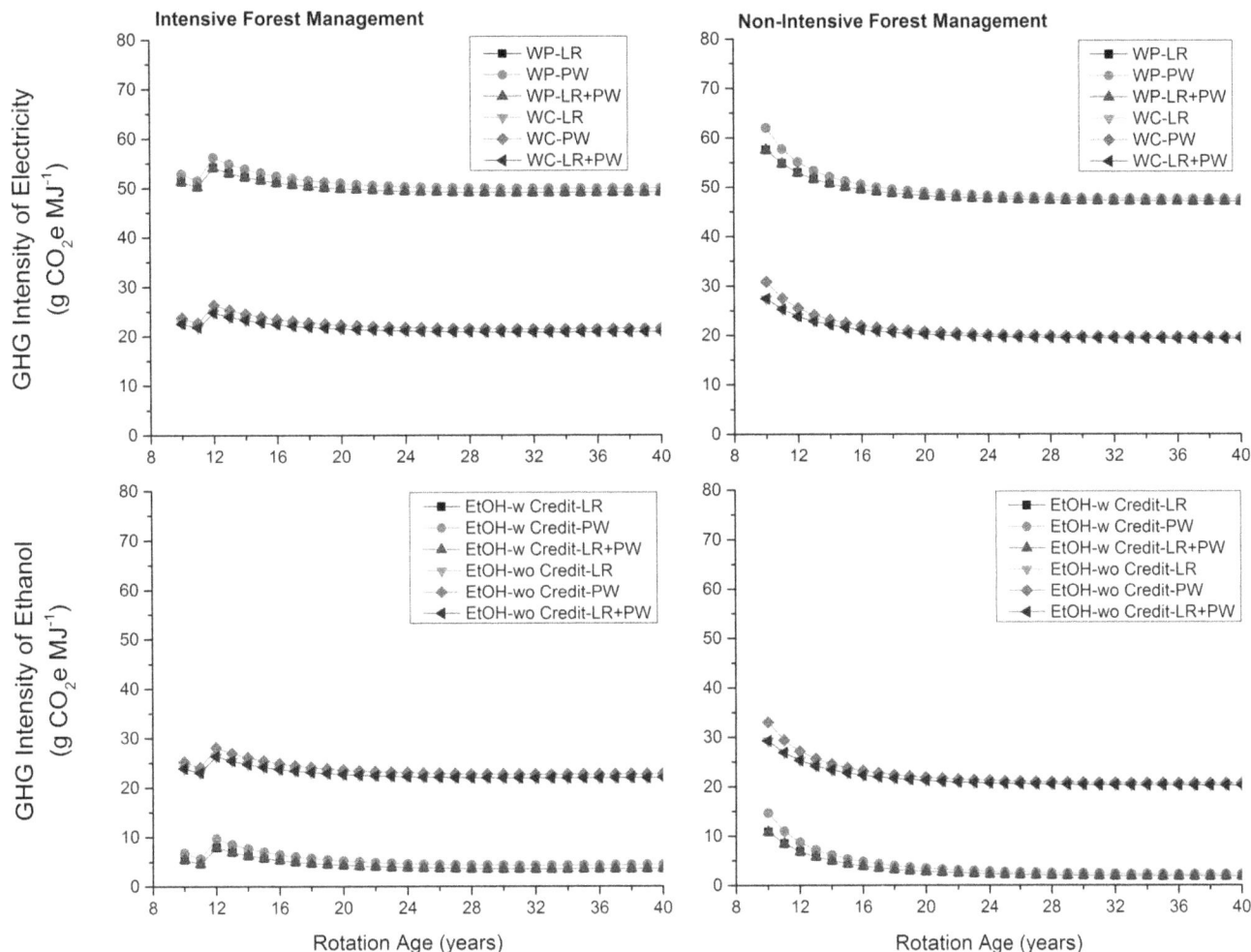

Figure 6. GHG intensity of energy products at the production level. LR: logging residues; PW: pulpwood; WP: wood pellets; WC: wood chips; w: with income from cogenerated electricity; wo: without income from cogenerated electricity.

feedstock. Similarly, we have not allocated any GHG emissions related to biomass production to logging residues when they were not used as a feedstock. Parameters reported in Table S2 in File S1 were used for ascertaining production cost of a MJ of generated electricity and produced ethanol.

Abatement Cost

We used Equation (10) to estimate the abatement cost of a metric ton of GHG emission for both bioenergy products.

$$\text{Abatement Cost} = \frac{\text{Unit cost of bioenergy product} - \text{Unit cost of fossil} - \text{based energy product}}{\text{GHG intensity of fossil} - \text{based energy product} - \text{GHG intensity of bioenergy product}} \quad (10)$$

The units of numerator and denominator portions of the above equation were ¢ km^{-1} and g CO$_2$e km^{-1}, respectively. The fuel economies of an electric and flex-fuel vehicles were taken as 1.4 km MJ^{-1} and 0.35 km MJ^{-1}, respectively [26]. The levelized unit production cost of electricity generated from coal and natural gas was taken as 2.78 ¢ MJ^{-1} and 1.87 ¢ MJ^{-1}, respectively [40]. Levelized electricity generation costs for electricity derived from biomass, coal, and natural gas are based on new generation

sources for 2018 expressed in 2011 dollars. The wholesale price of gasoline was taken as 2.56 ¢ MJ^{-1} [41]. The GHG intensity of electricity generated from coal and natural gas was taken as 343.1 g CO$_2$e MJ^{-1} and 178.61 g CO$_2$e MJ^{-1}, respectively [20]. The GHG intensity of gasoline was taken as 94 g CO$_2$e MJ^{-1} [35].

Results

The availability of large-diameter timber products (sawtimber and chip-n-saw) was smaller at initial harvest ages relative to small-diameter timber products (pulpwood and logging residues). However, availability of large-diameter timber products increased as trees gained girth and height with time (Figure 1). The availability of logging residues was maximum at harvest ages 33 (84.2 Mg ha^{-1}) and 39 (72.4 Mg ha^{-1}) years for intensive and non-intensive forest management, respectively. The availability of pulpwood was highest at harvest ages 13 (121.4 Mg ha^{-1}) and 18 (124.8 Mg ha^{-1}) years for intensive and non-intensive forest management, respectively. The combined availability of pulpwood and logging residues reached to a maximum value of 179.6 and 172.8 Mg ha^{-1} at plantation ages 21 and 22 years under intensive and non-intensive forest management scenarios, respectively.

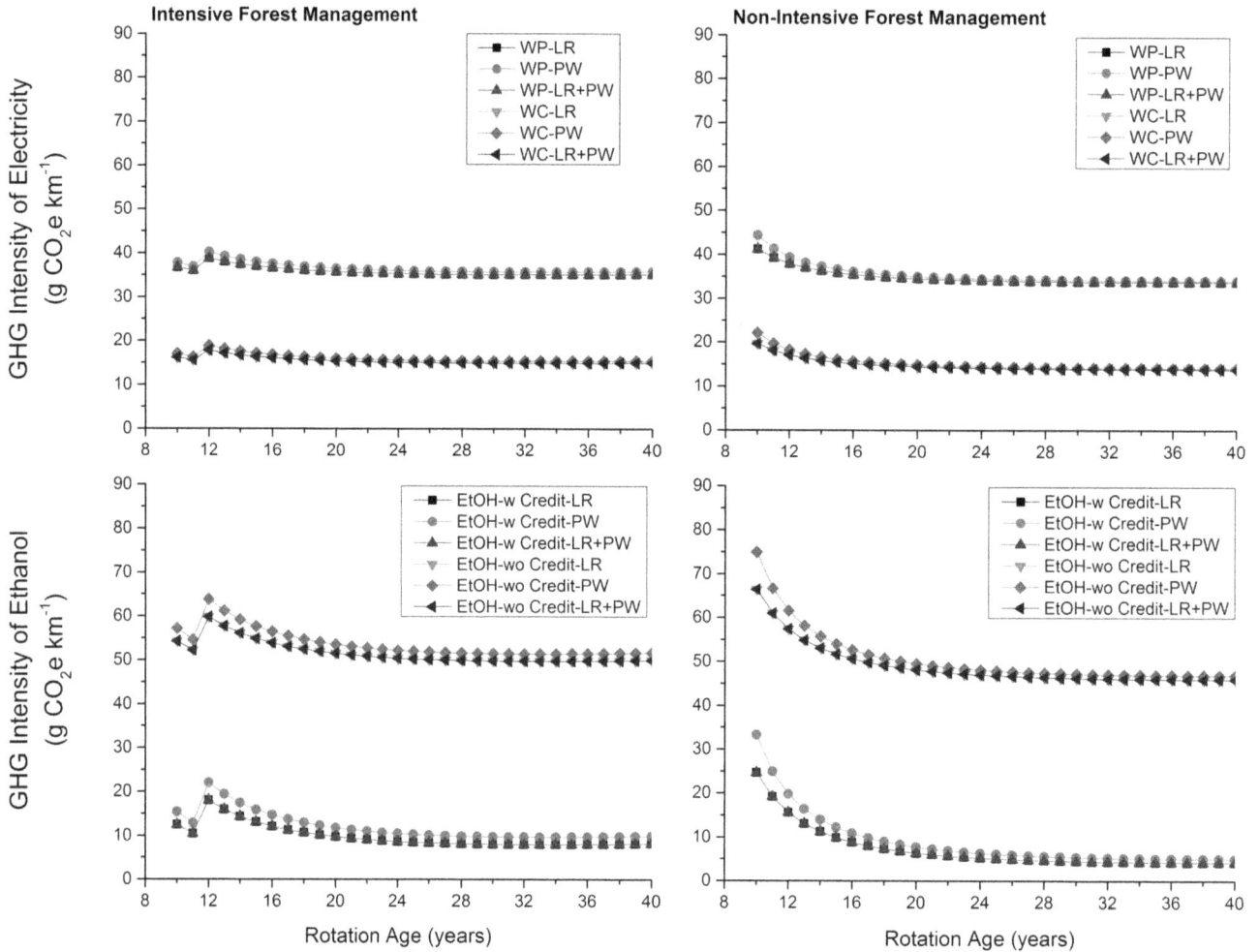

Figure 7. GHG intensity of energy products at the consumption level. LR: logging residues; PW: pulpwood; WP: wood pellets; WC: wood chips; w: with income from cogenerated electricity; wo: without income from cogenerated electricity.

Total availability of logging residues was always higher under intensive than non-intensive forest management at all harvest ages. Same case was observed with the combined availability of logging residues and pulpwood. However, total availability of pulpwood was only higher under intensive than non-intensive forest management when harvest age was lower than 16 years.

Under intensive and non-intensive forest management choices, LEVs were highest at 21st and 26th year of plantation, respectively (Figure 2). Thus, optimal rotation ages for intensive and non-intensive forest management choices were 21 and 26 years, respectively. Additional income from logging residues increased the LEV by 15 and 28 percentage points for intensive and non-intensive forest management choices at optimal rotation ages, respectively. As expected, opportunity cost increased with an increase or a decrease in the harvest age from the optimal rotation age. Quantities of total electricity generated and ethanol produced were proportional to the feedstock availability (Figure 3).

The cost of electricity generated from wood pellets was consistently higher (about 1.0 to 2.5 ¢ MJ^{-1}) than the cost of electricity generated using wood chips across same feedstocks mostly due to higher production and transportation costs of wood pellets (Figure 4). The cost of ethanol produced without any income from co-generated electricity was higher by 0.7 ¢ MJ^{-1} than the cost of ethanol produced with income from co-generated

electricity across same feedstocks. Across energy pathways, the cost of per MJ of energy obtained in the form of ethanol without any income from co-generated electricity was highest followed by electricity from wood pellets, ethanol with income from co-generated electricity, and electricity from wood chips. Unit production costs were comparable across feedstocks and choice of forest management especially after 12th year of plantation. At the consumption level, the cost of a km traveled using electricity produced with wood pellets was higher than that of a km traveled with electricity generated from wood chips (0.7 to 1.8 ¢ km^{-1}) across feedstocks (Figure 5). The cost of a km with ethanol produced in the presence of income from co-generated electricity was lower than the cost of a km with ethanol produced in the absence of income from co-generated electricity by 1.7 ¢ km^{-1}. A comparison across energy pathways revealed that a km of travel was much cheaper for an electric vehicle than a flex-fuel vehicle ranging from 5.6 ¢ km^{-1} and 17.4 ¢ km^{-1} depending upon whether wood pellets or wood chips were used for electricity generation (Table 1). This was mostly due to high fuel economy of electric vehicles than flex fuel vehicles.

The GHG intensity of electricity generated from wood pellets was highest whereas the GHG intensity of ethanol produced in presence of GHG credits due to supply of co-generated electricity to the grid was lowest at the production level (Figure 6). The GHG

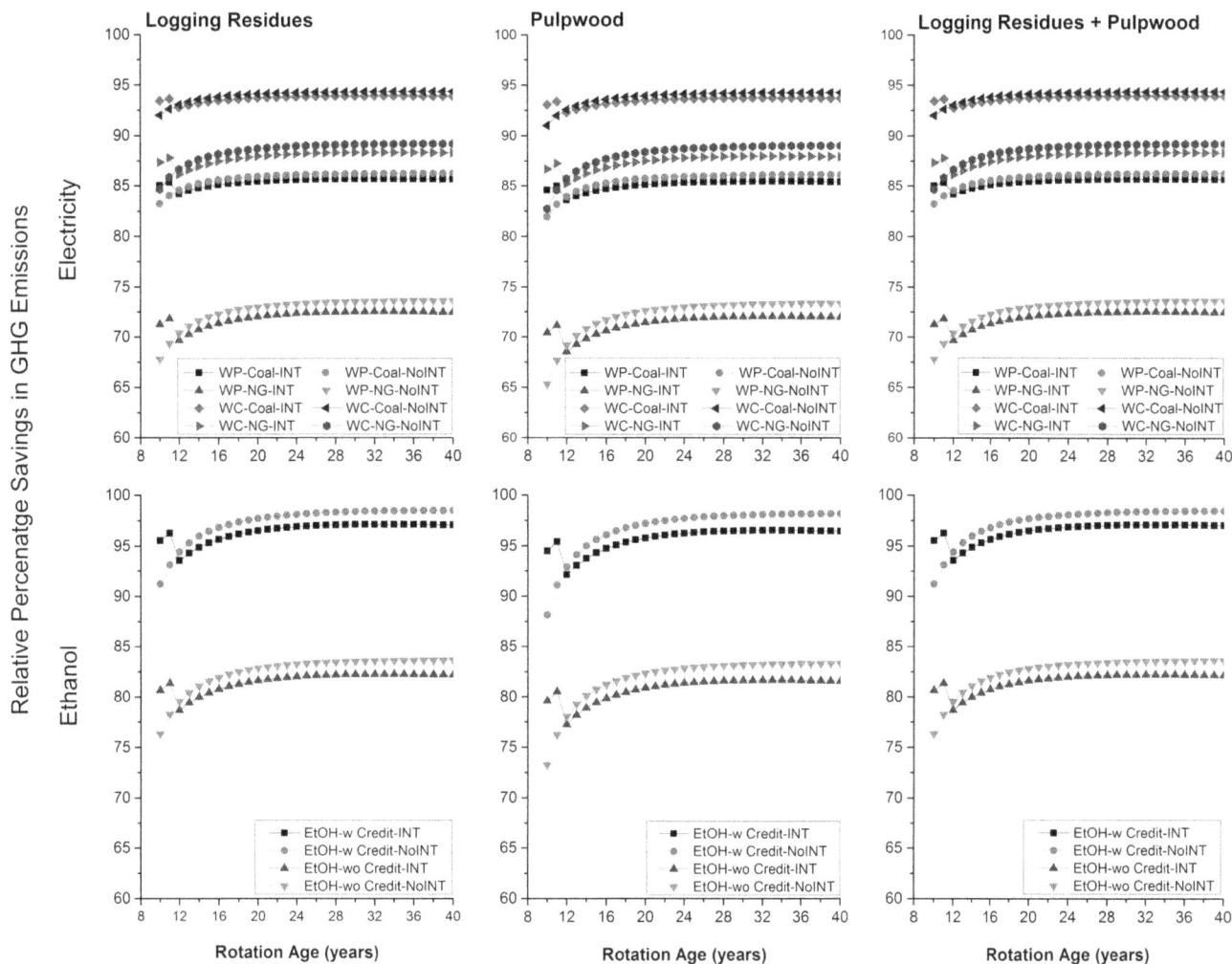

Figure 8. Relative percentage savings in GHG emissions. LR: logging residues; PW: pulpwood; WP: wood pellets; WC: wood chips; w: with income from cogenerated electricity; wo: without income from cogenerated electricity.

intensities of electricity generated from wood chips and ethanol produced in absence of any GHG credits were comparable at the production level (Table 1). At the consumption level, the GHG intensity of ethanol produced in the absence of any GHG credits was highest followed by electricity generated using wood pellets, electricity generated from wood chips, and ethanol produced in the presence of GHG credits (Figure 7). Percentage savings in GHG emissions relative to the electricity generated from coal and natural gas on per km traveled across feedstocks remained almost same (Figure 8). This was also the case for the produced ethanol. For generated electricity, relative percentage savings were higher (about 8% and 15% relative to coal and natural gas, respectively) when wood chips were used as a feedstock than wood pellets. Similarly, relative percentage savings were higher (about 15%) when GHG credits from co-generated electricity were considered. Across forest management choices, percentage savings in GHG emissions for non-intensive than intensive forest management were higher by about 2% only.

For generated electricity and produced ethanol, the abatement cost of GHG emissions did not vary much across feedstocks (Figure 9). Based on lowest abatement cost, a minimum carbon tax of $ 7.7 Mg CO_2e^{-1} or $ 73 Mg CO_2e^{-1} would be required to promote production of electricity from wood chips with respect to

electricity generated using coal and natural gas, respectively (Table 2). A minimum carbon tax of $ 42.5 Mg CO_2e^{-1} or $ 165 Mg CO_2e^{-1} would be required to promote production of electricity from wood pellets with respect to electricity generated using coal and natural gas, respectively. Similarly, a minimum carbon tax of $ 31 Mg CO_2e^{-1} or $ 108 Mg CO_2e^{-1} would be required to promote wood-based ethanol depending upon whether or not income and GHG credits from co-generated electricity at the time of ethanol production were considered. The abatement cost was higher under non-intensive than intensive forest management before harvest age of 24 years but for harvest ages 24 years and greater, the abatement cost was higher under intensive than non-intensive forest management. For generated electricity, the abatement cost was at least $ 34.8 Mg CO_2e^{-1} and $ 92.3 Mg CO_2e^{-1} less when wood chips were used as a fuel than wood pellets with respect to electricity generated using coal and natural gas, respectively. Relative abatement cost was at least $ 70 Mg CO_2e^{-1} less for ethanol produced in presence of income and GHG credits due to co-generated electricity than in absence of them.

Table 2. Range of abating GHG emissions for selected energy pathways.

		Abatement Cost ($Mg⁻¹ CO2e) (electricity relative to coal)						Abatement Cost ($Mg⁻¹ CO2e) (electricity relative to natural gas)						Abatement Cost ($Mg⁻¹ CO2e) (ethanol relative to gasoline)					
		Intensive			Non-Intensive			Intensive			Non-Intensive			Intensive			Non-Intensive		
		LR	PW	LR+PW	LR	PW	LR+PW	LR	PW	LR+PW	LR	PW	LR+PW	LR	PW	LR+PW	LR	PW	LR+PW
Electricity from Wood Pellets	Min	42.8	51.1	47.6	42.5	50.6	47.5	168.3	188.3	179.3	165.4	184.3	176.7	-	-	-	-	-	-
	Max	187.8	96.4	81.1	280.5	137.0	111.0	339.5	144.6	112.6	549.0	239.4	179.2	-	-	-	-	-	-
Electricity from Wood Chips	Min	7.7	13.5	11.1	7.7	13.4	11.3	73.9	86.1	80.9	73.0	85.0	80.4	-	-	-	-	-	-
	Max	110.4	45.8	34.8	174.5	72.7	55.1	285.3	151.9	130.0	424.9	215.6	175.5	-	-	-	-	-	-
Ethanol with co-generated electricity	Min	-	-	-	-	-	-	-	-	-	-	-	-	31.1	48.7	41.3	30.6	47.7	41.2
	Max	-	-	-	-	-	-	-	-	-	-	-	-	339.5	144.6	112.6	549.0	239.4	179.2
Ethanol without co-generated electricity	Min	-	-	-	-	-	-	-	-	-	-	-	-	110.3	131.8	122.5	108.2	128.8	120.7
	Max	-	-	-	-	-	-	-	-	-	-	-	-	476.8	244.8	207.9	735.0	370.3	293.0

LR: logging residues; PW: pulpwood; Intensive: intensive forest management; Non-intensive: non-intensive forest management.

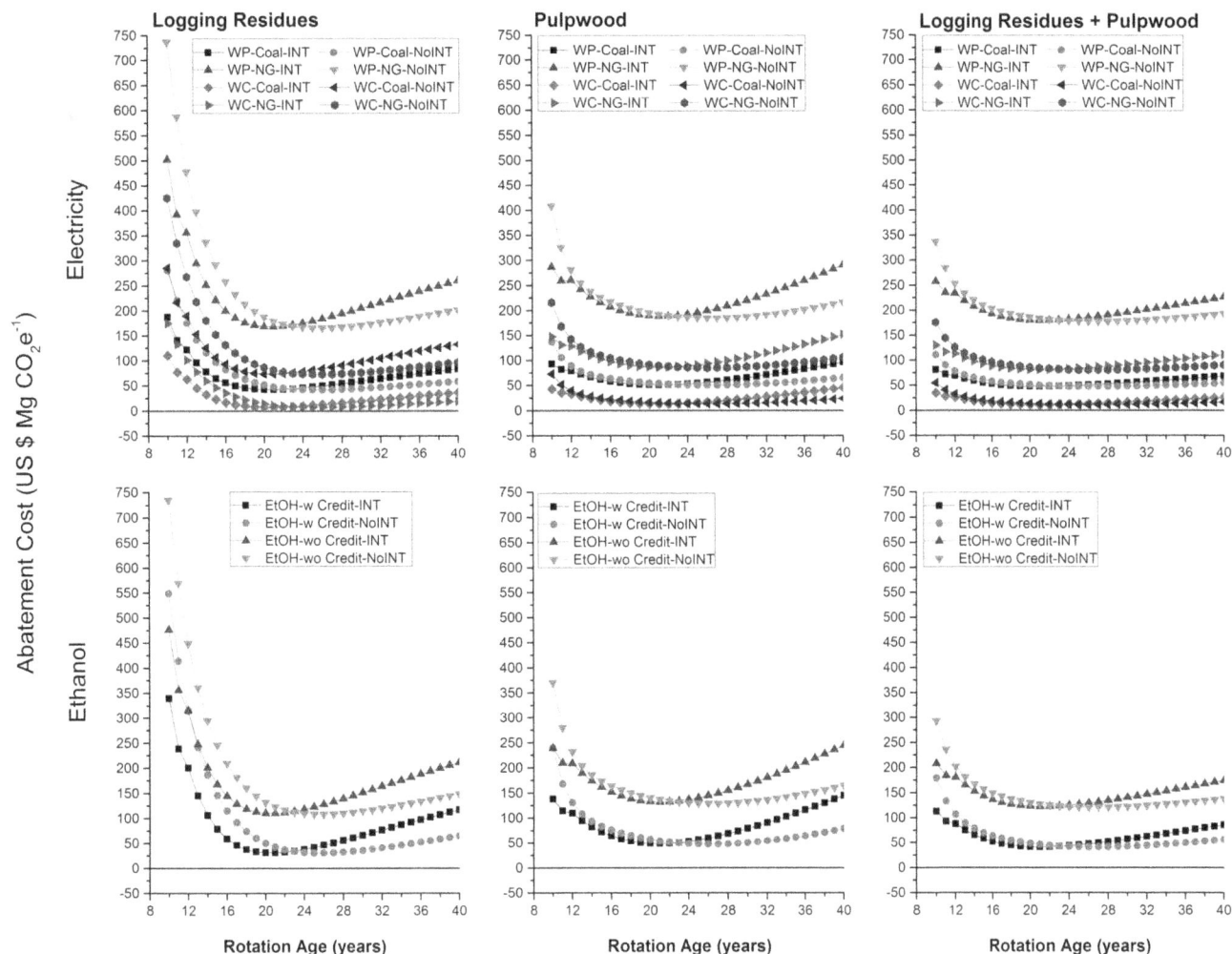

Figure 9. Abatement cost of GHG emissions with respect to corresponding fossil fuel-based energy products. LR: logging residues; PW: pulpwood; WP: wood pellets; WC: wood chips; w: with income from cogenerated electricity; wo: without income from cogenerated electricity.

Discussion and Conclusions

The use of wood chips instead of wood pellets for electricity generation was a better option both in terms of unit cost and environmental performance in the US. This was mostly due to additional costs and GHG emissions related with the production and transportation of wood pellets. An abatement cost of electricity generated using woody feedstocks varied decisively depending upon the selected baseline of electricity generated from fossil fuels. Cost of abating GHG emissions by electricity produced from either wood pellets or wood chips was much lower when it replaces coal-based electricity than natural gas-based electricity. Income and GHG credits accrued due to the supply of co-generated electricity at the time of ethanol production played a critical role in determining unit cost and GHG intensity of produced ethanol. This implies that industrial operations at an ethanol mill should be optimized so that a certain portion of co-generated electricity is supplied to the grid to earn extra income and GHG credits.

Cost of driving a km of an electric vehicle using electricity generated from wood chips was cheaper than a comparative flex-fuel vehicle utilizing ethanol derived from same woody feedstocks. Similarly, the GHG intensity of covering a km of distance by an electric vehicle was less than a comparative flex-fuel vehicle running on ethanol derived in absence of any co-generated electricity. The GHG intensity was higher for a km of distance covered by electric vehicle utilizing electricity generated using wood chips or wood pellets than a km of distance covered by flex-fuel vehicle using ethanol produced in presence of GHG credits related to co-generated electricity. Overall this implies that use of an electric vehicle running on electricity derived from wood chips should be preferred for simultaneously maximizing environmental and economic efficiencies. However, the abatement cost of doing so could range from \$7 to \$425 Mg CO_2e^{-1} depending upon the selected baseline of electricity generated from fossil fuels and the harvest age. We also found that the minimum abatement cost of GHG emissions for electricity derived from wood pellets (with respect to coal-based electricity) and ethanol derived in presence of co-generated electricity were close to each other especially for rotation ages which were near to optimal rotation ages.

The opportunity cost related with a change in rotation age from the optimal rotation age was a significant determinant of unit production cost of wood-based energy products. A departure from the optimal rotation age increased the unit production cost of wood-based energy products implying that a significant change in rotation age from current rotation ages would increase the prices of wood-based energy products. The unit production cost and environmental performance of wood-based energy products did not vary across feedstocks. Therefore, logging residues and pulpwood can be used as individual feedstocks on their own for manufacturing of wood-based energy products. However, it is preferable to use both pulpwood and logging residues as a single feedstock from the perspective of land-use efficiency [21]. Relative savings in GHG emissions were about 2% higher under non-intensive than intensive forest management starting from 12th year of plantation age implying that feedstocks derived from both intensive and non-intensive forest management could be used for wood-based bioenergy development without any significant drop in relative savings of GHG emissions.

This study suggests that the GHG intensity of wood-based energy products is less than the GHG intensity of corresponding fossil-fuel energy products. However, the unit production cost of wood-based energy products is higher than the corresponding fossil-fuel energy products depending upon the harvest age. This implies that financial support is required to promote production of wood-based energy products. This financial support could be in the form carbon tax on corresponding fossil fuel-based energy products. Other mechanism like subsidies/carbon markets should also be explored.

We have not considered carbon sequestered in soils in this study as carbon sequestered on reforested lands remain very stable with respect to time [42]. We have not considered carbon sequestered in other pools (live trees, dead trees, debris, and coarse roots) as well. We acknowledge this as a limitation of the existing study as a change in the rotation age will affect both these carbon pools with respect to time. A need exists to integrate the model developed in

this study with national or regional economic-wide equilibrium models [43,44] to assess the price dynamics of energy products derived from woody feedstocks with respect to the corresponding fossil-fuel energy products. This will also give an estimate of an opportunity cost related to diversion of pulpwood for bioenergy development than paper-based products. Moreover, we have primarily focused on variability in availability of feedstocks in this study. A need exists to capture variability on production technologies as well including other energy products like biodiesel and vehicle types. Finally, we have not considered biogenic emissions due to consumption of wood-based energy products as quantities of carbon at the landscape level under continuous forestry assumption does not change over time. We hope that this study will significantly benefit future research exploring carbon benefits of bioenergy development in the US and beyond. We are also hopeful that this study will provide policy makers an understanding about possible pathways and potential incentives needed to promote bioenergy development in the US.

Supporting Information

File S1   Supporting tables.

Author Contributions

Conceived and designed the experiments: PD MK. Performed the experiments: PD. Analyzed the data: PD. Contributed reagents/materials/analysis tools: PD. Wrote the paper: PD MK.

References

1. USEPA (2013) Inventory of U.S. Greenhouse Gas Emissions and Sinks: 1990–2011. United States Environment Protection Agency. Washington, DC.
2. DSIRE (2013) Database of State Incentives for Renewable & Efficiency. United States Department of Energy/North Carolina Solar Center. Raleigh, NC.
3. USEIA (2012) Annual Energy Outlook 2012 with Projections to 2035. United States Energy Information Administration. Washington, DC.
4. US Congress (2007) The Energy Independence and Security Act (EISA) of 2007.
5. ORNL (2011) U.S. Billion-Ton Update: Biomass Supply for a Bioenergy and Bioproducts Industry. Oak Ridge National Laboratory. Oak Ridge, TN.
6. Dwivedi P, Bailis R, Bush TG, Marinescu M (2011) Quantifying GWI of wood pellet production in the southern United States and its subsequent utilization for electricity production in the Netherlands/Florida. BioEnergy Res 4: 180–192.
7. Nuss P, Gardner KH, Jambeck JR (2013) Comparative life cycle assessment (LCA) of construction and demolition (C&D) derived biomass and U.S. Northeast forest residuals gasification for electricity production. Environ Sci Technol 47: 3463–3471. Available: http://dx.doi.org/10.1021/es304312f.
8. Heller MC, Keoleian GA, Mann MK, Volk TA (2004) Life cycle energy and environmental benefits of generating electricity from willow biomass. Renew Energy 29: 1023–1042. Available: http://www.sciencedirect.com/science/article/pii/S0960148103003914.
9. Robinson AL, Rhodes JS, Keith DW (2003) Assessment of potential carbon dioxide reductions due to biomass−coal cofiring in the United States. Environ Sci Technol 37: 5081–5089. Available: http://dx.doi.org/10.1021/es034367q.
10. Mann M, Spath P (2001) A life cycle assessment of biomass cofiring in a coal-fired power plant. Clean Prod Process 3: 81–91.
11. Tharakan PJ, Volk TA, Lindsey CA, Abrahamson LP, White EH (2005) Evaluating the impact of three incentive programs on the economics of cofiring willow biomass with coal in New York State. Energy Policy 33: 337–347. Available: http://www.sciencedirect.com/science/article/pii/S0301421503002453.
12. Pirraglia A, Gonzalez R, Denig J, Saloni D (2013) Technical and economic modeling for the production of torrefied lignocellulosic biomass for the U.S. densified fuel industry. BioEnergy Res 6: 263–275. Available: http://dx.doi.org/10.1007/s12155-012-9255-6.
13. Dwivedi P, Bailis R, Alavalapati JRR, Nesbit T (2012) Global warming impact of E85 fuel derived from forest biomass: A case study from Southern USA. BioEnergy Res 5: 470–480.
14. Nesbit T, Alavalapati J, Dwivedi P, Marinescu M (2011) Economics of ethanol production using feedstock from slash pine (Pinus elliottii) plantations in the southern United States. South J Appl For 35: 61–66.
15. Kadam KL, Wooley RJ, Aden A, Nguyen QA, Yancey MA, et al. (2000) Softwood forest thinnings as a biomass source for ethanol production: A

feasibility study for California. Biotechnol Prog 16: 947–957. Available: http://dx.doi.org/10.1021/bp000127s.
16. Budsberg E, Rastogi M, Puettmann M, Caputo J, Balogh S, et al. (2012) Life-cycle assessment for the production of bioethanol from willow biomass crops via biochemical conversion. For Prod J 62: 305–313.
17. Daystar J, Reeb C, Venditti R, Gonzalez R, Puettmann M (2012) Life-cycle assessment of bioethanol from pine residues via indirect biomass gasification to mixed alcohols. For Prod J 62: 314–325.
18. Gonzalez R, Treasure T, Phillips R, Jameel H, Saloni D, et al. (2011) Converting Eucalyptus biomass into ethanol: Financial and sensitivity analysis in a co-current dilute acid process. Part II. Biomass and Bioenergy 35: 767–772. Available: http://www.sciencedirect.com/science/article/pii/S0961953410003776.
19. Wang ZJ, Zhu JY, Zalesny RS Jr, Chen KF (2012) Ethanol production from poplar wood through enzymatic saccharification and fermentation by dilute acid and SPORL pretreatments. Fuel 95: 606–614. Available: http://www.sciencedirect.com/science/article/pii/S0016236111008015.
20. Lemoine D, Plevin R, Cohn A, Jones A, Brandt A, et al. (2010) The climate impacts of bioenergy systems on market and regulatory policy contexts. Environ Sci Technol 44: 7347–7350.
21. Campbell JE, Lobell DB, Field CB (2009) Greater transportation energy and GHG offsets from bioelectricity than ethanol. Science 324: 1055–1057. Available: http://www.ncbi.nlm.nih.gov/pubmed/19423776. Accessed 8 November 2013.
22. Wang M, Elgowainy A (2013) Life-cycle analysis of biofuels and electricity for transportation use. In: Jawahir I, Sikdar S, Hunag Y, editors. Treatise on Sustainability Science and Engineering. Springer. 231–257.
23. Peterson SB, Whitacre JF, Apt J (2010) The economics of using plug-in hybrid electric vehicle battery packs for grid storage. J Power Sources 195: 2377–2384. Available: http://www.sciencedirect.com/science/article/pii/S0378775309017303.
24. Farrell DML, DMK AE (2008) An innovation and policy agenda for commercially competitive plug-in hybrid electric vehicles. Environ Res Lett 3: 14003. Available: http://stacks.iop.org/1748-9326/3/i=1/a=014003.
25. Valatin G (2012) Marginal Abatement Cost Curves for UK Forestry. Edinburgh, United Kingdom.
26. EERE (2013) Alternative Fuels Data Center - Vehicle Cost Calculator. Energy Efficiency and Renewable Energy, United States Department of Energy. Washington, DC. Available: http://www.afdc.energy.gov/calc/.
27. Spelter H, Toth D (2009) North America's Wood Pellet Sector. Forests Product Laboratory. Madison, WI.
28. Smith W, Miles P, Perry C, Pugh S (2009) Forest Resources of the United States, 2007: A Technical Document Supporting the Forest Service 2010 RPA

Assessment. United States Department of Agriculture Forest Forest Service. Washington, DC.

29. Yin R, Pienaar L, Aronow M (1998) The productivity and profitability of fiber farming. J For 96: 13–18.

30. Jenkins J, Chojnacky D, Heath L, Birdsey R (2003) National scale biomass estimators for United States tree species. For Sci 49: 12–35.

31. Bridgwater AV, Toft AJ, Brammer JG (2002) A techno-economic comparison of power production by biomass fast pyrolysis with gasification and combustion. Renew Sustain Energy Rev 6: 181–246. Available: http://linkinghub.elsevier.com/retrieve/pii/S1364032101000107.

32. USEIA (2012) How much electricity is lost in transmission and distribution in the United States? United States Energy Iinformation Administration. Washington, DC. Available: http://www.eia.gov/tools/faqs/faq.cfm?id = 105&t = 3.

33. American_Renewables (2013) Gainesville Renewable Energy Center (GREC). Available: Gainesville Renewable Energy Center (GREC). Accessed 12 December 2013.

34. Lundgren K, Morales A (2012) Biggest English Polluter Spends $1 Billion to Burn Wood. www.bloomberg.com. Available: http://www.bloomberg.com/news/2012-09-25/biggest-english-polluter-spends-1-billion-to-burn-wood-energy.html.

35. Wang M (2001) Development and Use of GREET 1.6 Fuel-Cycle Model for Transportation Fuels and Vehicle Technologies. Argonne, IL.

36. PRé-Consultants (2013) US LCI Database Simapro LCA Software.

37. WDNR (2010) Forest Biomass and Air Emissions. Washington Department of Natural Resources. Olympia, WA.

38. Humbird D, Davis R, Tao L, Kinchin C, Hsu D, et al. (2011) Process Design and Economics for Biochemical Conversion of Lignocellulosic Biomass to Ethanol: Dilute-Acid Pretreatment and Enzymatic Hydrolysis of Corn Stover. National Renewable Energy Laboratory. Golden, CO.

39. Dwivedi P, Alavalapati JRR, Susaeta A, Stainback A (2009) Impact of carbon value on the profitability of slash pine plantations in the southern United States: an integrated life cycle and Faustmann analysis. Can J For Res 39: 990–1000. Available: http://www.nrcresearchpress.com/doi/abs/10.1139/X09-023. Accessed 25 November 2013.

40. USEIA (2012) Levelized Cost of New Generation Resoruces. United States Energy Information Adminstration. Washington, DC. Available: http://www.eia.gov/forecasts/aeo/er/electricity_generation.cfm.

41. USEIA (2013) Refiner Gasoline Prices by Grade and Sales Type. United States Energy Information Administration. Washington, DC.

42. Davis SC, Dietze M, DeLucia E, Field C, Hamburg SP, et al. (2012) Harvesting carbon from eastern US Forests: Opportunities and impacts of an expanding bioenergy industry. Forests 3: 370–397. Available: http://www.mdpi.com/1999-4907/3/2/370.

43. Abt KL, Abt RC, Galik C (2012) Effect of bioenergy demands and supply response on markets, carbon, and land use. For Sci 58: 523–539.

44. Daigneault A, Sohngen B, Sedjo R (2012) Economic approach to assess the forest carbon implications of biomass energy. Environ Sci Technol 46: 5664–5671. Available: http://www.ncbi.nlm.nih.gov/pubmed/22515911.

12

A Rare *Phaeodactylum tricornutum* Cruciform Morphotype: Culture Conditions, Transformation and Unique Fatty Acid Characteristics

Liyan He[1,2], **Xiaotian Han**[1], **Zhiming Yu**[1]*

1 Key Laboratory of Marine Ecology and Environmental Sciences, Institute of Oceanology, Chinese Academy of Sciences, Qingdao, China, **2** University of Chinese Academy of Sciences, Beijing, China

Abstract

A rare *Phaeodactylum tricornutum* cruciform morphotype was obtained and stabilized with a proportion of more than 31.3% in L1 medium and is reported for the first time. Long-term culture and observation showed that the cruciform morphotype was capable of transforming to the oval form following the degeneration of arms by two processes. After three months of culture, four morphotypes existed in a relatively stable proportion in culture for six months (10.5% for oval, 11.3% for fusiform, 37.2% for triradiate and 41.0% for cruciform). Low temperature was particularly beneficial for cruciform cell formation. As the culture temperature decreased from 25°C to 10°C, the percentage of the cruciform morphotype increased from 39.1% to 55.3% approximately. The abundant cruciform cells endowed this strain with unique fatty acid characteristics. The strain cultured at 15°C showed both maximum content of neutral lipid in a single cell and total yield. The maximum content of fatty acid methyl esters was C16:1 for *Phaeodactylum tricornutum* cultured at four temperatures (43.82% to 50.82%), followed by C16:0 (20.47% to 22.65%). Unique fatty acid composition endowed this strain with excellent quality for biodiesel production.

Editor: Hector Escriva, Laboratoire Arago, France

Funding: This work was financially supported by the National Natural Science Foundation of China (2011CB2009001), the State Oceanic Administration Project (GHME2001SW02) and the Funds for Creative Research Groups of China (41121064). The funders had no role in study design, data collection and analysis, decision to publish, or preparation of the manuscript.

Competing Interests: The authors have declared that no competing interests exist.

* E-mail: zyu@qdio.ac.cn

Introduction

Phaeodactylum tricornutum has been widely studied and reported as a model diatom [1–5] and is generally recognized as existing in three morphotypes: oval, fusiform and triradiate [6–10]. The pleiomorphic characteristics of the microalgae were first described by Douglas P. Wilson in 1946 and the organism was capable of producing four morphotypes (oval, fusiform, triradiate and cruciform) [11]. Only the first three morphotypes are common, while cruciform cells are rare and regarded as irregular. Since 1946, cruciform cells have rarely been reported. Subsequent studies on *Phaeodactylum tricornutum* have mainly concerned strains with one to three of the common morphotypes. And unlike the intensive studies on the three common *Phaeodactylum tricornutum* morphotypes, cruciform cells have been rarely reported as the difficulty of cultivation.

Phaeodactylum tricornutum, which is a unique diatom, aroused interest in the field of biodiesel production when the whole genome became available [2,5,12,13]. However, previous studies have mainly been based on the three common morphotypes: oval, fusiform and triradiate. The differences between the three common morphotypes are first seen in cell structure. Both atomic force microscopy (AFM) and scanning electron microscopy (SEM) have been employed to investigate the ultrastructure of the three common morphotypes [7,10]. Topographic imaging showed that the oval cells possess an outer layer of extracellular polymers and

rougher surface than those of fusiform and triradiate cells and spatially resolved force-indentation curves analysis showed that cell wall composition was different in the three morphotypes, and oval cells are the only morphotype to possess silicon [10]. The girdle regions of the fusiform and ovoid forms and the arms of the triradiate form were found to be localized in organelles [10]. Morphological diversity is important in microorganisms in order to respond to changes in environmental conditions. The transformation from one morphotype to another can be triggered by variations in culture conditions such as temperature, salinity and state of media, although the effects of temperature are indistinctive [6,7]. Hyposaline conditions can accelerate the conversion of fusiform and triradiate morphotypes to oval and round morphotypes [7]. To some extent, the relative content of the different morphotypes is a reflection of cell physiological characteristics and the environment. This is because differences in cell composition also exist between different morphotypes. Lipid and protein dry weight as well as exopolysaccharide content vary in different morphotypes [14]. The fatty acid composition of C14:0, C16:2 and C16:3 contents were significantly greater in the fusiform morphotype compared with the oval morphotype [9]. The antibacterial activity against *Staphylococcus aureus* was also different in fusiform-enriched cultures and was nearly twice that of 100% oval cells [9].With the characteristics of fast growth, easy culture, high lipid content and excellent fatty acid composition, *Phaeodacty-*

lum tricornutum has been widely used as potential biofuel feedstocks. Lipid metabolism in microalgae is regulated by many factors such as nutrients, temperature and light [15,16]. Temperature is a key factor which can affect cell growth and metabolic rates through nutrient absorbance, enzymatic activities and alterations in signal pathways. Previous studies have shown that the contents of eicosapentaenoic acid and polyunsaturated fatty acids were higher at lower temperatures [15,17].

As the only representative of the suborder Phaeodactylineae, family Phaeodactylacea, genus *Phaeodactylum* [14], new insights into the polymorphism of *Phaeodactylum tricornutum*, especially the description of the forth morphotype, have important ecological significance. In this study, we successfully cultured the fourth morphotype of *Phaeodactylum tricornutum* – the cruciform morphotype in a high proportion which is reported for the first time. We also demonstrated the transformation from cruciform to other common morphotypes and clarified the fatty acid characteristics of this unique strain at different temperatures.

Materials and Methods

Culture conditions

The *Phaeodactylum tricornutum* strain CCMM 2004 (maintained by the Institute of Oceanography, Chinese Academy of Sciences) has been grown in f/2 medium at $18\pm0.5°C$ for several years. Subculturing was conducted every month to avoid possible nutrient limitation. Before this study was conducted, the particular strain was found to possess cruciform morphotype. In this study, the strain was grown in 250 mL Erlenmeyer flasks, each containing 150 mL sterilized L1 medium [18] with an initial salinity of 30 at $15\pm0.5°C$. The light density was 4000 Lux with a 12:12 h light: dark cycle. The initial inoculation density was appropriately 5×10^4 cells/mL and all cultures were shaken twice a day. From January 12, 2013 to October 12, 2013, two new cultures were added each month under the same conditions. All cultures were kept without nutrients replenished.

Microscopic observation

The morphological characteristics of *Phaeodactylum tricornutum* CCMM 2004 were observed each month using a phase contrast microscope (Olympus IX71, Japan) equipped with a DP73 digital camera. Blood counting chamber was employed for quantifying each morphotype. For each sample, approximately 1000 cells were counted for morphological statistics after mixed evenly. Samples for SEM (Hitachi S-4800, Japan) observation were prepared using a freeze dehydration procedure [19].

PCR amplification of the 18S rDNA

For genomic DNA extraction, 10 mL cultures were collected by centrifugation at 4000 rpm using a Plant DNA Kit (Tiangen). PCR primers for 18S rDNA amplification are shown in Table 1 and were designed based on the sequences of *Phaeodactylum tricornutum* in NCBI (DQ402479, GQ452863 and HQ912556). Genomic DNA and PCR products were electrophoresed in a 1.2% (w/v) agarose gel buffered by TBE for 0.5 h before staining with ethidium bromide solution. 18S rDNA was sequenced by the Genomic Platform of IOCAS.

Growth and biomass estimation

Four temperatures (10°C, 15°C, 20°C and 25°C) were used in this study to determine the fatty acids produced by *Phaeodactylum tricornutum*. The growth of *Phaeodactylum tricornutum* cultured at different temperatures was monitored each day by reading the fluorescence value obtained by the TD700 Laboratory fluores-

cence spectrometer. Biomass was estimated by dry weight according to Phukan et al. [20].

Relative content of neutral lipids in a single cell determination

For intracellular neutral lipids determination, the fluorescence dye, BODIPY 505/515, was diluted in DMSO to achieve a final concentration of 100 mM as the stock solution. 10 μL stock solution was added to 1 mL *Phaeodactylum tricornutum* cell suspension. Observations were conducted using a fluorescent microscope (Nikon Eclipse 50i, Japan) after staining for 2 min. Flow cytometry was employed to determine green fluorescence of BODIPY-stained cells, cell size and morphology and red autofluorescence caused by chlorophyll, using a BD FACSVantage SE flow cytometer (Becton Dickinson, USA) equipped with an air-cooled 488 nm argon-ion laser. The optical system in the BD FACSVantage SE flow cytometer collects green light in the FL1 channel and red light in the FL3 channel. For flow cytometry determination, microalgal solutions were diluted to a density of 1×10^4 cells/mL before staining.

Lipid content and composition analysis

Approximately 3 g centrifuged fresh algal pellets from each sample were collected for lipid content and composition analysis. The samples were first dried by a lyophilizer. Lipid extraction was performed using the methods of Bligh and Dyer [21] for subsequent lipid analysis. An Agilent 7890N GC/5975N MS (Agilent Technologies Inc., USA) was employed for analysis of fatty acid methyl esters (FAMEs).

Calculation of FAME properties

Four indices including cetane number (*CN*), iodine number (*IN*), linolenic acid content and the content of FAME with ≥4 double bonds were introduced to calculate FAME properties. The equations for the former two indices were taken from Lapuerta et al [22].

Cetane number (*CN*)

$$CN = -21.157 + (7.965 - 1.785db + 0.235db^2)n - 0.099n^2 \quad (1)$$

Iodine number (*IN*)

$$IN = 100db_{oil}\frac{W_{I_2}}{W_{oil}} \quad (2)$$

Statistical analysis

The SAS 9.2 software system (SAS Institute Inc., Cary, NC, USA) for statistical analysis was employed in this study. One-way analysis of variance (ANOVA) was used to evaluate the differences in cell number, biomass and FAME contents following the treatments [23]. Correlation analysis was conducted using Pearson correlation coefficients. P-values <0.05 were considered significant.

Results and Discussion

Cell transformation between the four morphotypes

Following observation and counting over a nine-month period, the proportions of the four morphotypes were calculated and are shown in Fig 1. During the first two months of growth, only

Table 1. Primers for amplification of 18S rDNA from *Phaeodactylum tricornutum* CCMM 2004.

Name	Primer Sequence (5'-3')
18SF	AACCTGGTTGATCCTGCCAGT
18SR	GATCCTTCYGCAGGTTCACCTAC

cruciform, triadiate and fusiform morphotypes were found in the culture. During the first month, cell numbers of these three morphotypes were similar with 31.3%, 33.7% and 34.9% of cruciform, triadiate and fusiform, respectively. After two months, the proportions of cruciform, triadiate and fusiform were 48.9%, 37.8% and 13.3%, respectively. The fusiform morphotype reduced sharply during this stage with a 21.6% decrease, and the cruciform morphotype increased by 17.6%. The oval morphotype was absent for three months. The protortion of the oval form was 5.0% of the total cell count at that time. Subsequently, the four morphotypes showed relatively stable proportions in culture during the following six months (10.5% for oval, 11.3% for fusiform, 37.2% for triadiate and 41.0% for cruciform). During the nine months' observation, proportion of triadiate cells ranged from 33.3% to 39.5% and showed a consistency compared with the other three morphotypes. While a *Phaeodactylum tricornutum* strain CCAP 1052/1A with 100% triadiate forms changed to a 50:50 triadiate: fusiform ratio after cultured for one month and this ratio was maintained for four years [6]. Some similar researches also showed that stability of triadiate morphotype was lower than other morphotypes [24]. This study may be a special case compared with previous studies because other *Phaeodactylum tricornutum* strains often have one or no more than two preferential morphotypes in laboratory culture. Abundant morphotype compositions and complicated interactions between them may explain morphotype transformations of this strain.

Previous studies have focused on the morphology of *Phaeodactylum tricornutum*, and the results showed three morphotypes including fusiform, triadiate and ovoid [6–8,10]. In the present study another rare morphotype was detected, and the proportion of the rare cruciform morphotype was greater than 40%

depending on the culture conditions (Fig. 1). Another difference between this study and previous studies was the presence of the oval form, which was not observed in this study until after three months of culture. It is likely that this morphotype was the result of the transformation of other morphotypes. Long term observation results on intermediate morphotypes may support for the above speculation. Fig. 2 shows the specific transformation processes and the changes from cruciform, triadiate and fusiform to oval. Generally, this transformation involves four stages. Stage 1 is the initial period, and the three morphotypes show the characteristics of a thin form indicating that the content of intracellular substances is poor. In stage2, the cells are plump indicating many intracellular substances. In stage 3, the overall outline of the cells begins to change. In stage 4, the arms of the three morphotypes gradually disappear and the oval form is observed. In cruciform cells, two processes related to the formation of oval cells have been found. In the first process, four arms degenerate and disappear one by one. Thus, several intermediate forms exist during this process. In the second process, four arms degenerate and disappear around the same time. For triadiate and fusiform cells, only one process has been found to result in the oval form, respectively, and this process is similar to the second process found in the cruciform morphotype.

Combined with the study results of previous researchers [6,7,24], a new map of *Phaeodactylum tricornutum* morphotype transformation is shown in Fig. 3. Of these transformations, pathway 7 observed in this study is a new discovery. It shows that cruciform cells possess similar characteristics to fusiform and triadiate cells and degenerate arms to form oval cells.

Besides the confirmed pathway, other transformations, such as cruciform to triadiate, cruciform to fusiform and the reverse transformations, may also exist. This speculation is based on several intermediate forms shown in Fig 4. Imagine A and B show two complexes of fusiform and cruciform cells with different connecting locations. Through the conjunct arm, two different morphotype cells link together. Although it is difficult to decide whether the fusiform or the cruciform pre-exist, transformations between the two forms exist. It is likely that one arm of the cruciform cell elongates and develops into a fusiform cell. A similar process may occur in triadiate and cruciform cells (Fig. 4C), because the two cells are connected by one arm. Image D and E in Fig. 4 also show the possible transformation between triadiate and cruciform cells, where a triadiate cell develops one more arm to

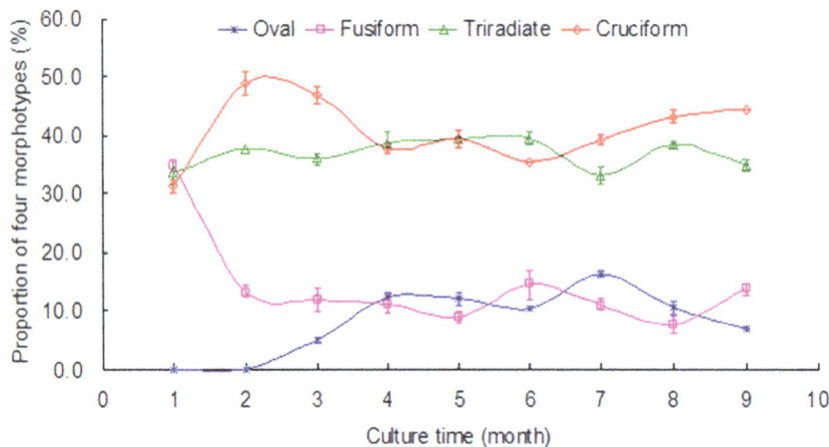

Figure 1. Proportions of the four morphotypes during long-term culture. Data are averages of duplicate measurements. Error bars represent standard deviation.

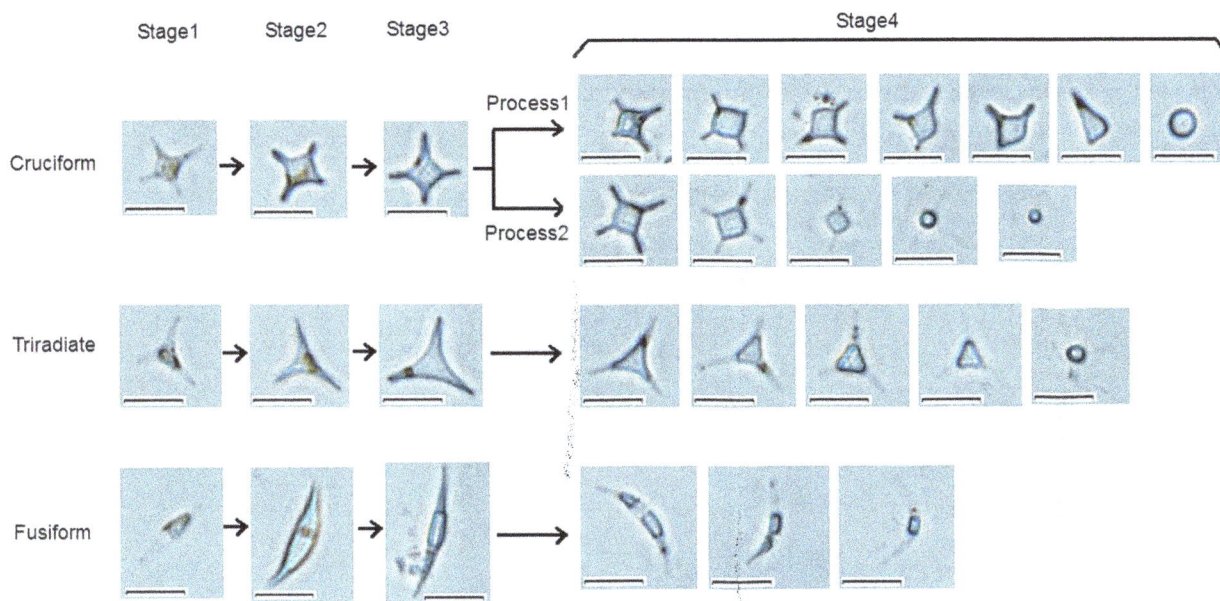

Figure 2. Transformation from cruciform, triradiate and fusiform to oval morphotype. This transformation involves four stages. In stage 1, the three morphotypes show the characteristics of a thin form indicating that the content of intracellular substances is poor. In stage2, the cells are plump indication many intracellular substances. In stage 3, the overall outline of the cells begins to change. In stage 4, the arms of the three morphotypes gradually disappear and the oval form is observed. Scale bars = 10 μm.

form a cruciform cell. From the above descriptions, the "conjunct arm" plays a significant role in cell transformation. Using scanning electron microscopy, the "conjunct arm" is clearly shown in Fig. 4F.

Sequence alignment results of 18S rDNA

After centrifugation and collection of microalgal cells, genomic DNA of *Phaeodactylum tricornutum* CCMM2004 cultured at 15°C containing all three morphotypes was extracted. Based on the template and microalgal 18S rDNA universal primers, a PCR product of approximately 1700 bp was obtained by electrophoresis. Further sequencing results showed no interference peak which indicated the purity of the strain. 1674 bp effective 18S rDNA sequences were obtained for *Phaeodactylum tricornutum* CCMM2004. The NCBI Blast nucleotide sequence showed 100% similarity with the *Phaeodactylum tricornutum* 18S rRNA sequence (GenBank accession no. CQ452863).

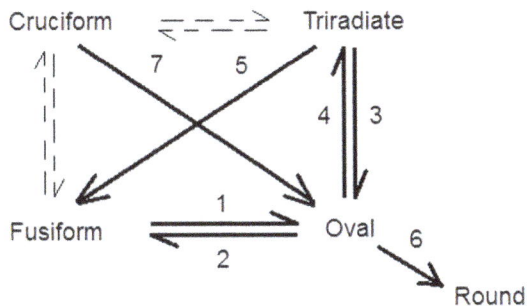

Figure 3. Transformations between the four morphotypes. Transformations 1–6 are study results of previous researchers. Pathway 7 observed in this study is a new discovery.

Morphological characteristics of *Phaeodactylum tricornutum* cultured at different temperatures

As shown in Fig. 5, the proportions of the three morphotypes were different depending on the culture temperature. The cruciform morphotype showed the greatest proportion and fusiform cells showed the smallest proportion. As the temperature increased from 10°C to 20°C, the percentage of cruciform cells decreased gradually from 55.3% to 48.3%, and the proportion of both fusiform and triradiate morphotypes increased. Fusiform cells increased from 8.3% to 15.0% and the triradiate morphotype increased from 36.5% to 36.7%. At 25°C, the proportions of cruciform and fusiform cells were 39.1% and 22.8%, decreasing by 9.2% and increasing by 7.8%, respectively, compared with that at 20°C. Generally, the percentage of triradiate cells showed maximum stability. Lower temperature was beneficial for cruciform cells and higher temperature facilitated the formation of fusiform cells. In this study, low temperature facilitated the formation of the cruciform morphotype. The percentage of this morphotype at 10°C was 16.2% more than that at 25°C. That may be a defence mechanism of *Phaeodactylum tricornutum* CCMM 2004 to different temperatures.

Culture temperature plays an important role in *Phaeodactylum tricornutum* morphotype characteristics. For strain CCMP633, the proportion of the oval morphotype was 60%–75% at 15°C –19°C, however, the proportion of fusiform cells was 80%–95% at 25°C – 28°C [8]. Oval cell growth increased notably in strain Pt1 8.6F when the culture temperature was altered from 19°C to 15°C compared to that at 28°C [7]. The proportions of the different morphotypes of *Phaeodactylum tricornutum* primarily depend on the strain itself to a great extent. Approximately 95%–100% of the *Phaeodactylum tricornutum* strains CCMP632, CCAP 1052/1A, CCAP 1052/6, CCMP630, CCMP631, CCMP1327 and MACC B228 were fusiform cells, whereas 80%–85% of *Phaeodactylum tricornutum* strain NEPCC 640 were triradiate cells [8]. When cultured in standard temperature conditions (19°C), 80% of Pt1

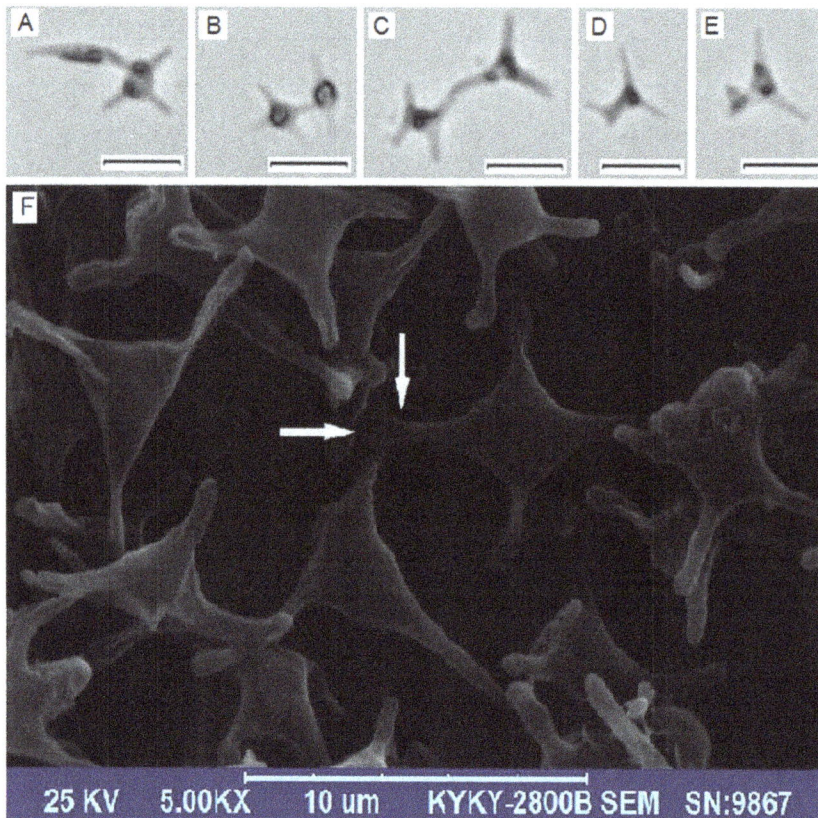

Figure 4. Intermediate forms in the culture. A–E are DIC photos. Scale bars = 10 μm. F is SEM result and the white arrows direct to the "conjunct arm".

8.6F were fusiform cells whereas 60% of Pt3O were oval cells [7]. Temperature has a significant influence on the composition of microalgae. Available evidence shows that the contents and composition of FAME also varied in different morphotypes of *Phaeodactylum tricornutum* [9].

Growth and biomass

Phaeodactylum tricornutum cells are capable of growing at the four temperatures studied in this report. The color of the *Phaeodactylum tricornutum* culture solution at four temperatures was also different which suggested a difference in metabolite composition and

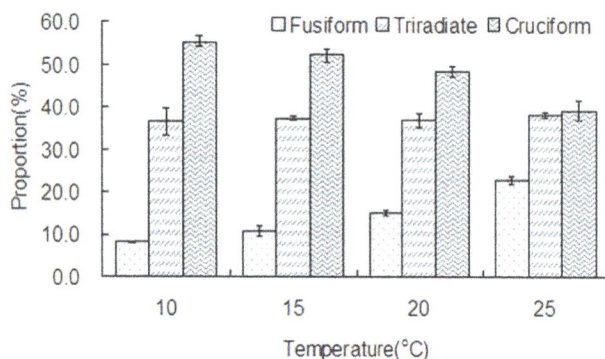

Figure 5. Proportion of the three morphotypes at different temperatures. Data are averages of duplicate measurements. Error bars represent standard deviation.

contents. The biomass yields of *Phaeodactylum tricornutum* cultured at four temperatures are shown in Fig. 6. Biomass yields at 10°C, 15°C and 20°C were about 0.121–0.136 g/L, and were less than that at 25°C (0.161 g/L). The temperature range for the growth of *Phaeodactylum tricornutum* is relatively wide [15]. This strain grows well at low temperature.

Fluorescence staining and flow cytometry determination

The refractive index of fat particles is different from that of the cytoplasm. Although the lipid dye, Nile Red, has been used for over 20 years, it has some disadvantages, which can not be overlooked. Nile Red is unable to efficiently penetrate microalgae with thick or rigid cell walls, adding to the difficulties in effective staining [25]. In addition, even for microalgae without cell walls, the stained lipids are seen as yellow and similar to chlorophyll autofluorescence red, which makes observation and quantification difficult. BODIPY 505/515 is a new fluorescence dye with a small fluorescence Stokes shift, large extinction coefficient and high fluorescence quantum yield for lipid staining [26]. It overcomes the insufficiencies of Nile Red and has been successfully used in many microalgal species [26,27]. BODIPY-stained *Phaeodactylum tricornutum* cells showed red and specific green fluorescence when excited by blue light (Fig. 7) which is similar with previous studies on *Chlorella vulgaris* and *Chaetoceros calcitrans* [28]. The red fluorescence represents chlorophyll and the green fluorescence represents cellular lipids. Differences between them can be easily distinguished. In Fig. 7, triradiate cells showed the maximum green fluorescence intensity, followed by fusiform cells, and cruciform cells showed minimum fluorescence intensity. Different

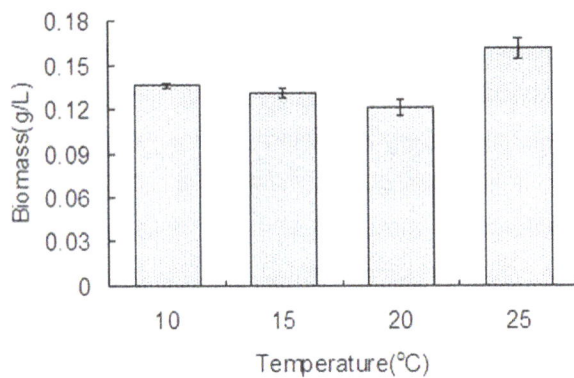

Figure 6. Biomass of *Phaeodactylum tricornutum* **cultured at different temperatures.** Data are averages of duplicate measurements. Error bars represent standard deviation.

green fluorescence intensities in the three morphotypes indicated differences in lipid contents. Besides lipid contents, differences between different morphotypes of *Phaeodactylum tricornutum* also exist in other cellular characteristics. Based on the difference in extracellular polymeric substance secretion, Stanley et al. showed that oval cells were capable of adhering to both hydrophilic acid-washed glass and polydimethylsiloxane elastomer, while fusiform cells were unable to adhere to either surface [29].

Phaeodactylum tricornutum cells cultured at four temperatures were detected by flow cytometry. For each sample, 30000 events were carried out for statistical analysis. Distributions of cell shapes and sizes were similar in the four different treatment groups. Cells at 10°C showed more dispersion. Chlorophyll autofluorescence was reflected by FL3. The values obtained at 10°C were less than those obtained at the other three temperatures indicating less chlorophyll content in *Phaeodactylum tricornutum* at low temperature. The SSC, FSC and FL3 data showed that *Phaeodactylum tricornutum* cells grew well in the four groups which is similar with research results of Jiang et al. [15], both indicating wide temperature adaptability. After staining with BODIPY 505/515, FL1 values of *Phaeodactylum tricornutum* cells cultured at 15°C showed the greatest increase, rising from 1.04 to 151.25 (Table 1). The FL1 values of stained cells cultured at 10°C, 20°C and 25°C also showed an average increase of 72.90, 45.01 and 53.76 respectively (Table 2). These results suggest that 15°C is the optimal temperature and 20°C is relatively unfavourable for lipid accumulation in *Phaeodactylum tricornutum*. Flow cytometry can analyse the chemical composition

of biomass by fluorescence labelling [30]. The combination of flow cytometry and fluorescence staining is a new method for the determination of cellular lipid. The fluorescence dyes employed for staining are primarily Nile red and BODIPY. The lipid contents of many microalgal species have been determined using this method [26,27]. After staining with 1 μM BODIPY 505/515, the mean fluorescence value of *Chrysochromulina sp.* was 227 which was 100 times more than unstained cells [26]. The amount of fluorescence enhancement is related to the microalgal species. A fluorescence enhancement of 50–524 times in different microalgae was observed in a previous study [26,27]. The staining efficiency of microalgal cellular lipids was thought to be related to dye concentration, staining time and cell concentration.

Fatty acid content and composition

The composition and percentage of fatty acids in microalgae can be regulated by altering the culture temperature. In this study, the fatty acid profiles of *Phaeodactylum tricornutum* cultured at four temperatures are shown in Table 3. The maximum lipid contents at four temperatures were all C16:1. As the temperature increased from 10°C to 25°C, the percentage of C16:1 increased from $43.82 \pm 0.16\%$ to $50.82 \pm 0.21\%$. These results are similarly to those of Jiang et al. [15] who showed that lipid contents in *Phaeodactylum tricornutum* 2038, with the exception of the percentage of C16:1, were all about 10% less than their counterpart in the present study. C16:1 content in *Phaeodactylum tricornutum* CCAP1055/1 can reach $46.35 \pm 2.3\%$ which is the maximum content [31]. The content of C16:0 was also high at approximately 20%. The sum of C16:1 and C16:0 can reach 66.47% and was a maximum of 71.71% in this study. Saturated fatty acid contents range from 41.72% to 44.68% which is beneficial for biodiesel cetane number and oxidative stability [22]. Over 50 years ago, researchers discovered that different morphotypes of *Phaeodactylum tricornutum* differed not only in their shapes, but also in their biomacromolecule contents. Two *Phaeodactylum tricornutum* cultures dominated by fusiform and oval cells, respectively, showed different protein dry weight (41% and 34%, respectively), lipid dry weight (34% and 24%, respectively) and exopolysaccharide content (<1% and 16%, respectively) [14]. In recent years, the contents of the fatty acids, EPA, HTA and PA extracted from fusiform cells were higher than those in oval cells (42.7 and 36.6×10^{-13} g, respectively) [9]. Research on *Phaeodactylum tricornutum* 2038 showed that the percentage of PUFAs in total fatty acids decreased gradually from $31.9 \pm 1.0\%$ to $21.8 \pm 0.4\%$ as the culture temperature increased from 10°C to 25°C [15]. In the present study, as shown in Fig. 5, in the three growth periods,

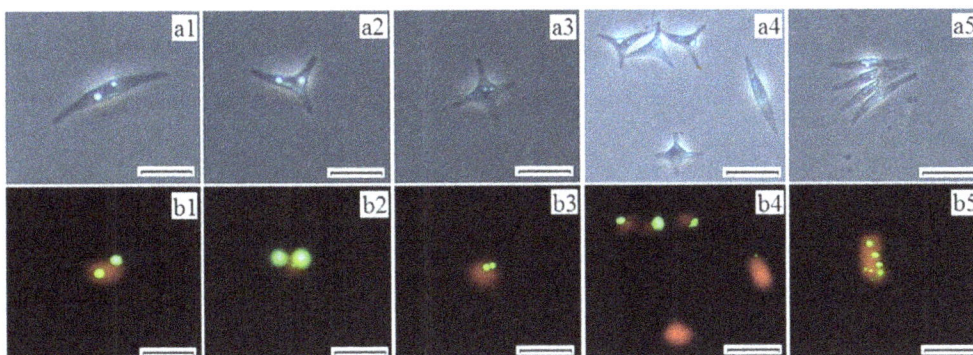

Figure 7. Images of *Phaeodactylum tricornutum* **cells stained with BODIPY 505/515.** a1, a2, a3, a4 and a5 are images at white light, while b1, b2, b3, b4 and b5 are corresponding images at the exciting light. Scale bars = 10 μm.

Table 2. Histogram statistical data for changes in fluorescence value.

Temperature(°C)	M1		M2		M1–M2
	Mean	% Total	Mean	% Total	
10	1.05	99.44	73.95	99.15	72.90
15	1.04	99.46	151.25	99.08	150.21
20	1.04	99.48	46.05	99.38	45.01
25	1.04	99.55	54.80	99.99	53.76

the percentage of PUFAs in total fatty acids fell to a minimum mean value of 19.04% at 10°C which was lower than that at 15°C, 20°C and 25°C (25.90%, 26.12% and 24.03%, respectively). With regard to the contribution of C16:1 and C18:1, the percentage of MUFA in total fatty acids in all treatments was maximum. Based on previous research results and those of the present study, we consider that the unique fatty acid characteristics of *Phaeodactylum tricornutum* may be caused by the presence of abundant cruciform cells.

Content of C16:1 was greatest in all FAMEs and it showed a significant correlation with *Phaeodactylum tricornutum* morphotypes. Specifically, the percentage of fusiform and triradiate morphotypes showed significant positive correlations (P<0.01 and P<0.05, respectively) with the content of C16:1. A significant negative correlation (P<0.01) was observed between the percentage of the cruciform morphotype and the content of C16:1. A previous study showed similar results. Research on *Phaeodactylum tricornutum* SAG1090-6 enriched oval and fusiform cells showed that C14:0, C16:2 n-4 and C16:3 n-4 contents in fusiform cells were significantly greater (P<0.05) than in oval cells [9].

Characteristics of biodiesel produced from *Phaeodactylum tricornutum*

In the European Union, Standard EN 14214 was drafted for biodiesel use in vehicles. ASTM Biodiesel Standard D 6751 was also drafted in the United States. Of the indices included, the

cetane number is the parameter used to evaluate diesel ignition quality. In this study, the cetane numbers for biodiesel at the four temperatures are shown in Table 4, and range from 64.3 to 65.2. These values conform to both EN 14214 (51 min) and D6751 (47 min) [32]. As a measure of total unsaturated oil, the iodine index was also employed in this study. The iodine indices at the four temperatures were below the maximum standard of 120 (EN 14214). Other indices included in the above standards involved restrictions on the fatty acid profile such as linolenic acid content, content of FAME with ≥4 double bonds and kinematics viscosity. The former indices were introduced to restrict excess double bonds which are easily oxidized. Kinematics viscosity is affected by both saturated and short-chain fatty acid contents.

Conclusion

This is the first study to induce a *Phaeodactylum tricornutum* strain rich in the rare cruciform morphotype (maximum of 55.3±1.2% in this study). Long-term culture and observation showed that cruciform cells were capable of transforming to oval cells following the degeneration of arms by two processes: four arms degenerate and disappear one by one or around the same time. This new discovery deepens the understanding of *Phaeodactylum tricornutum* polymorphism. This study also demonstrated the unique fatty acid profiles of the strain cultured at different temperatures, and these characteristics were partially because of abundant cruciform cells.

Table 3. Fatty acid composition of *Phaeodactylum tricornutum* cultured at different temperatures.

FAME	Temperature (°C)			
	10	15	20	25
C8:0	4.71±0.02	5.12±0.01	4.01±0.00	4.29±0.03
C14:0	6.12±0.03	6.45±0.02	7.65±0.07	7.47±0.05
C16:1	43.82±0.16	45.86±0.09	47.34±0.17	50.82±0.21
C16:0	22.65±0.09	20.47±0.07	21.56±0.02	20.89±0.11
C18:3	1.79±0.02	1.66±0.00	1.04±0.01	0.90±0.02
C18:2	1.76±0.03	1.82±0.06	2.44±0.01	2.00±0.05
C18:1 (cri-9)	6.69±0.08	6.72±0.04	5.23±0.02	3.82±0.07
C18:1 (trans-9)	1.26±0.03	0.82±0.01	0.97±0.03	0.74±0.02
C18:0	2.30±0.02	2.06±0.06	1.88±0.05	1.95±0.01
C19:0	8.90±0.04	9.02±0.03	7.88±0.05	7.12±0.03
SFA	44.68±0.12	43.12±0.09	42.98±0.15	41.72±0.18
PUFA	3.55±0.07	3.48±0.09	3.48±0.02	2.90±0.04
C14-C18	86.39±0.31	85.86±0.27	88.11±0.35	88.59±0.24

Table 4. Biodiesel properties at the four culture temperatures.

Properties	10°C	15°C	20°C	25°C	Standard[a]
Cetane number	65.2	64.6	64.8	64.3	51 min
Iodine index	60.3	61.9	61.4	62.2	120 max
Linolenic acid content	1.79	1.66	1.04	0.90	12.0 max
Content of FAME with ≥4 double bonds	0	0	0	0	1 max

[a]The standard here refers to EN 14214.

Moreover, the increase in cruciform morphotype at low temperature reflects the adaptability of this strain to environmental change. These findings provide a platform for further research on the cold-resistance mechanism of this strain.

Acknowledgments

We would like to thank Dr. Baijuan Yang for fatty acid data, Nana Zong for SEM photos. We also thank Prof. Yingzhong Tang, the editors and anonymous reviewers for the comment and suggestion on the manuscript.

Author Contributions

Conceived and designed the experiments: LH XH ZY. Performed the experiments: LH. Analyzed the data: LH XH ZY. Contributed reagents/materials/analysis tools: XH ZY. Wrote the paper: LH.

References

1. Scala S, Carels N, Falciatore A, Chiusano ML, Bowler C (2002) Genome properties of the diatom Phaeodactylum tricornutum. Plant Physiology 129: 993–1002.
2. Montsant A, Jabbari K, Maheswari U, Bowler C (2005) Comparative genomics of the pennate diatom Phaeodactylum tricornutum. Plant Physiology 137: 500–513.
3. Fabris M, Matthijs M, Rombauts S, Vyverman W, Goossens A, et al. (2012) The metabolic blueprint of Phaeodactylum tricornutum reveals a eukaryotic Entner-Doudoroff glycolytic pathway. Plant Journal 70: 1004–1014.
4. Chauton MS, Winge P, Brembu T, Vadstein O, Bones AM (2013) Gene Regulation of Carbon Fixation, Storage, and Utilization in the Diatom Phaeodactylum tricornutum Acclimated to Light/Dark Cycles. Plant Physiology 161: 1034–1048.
5. Bowler C, Allen AE, Badger JH, Grimwood J, Jabbari K, et al. (2008) The Phaeodactylum genome reveals the evolutionary history of diatom genomes. Nature 456: 239–244.
6. Bartual A, Villazan B, Brun FG (2011) Monitoring the long-term stability of pelagic morphotypes in the model diatom Phaeodactylum tricornutum. Diatom Research 26: 243–253.
7. De Martino A, Bartual A, Willis A, Meichenin A, Villazan B, et al. (2011) Physiological and Molecular Evidence that Environmental Changes Elicit Morphological Interconversion in the Model Diatom Phaeodactylum tricornutum. Protist 162: 462–481.
8. De Martino A, Meichenin A, Shi J, Pan KH, Bowler C (2007) Genetic and phenotypic characterization of Phaeodactylum tricornutum (Bacillariophyceae) accessions. Journal of Phycology 43: 992–1009.
9. Desbois AP, Walton M, Smith VJ (2010) Differential antibacterial activities of fusiform and oval morphotypes of Phaeodactylum tricornutum (Bacillariophyceae). Journal of the Marine Biological Association of the United Kingdom 90: 769–774.
10. Francius G, Tesson B, Dague E, Martin-Jezequel V, Dufrene YF (2008) Nanostructure and nanomechanics of live Phaeodactylum tricornutum morphotypes. Environmental Microbiology 10: 1344–1356.
11. Wilson DP (1946) The triradiate and other forms of nitzschia closterium (Ehrenberg) WM.Smith,forma minutissima of Allen and Nelson. Journal of the Marine Biological Association of the United Kingdom: 235–270.
12. Radakovits R, Eduafo PM, Posewitz MC (2011) Genetic engineering of fatty acid chain length in Phaeodactylum tricornutum. Metabolic Engineering 13: 89–95.
13. Valenzuela J, Mazurie A, Carlson RP, Gerlach R, Cooksey KE, et al. (2012) Potential role of multiple carbon fixation pathways during lipid accumulation in Phaeodactylum tricornutum. Biotechnology for Biofuels 5: 40.
14. Lewin JC, Lewin RA, Philpott DE (1958) Observations on Phaeodactylum-Tricornutum. Journal of General Microbiology 18: 418–426.
15. Jiang H, Gao K (2004) Effects of Lowering Temperature during Culture on the Production of Polyunsaturated Fatty Acids in the Marine Diatom Phaeodactylum Tricornutum (Bacillariophyceae)1. Journal of Phycology 40: 651–654.
16. Garcia MCC, Miron AS, Sevilla JMF, Grima EM, Camacho FG (2005) Mixotrophic growth of the microalga Phaeodactylum tricornutum - Influence of different nitrogen and organic carbon sources on productivity and biomass composition. Process Biochemistry 40: 297–305.
17. Liu CP, Lin LP (2005) Morphology and eicosapentaenoic acid production by Monodus subterraneus UTEX 151. Micron 36: 545–550.
18. Guillard RRL, Hargraves PE (1993) Stichochrysis-Immobilis Is a Diatom, Not a Chyrsophyte. Phycologia 32: 234–236.
19. Veltkamp CJ, Chubb JC, Birch SP, Eaton JW (1994) A Simple Freeze Dehydration Method for Studying Epiphytic and Epizoic Communities Using the Scanning Electron-Microscope. Hydrobiologia 288: 33–38.
20. Phukan MM, Chutia RS, Konwar BK, Kataki R (2011) Microalgae Chlorella as a potential bio-energy feedstock. Applied Energy 88: 3307–3312.
21. Bligh EG, Dyer WJ (1959) A Rapid Method of Total Lipid Extraction and Purification. Canadian Journal of Biochemistry and Physiology 37: 911–917.
22. Lapuerta M, Rodriguez-Fernandez J, de Mora EF (2009) Correlation for the estimation of the cetane number of biodiesel fuels and implications on the iodine number. Energy Policy 37: 4337–4344.
23. Bartual A, Gálvez JA, Ojeda F (2008) Phenotypic response of the diatom Phaeodactylum tricornutum Bohlin to experimental changes in the inorganic carbon system. Botanica Marina 51: 350–359.
24. Tesson B, Gaillard C, Martin-Jézéquel V (2009) Insights into the polymorphism of the diatom Phaeodactylum tricornutum Bohlin. Botanica Marina 52: 104–116.
25. Chen W, Sommerfeld M, Hu Q (2011) Microwave-assisted nile red method for in vivo quantification of neutral lipids in microalgae. Bioresour Technol 102: 135–141.
26. Cooper MS, Hardin WR, Petersen TW, Cattolico RA (2010) Visualizing "green oil" in live algal cells. Journal of Bioscience and Bioengineering 109: 198–201.
27. Brennan L, Blanco Fernandez A, Mostaert AS, Owende P (2012) Enhancement of BODIPY505/515 lipid fluorescence method for applications in biofuel-directed microalgae production. J Microbiol Methods 90: 137–143.
28. Govender T, Ramanna L, Rawat I, Bux F (2012) BODIPY staining, an alternative to the Nile Red fluorescence method for the evaluation of intracellular lipids in microalgae. Bioresour Technol 114: 507–511.
29. Stanley MS, Callow JA (2007) Whole cell adhesion strength of morphotypes and isolates of Phaeodactylum tricornutum (Bacillariophyceae). European Journal of Phycology 42: 191–197.
30. Hyka P, Lickova S, Pribyl P, Melzoch K, Kovar K (2012) Flow cytometry for the development of biotechnological processes with microalgae. Biotechnol Adv 31: 2–16.
31. Zendejas FJ, Benke PI, Lane PD, Simmons BA, Lane TW (2012) Characterization of the acylglycerols and resulting biodiesel derived from vegetable oil and microalgae (Thalassiosira pseudonana and Phaeodactylum tricornutum). Biotechnology and Bioengineering 109: 1146–1154.
32. Knothe G (2006) Analyzing biodiesel: Standards and other methods. Journal of the American Oil Chemists Society 83: 823–833.

13

Energy, Water and Fish: Biodiversity Impacts of Energy-Sector Water Demand in the United States Depend on Efficiency and Policy Measures

Robert I. McDonald[1]*, Julian D. Olden[2], Jeffrey J. Opperman[3], William M. Miller[4], Joseph Fargione[5], Carmen Revenga[6], Jonathan V. Higgins[7], Jimmie Powell[1]

1 Worldwide Office, The Nature Conservancy, Arlington, Virginia, United States of America, 2 School of Aquatic & Fishery Sciences, University of Washington, Seattle, Washington, United States of America, 3 Freshwater Focal Area Program, The Nature Conservancy, Chargin Falls, Ohio, United States of America, 4 Department of Chemical and Biological Engineering, Northwestern University, Evanston, Illinois, United States of America, 5 North America Region, The Nature Conservancy, Minneapolis, Minnesota, United States of America, 6 Marine Focal Area Program, The Nature Conservancy, Arlington, Virginia, United States of America, 7 Freshwater Focal Area Program, The Nature Conservancy, Chicago, Illinois, United States of America

Abstract

Rising energy consumption in coming decades, combined with a changing energy mix, have the potential to increase the impact of energy sector water use on freshwater biodiversity. We forecast changes in future water use based on various energy scenarios and examine implications for freshwater ecosystems. Annual water withdrawn/manipulated would increase by 18–24%, going from 1,993,000–2,628,000 Mm^3 in 2010 to 2,359,000–3,271,000 Mm^3 in 2035 under the Reference Case of the Energy Information Administration (EIA). Water consumption would more rapidly increase by 26% due to increased biofuel production, going from 16,700–46,400 Mm^3 consumption in 2010 to 21,000–58,400 Mm^3 consumption in 2035. Regionally, water use in the Southwest and Southeast may increase, with anticipated decreases in water use in some areas of the Midwest and Northeast. Policies that promote energy efficiency or conservation in the electric sector would reduce water withdrawn/manipulated by 27–36 m^3GJ^{-1} (0.1–0.5 m^3GJ^{-1} consumption), while such policies in the liquid fuel sector would reduce withdrawal/manipulation by 0.4–0.7 m^3GJ^{-1} (0.2–0.3 m^3GJ^{-1} consumption). The greatest energy sector withdrawal/manipulation are for hydropower and thermoelectric cooling, although potential new EPA rules that would require recirculating cooling for thermoelectric plants would reduce withdrawal/manipulation by 441,000 Mm^3 (20,300 Mm^3 consumption). The greatest consumptive energy sector use is evaporation from hydroelectric reservoirs, followed by irrigation water for biofuel feedstocks and water used for electricity generation from coal. Historical water use by the energy sector is related to patterns of fish species endangerment, where water resource regions with a greater fraction of available surface water withdrawn by hydropower or consumed by the energy sector correlated with higher probabilities of imperilment. Since future increases in energy-sector surface water use will occur in areas of high fish endemism (e.g., Southeast), additional management and policy actions will be needed to minimize further species imperilment.

Editor: Chenyu Du, University of Nottingham, United Kingdom

Funding: Most authors had no specific research grants supporting this project. RIM, JJO, JF, CR, and JVH were supported institutionally by the Nature Conservancy, which is funded by its members and donors. JDO and WMM were supported institutionally by the University of Washington and Northwestern University, respectively. JP's work on this project was supported by a grant from Royal Dutch Shell that supports research into energy production and its impacts on water. The funders had no role in study design, data collection and analysis, decision to publish, or preparation of the manuscript.

Competing Interests: The authors have declared that no competing interests exist.

* E-mail: rob_mcdonald@tnc.org

Introduction

In the United States (US), the energy sector is responsible for more than half of all water withdrawals [1]. Only a fraction of this water is consumed, with the remainder returned to the hydrologic system after likely modification of its physical (e.g., flow regimes, temperature) and chemical (e.g., dissolved oxygen) properties. With continued population growth and economic development in the US, total energy consumption is expected to increase over the coming decades. At the same time, the combination of energy sources used by Americans is changing, driven by shifts in the availability, cost effectiveness, and investment in emerging and traditional technologies. For example, new techniques for extract-

ing natural gas from shale have increased supply and decreased the price of natural gas, affecting investments in new production in many other energy technologies, as well as increasing the water used for extracting natural gas [2,3]. Moreover, concerns about energy security and the environmental impacts of energy production are leading policymakers to change the incentives and regulations that govern the energy sector. For instance, the US has made significant investments in subsidizing biofuel production, incentivizing new renewable electric generation capacity, and funding research into developing commercially viable technologies for carbon capture and storage (CCS) of emissions from fossil fuels, particularly coal [4]. Another example is potential new EPA regulations under Section 316(b) of the Clean

Water Act that may force some thermoelectric plants to switch from current once-through cooling to recirculating cooling. Section 316(b) requires that facilities use the best available cooling technology to minimize environmental impacts, which in some cases may require facilities that currently use once-through cooling to transition to using recirculating cooling.

The future expansion of energy consumption and changes in the use of different sources could cause major changes in energy sector water use. Water withdrawal and consumption by the energy sector may increase in some areas [5], altering water quality and quantity in freshwater ecosystems [6,7], and further threatening an already imperiled fauna. In the US, freshwater taxa already have a greater proportion of their species imperiled than terrestrial taxa [8], and are expected to disappear at a rate five times that of terrestrial fauna in the future [9]. The potential for future energy sector withdrawals to worsen the current freshwater biodiversity crisis appears high, but remains uncertain. The central goal of this paper is to explore the relationships between energy policies on water use and the implications of these energy-related water impacts on freshwater ecosystems, specifically freshwater fishes. Freshwater fishes are a useful indicator taxa for freshwater biodiversity more broadly: they are widely distributed across the US, their abundance and distribution patterns commonly reflect impacts to many other components of freshwater biodiversity [10], they respond to changes in water consumption and withdrawal [11], and they contribute valuable goods and services to human society [12].

A number of recent studies have looked at how changes in the energy sector will affect water withdrawals or consumption [5,13–24]. In this paper, we have the following objectives:

1. Synthesize information on the water-use intensity (m^3GJ^{-1}) of various energy production techniques, using high and low values of water-use intensity to provide a realistic range for each technique;

2. Present scenarios of future water withdrawal and consumption by the energy sector; and

3. Compare energy sector water-use with current patterns of threats to endangered fish species by major water resource region (Figure 1).

We present three scenarios of water use, based upon energy production scenarios developed by the US Energy Information Administration (EIA).

● **Reference Case**- The "business as usual" scenario, depicting the future of US energy markets under current policies, with baseline assumptions about the rate of economic growth, world oil price, and the development of new technologies.

● **Greenhouse Gas (GHG) Price Case**- Similar to the Reference Case scenario, but a price is set on greenhouse gas emissions throughout the entire US economy.

● **Extended Policies Case**- Similar to the Reference Case scenario, but additional government regulations foster improved electricity and liquid fuel efficiency.

Materials and Methods

We first estimated water-use intensity for energy production techniques, drawing heavily from several of the published reviews of this topic [17–19,22–24]. We then obtained scenarios of future energy demand and multiplied those by water-use intensity to estimate total water use. We partitioned the new energy production and water use among major water resource regions, using geospatial information on the supply of needed natural resources and demand for energy. Finally, we statistically compared the water withdrawal and consumption by energy type in each of the water resource regions with the probability of a fish species being imperiled, and then used this statistical model to estimate the likely impact future energy sector water use would

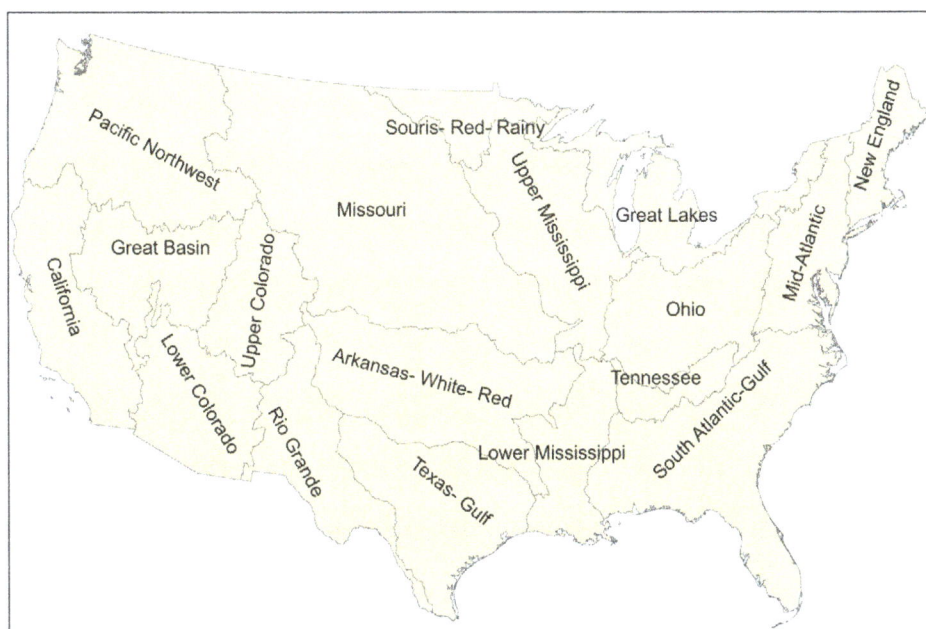

Figure 1. Water resource regions. The 18 water resource regions of the United States, as defined by the 2-digit Hydrologic Unit Codes (HUC) of the USGS.

have on freshwater fishes as an indicator of freshwater biodiversity more broadly.

Calculating water requirements

We recognized 12 energy production techniques (solar photovoltaic, solar thermal, wind, geothermal, biopower, hydropower, coal, nuclear, natural gas, power generated from municipal waste, biofuels, and petroleum), based upon the sectors used in the Annual Energy Outlook [4]. Except where noted, we define each energy production technique identically in scope to the Annual Energy Outlook. For certain energy types, we recognize multiple different techniques for producing energy that have very different water footprints (e.g., a coal power plant that is once-through cooled versus one that uses closed-loop cooling).

In this paper, we define "water use" as any use of surface water or groundwater to produce energy, including water used for hydropower production and water used to irrigate bioenergy feedstocks. Note that other papers have categorized water use into three categories: "blue", "green," and "gray"[25]. Our definition of "water use" is roughly analogous to "blue water," defined as the amount of water taken from groundwater or surface water that does not return to the catchment from which it is withdrawn. We do not estimate "green" water use (i.e., the amount of rainwater used by crops) or "grey" water use (i.e., the amount of water needed to safely dilute pollutants or impurities) (cf. [14,15,16]).

We divide "water use" into two subcategories: water withdrawn/manipulated, the removal of water from a surface or groundwater source; and water consumption, the portion of water withdrawal that is not returned to the environment but is consumed by the process of energy production. This consumption can take several forms, including evaporation (e.g., in the cooling loop of a thermoelectric plant or from a reservoir), transpiration (e.g., irrigation water applied to energy crops), or incorporation into a product, byproduct or material of production (e.g., water used in biofuel production). Note that our definition of "withdrawal/manipulation" includes water that is removed from a river system only briefly, as water passed through a hydropower turbine located at a dam site. By classifying water used by hydropower as water manipulated, we are using a terminology that differs from the United States Geological Survey, which presents statistics on hydropower water use separate from water use for thermoelectric plant cooling. We adopted this different terminology because one of our primary goals in this paper is to present a full picture of the water used for energy production, including how that water use affects freshwater ecosystems. Dams can affect a river's flow regime, connectivity and water quality and are among of the leading sources of threat for aquatic species. Therefore, quantifying the volume of water that is run through hydropower dams' turbines (i.e., manipulated) provides relevant information on how water management by hydropower affects fish species.

We recognize two major parts of the production process in our estimate of water-use intensity (m^3 of water per GJ of energy): material/resource acquisition and processing (e.g., mining coal and preparing it for use) (Table 1); and electricity generation (Table 2). Note that the endpoint of our production analysis is the delivered energy embodied in electricity or liquid fuels. We do not account for energy that is lost during energy consumption, for example, when electricity is converted to heat for a home or when liquid fuels are converted to a car's kinetic energy. Similarly, the beginning of our analysis of the production process is material/resource acquisition and processing. We do not account for water or energy, for example, used to create the steel in the machinery that mines the coal.

For each part of the production process, we defined a high- and low-end estimate of water-use intensity (for both withdrawal/manipulation and consumption). Generally, the high-end estimate gives current levels of water-use intensity as reported in the literature, whereas the low-end estimate of water-use intensity is the lowest level reported in the literature or the lowest level likely with future technological changes (cf. [26]). This approach with a high and low estimate is meant to represent the uncertainty about future water-use intensity. It does not consider the potential impact of catastrophic events, such as an oil spill, on water resources.

There is a large variation in water-use intensity, in terms of withdrawal/manipulation, between thermoelectric plants that use once-through cooling and those that use recirculating cooling (Table 1). To reflect this fact, we have calculated at a regional level the proportion of plants that use these various technologies, using a comprehensive commercial database of the plants within the US created by the Ventyx Corporation [27]. The Ventyx data is useful because it has the spatial locations of all significant electricity producing facilities in the United States, allowing us to accurately determine what watershed the facility is located in. Our high-end estimates for 2035 assume that all existing plants continue with current cooling technology, but that all new plants use recirculating cooling technology, consistent with the general trend for recirculating cooling plants to become a greater proportion of the nation's thermoelectric plants. Our low-end estimates for 2035 assume that, between 2010 and 2030, all existing plants that use once-through cooling are converted to recirculating cooling systems. Such a transition would be expensive, and would likely only come about through some regulatory requirement, such as changes to regulations developed by the EPA around Section 316(b) of the Clean Water Act. It is important to note that existing regulations around Section 316(b) are unlikely to require such a large change in cooling technologies, and that such a change is not necessarily cost-effective or beneficial to biodiversity in all situations. However, we have included this scenario in our low-end estimates for 2035 to show the potential water withdrawal/manipulation reduction if existing plants are slowly transitioned to recirculating cooling systems.

We have included water used for the irrigation of bioenergy feedstocks in our calculations of material acquisition for biopower and biofuel production. There are three major feedstocks considered in our analysis: corn, soybean, and cellulosic.

Corn and soybean are the two feedstocks currently used for commercial biofuel production, and for these two our high-end estimate is simply a function of the current average state-level irrigation rate of the crops (m^3/tonne). The advantage of using state-level estimates is that it accounts for the considerable variation in irrigation rates between corn raised in, for instance, Indiana (primarily rain-fed) and Nebraska (a significant portion irrigated). Our low-end estimate for corn and soybean in 2035 assumes continued gains in the yield of these crops with no additional inputs of water required, consistent with historical trends, plus a transition away from gravity-fed irrigation toward more efficient sprinkler systems, also consistent with historical trends. We note that rainfed biofuel crops also transpire significant amounts of water [28], which could otherwise be used for other purposes, such as food production; however, this "green water" use is beyond the scope of our paper.

The third major feedstock we consider is generic biomass used for either cellulosic ethanol or for biopower.Here, we recognized five sources of biomass, consistent with NREL research into price-supply curves for each of these sources [29]: urban waste wood; mill waste wood; forestry residuals; agricultural residues, and dedicated biomass crops. All are assumed to have no additional

Table 1. Statistics for material resources acquisition/processing.

Type	Withdrawal/manipulation (m³GJ⁻¹)		Consumption (m³GJ⁻¹)		Water-intensity varies by	EIA forecasts by	Notes
	Low	*High*	*Low*	*High*			
Solar PV	0.486	0.549	0.061	0.161	National	Elec. Producing regions	High withdrawal value: 60% mono-SI, 40% multi-SI. Proportional split based on [42]. Low withdrawal value: 48% mono-SI, 32% multi-SI, 20% CdTe representing hypothetical future split. Water use values from [23]. Consumption: High and low values from [43].
Solar Thermal	0.0825	0.36	0.024	0.105	National	Elec. Producing regions	High values: 0.105 m³GJ⁻¹ consumption for plant construction and O&M [43]. Low values: 0.024 m³GJ⁻¹ consumption for plant construction and O&M [43]. Adjustment for withdrawal based on observed rates for solar power [43] and the ratios between withdrawal and consumption for construction shown above for the high water use number for solar PV.
Wind	0.047	0.089	0.011	0.019	National	Elec. Producing regions	Withdrawal: Taken from [23]. High value is Denmark highest example, low value is Denmark lowest example. Consumption: Taken from [44],Table 6.
Geothermal	0.003	0.031	0.001	0.011	National	Elec. Producing regions	Consumption data from Table 4-3 in [45]. Withdrawals assumed 3x bigger.
Coal	0.028	1.21	0.003	0.328	National	Coal producing regions	Withdrawal: [23], low number is for western surface mining and train transportation, high number is for eastern underground mining and slurry pipeline. Withdrawal: [23], low number is for surface mining and train transportation, high number is for underground mining and slurry pipeline.
Nuclear	0.083	0.392	0.047	0.159	National	Elec. Producing regions	Withdrawal: [23], low number is for centrifuge enrichment, top number is for diffusion enrichment. Consumption: Range taken from [17]. Brackets other studies reported values. There is variation by mining type and method of enrichment, which is captured within this range.
Natural Gas	0.033	0.153	0.025	0.036	National	Natural gas producing regions	Withdrawal: [23], low number for offshore extraction, high number on-shore (plus other components). Consumption: Low value from [23], high number assumed 50% above their number. Interestingly, our high estimates are above some estimates of the extra use for shale gas- like [2] estimated 5–29 gallons per MWh, which gives 0.0189–0.110 m³GJ⁻¹.

Table 1. Cont.

Type	Withdrawal/manipulation (m^3GJ^{-1})		Consumption (m^3GJ^{-1})		Water-intensity varies by	EIA forecasts by	Notes
	Low	High	Low	High			
Hydropower	0.00	0.00	0.00	0.00	National	Elec. Producing regions	Dam construction assumed trivial relative to water use for electricity production.
Municipal Waste	0.00	0.00	0.00	0.00	National	Elec. Producing regions	Assumed municipal waste streams would have been created anyway, so no water for waste creation.
Petroleum	0.22	0.27	0.07	0.21	National	Petroleum producing regions	[46] lists average injection water withdrawal as 8 gal H20/gal oil, with consumption varying from 2.1-5.4. Processing can vary from 0.5-2.5, and we assume all processing withdrawals are consumed.
Biopower	0.00	0.00	0.00	0.00	National except for energy crops, which are state-level	Biomass market	Assumed zero except for energy crops, because the waste would have been collected and stored anyway. Assumed all rain-fed for energy crop biomass market.
Biofuel- corn	19.4	24.3	16.4	19.7	State-level [47]	Biomass market	High number is existing average, averaging across irrigated and non-irrigated acres. Low number is estimated value for 2035, with an increase in yield and a full-switch to pressure irrigation (and thus no gravity irrigation).
Biofuel- soybean	58.3	71.7	49.6	53.6	State-level [47]	Biomass market	High number is existing average, averaging across irrigated and non-irrigated acres. Low number is estimated value for 2035, with an increase in yield and a full-switch to pressure irrigation (and thus no gravity irrigation).
Biofuel- cellulosic	0.0	0.0	0.0	0.0	State-level [47]	Biomass market	Assumed all rain-fed. See text for details.

water-use involved with their creation and use for energy purposes. For instance, forestry residues would exist anyway without a market for biomass, and so it is assumed that there is no extra "blue" water involved with their use in bioenergy projects. For dedicated biomass crops, we assume there is no irrigation involved in production (i.e., the crops are all rain-fed). This assumption is consistent with that adopted in the US Department of Energy's Billion-Ton Update study [30], and seems plausible given that the relatively low profit margins considered likely for cellulosic feedstock production may make intensive production with irrigation cost prohibitive.

Finally, to estimate the amount of water saved when a unit of energy is not consumed, due to either efficiency gains or reductions in demand, we calculated the average water withdrawal/manipulation and water consumption per unit energy for both the liquid fuel sector and for the electricity sector. To derive these values, we first calculated total energy consumption for each sector in 2010 and divided by the total water use for each sector.

Scenarios of future energy use

Our energy scenarios are taken from the EIA's Annual Energy Outlook (AEO) 2011 [4]. These scenarios were calculated by EIA's National Energy Modeling System, a comprehensive econometric model of US energy production, imports, and consumption. For each energy production technology, each EIA scenario projects energy produced and (for electricity) generation

capacity, from now until 2035, by subregion. For electricity-producing technologies, 22 electric market subregions, whose boundaries delineate areas of the US electric grid that are relatively disconnected from one another, are used. Additionally, projections of old plant capacity that will be retired and new plant capacity that will be created are available. For coal, oil, and natural gas production, resource extraction is also listed by major geographic regions of the US. For each technology, we applied our high and low water-use intensity estimates for each subregion, taking into account that the material/acquisition and processing component and the electric generation component may occur in different subregions.

We present three water use scenarios, based upon energy production scenarios developed by the EIA. Much more information on these standards available in the AEO [4].

- **Reference Case**- This scenario incorporates baseline economic growth (2.7% per year economic growth between 2009 and 2035), increasing crude oil prices (rising to $125 per barrel), and assumes that the Renewable Fuel Standard target will be met in the immediate future.

- **Greenhouse Gas (GHG) Price Case**- Similar to the Reference Case scenario but it applies a price for CO_2 emissions throughout the economy. The CO_2 price starts at $25 per ton beginning in 2013 and increases to $75 per ton by 2035.

Table 2. Statistics for electricity generation.

Type	Withdrawal/manipulation (m³GJ⁻¹)		Consumption (m³GJ⁻¹)		Water-intensity varies by	EIA forecasts by	Notes
	Low	*High*	*Low*	*High*			
Solar PV	0.004	0.021	0.004	0.021	National	Elec. Producing regions	[43] and [23]
Solar Thermal	0.58	1.06	0.58	1.06	National	Elec. Producing regions	Withdrawal and consumption: range in [23] of commercial operational technologies (excluding dish sterling and dry cooling).
Wind	0	0.001	0	0.001	National	Elec. Producing regions	[23]. Assumed all water consumed.
Geothermal	1.89	12.4	0.66	1.89	Varies by mix of open and recirculating cool in each elec. Producing region	Elec. Producing regions	[23]
Coal: once-through cooling	21.1	52.6	0.35	1.23	Varies by mix of open and recirculating cool in each elec. Producing region	Elec. Producing regions	Withdrawal: [17]. Consumption: Lower is [17] example of open loop, upper is highest value in [23].
Coal: recirculating cooling	0.35	1.23	0.31	1.23	Varies by mix of open and recirculating cool in each elec. Producing region	Elec. Producing regions	Withdrawal: lower is from closed loop tower example in [17], upper is from wet tower, subcritical example in [23]. Consumption: Lower is [17] example of IGCC, dry fed with cooling tower, upper is highest value in [17].
Nuclear: once-through cooling	26.3	63.1	0.42	0.94	Varies by mix of open and recirculating cool in each elec. Producing region	Elec. Producing regions	Withdrawal: [17]. Numerous other sources fell in this range. Consumption: Lower value from [48]. Upper value from maximum consumption of any plant in [23]. Note that some nuclear power plants use saline or brackish water for cooling, which may inflate average water-use statistics cited in the literature.
Nuclear: recirculating cooling	0.59	1.19	0.45	0.94			Withdrawal: [17]. Numerous other sources fell in this range. Consumption: Lower value from [17]. Upper value from maximum consumption of any plant in [23].
Natural Gas: once-through cooling	7.89	52.6	0.11	0.35	Varies by mix of open and recirculating cool in each elec. Producing region	Elec. Producing regions	Withdrawal:[17], low end natural gas CC with open loop, high end [17] for generic steam plant, open loop. Consumption: low end [17] for NGCC open loop, high end [17] for generic steam plant, open loop.
Natural Gas: recirculating cooling	0.25	0.66	0.20	0.54			Withdrawal: Low value from [17], NGCC closed loop tower, upper value from [17] closed loop generic steam plant. Consumption: low from [17], NGCC, closed loop, upper value from [17]closed loop generic steam plant.
Hydropower	1,811 (US mean)	2,173 (US mean)	4.6 (US mean)	14.1 (US mean)	State-level	Elec. Producing regions	Manipulation: Calculated from head of dams listed in the National Inventory of Dams [36]. For average head in each electricity producing region, high number is 75% turbine efficiency, low number is 90% turbine efficiency. Consumption: NREL data [49].

Table 2. Cont.

Type	Withdrawal/manipulation (m³GJ⁻¹)		Consumption (m³GJ⁻¹)		Water-intensity varies by	EIA forecasts by	Notes
	Low	High	Low	High			
Municipal Waste	6.6	16.7	0.1	0.5	National	Elec. Producing regions	[40] generic thermoelectric plant numbers
Petroleum: once-through cooling	21.1	52.6	0.35	0.52	Varies by mix of open and recirculating cool in each elec. Producing region	Elec. Producing regions	Withdrawal: [17], generic open-loop thermoelectric plant. Consumption: lower [17], generic open-loop thermoelectric plant. Upper assumed 50% higher.
Petroleum: recirculating cooling	0.35	0.66	0.35	0.52			Withdrawal: [17] generic closed loop thermoelectric (tower). Consumption: [17] generic closed loop thermoelectric plant.
Biopower	6.6	16.7	0.1	0.5	Varies by mix of open and recirculating cool in each elec. Producing region	Elec. Producing regions	[23] generic thermoelectric plant numbers

- **Extended Policies Case**- This scenario differs from the Reference Case in that additional government regulations foster improved electricity and liquid fuel efficiency. It assumes new light duty vehicle CAFE standards (to 46 miles per gallon by 2025) and tailpipe emissions standards, and includes additional rounds of efficiency standards for currently covered products, as well as new standards for products not yet covered.

Where energy sector water impact occurs

The next phase of the analysis involved apportioning the water use predictions by various subregions among 18 major water resource regions for the contiguous US, as defined by the USGS Hydrologic Unit Code (HUC) system. Our general strategy was to use higher-resolution information on where material acquisition/resource processing or electricity generation occurs to partition the water use as accurately as possible. For projections into the future, we used information on both current and proposed energy facilities contained in the Ventyx database to partition the water use.

For material acquisition/resource processing, the method used varied by energy technique (Table 1). For technologies that involved resource extraction (coal, natural gas, petroleum, uranium), we used maps of production areas available from the EIA. Bioenergy irrigation was partitioned using a map of irrigated area in the US [31]. Biofuel processing was partitioned using information on the location of current and proposed biofuel production facilities from the Ventyx database. For other energy production techniques, material acquisition/resource processing is small relative to water use for electricity generation, and it was partitioned proportional to electric generation (see below), essentially assuming that most material acquisition/resource processing occurs in the same major water resource regions where this electricity is generated.

For electricity generation, we used the Ventyx database to calculate, for each energy production technique, the total electricity generation in each major water resource region. Since the location, as well as the capacity (MW), of most facilities is known with great precision, it is possible to calculate this accurately. Water-use was then partitioned among major water resource regions using this calculation.

Energy sector water use and freshwater biodiversity

Data on the status, source of imperilment, and geographic range of US freshwater fish species were taken from NatureServe [32], as updated in Mims et al. [33]. Fish species were classified as imperiled according to NatureServe conservation ranking categories of G1 (defined as at very high risk of extinction due to extreme rarity, such as 5 or fewer populations, very steep declines, or other factors) and G2 (defined as at high risk of extinction or elimination due to very restricted range, very few populations, steep declines, or other factors) [34]. Next, for all 239 imperiled freshwater fish species detailed text descriptions of the source of imperilment were used to assign threats to species from 9 major threat categories: dams/impoundments; invasive/introduced species; altered hydrologic flow/channelization; overharvesting/overfishing; pollution/water quality; sedimentation/turbidity/siltation; excess water consumption/withdrawal; and hybridization. Most species had more than one threat listed, with dams/impoundments being most commonly listed (36.0% of species), followed by pollution/water quality threat (32.6% of species).

While water used in cooling of thermoelectric power plants is overwhelmingly from surface water, water used for irrigation is frequently obtained from groundwater [1]. The effects of groundwater use on freshwater fish species are often different than those of surface water use. Although there are linkages between groundwater use and the quantity and quality of surface water, these links are complex and depend on the hydrology of the river basin and underlying aquifer. In order to make our estimates of water use more biologically meaningful, we used county-level information on the proportion of withdrawals from groundwater and surface water to split water use into its surface and groundwater components. For this calculation, all thermoelectric cooling water use was split into surface and groundwater components based upon the proportion of freshwater withdrawals from groundwater reported for the power sector, while irrigation water use was split into surface and groundwater components based upon the proportion of freshwater withdrawals reported from groundwater for the agriculture sector [1]. For our statistical analysis, we have only used information on the surface component of withdrawal or consumption, since the amount of groundwater used by the energy sector is a relatively small portion of the total groundwater withdrawal in most basins and hence seemed unlikely to be statistically related to fish imperilment. Moreover, the amount of water available in groundwater basins is often unknown or poorly characterized, making the normalization (see below) of groundwater use difficult.

Next, we normalized surface water use by available water, to obtain the proportion of water used for energy production in each subregion. Specifically, we divided three metrics of water-use (hydropower water manipulation, non-hydropower energy sector surface water withdrawals, and energy sector surface water consumption) by average annual surface water availability in each water resource region [35] (Table 3). Hydropower water manipulation was treated separately from non-hydropower energy sector surface water withdrawal for this calculation to see if there were different patterns for the two subcomponents of withdrawal/manipulation. Finally, we calculated the range-area-weighted average of normalized water use for each fish species:

where R_i is the total range size (area) of the species in major hydrologic region i, and U_i is the normalized water use in major hydrologic region i.

We tested two related hypotheses using logistic regression analysis. First, we tested to see if expert evaluation of the threats facing each species is consistent with our metrics of water use, simply examining whether fish species with a particular reported threat had higher relative water-use on the appropriate metric (e.g, species threatened by dams and hydropower water use). Second, we tested to see if the probability of fish species imperilment is positively correlated with one of our three metrics of water-use (hydropower water manipulation, non-hydropower energy sector surface water withdrawals, and energy sector surface water consumption), after accounting for species range size. For each logistic regression analysis, we first added the term for species

range, then the term for normalized water use, and then tested for any interaction terms. At each step, the significance of each addition was tested using likelihood ratio tests. When comparing between models using different metrics of normalized water use (i.e., not nested models), we used Akaike's Information Criterion (AIC). To improve normality of variables and meet the assumptions of logistic regression, our metrics of normalized water-use and species area were log-transformed.

Results

Energy production and consumption

Current domestic energy production is dominated by coal and natural gas [4], with significant contributions from oil production and nuclear energy (Figure 2A). By 2035, the Reference Case predicts an increase in production from all sources, particularly natural gas. There would also be a large increase in biofuel production, driven by US federal policy. The GHG Price Case predicts much less production from coal, and an increase in wind and biofuels, relative to the Reference Case. The Extended Policies Case predicts a similar combination of energy sources as the Reference Case, but with less overall production, as increased energy efficiency would reduce aggregate demand for energy.

Coal, natural gas, and nuclear dominate current domestic electricity generation (Figure 2B). According to projections under the Reference Case, electricity generation increases in most sectors, including the renewable technologies of wind and biomass. The Extended Policies Case includes a composition of energy production sources similar to the Reference Case, but with less total domestic electricity production due to the decreased demand from efficiency measures. The GHG Price Case predicts reduced electricity generation from coal and greatly increased generation from wind and biomass relative to the Reference Case.

Water use intensity

The water withdrawal/manipulation intensity of energy production techniques can vary over five orders of magnitude (Figure 3A). The largest water withdrawal/manipulation intensity is for hydropower. The national average water withdrawal/manipulation intensity for hydropower, weighted by electricity-production, is estimated as $1{,}810$–$2{,}170$ $m^3 GJ^{-1}$ (ranges for hydropower indicate uncertainty about the efficiency of turbines). However, the water withdrawal/manipulation intensity for hydropower varies greatly for different electricity producing regions. This large regional variation is primarily due to variation in hydrologic head [36]: electricity producing regions with dams with a large hydrologic head will have relatively smaller water withdrawal/manipulation intensities. For instance, the "Western Electricity Coordinating Council/Southwest" electricity producing region, which includes most of Arizona and New Mexico, has a large weighted average head, driven by big dams like the Hoover Dam and Glen Canyon Dam, and thus has a low estimated water withdrawal/manipulation intensity of 544–652 $m^3 GJ^{-1}$. By contrast, the "Florida Reliability Coordinating Council" region has a small weighted average head, driven by the lack of topographic relief in Florida, and thus has a high estimated water withdrawal/manipulation intensity of $14{,}900$–$17{,}900$ $m^3 GJ^{-1}$.

For thermoelectric power production (e.g., from coal, natural gas and nuclear energy sources), the major difference in water withdrawal/manipulation intensity is between once-through cooling (high water withdrawal/manipulation intensity) and recirculating cooling (low water withdrawal/manipulation intensity). Nuclear power has higher average water withdrawal/manipulation intensity because a higher proportion of nuclear plants are

Table 3. Average water availability by major hydrologic region.

Water Resource Region	Average flow, million m³/yr (1901–2009)
New England	97,100
Mid-Atlantic	133,600
South Atlantic-Gulf	306,400
Great Lakes	143,200
Ohio	188,900
Tennessee	58,000
Upper Mississippi	96,700
Lower Mississippi	111,300
Souris-Red-Rainy	8,100
Missouri	76,900
Arkansas-White-Red	41,100
Texas-Gulf	24,100
Rio Grande	6,200
Upper Colorado	16,700
Lower Colorado	5,600
Great Basin	15,600
Pacific Northwest	223,700
California	117,500

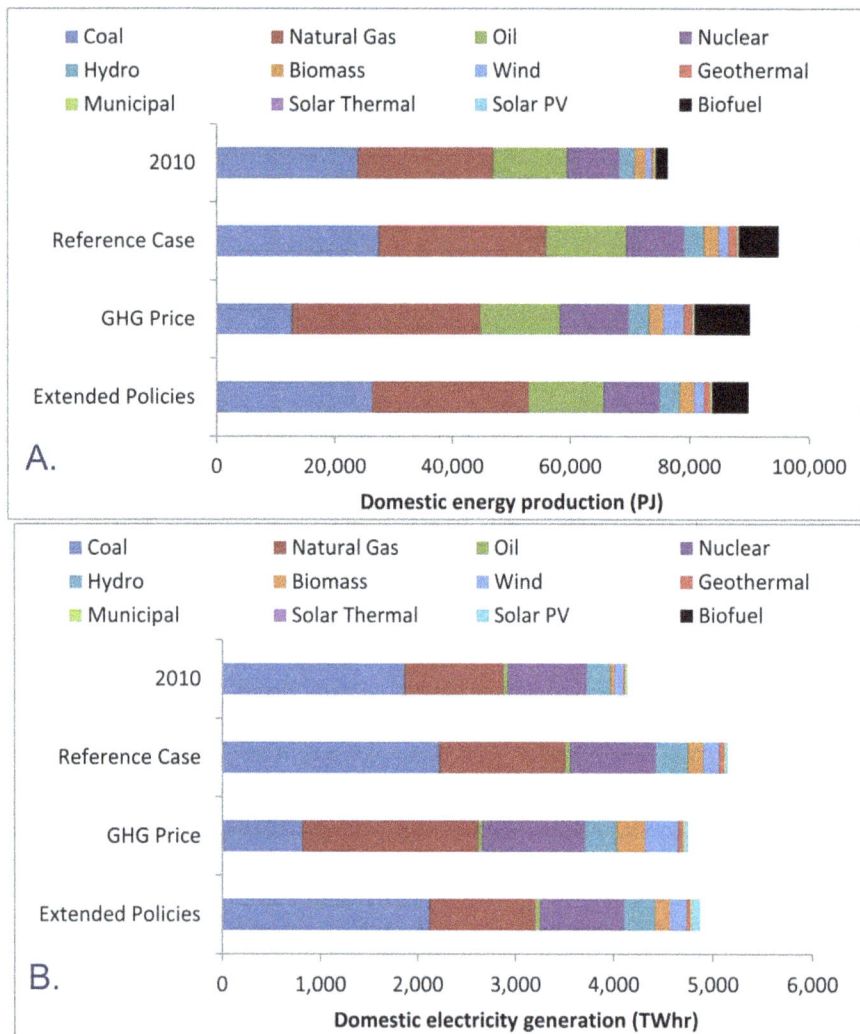

Figure 2. United States domestic energy creation. US annual energy production (A) and electricity generation (B), in 2010 and in 2035 for three scenarios of future energy policy. Annual energy production is shown in petajoules and electricity generation is shown in terawatt-hours.

once-through cooling than other types of thermoelectric plants. Natural gas power has lower average water withdrawal/manipulation intensity because combined cycle gas turbine plants (the dominant natural gas power plant type) generally use less cooling water. Finally, renewable energy production technologies, such as solar and wind, have among the lowest water withdrawal/manipulation intensities of the technologies assessed.

Energy conservation (reduced energy consumption caused by increases in energy efficiency or reduced demand) would reduce US water withdrawal/manipulation. This effect is greatest for the electric sector, where every 1 GJ of electricity conserved would save 27–36 m^3 of water withdrawal/manipulation (midpoint of range, 31.1 m^{3}GJ^{-1}). By contrast, the effect would be less for the liquid fuel sector (0.4–0.7 m^{3}GJ^{-1} of liquid fuels saved, midpoint estimate 0.5 m^{3}GJ^{-1}). This results in part because so much of the energy in the US liquid fuel sector comes from petroleum, which is extracted abroad and hence does not figure into the calculations of US water withdrawal/manipulation. Another reason for this trend is that the use of cooling water in many thermoelectric plants is so high relative to the water used in the extraction of petroleum.

Water-consumption intensity trends among sectors have somewhat similar patterns to those for water withdrawal/manipulation

intensity (Figure 3B). Hydroelectric power demonstrates among the highest water consumption intensities due primarily to evaporation from reservoirs. The weighted-average water consumption intensity for the US is 4.6–14.1 m^{3}GJ^{-1} (midpoint estimate 9.3 m^{3}GJ^{-1}). Note that some hydropower dams, including many of the largest, are multipurpose dams that also provide water supply or flood control benefits. Moreover, the amount of water evaporated off reservoirs per unit power produced varies among electricity producing region, depending on the climate (arid climates have more evaporation than humid climates) and the configuration of reservoirs (wide shallow reservoirs have more evaporation than narrow deep reservoirs). For instance, the "Texas Regional Entity" electricity producing region, which includes most of south Texas, has the largest estimated water consumption intensity of 47–144 m^{3}GJ^{-1} (midpoint estimate 96 m^{3}GJ^{-1}), driven by a dry climate and relatively little topographic relief, which implies wide flat reservoirs. By contrast, the "Midwest Reliability Council -West" region, which includes places like Minnesota, has the lowest estimated water-consumption intensity of 0.7–2.3 m^{3}GJ^{-1} (midpoint estimate 1.5 m^{3}GJ^{-1}), presumably because of the low evaporation rates off reservoirs in this relatively cold and humid climate.

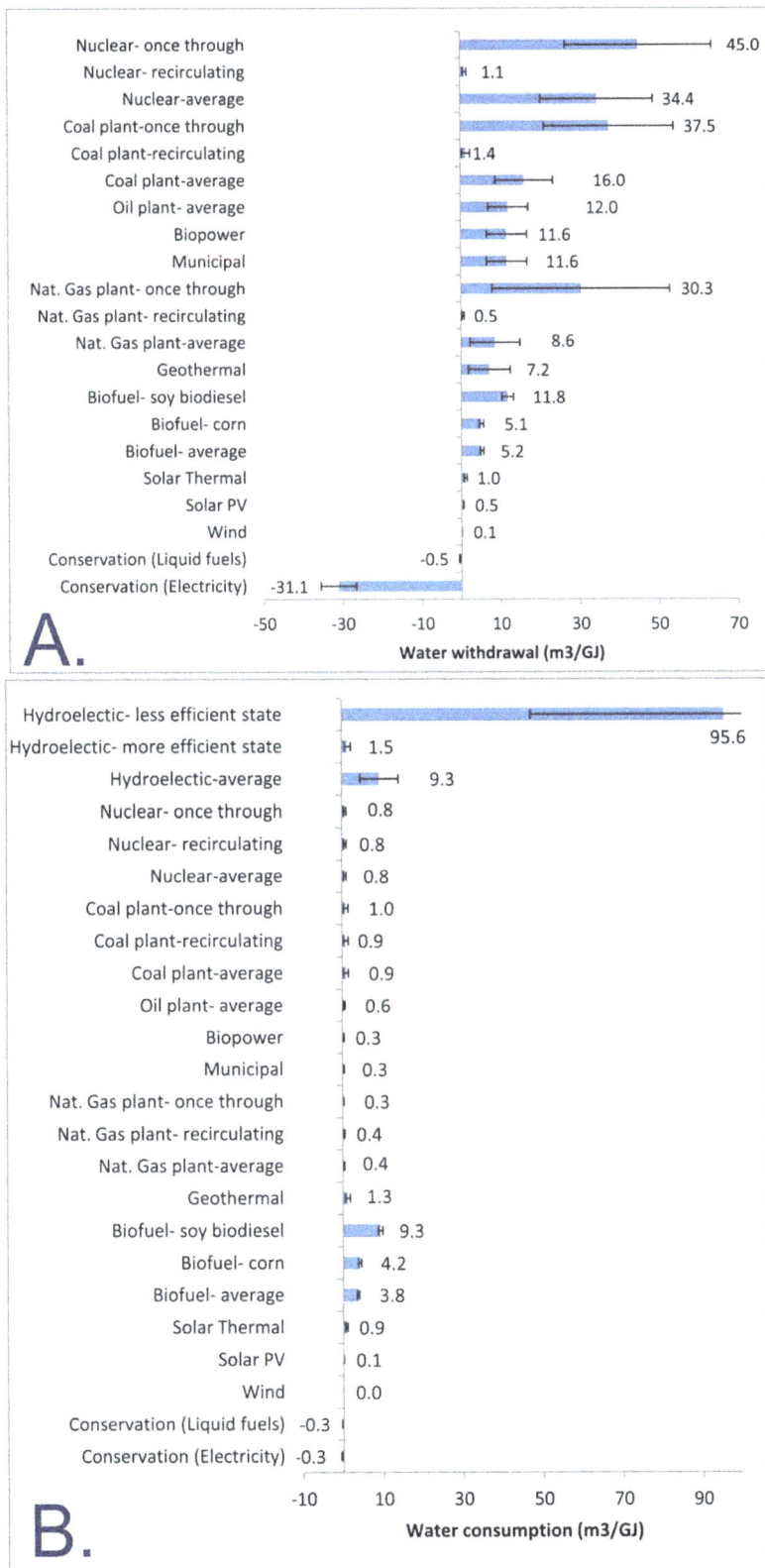

Figure 3. Water-use intensity of energy technologies. Water-use intensity (m^3GJ^{-1}) of US domestic energy production or energy conservation, in terms of water withdrawal (A) or water consumption (B). These water-use intensity estimates include water for material acquisition and processing, as well as for electricity generation where applicable. Errors bars indicate the range of our low and high water-use intensity estimates. The value labeled is the midpoint between these high and low estimates. The effect of energy conservation is shown using the energy mix in 2010. For hydropower, for display purposes typical consumption values are shown for more efficient and less efficient regions. Because hydropower water manipulation is more than an order of magnitude greater than water withdrawals for other technologies, hydropower is omitted in the top panel (A).

Biofuel production also has high water consumption intensities, due to the high fraction of irrigation water that is either lost to evapotranspiration or incorporated into plant biomass. Compared with the large differences in water withdrawal/manipulation intensities, there is little difference in water consumption intensities between once-through and recirculating cooling thermoelectric plants. The large differences in the intensity of water withdrawal/manipulation are offset because the vast majority of water used in once-through cooling is returned rather than consumed. Geothermal and solar thermal have similar water consumption intensities to fossil fuel technologies. However, solar PV and wind have much lower water consumption intensity.

Energy conservation would also reduce US water consumption. This effect would be similar in size for the liquid fuel sector, where every 1 GJ conserved would save 0.2–0.3 m^3 of water consumption (midpoint of range, 0.25 m^{3}GJ^{-1}) and for the electricity sector (0.1–0.5 m^{3}GJ^{-1} of electricity saved, midpoint estimate 0.3 m^{3}GJ^{-1}). The effect of energy conservation of liquid fuels on US water consumption is high, relative to the situation with US water withdrawal/manipulation, because a fraction of liquid fuels come from biofuels, and irrigation water used for biofuel production has a much larger consumption of water per unit energy than other energy production techniques.

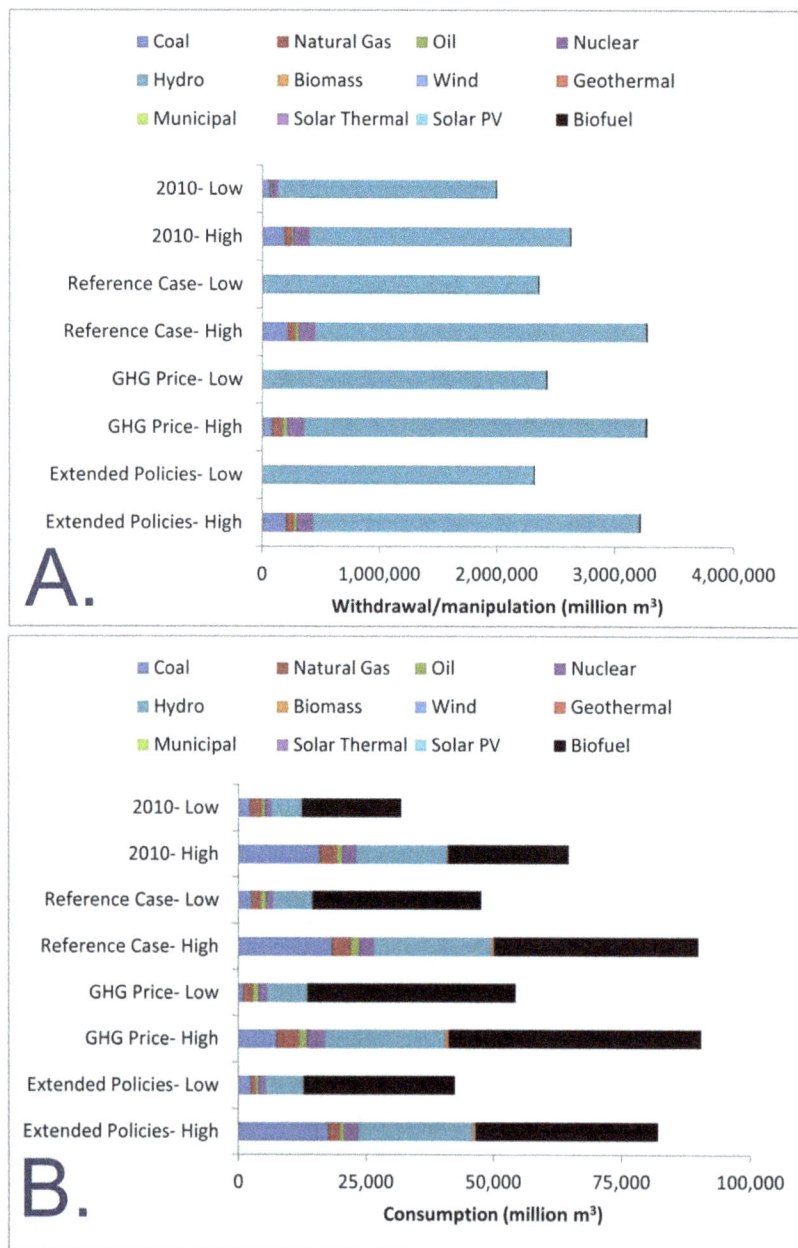

Figure 4. Water use with different energy policies. Water withdrawn (A) and consumed (B) for US domestic energy production, in Mm^3, in 2010 and in 2035 for three scenarios of future energy policy. For each scenario, we show the value implied by our low and high water-use intensity estimates (Figure 3). Note the different scales between the two graphs.

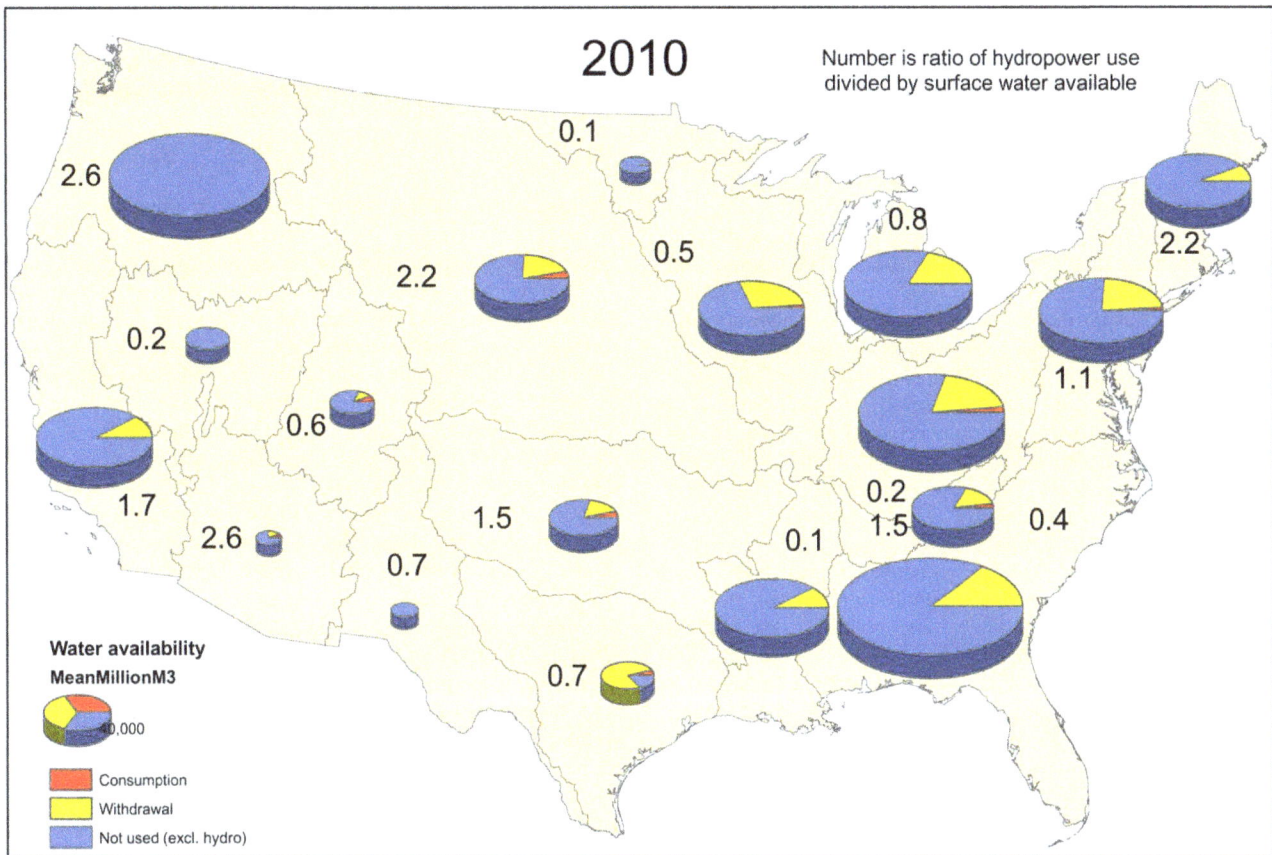

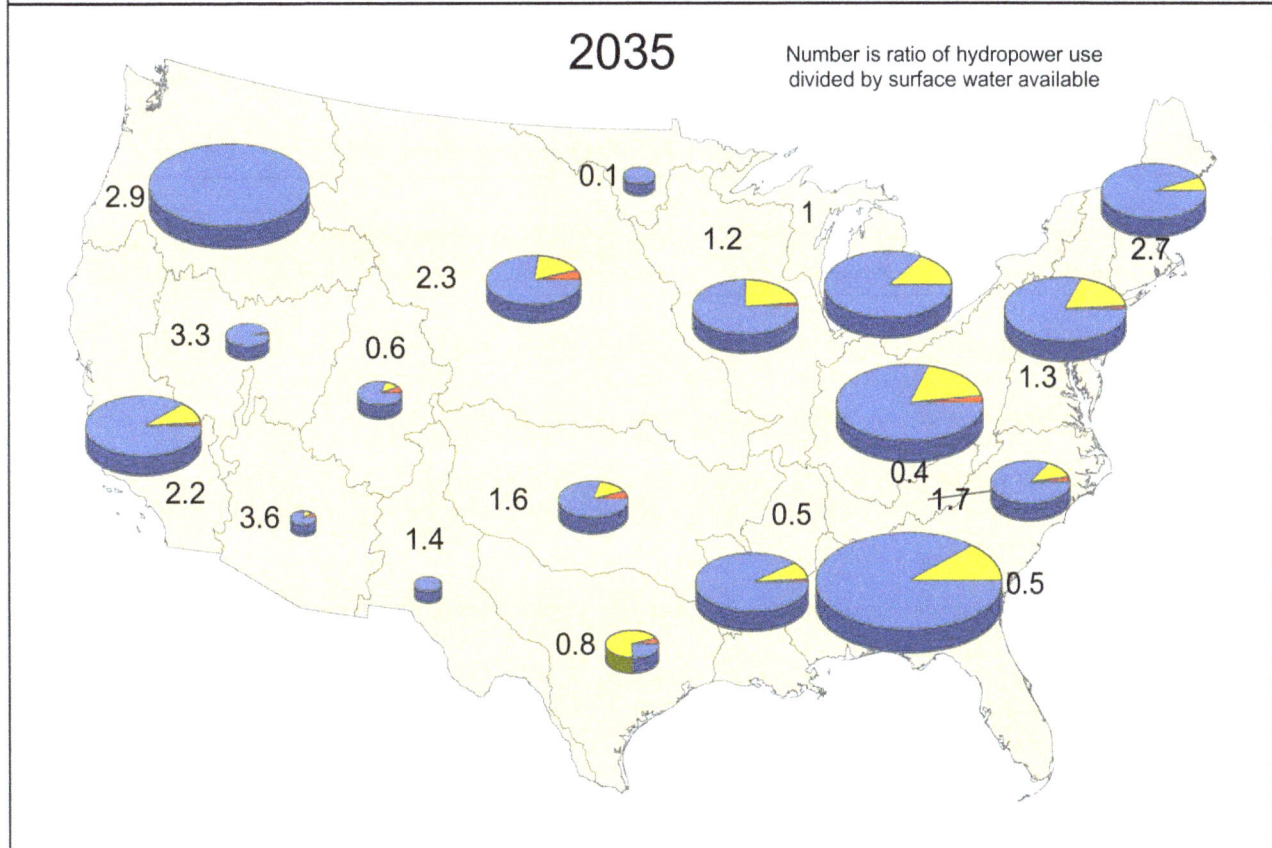

Figure 5. Water use by major water resource regions. Water use by the energy sector in major water resource regions in 2010 (A) and 2035 (B), under the Reference Case. The size of the pie chart indicates the total water available (mean Mm^3 per year) in major water resource regions. The pie chart is divided into three colors, based on energy sector water use (excluding hydropower production). Water not used by the energy sector is shown in blue, while water withdrawn but not consumed is shown in yellow, and water withdrawn and consumed is shown in red. Then, the number in each region indicates the amount of water used specifically for hydropower production divided by total water available.

Energy sector water use

Hydropower currently accounts for the largest total withdrawal/manipulation ($1,851,000-2,222,000$ Mm^3) by far (Figure 4A), followed by coal ($63,500-186,000$ Mm^3) and nuclear ($58,200-140,000$ Mm^3). Note that for future scenarios our high-intensity number assumes a slow shift in technology as new thermoelectric plants use recirculating cooling and old plants gradually cease operations. By contrast, our low-intensity number describes a future in which EPA regulations or incentives drive all older plants to use recirculating cooling by 2035. By 2035, the Reference Case predicts that overall water withdrawal/manipulation will increase, because of additional hydroelectric, biofuel, and nuclear energy production.

The rank ordering of energy technologies is different with regards to current water consumption (Figure 4B), led by hydropower ($5,830-17,800$ Mm^3), and followed by biofuels ($4,300-5,300$ Mm^3), coal ($2,240-15,800$ Mm^3) and natural gas ($2,240-3,410$ Mm^3). Any shift toward a greater fraction of thermoelectric plants using recirculating cooling, rather than once-through cooling, does not greatly affect total consumption, since consumption water-use intensities for the two technologies are similar. The GHG Price Case predicts similar water consumption to the Reference Case, while the Extended Policies case predicts less water consumption due to decreased energy consumption.

National-level statistics of energy sector water use mask significant variation among major water use regions (Figure 5A).

The largest hydroelectric water withdrawal/manipulation, relative to the average annual water availability, is in the New England, Missouri, Lower Colorado, and Pacific Northwest regions, where the average water molecule has gone through more than two hydroelectric turbines by the time it flows to the ocean. By contrast, hydropower withdrawal/manipulation are a small fraction of the available water in the Great Basin, Ohio, Lower Mississippi, and Souris-Red-Rainy regions. Estimated hydroelectric water withdrawal/manipulation in million m^3 is shown in Table 4. Under the Reference Case, some water resource regions such as the Great Basin, Lower Colorado, Upper Mississippi, and Rio Grande regions are projected to have a substantial increase in hydroelectric power (Figure 5B). The other two scenarios predict similar spatial patterns as the Reference Case.

As a share of available water, the largest withdrawal (excluding hydropower use) is in the Texas-Gulf and Lower Colorado water resource regions (Figure 5A). The lowest withdrawal relative to availability is in the Pacific Northwest, Great Basin, and Souris-Red-Rainy water resource regions. The vast majority of all withdrawals (excluding hydropower use) are for thermoelectric plant cooling, although the Missouri River basin also has a significant amount of water withdrawn for irrigation for bioenergy production (Table 4). By 2035, the Reference Case predicts an increase in energy sector withdrawal (Figure 5B), driven by irrigation water for energy crops, particularly corn for ethanol production. At the same time, there may be a decrease in withdrawals by 2035 in some regions like the Texas-Gulf and

Table 4. Water withdrawal/manipulation and consumption by major hydrologic region.

Water Resource Region	Withdrawal/manipulation (million m³)			Consumption (million m³)		
	Thermoelectric	Hydropower	Biomass production	Thermoelectric	Hydropower	Biomass production
New England	4,340-11,216	213,207-255,848	0	116-310	96-294	0
Mid-Atlantic	17,237-46,934	140,466-168,559	12-17	635-1,783	172-525	11-14
South Atlantic-Gulf	23,497-60,122	123,343-148,011	57-69	789-2,604	423-1,294	49-56
Great Lakes	14,512-38,488	116,730-140,076	28-38	403-1,340	222-678	25-31
Ohio	20,255-57,933	45,794-54,953	16-22	1,163-4,797	235-717	14-20
Tennessee	5,981-14,830	84,374-101,249	3	162-450	1,101-3,366	3
Upper Mississippi	15,895-42,835	47,648-57,178	18-26	441-1,905	22-67	17-23
Lower Mississippi	4,745-17,421	10,601-12,722	71-102	164-958	57-175	61-81
Souris-Red-Rainy	48-127	1,128-354	10-13	6	1	8-11
Missouri	7,752-26,302	165,476-198,572	1,331-1,716	351-2,557	139-426	1,156-1,395
Arkansas-White-Red	3,709-12,868	63,121-75,745	213-270	219-1,493	690-2,110	184-219
Texas-Gulf	8,936-31,733	17,192-20,630	124-156	327-1,569	78-237	108-127
Rio Grande	3-29	4,613-5,536	46-63	8	30-91	40-51
Upper Colorado	295-5,402	10,127-12,153	100-141	161-1,617	238-26	86-114
Lower Colorado	855-2,353	14,333-17,199	9-13	52-209	510-1,558	8-10
Great Basin	33-228	3,696-4,436	18-26	18-84	10-30	15-21
Pacific Northwest	806-2,419	587,454-04,944	146-182	90-292	986-3,016	127-148
California	5,129-19,232	202,466-42,959	63-82	214-504	818-2,502	55-67

Lower Colorado water resource regions, depending on the rate of transition from once-through to recirculating cooling.

Relative to available water, consumption by the energy sector is generally low (Figure 5A). The sources of water consumption vary significantly among water resource regions (Table 4). In the majority of water resource regions, consumption by thermoelectric power plants is the dominant source of consumption. However, in a few water resource regions, like the Tennessee and the Lower Colorado, water consumed by hydropower (i.e., evaporation off reservoirs) is the dominant source of water consumption. Water consumed in bioenergy production is a significant part of total energy sector water consumption in the Missouri water resource region. By 2035, the Reference Case predicts an increase in biofuel water consumption (Figure 5B), primarily in the Missouri and Upper Mississippi water resource regions, and an increase in hydroelectric consumption, which is largest in the Great Basin and Lower Colorado water resource regions.

Implications for fish species

Our metric of greater normalized hydroelectric water manipulation (i.e., hydropower use through turbines divided by water availability) appears consistent with expert evaluation of the threat dams pose to fish species. Fish for which "dams/impoundments" were reported as a threat had an average normalized hydroelectric water manipulation of 0.85, whereas those fish species where it was not reported as a threat had an average normalized hydroelectric water manipulation of 0.76.

Water resource regions with greater normalized hydroelectric water manipulation have a greater proportion of imperiled fish species, after controlling for species range size and its interaction with normalized hydroelectric water manipulation ($\chi^2 = 255.97$, df $= 3$, P$<$0.001, Table 5). The effect of the interaction term in this best-fit model can best be seen graphically (Figure 6). Species with smaller ranges have a higher probability of being imperiled at any level of normalized hydroelectric water manipulation than do species with larger ranges. Moreover, the probability of imperilment for species with small ranges increase more rapidly with increases in normalized hydroelectric water manipulation than do species with large ranges.

Similarly, our metric of greater normalized energy sector consumption (i.e., consumption divided by water availability) appears consistent with expert evaluation of the threat posed to fish species. Fish for which "excess withdrawals/consumption" were reported as a threat had an average normalized energy sector

water consumption of 0.011, whereas those fish species where it was not reported as a threat had an average normalized energy sector water consumption of 0.008.

Water resource regions with greater normalized energy sector water consumption have a greater proportion of imperiled fish species, after controlling for species range size and its interaction with normalized energy sector water consumption ($\chi^2 = 268.78$, df $= 3$, P$<$0.001, Table 6). The effect of the interaction term in this best-fit model can best be seen graphically (Figure 7). Species with smaller ranges have a higher probability of being imperiled at any level of normalized energy sector water consumption than do species with larger ranges. Moreover, species with small ranges increase more rapidly with increases in normalized energy sector water consumption than do species with large ranges.

Care must be taken when extrapolating these correlative patterns to the future, because our regression model was fit to cross-sectional data not panel data. However, some insight into which types of species are most likely to be imperiled in the future can be gained by examining a projection of regression results. In general, species with small ranges that are in water resource regions with a projected increase in energy-sector water consumption are most at risk. Under the Reference Case, water resource regions with an increase in consumption include the Lower Mississippi, the Texas-Gulf, and the Lower Colorado. Note that a more realistic evaluation of the effect of new energy sector water use on a particular fish species would require an analysis of the new energy development within that species range, as well as knowledge of the species-specific sensitivity to alterations to hydrological conditions associated with energy sector water use.

Discussion

Our study revealed strong correlative relationships between current energy-sector water use and the likelihood of fish species imperilment in the US. If this association holds into the future then we expect that projected increases in energy-sector water use have the potential to further threaten freshwater fish. Since federal and state energy policies affect the combination of technologies used to produce energy, which in turn affects the amount of water withdrawn and consumed for energy production, energy policy decisions plays an important role in the conservation of freshwater fish biodiversity into the future. Below, we discuss a few key aspects of energy policy that affect how much energy sector water use occurs, and then explore their implications for fish species imperilment.

Table 5. Probability of a fish species being imperiled as a function of normalized hydropower manipulation.

Predictor	β	SE β	Wald's χ^2	df	P	e^β (odds-ratio)
Intercept	9.97	0.97	104.92	1	<0.001	NA
LN(Normalized hydropower water manipulation)	−3.47	1.23	7.94	1	0.0048	0.031
LN(Total Range)	−1.06	0.095	126.32	1	<0.001	0.347
Interaction	0.37	0.12	9.05	1	0.0026	1.45
Test			χ^2	df	P	
Overall model evaluation:						
Likelihood ratio test			255.97	3	<0.001	
Score test			210.63	3	<0.001	
Wald test			133.17	3	<0.001	

Note: Kendall's Tau-a = 0.240. Goodman-Kruskal Gamma = 0.763. Somers's Dxy = 0.762, c-statistic = 88.1%.

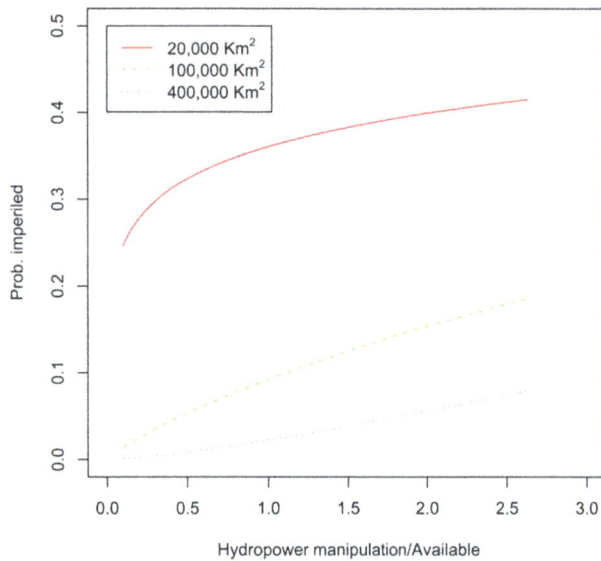

Figure 6. Fish imperilment as a function of hydropower water manipulation. Probability of a fish species being imperiled, as a function of the normalized hydropower water manipulation (i.e., water used in turbines/available). Curves are shown for three range sizes (km^2), corresponding to the 25, 50, and 75th percentile of fish species range sizes.

Energy conservation

Policies that limit total energy consumption, either through increased energy efficiency or incentives to reduce energy use, conserve water. Energy efficiency gains are greatest in the Extended Policies Case, where strict energy efficiency standards for buildings and appliances and tighter Corporate Average Fuel Economy (CAFE) standards would reduce energy production by 5,000 PJ below the Reference Case. Our calculations suggest this would save 33,900–54,300 Mm^3 of annual withdrawal/manipulation (2,510–4,650 Mm^3 of consumption). In general, policies that limit use of liquid fuels would cause a slightly greater reduction in water consumption per unit energy (m^{3}GJ^{-1}) than would policies that limit electricity use, largely because the consumptive water-use intensity of corn ethanol production is so high.

Climate policy

The effect of climate policy on energy-sector water use is complicated and varies by energy-production technology. The GHG Price Case would reduce total energy production by 4,750 PJ below the Reference Case, as higher electricity prices from fossil fuel sources would drive reductions in energy use. The GHG Price Case would also reduce water use by accelerating the retirement of old thermoelectric plants that disproportionately use once-through cooling. It would also shift some electricity production to renewable technologies that have relatively low water-use intensities. This would reduce withdrawals by thermoelectric plants by 1,740–91,000 Mm^3 (1,180–9,000 Mm^3 of consumption). Our study did not consider the effect of CCS technology on water use. Chandell et al. [13] estimated that CCS could increase water-use intensity by 25%, but their results are in agreement with our finding that overall climate policy would decrease water use by thermoelectric plants. However, the effects of climate policy on water-use in the liquid fuels sector are less clear. Increased biofuel production under the GHG Price Case means that withdrawals and consumption for this sector would increase by an even greater amount than the savings in the thermoelectric sector. The net effect is that the GHG Price Case withdrawal/manipulation would be approximately equal in 2035 to those in the Reference Case, as decreased withdrawals by thermoelectric plants are offset by increased biofuel production.

Bioenergy production

All of our scenarios assume full implementation of the strong incentives for biofuel production mandated by the Energy Independence and Security Act of 2007 (EISA). Increased biofuel from corn and soybeans will increase the amount of water used for energy in the US, since both crops are occasionally irrigated. Note that our analysis assumed that new feedstocks for cellulosic ethanol are entirely rain-fed. If, however, feedstocks for cellulosic ethanol require irrigation, at least in some places in the US, then the water required for biofuel production may increase significantly.

Thermoelectric plant cooling

The switch of thermoelectric plants from once-through cooling to recirculating cooling could substantially reduce water withdrawals. The biggest unknown here is how fast this shift will be, which is a function of market dynamics (i.e., how fast existing once-through cooled plants are retired) and policy regulation (i.e., if EPA regulations encourage existing plants to convert to

Table 6. Probability of a fish species being imperiled as a function of normalized energy sector water consumption.

Predictor	β	SE β	Wald's χ^2	df	P	e^β (odds-ratio)
Intercept	−9.92	4.75	4.37	1	0.036	NA
LN(Normalized energy sector surface water consumption)	−4.44	1.07	17.30	1	<0.001	0.012
LN(Total Range)	1.08	0.47	5.20	1	0.023	2.94
Interaction	0.48	0.11	19.92	1	<0.001	1.6
Test			**χ^2**	**df**	**P**	
Overall model evaluation:						
Likelihood ratio test			268.78	3	<0.001	
Score test			221.34	3	<0.001	
Wald test			128.97	3	<0.001	

Note: Kendall's Tau-a = 0.246. Goodman-Kruskal Gamma = 0.784. Somers's Dxy = 0.783, c-statistic = 89.1%.

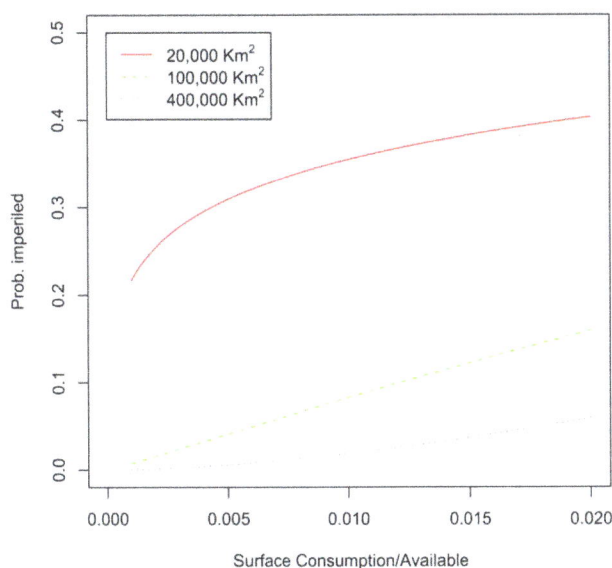

Figure 7. Fish imperilment as a function of surface water consumption. Probability of a fish species being imperiled, as a function of the normalized energy sector surface water consumption (i.e., consumption/available). Curves are shown for three range sizes (km^2), corresponding to the 25, 50, and 75th percentile of fish species range sizes.

recirculating cooling). Our calculations show that, in the Reference Case, conversion of all existing plants to recirculating cooling over a 20-year period would reduce annual water withdrawals in 2035 by 441,000 Mm^3, relative to a continuation of the current gradual shift as existing once-through cooling plants are retired and new recirculating cooling ones come online. However, the switch from once-through cooling to recirculating cooling would not significantly reduce water consumption, which appears to have a greater impact, and is more predictive of native fish imperilment compared to water withdrawal. Moreover, the greatest reductions in withdrawals from the switch from once-through cooling to recirculating would be in the Northeast and Midwest, areas that have relatively low levels of fish endemism and imperilment.

Implications for fish imperilment

The potential future impacts of energy-sector water use vary significantly by water resource region. Our statistical results show

that fish species most likely to be imperiled have small ranges and are in water resource regions with high energy-sector normalized water consumption. The Southwestern US has both factors, with a large number of species with small ranges and high normalized water consumption, reinforcing the threat of water development on arid and semi-arid fish [37]. The Southeastern US also has high levels of species endemism and relatively high normalized water consumption. One implication of our results is that irrigation water for biofuel feedstocks may impact fish in places such as the Missouri and Arkansas-White-Red water resource regions. However, it is worth remembering that while our analysis looked at aggregate energy-sector consumption, different kinds of water consumption may have very different effects on fish. Water consumption in thermoelectric plant cooling likely has different ecological effects on particular fish species than water consumption in hydroelectric power consumption or during irrigation of bioenergy crops. More species-specific and basin-specific analyses are needed to understand the components of energy sector water consumption that are of most concern for species imperilment.

Conclusions

Our results emphasize that policy decisions about energy are also decisions about water use, and that water sustainability and the health of freshwater ecosystems should be fully considered among the many factors that drive energy policy. The per-unit energy impacts of different energy technologies on water use, land use [26], and GHG emissions [38–41] vary dramatically. Energy policy involves complex tradeoffs among different aspects of freshwater ecosystems, economic needs, food security, climate change, and energy security. Nevertheless, certain technologies seem to have limited adverse impacts regardless of future scenarios. Energy efficiency, most notably, saves water and land [26] while reducing GHG emissions, resulting in a more sustainable future for biodiversity and energy production.

Acknowledgments

We thank Joe Kiesecker and three anonymous reviewers for comments on earlier drafts of this manuscript. J. Slaats provided GIS support for this project. Staff at EIA were very helpful in answering technical questions. We thank all of the organizations that created data that made this analysis possible, including: NETL, DOE NREL, USGS, and Ventyx Corporation.

Author Contributions

Conceived and designed the experiments: RIM JJO WMM JF CR JVH JP. Performed the experiments: RIM JDO JJO. Analyzed the data: RIM JDO JJO WMM JF. Wrote the paper: RIM JDO JJO WMM JF JVH.

References

1. Kenny J, Barber N, Hutson S, Linsey K, Lovelace J, et al. (2009) Estimated Use of Water in the United States in 2005. Reston, VA: U.S. Geological Survey.
2. Andrews A, Folger P, Humphries M, Copeland C, Tiemann M, et al. (2009) Unconventional gas shales: Development, technology, and policy issues. Washington, DC: Congressional Research Service.
3. Entrekin S, Evans-White M, Johnson B, Hagenbuch E (2011) Rapid expansion of natural gas development poses a threat to surface waters. Frontiers in Ecology and the Environment 9: 503–511.
4. EIA (2011) Annual Energy Outlook. Washington, DC: U.S. Energy Information Administration. Available: http://205.254.135.7/forecasts/aeo/pdf/0383(2011).pdf. Accessed 2011 Sept 19.
5. Cooley H, Fulton J, Gleick P (2011) Water for energy: Future water needs for electricity in the intermountain West. Oakland. CA: Pacific Institute.
6. Olden JD, Naiman RJ (2010) Incorporating thermal regimes into environmental flows assessments: modifying dam operations to restore freshwater ecosystem integrity. Freshwater Biology 55: 86–107.
7. Poff NL, Olden JD, Merritt DM, Pepin DM (2007) Homogenization of regional river dynamics by dams and global biodiversity implications. Proceedings of the National Academy of Sciences of the United States of America 104: 5732–5737.

8. Wilcove D, Master L (2005) How many endangered species are there in the United States? Frontiers in Ecology and the Environment 3: 414–420.
9. Ricciardi A, Rasmussen J (1999) Extinction rates of North American freshwater fauna. Conservation Biology 13: 1220–1222.
10. Higgins JV, Duigan C (2009) So much to do, so little time: identifying priorities for freshwater biodiversity conservation in the United States and Britain. In: Boon PJ, Pringle CM, editors. Assessing the conservation value of fresh waters-an international perspective. Cambridge, UK: Cambridge University Press. pp. 293.
11. Richter B, Baumgartner J, Braun DP, Powell J (1998) A spatial assessment of hydrologic alteration with a river network. Regulated Rivers Research & Management 14: 329–340.
12. Holmlund CM, Hammer M (1999) Ecosystem services generated by fish populations. Ecological Economics 29: 253–268.
13. Chandel M, Pratson L, Jackson RB (2011) The potential impacts of climate-change policy on freshwater use in thermoelectric power generation. Energy Policy 39: 6234–6242.
14. Gerbens-Leenes PW, Hoekstra A, van der Meer T (2009) The water footprint of energy from biomass: A quantitative assessment and consequences of an

increasing share of bio-energy in energy supply. Ecological Economics 68: 1052–1060.

15. Gerbens-Leenes PW, Hoekstra A, Van der Meer T (2009) The water footprint of bioenergy. Proceedings of the National Academy of Sciences 106: 10219–10223.

16. Gerbens-Leenes PW, Hoekstra A (2011) The water footprint of biofuel-based transport. Energy and Environmental Science 4: 2658–2668.

17. DOE (2006) Energy Demands on Water Resources: Report to Congress on the Interdependency of Energy and Water. Washington, DC: U.S. Department of Energy.

18. DOE NETL (2008) Estimating Freshwater Needs to Meet Future Thermoelectric Generation Requirements. Washington, DC: U.S. Department of Energy and National Energy Technology Laboratory.

19. EPRI (2002) Water & Sustainability (Volume 3): U.S. Water Consumption for Power Production–The Next Half Century. Palo Alto, CA: Electric Power Research Institute.

20. Mulder K, Hagens N, Fisher B (2010) Burning water: A comparative analysis of the energy return on water invested. Ambio 39: 30–39.

21. Carter NT (2010) Energy's water demand: trends, vulnerabilities, and management. Washington, D.C.: Congressional Research Service.

22. DOE (2009) Water Requirements for Existing and Emerging Thermoelectric Plant Technologies. Washington, DC: U.S. Department of Energy.

23. Fthenakis V, Kim HC (2010) Life-cycle uses of water in U.S. electricity generation. Renewable and Sustainable Energy Reviews 14: 2039–2048.

24. Mielke E, Anandon L, Narayanamurtia V (2010) Water consumption of energy resource extraction, processing, and conversion. Cambridge, MA: Belfer Center for Science and International Affairs, Harvard Kennedy School.

25. Hoekstra A, Chapagain A, Aldaya M, Mekonnen M (2011) Water footprint assessment manual: Setting the global standard. London, UK: Earthscan.

26. McDonald RI, Fargione J, Kiesecker J, Miller W, Powell J (2009) Energy Sprawl or Energy Efficiency: Tradeoffs in U.S. Climate Policy Effects on Natural Habitats. PLoS One.

27. Ventyx Corporation (2011) Velocity Suite: EV Energy Map. Available: http://www.ventyx.com/velocity/ev-energy-map.asp. Accessed: 2011 March 29.

28. Gerbens-Leenes PW, Van Lienden AR, Hoekstra A, Van der Meer T (2012) Biofuel scenarios in a water perspective: The global blue and green water footprint of road transport in 2030. Global Environmental Change 22: 764–775.

29. Bain R, Amos W, Downing M, Perlack R (2003) Biopower technical assessment: state of the industry and technology. Washington, DC: U.S. Department of Energy. NREL/TP-510-33123 NREL/TP-510-33123.

30. DOE (2011) U.S. Billion-Ton Update: Biomass Supply for a Bioenergy and Bioproducts Industry; Perlack R, Stokes B, editors. Oak Ridge, TN: U.S. Department of Energy, ORNL/TM-2011/224.

31. Brown J, Maxwell S, Pervez S, Wardlow B, Callahan K (2008) National Irrigated Lands Mapping via an Automated Remote Sensing Based Methodology. Proceedings of the 88th Annual Meeting of the American Meteorological Association. New Orleans, LA: American Meteorological Association.

32. NatureServe (2009) NatureServe Explorer: an online encyclopedia of life. Version 7.1. Available: http://www.natureserve.org/explorer/. Accessed 2009. Arlington, VA: NatureServe.

33. Mims MC, Olden JD, Shattuck ZR, Poff NL (2010) Life history trait diversity of native freshwater fishes in North America. Ecology of Freshwater Fish 19: 390–400.

34. Faber-Langendoen D, Master L, Nichols J, Snow K, Tomaino A, et al. (2009) NatureServe Conservation Status Assessments: Methodology for Assigning Ranks. Arlington, VA: NatureServe.

35. USGS (2011) Estimates of HUC runoff from 1901–2009. Washington, DC.

36. USACE (2010) National Inventory of Dams. Washington, DC: US Army Corps of Engineers.

37. Sabo JL, Sinha T, Bowling LC, Schoups G, Wallender W, et al. (2010) Reclaiming freshwater sustainability in the Cadillac Desert. Proceedings of the National Academy of Sciences 107: 21263-21269.

38. Raadal HL, Gagnon L, Modahl IS, Hanssen OJ (2011) Life cycle greenhouse gas (GHG) emissions from the generation of wind and hydro power. Renewable and Sustainable Energy Reviews 15: 3417–3422.

39. Lenzen M (2008) Life cycle energy and greenhouse gas emissions of nuclear energy: A review. Energy conversion and management 49: 2178–2199.

40. Fthenakis V, Kim HC (2007) Greenhouse-gas emissions from solar electric- and nuclear power: A life-cycle study. Energy Policy 35: 2549–2557.

41. Sterner M, Fritsche U (in press) Greenhouse gas balances and mitigation costs of 70 modern Germany-focused and 4 traditional biomass pathways including land-use change effects. Biomass and bioenergy doi:/10.1016/j.biombioe.2011.08.024.

42. Aabakken J (2006) Power Technologies Energy Data Book (NREL/TP-620-39728). Golden, Colorado: National Renewable Energy Laboratory, U.S. Department of Energy.

43. Harto C, Meyers R, Williams E (2010) Life cycle water use of low-carbon transport fuels. Energy Policy 38: 4933–4944.

44. Vestas Wind Systems A/S (2006) Life cycle assessment of electricity delivered from an onshore power plant based on Vestas V82-1.65 MW turbines. Randers, Denmark: Vestas Wind Systems A/S.

45. Macknick J, Newmark R, Heath G, Hallet KC (2011) A review of operational water consumption and withdrawal factors for electricity generating technologies (NREL/TP-6A20-50900). Golden, Colorado: National Renewable Energy Laboratory, U.S. Department of Energy.

46. Wu M, Mintz M, Wang M, Arora S (2009) Consumptive water use in the production of ethanol and petroleum gasoline (ANL/ESD/09-01). Argonne, IL: Argonne National Laboratory, U.S. Department of Energy.

47. USDA (2008) Farm and Ranch Irrigation Survey. Washington, DC: U.S. Department of Agriculture.

48. NEI (2010) Water use and nuclear power plants. Washington, DC: Nuclear Energy Institute.

49. Torcellini P, Long N, Judkoff R (2003) Consumptive water use for U.S. power production (NREL/TP-550-33905). Golden, Colorado: National Renewable Energy Laboratory, U.S. Department of Energy.

14

De novo Biosynthesis of Biodiesel by *Escherichia coli* in Optimized Fed-Batch Cultivation

Yangkai Duan[⑨]**, Zhi Zhu**[⑨]**, Ke Cai, Xiaoming Tan, Xuefeng Lu***

Key Laboratory of Biofuels, Qingdao Institute of Bioenergy and Bioprocess Technology, Chinese Academy of Sciences, Qingdao, China

Abstract

Biodiesel is a renewable alternative to petroleum diesel fuel that can contribute to carbon dioxide emission reduction and energy supply. Biodiesel is composed of fatty acid alkyl esters, including fatty acid methyl esters (FAMEs) and fatty acid ethyl esters (FAEEs), and is currently produced through the transesterification reaction of methanol (or ethanol) and triacylglycerols (TAGs). TAGs are mainly obtained from oilseed plants and microalgae. A sustainable supply of TAGs is a major bottleneck for current biodiesel production. Here we report the *de novo* biosynthesis of FAEEs from glucose, which can be derived from lignocellulosic biomass, in genetically engineered *Escherichia coli* by introduction of the ethanol-producing pathway from *Zymomonas mobilis*, genetic manipulation to increase the pool of fatty acyl-CoA, and heterologous expression of acyl-coenzyme A: diacylglycerol acyltransferase from *Acinetobacter baylyi*. An optimized fed-batch microbial fermentation of the modified *E. coli* strain yielded a titer of 922 mg L^{-1} FAEEs that consisted primarily of ethyl palmitate, -oleate, -myristate and -palmitoleate.

Editor: Arnold Driessen, University of Groningen, Netherlands

Funding: This study was supported by the "100-Talent Program of the Chinese Academy of Sciences" foundation (Grant O091001110A). The funders had no role in study design, data collection and analysis, decision to publish, or preparation of the manuscript.

Competing Interests: The authors have declared that no competing interests exist.

* E-mail: lvxf@qibebt.ac.cn

⑨ These authors contributed equally to this work.

Introduction

In order to meet the rapidly growing demand for transportation fuel and to achieve reduction of carbon dioxide emissions, development of renewable energy sources has become more and more urgent. Biodiesel, as one type of renewable energy, is an ideal substitute for petroleum-based diesel fuel and is usually made from plant oils or animal fats (triacylglycerides) by transesterification with methanol or ethanol resulting in fatty acid methyl esters (FAMEs) and fatty acid ethyl esters (FAEEs). However, the limited supply of bioresources to obtain triacylglycerides (TAGs) is becoming a major bottleneck for biodiesel production. The main reason is that vegetable oil feedstocks are also food sources and their planting is geographically limited. Microalgae are currently viewed as one of the most promising TAG feedstocks for biodiesel production. Although the productivity of these photosynthetic microorganisms greatly exceeds that of agricultural oleaginous crops, current microalgae production using the best available strains and cultivation methods has not yet become economically feasible for biodiesel production [1].

Lignocellulosic biomass is a relatively sustainable bioresource to make biofuels. In the past few decades, tremendous research and development efforts on cellulosic ethanol have resulted in significant advances and many technical problems have been solved [2,3]. However, due to its low energy density, high vapor pressure and corrosiveness, bioethanol is not an ideal alternative to petroleum-derived fuels. Moreover, the high solubility of ethanol also results in toxicity to the microbes used to produce it [4]. Recently, technical routes have been developed to produce

novel biofuel products with higher energy density and hydrophobic properties from lignocellulose-derived sugars. Liquid alkanes chemically identical to petroleum fuels can be made by both biosynthesis with genetically modified *E. coli* [5] and synthetic catalytic conversion [6]. Biosynthetic production of C4–C8 alcohols with straight or branched chains through manipulation of amino acid metabolic pathways in *E. coli* has been reported [7]. Thus, the production of novel biofuels from lignocellulosic biomass-derived sugars is a promising alternative to bioethanol.

Bioresource technologies for biodiesel production from lignocellulosic biomass, distinct from approaches using the oily fraction of biomass, can be developed through constructing non-native biosynthetic pathways of biodiesel molecules in microbial hosts. Direct microbial production of FAEEs in engineered *E. coli* was first reported by co-expressing genes coding enzymes for ethanol production from *Zymomonas mobilis* and the WS/DGAT gene encoding acyl-coenzyme A: diacylglycerol acyltransferase from *Acinetobacter baylyi* strain ADP1 [8]. However, biosynthesis of FAEEs in that work relied on the supplementation of exogenous fatty acids. In last couple of years, research focusing on overproduction of fatty acids by genetically engineering the fatty acid metabolic network towards providing precursors of fatty acid-based biofuels such as FAMEs, FAEEs, fatty alcohols, and fatty alkanes, has been extensively investigated in *E. coli*. The titer of 2.5 g L^{-1} day^{-1} free fatty acids was produced by an *E. coli* mutant strain in fed-batch fermentation with overexpression of acetyl-CoA carboxylase (ACC) and acyl-ACP thioesterase (TE) from *E. coli* as well as plant TE from *Cinnamomum camphorum*, and with deletion of *fadD*, which codes for fatty acyl-CoA synthase, the first step of fatty acid degradation [9]. An improved yield of fatty acids to a titer of $4.5 \text{ g L}^{-1} \text{ day}^{-1}$ has been demonstrated by the same group [10]. A very recent research publication in Nature has demonstrated *de novo* biosynthesis of FAEEs in *E. coli* by co-overproduction of ethanol and fatty acids in genetically engineered *E. coli*. In this work, several genetic engineering strategies were developed to improve fatty acid production of the ethanol-producing mutant strain to increase the yield of FAEEs, including overexpression of thioesterases, heterologous introduction of *fadD* gene from *Saccharomyces cerevisiae*, and the deletion of the *fadE* gene the encodes acyl-CoA dehydrogenase, leading to a yield of FAEEs of up to 674 mg L^{-1}. Researchers have also demonstrated the feasibility of *in vivo* production of FAEEs from hemicellulose achieved *via* genetic engineering of the endoxylanase catalytic domain from *Clostridium stercorarium* and the xylanase from *Bacteroides ovatus* [11].

In our study, six distinct genetic alterations have been introduced into an *E. coli* BL21 (DE3) host strain (Fig. 1): (1) heterologous expression of the genes *pdc* and *adhB* from the ethanol-producing pathway in *Zymomonas mobilis*, coding pyruvate decarboxylase and alcohol dehydrogenase, respectively; (2) overexpression of *accBACD* genes coding acetyl-CoA carboxylase, which converts acetyl-CoA to malonyl-CoA, the first step of fatty acid biosynthesis that is proposed to be the rate-limiting step; (3) overexpression of the modified *tesA'* gene from *E. coli*, coding a leaderless version of thioesterase; (4) knockout of the *fadE* gene, coding acyl-CoA dehydrogenase that dehydrogenates fatty acyl-CoA, the second step of fatty acid degradation; (5) overexpression of the *fadD* gene from *E. coli*, coding a fatty acyl-CoA ligase that catalyzes the conversion of free fatty acids to fatty acyl-CoA; and (6) heterologous expression of the *atfA* gene from *A. baylyi*, coding the wax ester synthase/acyl-coenzyme A: diacylglycerol acyltransferase (WS/DGAT). Here, we also evaluated FAEE production in a scaled-up fed-batch fermentation, and optimized the nutritional and environmental conditions to improve the yield of FAEEs.

Compared to Steen *et al.*'s work on FAEE production reported in Nature [11], three significant differences were made in the current study. First, the *E. coli* mutant strain producing FAEE with over-expression of acetyl-CoA carboxylase that has been proved to be able to increase the rate of fatty acid biosynthesis in *E. coli* [12] was constructed in our study. Second, the effect of *fadE* deletion on FAEE production in both shake flask and scale-up fed-batch fermentation experiments was specially examined and analyzed in the report here. Third, we also evaluated FAEE production in a scale-up fed-batch fermentation, and optimized the nutritional conditions to improve the yield of FAEEs.

Results and Discussion

Shake flask experiment for *E. coli* strains

With the aim to achieve *de novo* biosynthesis of FAEEs in *E. coli*, plasmid pXT11, which contains *pdc* and *adhB* from *Zymomonas mobilis* ZM4, *fadD* from *E. coli*, and *atfA* from *Acinetobacter baylyi* ADP1, was constructed. When the *E. coli* strain harboring pXT11 was induced by 0.5 mM IPTG, no trace of FAEEs was detected by GC-MS. This is consistent with the previously reported results that FAEEs biosynthesis was dependent on the presence of a certain amount of free fatty acids and that the native level of fatty acyl-CoA was not sufficient to make FAEEs [8]. pMSD8 and pMSD15, containing *E. coli accBCDA* and *tesA'* expressing cassettes respectively, were proven to be effective in overproduction of free fatty acids in *E. coli* [9,12]. In this study, pXT11, pMSD8, and pMSD15 were co-transformed into *E. coli* BL21 (DE3), and FAEEs

were produced with the concentration of 34.6 mg L^{-1} by culturing the cells in shaking flasks with glucose as sole carbon source. To further increase the yield of FAEEs, the *fadE* gene, which encodes acyl-CoA dehydrogenase, was deleted to block degradation of fatty acyl-CoA, and it was observed that more than double the concentration of FAEEs, 77.5 mg L^{-1}, was produced in the recombinant strain under the same culture conditions. This result indicates that overproduced fatty acyl-CoA, which is the substrate of acyl-coenzyme A: diacylglycerol acyltransferase that catalyzes the production of FAEEs, is essential for *de novo* biosynthesis of FAEEs in *E. coli*, and that the metabolic engineering strategy applied here is effective to achieve accumulation of fatty acyl-CoA.

The effect of initial culture medium on FAEE production

Fig. S1a shows the production of different FAEEs using *E. coli* mutant strain BL21 (ΔfadE)/pXT11/pMSD8/pMSD15 fermented with three initial culture media. The maximum concentrations of FAEEs were 735, 922, and 328 mg L^{-1} (Table 1) when the cells were fermented in LB, 2LB, or 2LB+Phosphates medium, respectively. Thus, 2LB medium was selected as the initial culture medium for further experiments. It suggests that higher levels of nutrients in the initial culture medium have a positive effect for FAEE production and that excessive phosphates have negative effect.

The effect of feeding conditions on FAEEs production

The effects of glucose concentration (75, 100 and 150 g/750 ml) in the fed-batch fermentation of strain BL21 (ΔfadE)/pMSD8/pMSD15/pXT11 with the feed rate of 0.22 ml min^{-1} on the FAEEs production were evaluated. FAEEs were generated at 922 mg L^{-1} with 100 g glucose/750 ml feeding culture, while 581 and 588 mg L^{-1} FAEEs were produced in 75 g and 150 g glucose/750 ml feeding culture, respectively (Table 1, Fig. S1b). Moreover, the total FAEE concentration was only 464 mg L^{-1} when the glucose concentration was increased to 200 g glucose/750 ml feeding culture and the feed rate was decreased to 0.11 ml min^{-1}, as shown in Table 1. This suggests that the concentration of glucose in the feeding culture and the feeding rate are highly related to FAEE production.

The effect of culture temperature on FAEE production

The best FAEEs-producing strain in this work, *E. coli* BL21 (ΔfadE)/pMSD8/pMSD15/pXT11, contains three plasmids with three different origins. Temperature could affect the stability of these plasmids and also affect the production of FAEEs. When the strain was cultured at 37°C, no detectable FAEEs were found by GC-MS (data not shown). However, FAEEs were produced at 922 mg L^{-1} by the same strain when cultured at 30°C, and 652 mg L^{-1} of FAEEs were produced when the culturing temperature was decreased to 25°C (Table 1, Fig. S1c).

The effect of induction time point on FAEE production

In order to test the impact of induction time on FAEE production, three different time points in the early exponential stage were tested for induction of gene expression. As shown in Fig. S1d, when the culture was induced at an OD_{600} of 4, of 312 mg L^{-1} of FAEEs were produced at 18 h post-induction, and the total FAEE concentration did not change significantly during the next 20 h. When induced at an OD_{600} of 16, the concentration of FAEEs reached 682 mg L^{-1} at 32 h after the induction. The maximum FAEE yield of 922 mg L^{-1} was achieved 45 h after induction when the culture was induced at an OD_{600} of 11 (Table 1).

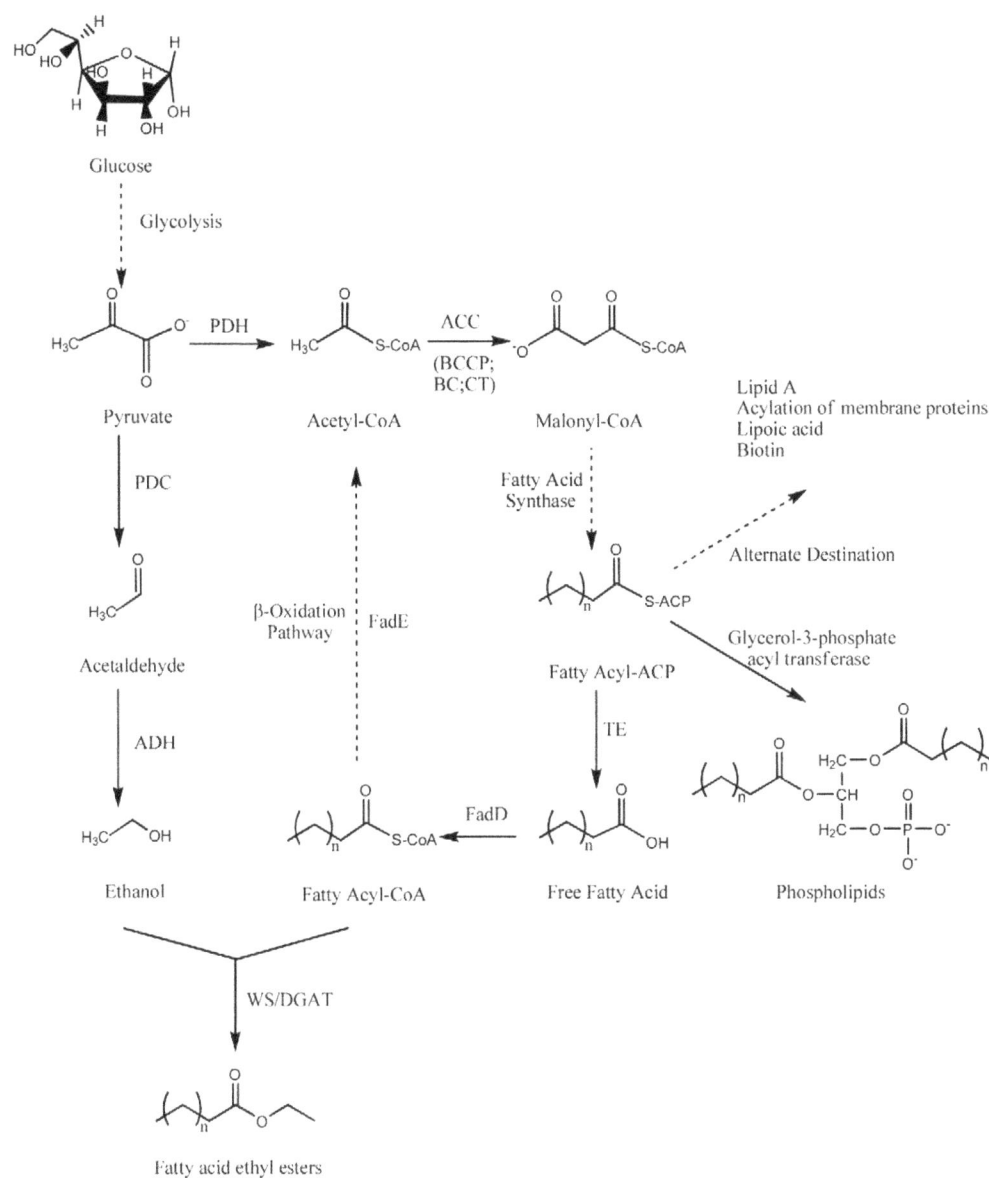

Figure 1. Constructed *de novo* biosynthetic pathway of fatty acid ethyl esters in *E. coli.* PDH: pyruvate dehydrogenase; ACC: acetyl-CoA carboxylase; BCCP: biotin carboxyl carrier protein; BC: biotin carboxylase; CT: carboxyltransferase; PDC: pyruvate decarboxylase; ADH: alcohol dehydrogenase; TE: thioesterase; FadD: fatty acyl-CoA synthase; FadE: acyl-CoA dehydrogenase; WS/DGAT: wax synthase/acyl-coenzyme A: diacylglycerol acyltransferase.

Production of FAEEs in a fed-batch fermentation under the optimized conditions

To evaluate the performance of the fatty acid ethyl ester-overproducing strain in a large scale, a 5-L fed-batch fermentation was performed with the optimized cultivation conditions described above, in which fed-batch fermentations were carried out at 30°C with 2LB as the initial culture medium and 100 g glucose/750 ml as the feeding culture at a feed rate of 0.22 ml min^{-1}, and the culture was induced at an OD_{600} of 11. Three recombinant *E. coli* strains, BL21 (DE3)/pXT11, BL21 (DE3)/pMSD8/pMSD15/ pXT11, and BL21 (ΔfadE)/pMSD8/pMSD15/pXT11, were grown, and the concentrations of cells, ethanol, and FAEEs were measured.

As shown in Fig. 2a, the growth of all three *E. coli* mutant strains was consistent with the logistic growth model. In the fed-batch

fermentation, there was nearly no lag phase and cells grew directly into the exponential period, followed by a steady period after around 65–70 h. The maximum optical densities of the three strains, BL21 (DE3)/pXT11, BL21 (DE3)/pMSD8/pMSD15/ pXT11, and BL21 (ΔfadE)/pMSD8/pMSD15/pXT11, reached around 24, 30 and 28 respectively. There is no significant difference among three cell growth profiles.

Ethanol production of the mutant strain BL21 (DE3)/pXT11, not producing FAEE, was clearly different than that in the two FAEE-producing strains (Fig. 2b). After induction, the concentration of ethanol increased slightly to around 1.3 g L^{-1} and then remained unchanged until the end of fermentation (Fig. 2b). It is reasonable that there is no ethanol consumption since there is no WS/DGAT enzyme in this strain to convert ethanol and no FAEE production in the strain. The ethanol production profile of the

Table 1. FAEE production of *E. coli* mutant strain BL21 (ΔfadE)/pXT11/pMSD8/pMSD15 under varied fed-batch fermentation conditions described in Materials and Methods.

Categories of fermentation conditions	Varing conditions	Maximum production of FAEE (mg L⁻¹)
Initial culture medium	LB	735
	2LB	922
	2LB+phosphates	328
Feeding conditions	75 g/0.22 ml min⁻¹	588
	100 g/0.22 ml min⁻¹	922
	150 g/0.22 ml min⁻¹	581
	200 g/0.11 ml min⁻¹	464
Culture temperature	30°C	922
	25°C	652
Time for starting induction	4 hr	333
	11 hr	922
	16 hr	682

strain BL21 (DE3)/pMSD8/pMSD15/pXT11 was similar to that of the strain BL21 (ΔfadE)/pMSD8/pMSD15/pXT11 (Fig. 2b), and the maximum ethanol accumulations of both FAEE-producing strains were also similar (about 3 g L⁻¹). The decrease of ethanol accumulations of the strain BL21 (ΔfadE)/pMSD8/pMSD15/pXT11 occurred earlier than that from the strain BL21 (DE3)/pMSD8/pMSD15/pXT11, and it is consistent with the FAEE production shown in Fig. 2c.

Based on the results obtained from the shake flask experiments, it was expected that no detectable FAEE would be obtained after induction during the fed-batch fermentation of BL21 (DE3)/pXT11 (Fig. 2c). FAEE production during fermentation of BL21 (DE3)/pMSD8/pMSD15/pXT11 and BL21 (ΔfadE)/pMSD8/pMSD15/pXT11 increased along with the decrease in ethanol production, and reached a maximum of 477 and 922 mg L⁻¹ respectively, about 14- and 12-fold higher than the production in the corresponding shake flask experiments respectively. The fed-batch fermentation experiments also demonstrated the same doubling effect of the deletion of *fadE* on FAEE production as was observed in the shake flask experiments. To confirm FAEE production under the optimized fed-batch fermentation conditionsAnd thereby, two other parallelled fermentation experiments of fermentation were performed with the strain BL21 (ΔfadE)/pMSD8/pMSD15/pXT11 under the optimized culture conditions (Fig. 2d). The FAEE production, the glucose conversion efficiency and the specific productivity were calculated to be 818.50±94.64 mg L⁻¹, 24.56±2.84 mg FAEE/g glucose and 0.46±0.11 mg L⁻¹ OD⁻¹ h⁻¹ respectively.

Composition of FAEEs during the optimized fed-batch fermentation

Fig. 3 illustrates the compositions of FAEEs produced in the fed-batch fermentation cultures of the three *E. coli* mutant strains. No detectable FAEE was found in the culture of BL21 (DE3)/pXT11. The fermentation of BL21 (DE3)/pMSD8/pMSD15/pXT11 produced 151.6 mg L⁻¹ ethyl myristate (C14:0; 32%) as the major constituent, with 99.6 mg L⁻¹ ethyl oleate (C18:1; 21%), 80.2 mg L⁻¹ ethyl palmitate (C16:0; 17%), 76.3 mg L⁻¹ ethyl

palmitoleate (C16:1; 16%), 33.2 mg L⁻¹ ethyl laurate (C12:0; 7%), and 33.1 mg L⁻¹ ethyl myristoleate (C14:1; 7%) as the minor FAEEs observed. The fed-batch fermentation of BL21 (ΔfadE)/pMSD8/pMSD15/pXT11 strain under the same cultivation condition produced ethyl palmitate (16:0; 31.3%) and ethyl oleate (18:1; 31.4%) as the two major FAEE constituents, together with ethyl myristate (14:0, 24.1%), ethyl palmitoleate (16:1, 9.9%), and other two minor FAEE constituents, ethyl laurate (12:0) and ethyl myristoleate (14:1).

By comparison of the FAEE production of the two *E. coli* mutant strains, BL21 (DE3)/pMSD8/pMSD15/pXT11 and BL21 (ΔfadE)/pMSD8/pMSD15/pXT11, we found that blocking the degradation of fatty acyl-CoA through the deletion of the *fadE* gene encoding acyl-CoA dehydrogenase caused significant changes in the FAEE composition. Both mutant strains produced fatty acid ethyl esters with carbon chain length varying from 12 to 18. Neither of mutant strains produced ethyl stearate (18:0), and ethyl laurate and ethyl myristoleate were minor constituents for both mutants. After *fadE* deletion, major products were ethyl palmitate (16:0) and ethyl oleate (18:1) rather than ethyl myristate (14:0) in the mutant strain BL21 (DE3)/pMSD8/pMSD15/pXT11.

Conclusions

In this study, a *de novo* biosynthetic pathway yielding fatty acid ethyl esters was constructed by genetically engineering *E. coli*. The fed-batch microbial fermentation was optimized with a maximum production of 922 mg L⁻¹ FAEEs. Although the titer of FAEEs is low for further scaling-up, this work shows the feasibility and potential to utilize lignocellulosic biomass-derived sugars instead of oily biomass-derived TAGs to produce biodiesel. FAEE production could be significantly improved by increasing the fatty acid biosynthetic flux, balancing ethanol production and fatty acid synthesis, and engineering the WS/DGAT enzyme toward higher substrate specificity to ethanol and higher catalytic efficiency.

Materials and Methods

Enzymes, DNA Kits and Strains

Taq, Pfu DNA polymerase and T4 DNA ligase were purchased from Fermentas (Burlington, Canada) and all restriction enzymes were from Takara (Kyoto, Japan). Plasmid mini kits, PCR purification kits and gel extraction kits were ordered from Omega (Norcross, USA). *E. coli* strain BL21 (DE3) and DH5α were obtained from Takara (Kyoto, Japan).

Plasmid construction

Detailed information about the plasmids used in this study is shown in Table 2. Plasmids pMSD8 and pMSD15 were kindly provided by Dr. John Cronan at University of Illinois at Urbana-Champaign, and pXL67 and pXL72 were generous gifts from Dr. Chaitan Khosla at Stanford University (USA). The *E. coli fadD* gene was excised from pXL72 *via* digestion with NdeI and SpeI, and cloned into the same sites of pXL67, resulting in pXT3. *Pdc* and *adh* genes from *Zymomonas mobilis* ZM4, coding pyruvate decarboxylase and alcohol dehydrogenase, were amplified from genomic DNA with primers pdc-up/pdc-down and adh-up/adh-down, and cloned into pXL67 resulting pXT4 and pXT5 respectively. The XbaI-SpeI double digested DNA fragment of pXT3, containing the *fadD* gene, was inserted into the SpeI site of pXL67, resulting in pXT9. The XbaI-SpeI double-digested DNA fragment of pXT5, containing the *adh* gene, was inserted into the SpeI site of pXT4, resulting in pXT10. Finally, the XbaI-SpeI double-digested DNA fragment of pXT10, containing both *pdc* and *adh* gene, was inserted into the SpeI site of pXT9, resulting in

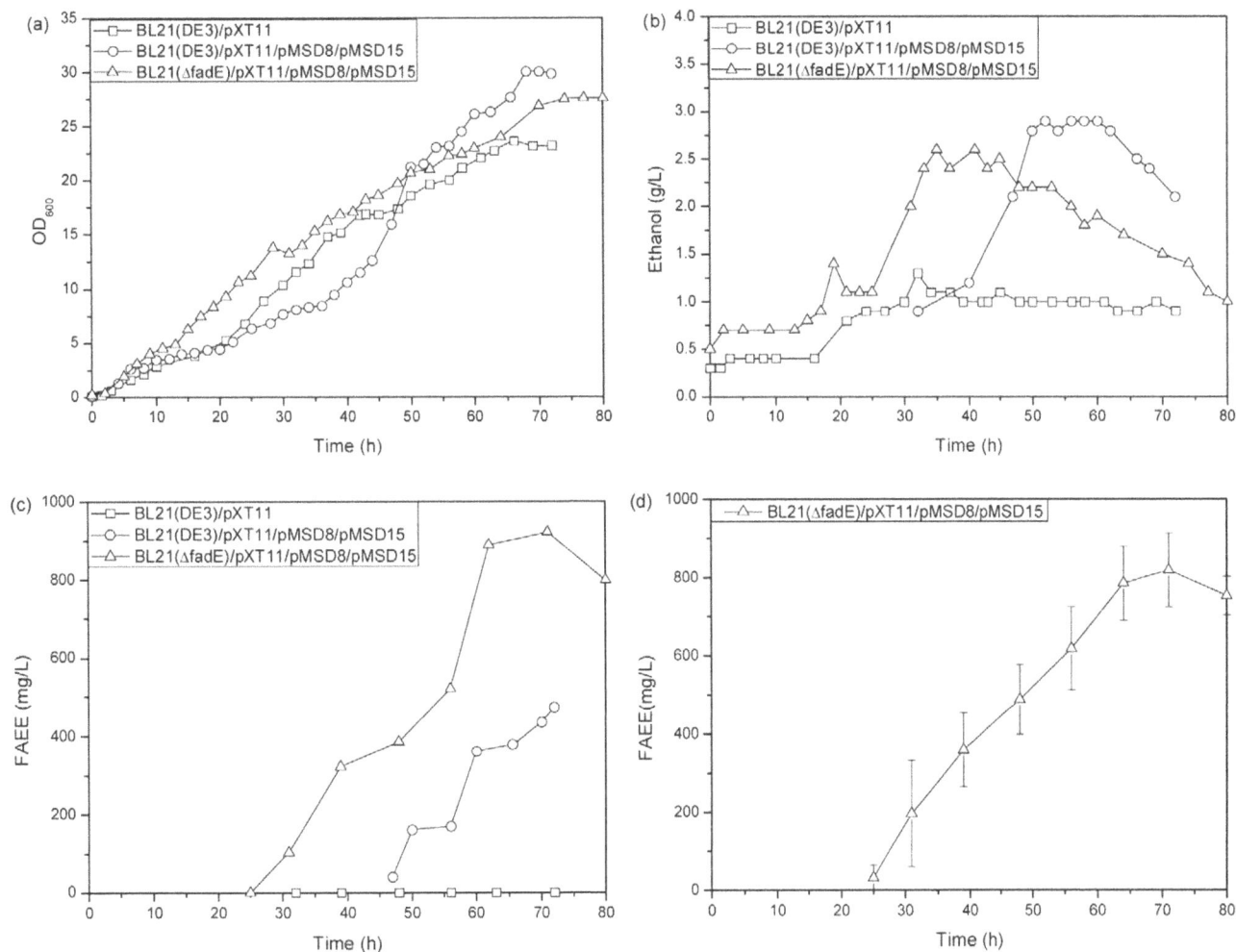

Figure 2. Analysis on fed-batch fermentations under the optimized conditions described in Materials and Methods. (a) Cell growth, (b) Ethanol production, (c) FAEE production of three *E. coli* mutant strains, BL21 (DE3)/pXT11 and BL21 (DE3)/pMSD8/pMSD15/pXT11 and BL21 (ΔfadE)/pMSD8/pMSD15/pXT11. (d) Three paralleled experiments for FAEE production of the strain BL21 (ΔfadE)/pMSD8/pMSD15/pXT11.

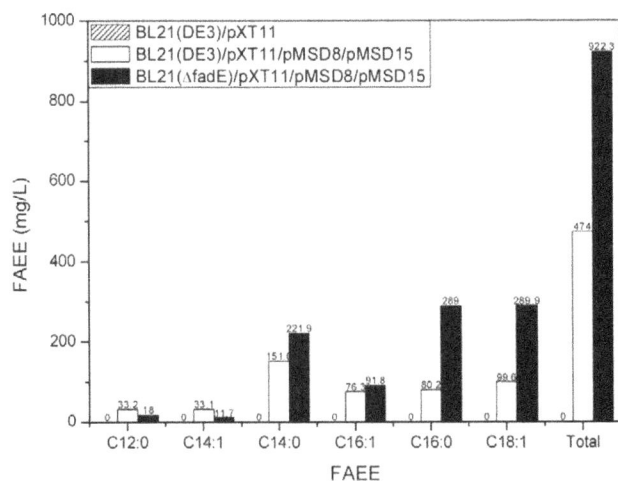

Figure 3. Composition of fatty acid ethyl esters with different carbon chain length and saturation degree during the optimized fed-batch fermentation of *E. coli* mutant strain BL21 (ΔfadE)/pMSD8/pMSD15/pXT11.

pXT11, in which four genes were assembled in the order of *atfA-fadD-pdc-adh* under control of the T7 promoter. Primers used here were listed in Table S1.

fadE gene deletion from the chromosome of *E. coli* strain BL21 (DE3)

Homologous replacement was used to delete the *fadE* gene from the chromosome of *E. coli* strain BL21 (DE3) as described by Datsenko and Wanner (Datsenko and Wanner, 2000). Briefly, primers ΔfadE-up/ΔfadE-down were used to amplify the kanamycin resistance cassette (FRT-*kan*-FRT) from the template plasmid pKD46 (Table S1). The purified PCR product was subjected to DpnI digestion and electroporated into *E. coli* BL21 (DE3) carrying pKD46 (a helper plasmid that expresses the λ-Red functions). To test the insertion of the FRT-*kan*-FRT cassette, colony PCR was performed with primers k1/fadE-C1, k2/fadE-C2 and fadE-C1/fadE-C2, using kanamycin resistance transformants as templates. Subsequently, the kanamycin resistance gene was eliminated by using pCP20, a helper plasmid expressing the flippase (FLP) recombinase, and the kanamycin-sensitive transformants were also confirmed by colony PCR with primers k1/fadE-C1, k2/fadE-C2 and fadE-C1/fadE-C2 (Table S1).

Table 2. Plasmids constructed and used in this study.

Plasmids	Relevant characteristic(s)	Source of reference(s)
pMSD8	Ap^r; pFN476 derivative containing *E. coli accBCDA* genes; T7 promoter	[12]
pMSD15	Cm^r; pACYA184 derivative containing *E. coli 'tesA* gene (without leading sequence); P$_{BAD}$ promoter	[12]
pXL67	Ap^r; pET28b derivative containing *Acinetobacter baylyi* ADP1 *atfA* gene; T7 promoter	Gift from Khosla
pXL72	Km^r; pCR-Blunt vector derivative containing *E. coli fadD* gene	Gift from Khosla
pKD46	Ap^r; Vector expressing the Red genes (γ, β, and *exo*) from phage λ under the arabinose-inducible *araB* promoter, *oriR101*, *repA101*(Ts)	[13]
pKD4	Ap^r Km^r; Template plasmid for *kan* flanked by FRT sequences, *oriR6K*	[13]
pCP20	Ap^r Cm^r; Helper plasmid, *FLP* λ *cl857* λ P$_R$ *repA*(Ts)	[14]
pXT9	Ap^r; pET28b derivative containing *atfA-fadD* genes; T7 promoter	This study
pXT10	Ap^r; pET28b derivative containing *pdc-adhB* genes; T7 promoter	This study
pXT11	Km^r; pET28b derivative containing *atfA-fadD-pdc-adhB* genes; T7 promoter	This study

Cell transformation

E. coli BL21 (DE3) and BL21 (ΔfadE) competent cells were transformed by plasmids pMSD8, pMSD15, and pXT11, and cells were selected on solid LB plates containing carbenicillin (50 µg ml^{-1}), kanamycin (50 µg ml^{-1}), and chloramphenicol (17 µg ml^{-1}).

Shake flask cultures

Recombinant strains of *E. coli* were streaked onto LB agar plates with antibiotics (25 mg L^{-1} amplicillin, 25 mg L^{-1} kanamycin, and 17 mg L^{-1} chloramphenicol) and incubated at 37°C for 12–20 h. Single colonies were picked and inoculated into 10 ml of LB medium in 50 ml flasks, and the flasks were incubated at 37°C in a rotary shaker at 200 rpm for 12 h. The cells were collected by centrifugation at 5000 rpm for 1 min, resuspended into 50 ml of sterilized LB with 5 g L^{-1} glucose in 250 ml flasks, and shaken until an OD$_{600}$ of 1.5–2 was reached. Arabinose was subsequently added into the culture to a final concentration of 0.4% for induction of the *araBAD* promoter. One hour later, IPTG was added to a final concentration of 0.5 M for induction of the T7 promoter. At about 20 h after induction, the culture was extracted for GC-MS analysis.

Fed-batch fermentation

Fed-batch fermentation was performed in a 5-L fermentor (Biostat Bplus, Sartorius) with a working volume of 3 L. Three different LB formulations were evaluated as the initial medium, as shown in Table 1. 2LB medium contained 2% w/v tryptone, 1% w/v yeast extract and 1% w/v sodium chloride, while LB medium contained 1% w/v tryptone, 0.5% w/v yeast extract, 1% w/v sodium chloride, and the 2LB+Phosphates medium was the same as 2LB with the addition of 8.7 g/L K$_2$HPO$_4$ and 4.2 g/L Na$_2$HPO$_4\cdot$12H$_2$O. Appropriate antibiotics were added (see above). The stirrer speed was adjusted to 400 rpm. Unless stated otherwise, the pH was controlled at 7.5 by automatic addition of 2 M hydrochloric acid. The flow rate of air was maintained at 1 VVM (air volume per broth volume per minute). Inoculum was 5% (v/v) of overnight cultures (see above). Starting at an OD$_{600}$ of 5–6, the culture was fed at a constant rate of 0.22 ml min^{-1} or 0.11 ml min^{-1} with sterilized glucose feed solutions containing 100 g yeast extract, 1.5 g magnesium sulfate, 0.75 g ammonium sulfate, and glucose at concentrations from 75 g to 200 g, all in a total volume of

750 ml (Table 1). Cells were induced at three different cultivation stages (OD$_{600}$ values of 4, 11 or 16) by arabinose (final concentration of 0.4%) for pMSD15. After one hour IPTG (final concentration of 0.5 mM) was added to induce genes coded on pMSD8 and pXT11. Fermentation broth samples ($\sim$20 ml) were collected at a series of time points and immediately kept at -80°C for fatty acid ethyl ester analysis.

Analysis method

For analysis of fatty acid ethyl esters, 5 ml culture was mixed thoroughly with 5 ml of organic solvent containing chloroform and methanol (the ratio of organic solvents is 2:1 by v/v), with 0.1 mg nonadecanoic acid methyl ester added as an internal standard. The organic phase was then collected and evaporated to dryness under a nitrogen atmosphere, and redissolved in 2 ml of n-hexane. Samples were analyzed by GC-MS (Thermo Scientific ITQ 1100^{TM} GC/MSn system, USA) using a single quadrupole MS with an electron impact ionization source. The TR-5 MS GC column was 30 m in length, with 0.25 mm ID and 25 m film thickness. The following temperature program was applied: 1 min at 40°C, 15 min ramp to 280°C, and constant at 280°C for 10 min. The quantification of fatty acid methyl esters was achieved by reference to the internal standard.

Ethanol concentration was determined using a biosensor (SBA-40C) from Biology Institute of Shandong Academy of Science (Jinan, China).

Supporting Information

Figure S1 FAEE production of *E. coli* mutant strain BL21 (ΔfadE)/pXT11/pMSD8/pMSD15 under varied fed-batch fermentation conditions: (a) varying initial culture medium; (b) varying feed glucose and feeding rate; (c) varying cultivation temperature; (d) varying induction time point.

Table S1 Strains and primers used in this study.

Acknowledgments

We would like to thank Dr. Chaitan Khosla and Dr. John Cronan for plasmid gifts, Dr. Kenneth Reardon for valuable discussion, and Dr. Wenna Guan and Miss Cong Wang for GC-MS technical assistance.

Author Contributions

Conceived and designed the experiments: XL. Performed the experiments: YD ZZ KC XT. Analyzed the data: XL YD ZZ KC. Contributed reagents/materials/analysis tools: YD ZZ KC. Wrote the paper: XL YD ZZ KC XT.

References

1. Wijffels RH, Barbosa MJ (2010) An outlook on microalgal biofuels. Science 329: 796–799.
2. Galbe M, Zacchi G (2007) Pretreatment of lignocellulosic materials for efficient bioethanol production. Adv Biochem Eng Biotechnol 108: 41–65.
3. Merino ST, Cherry J (2007) Progress and challenges in enzyme development for biomass utilization. Adv Biochem Eng Biotechnol 108: 95–120.
4. Somerville C (2007) Biofuels. Curr Biol 17: R115–119.
5. Schirmer A, Rude MA, Li X, Popova E, del Cardayre SB (2010) Microbial biosynthesis of alkanes. Science 329: 559–562.
6. Huber GW, Chheda JN, Barrett CJ, Dumesic JA (2005) Production of liquid alkanes by aqueous-phase processing of biomass-derived carbohydrates. Science 308: 1446–1450.
7. Zhang K, Sawaya MR, Eisenberg DS, Liao JC (2008) Expanding metabolism for biosynthesis of nonnatural alcohols. Proc Natl Acad Sci U S A 105: 20653–20658.
8. Kalscheuer R, Stolting T, Steinbuchel A (2006) Microdiesel: *Escherichia coli* engineered for fuel production. Microbiology 152: 2529–2536.
9. Lu X, Vora H, Khosla C (2008) Overproduction of free fatty acids in *E. coli*: implications for biodiesel production. Metab Eng 10: 333–339.
10. Liu T, Vora H, Khosla C (2010) Quantitative analysis and engineering of fatty acid biosynthesis in *E. coli*. Metab Eng 12: 378–386.
11. Steen EJ, Kang Y, Bokinsky G, Hu Z, Schirmer A, et al. (2010) Microbial production of fatty-acid-derived fuels and chemicals from plant biomass. Nature 463: 559–562.
12. Davis MS, Solbiati J, Cronan JE, Jr. (2000) Overproduction of acetyl-CoA carboxylase activity increases the rate of fatty acid biosynthesis in *Escherichia coli*. J Biol Chem 275: 28593–28598.
13. Datsenko KA, Wanner BL (2000) One-step inactivation of chromosomal genes in *Escherichia coli* K-12 using PCR products. Proc Natl Acad Sci U S A 97: 6640–6645.
14. Cherepanov PP, Wackernagel W (1995) Gene disruption in *Escherichia coli*: Tc^R and Km^R cassettes with the option of Flp-catalyzed excision of the antibiotic-resistance determinant. Gene 158: 9–14.

15

Microbial Succession during Thermophilic Digestion: The Potential of *Methanosarcina* sp.

Paul Illmer*, Christoph Reitschuler, Andreas Otto Wagner, Thomas Schwarzenauer, Philipp Lins

University Innsbruck, Institute of Microbiology, Innsbruck, Austria

Abstract

A distinct succession from a hydrolytic to a hydrogeno- and acetotrophic community was well documented by DGGE (denaturing gradient gel electrophoresis) and dHPLC (denaturing high performance liquid chromatography), and confirmed by qPCR (quantitative PCR) measurements and DNA sequence analyses. We could prove that *Methanosarcina thermophila* has been the most important key player during the investigated anaerobic digestion process. This organism was able to terminate a stagnation phase, most probable caused by a decreased pH and accumulated acetic acid following an initial hydrolytic stage. The lack in *Methanosarcina* sp. could not be compensated by high numbers of *Methanothermobacter* sp. or *Methanoculleus* sp., which were predominant during the initial or during the stagnation phase of the fermentation, respectively.

Editor: Melanie R. Mormile, Missouri University of Science and Technology, United States of America

Funding: There was no special funding except university resources. The funders had no role in study design, data collection and analysis, decision to publish, or preparation of the manuscript.

Competing Interests: The authors have declared that no competing interests exist.

* E-mail: Paul.Illmer@uibk.ac.at

Introduction

Irrespective of the disputed contribution of man to the global warming, the dramatic effects *per se* and the involvement of gases like CO_2 and CH_4 are unquestionable. Therefore it has (or should have) become an important global goal to reduce uncontrolled greenhouse gas emissions. One possibility to do so (and to fulfill the Kyoto Protocol) is to increase the portion of renewable energy sources like biogas. Therefore and before the background of a decreasing availability and increasing costs of fossil energy sources, the European Union has decided that by the year 2020 about 5% of the total energy budget should be derived from biogas production (De Vrieze et al, 2012; EC, 2011). No wonder that both, the number and capacity of biogas plants have steadily increased during the last decades [1].

Unfortunately, most of these plants are designed on the basis of empirical data and quite often unexpected and unexplainable fluctuations in fermenter performance occur [2,3] and even recent publications come to the conclusion that the engaged microorganisms still work within a 'black box' [4]. However, there has been significant progress in identifying and investigating microbial key players of anaerobic fermentations especially since culture independent techniques have become increasingly available in microbiology [1,5,5–9].

In a former investigation we could prove *Methanosarcina* sp. to be a key player during thermophilic biogas production – especially during the recovery after disturbed fermentations [7], a finding which corresponds with similar investigations [10,11]. It was possible to prove that inoculation with *Methanosarcina* sp. could successfully restart or at least accelerate the restoration process after a disturbance of the fermentation [12]. Despite this progress

a couple of questions remain unsolved, especially those connected with the microbial succession during optimal and malfunctioning.

Thus, within the present investigation we used different methods to characterize the microbial succession during a batch fermentation. These methods comprised both, fingerprint and analytical approaches and especially focused on the abundance of *Methanosarcina* sp., and its correspondence with the biogas production, as we assumed that a lack of *Methanosarcina* sp. might cause severe disturbance during thermophilic digestion.

Materials and Methods

Medium and medium preparation

The synthetic minimal medium described by [7] was used with the modifications that the concentration of $NaHCO_3$, carboxymethylcellulose (CMC) and peptone from casein was reduced to 4.2 g (50 mM), 2 g and 2 g per litre, respectively. The third complex carbon source was yeast extract (2 g L^{-1}), leading to a total carbon content of 3.18 g L^{-1} medium. The components were weighed in a bottle and dissolved with A. dest. resulting in 5 L medium. After autoclaving the hot medium ($>75°C$) was immediately transferred with a peristaltic pump into a lab-scale fermenter (see below) while flushing it with N_2/CO_2 (7/3) to reduce the contamination with O_2. Anaerobic conditions were controlled by the redox indicator resazurine and the pH value was set to 7.5 with HCl.

Inoculum, cultivation conditions and sampling

Sludge of a thermophilic anaerobic plug-flow reactor with an operating volume of 750 000 liters [2] was used as an inoculum. After the transport to the laboratory the sludge was diluted 1:5 (v/v) with boiled distilled water, which was flushed with pure N_2 for

10 min during cooling down. The diluted fermenter sludge (DFS) was shaken for 30 min at 100 rpm to allow homogenization. The DFS had a dry matter concentration of approximately 2.5%, which consisted of 65% organic matter. The dried sludge had a total carbon and nitrogen content of about 30% and 2.5%, respectively. For detailed chemical, physical and biological properties of the sludge it is referred to [2]. The 5 L medium was inoculated with 555 mL DFS representing a 1:10 inoculation.

The fermenter was maintained at $52 \pm 0.02°C$, moderately stirred with 50 rpm, and sampled 28 times during the whole investigation period of 65 days. At every sampling point 66 mL of the culture broth were withdrawn for subsequent analyses. As no fresh medium was provided due to the batch cultivation mode, a gradual decrease of the liquid volume was apparent with a final total withdrawal of approximately 1.85 L.

Fermenter system and determination of the gas production

As a fermentation system a software-controlled BIOSTAT Aplus (Sartorius, Germany) fermenter with an operation unit and a working volume of 5 L was chosen. The fermenter had a liquid and gas sampling port, ports for pH adjustment, and a port for gas sparging to get rid of remaining O_2. The produced biogas had to pass an exhaust cooler to reduce the loss of water vapor, which would be significant under thermophilic conditions. Afterwards the quantitative gas production was evaluated with a Rigamo MilliGascounter (Ritter GmbH, Germany) with a resolution of approximately 3.3 mL. Every time a switch occurred, the software calculated the cumulative gas production and gas production rate. The temperature was maintained by a cooling finger inside the fermenter and a heating blanket around the glass vessel, and the pH was measured online.

Biogas composition and VFAs

The analysis of the gas quality (H_2, CO_2, CH_4 and O_2 concentration) was done according to [12]. The samples for the determination of the volatile fatty acids (C_1–C_7) were prepared as previously described [12,13] but the operational settings of the HPLC system LC-20A prominence (Shimadzu) were slightly modified: oven temperature 65°C, flow rate 0.8 mL min^{-1}, mobile phase 5 mM H_2SO_4 and measurement of absorbance at 210 nm.

DNA extraction, end-point and quantitative PCR, DGGE and DNA sequencing

Out of selected samples (see below) 700 µL were extracted with a NucleoSpin Soil DNA extraction kit (Macherey-Nagel) and eluated in 50 µL buffer. Quantity and quality of extracted DNA were analyzed in duplicates with a NanoDropTM 2000c spectrophotometer (Thermo Scientific).

For archaea-specific end-point PCR/DGGE (denaturing gradient gel electrophoresis) analysis the primer pair 787F and 1059R (Arc) was applied [14] at which a GC-clamp was attached to the 5'-end of the forward primer [15]. The PCR reaction mix contained 200 µM dNTPs, 0.2 µM of each primer and 0.08% BSA (bovine serum albumin). For end-point amplification a Taq-DNA-Polymerase (BioThermTM) was used and finally 1 µL of template was added. Amplification conditions for archaea detection included an initial denaturation step (5 min, 95°C), 35 cycles of denaturation (45 s, 95°C), annealing (45 s, 57°C) and elongation (45 s, 72°C), and a final elongation step (7 min, 72°C). All PCR products were checked in a 1.5% agarose gel electrophoresis.

For methanogen specific end-point PCR/dHPLC general primers (109f/1492r) [16,17] and methanogen specific primers (O357f/O691r) [18] were used according to standard protocols. The reaction mixture contained 25 µL MyTaqTM 2×Mix PCR mixture, primers in a final concentration of 0.5 µM, 50 µg bovine serum albumin (aqueous solution, filter sterilized), and PCR grade water to achieve a final volume of 50 µL. Following PCR-programs were used for amplification of DNA: for 109f/1492r an initial denaturation step (10 min, 95°C), 35 cycles of denaturation (30 s, 95°C), annealing (30 s, 52°C) and elongation (45 s, 72°C) and a final elongation step (10 min, 72°C); for O357fGC/691r: an initial denaturation step (10 min, 95°C), 35 cycles of denaturation (30 s, 95°C), annealing (30 s, 49°C) and elongation (30 s, 72°C) and a final elongation step (7 min, 72°C).

All three parallels of the selected sampling points (day 0, 4, 18, 26 and 41) were analyzed in a DGGE and selected reference organisms were analyzed, to compare band patterns with the complex DNA samples to check for possible similarities. The DGGE protocol was altered based on the work of [19]. The acrylamide concentration in the gel was between 7 and 8%, while urea and formamide concentrations were set between 40 to 60%. For the separation an INGENYphorU electrophoresis system was used (60°C, 100 V for 16 h). Afterwards DNA bands were stained with silver nitrate. For the evaluation conserved gels were scanned and analyzed via GelCompare II software (Applied Maths). For quantification of distinct DNA bands densitometric curves of each lane were readout with ImageJ software (available at: http://rsb.info.nih.gov/ij/) after color separation and background subtraction. Afterwards, an averaged threshold was determined; peak areas were defined and set to relation to the sum of peak areas per lane. To gain more detailed qualitative information most representative samples were loaded on a new gel, bands were separated as described above and most abundant bands were isolated. DNA bands were stained with SYBR Gold Nucleic Acid Gel Stain (Invitrogen). Under UV light fluorescing bands were excised, suspended in A. d. and used as template in a further PCR with the archaea-specific primer pair 787F (without GC-clamp) and 1059R. Positive PCR products were purified with NucleoSpin Extract II (Macherey-Nagel) and sequenced by Eurofins MWG Operon. Passed sequences were processed with CLC DNA Workbench 5.6.1 (CLC bio) and aligned via NCBI Blast tool (http://blast.ncbi.nlm.nih.gov/).

For quantifying total archaea, representative for methanogenic archaea, the above mentioned primer pair Arc was applied in a quantitative PCR (qPCR) after evaluation of its specificity and applicability [20]. We used specific primers for both, for archaea and for methanogens but, although both results were very closely correlated, the latter primers were less reliable so that only the data for archaea are presented within this paper.

For amplification the SensiFAST SYBR No-ROX kit (Bioline) was used. The primer concentration was set to 0.2 µM per reaction. Amplification conditions were as following: 35 cycles of repeating denaturation- (20 s, 95°C), annealing- (20 s, 61°C), and elongation- (20 s, 72°C) steps. For quantification of cellulose-degrading microorganisms the primer set cel5 was applied with specifications according to [21]. A Corbett Life Science (Qiagen) Rotor-Gene 6000 system was used for measurements. PCR products were checked with melt curve analysis for specific amplification, absence of primer dimers and melting behavior of products.

dHPLC and DNA sequencing

dHPLC (denaturing high performance liquid chromatography) was basically carried out as described in [22] using an elution gradient from 50 to 56% buffer B in 24 min. To obtain additional information on the microbial methanogenic community, pure culture amplicons of various methanogenic archaea were used in order to match peaks with the same retention time. In cases of uncertainty but similar retention time, samples were spiked with pure culture amplicons of the nearest peak derived from a pure culture. If a second peak was found, the peak match was rejected. Additionally, peaks of dHPLC separation were collected (using a Shimadzu fraction collection system FRC-10A), liquid volatilized, and an additional PCR was carried out using methanogen specific primers. Subsequently, an aliquot was loaded on to dHPLC to test the presence of only one peak (in order to allow sequencing), else the collection procedure was repeated. When satisfying results were obtained (not possible for all peaks), amplicons were sequenced at MWG Operon (Germany). Sequence comparison and blast search were carried out using CLC Main Workbench 6.7 (CLC bio).

PLFA analysis

For the analysis of phospholipid fatty acids (PLFA), the samples were extracted following the method described by [23]. The extracts were separated into neutral-, glyco-, and polar lipids by solid-phase extraction on a Strata-Si Column (Phenomenex), and the polar fraction was subsequently transesterified via a modification of the method described by [24]. The PLFAs were analyzed by a GC 2010 (Shimadzu, Japan), equipped with a flame ionization detector (FID) and helium as carrier gas. A fused silica capillary column (Equity-1, 60 m, 0.25 mm inner diameter, 0.25 μm film thickness, Supelco) was used. The injector port was set to $250°C$ and the FID to $330°C$. The employed oven temperature program was: $100°C$ for 3 min, increase to $300°C$ at $3°C\ min^{-1}$ and hold for 15 min as described in [25].

Results and Discussion

Figure 1 shows the course of the batch fermentation with respect to the most important fermenter properties including quality and quantity of biogas. At each of the sampling days 66 mL of the culture broth was withdrawn and kept frozen at $-20°C$ till the whole process was finished. Afterwards, concentrations of VFAs were determined within all samples. On the basis of the process parameters, we decided to investigate samples from t = 0, 4, 18, 26 and 41, with molecular approaches and PLFA analysis in detail. At these days – indicated by dashed lines in Figure 1 – distinct changes in fermentation performance occurred, and thus, differences in microbiology should become obvious.

Start up

At the very beginning of the fermentation there was a distinct decrease in pH from about 7.5 to 6.6 connected with a sharp increase in the concentrations of H_2 and CO_2 in the headspace and the concentrations of acetic and butyric acid in the sludge. Altogether, this obviously reflects the high metabolic activity of hydrolytic and acetogenic microorganisms, resulting in about 40% CO_2, 17% H_2, 15 mM acetic acid and 4 mM butyric acid within only one single day. The production of appreciable amounts of iso-butyric acid and propionic acid took more time and concentrations reached approximately 1.5 mM and 3 mM at t = 1 and t = 4, respectively. At these levels the two acids remained remarkably constant till the second phase of high biogas production occurred.

PLFA analyses showed high initial concentrations of long polyunsaturated fatty acids, pointing to a high abundance of eukaryotic cells, possibly deriving from plant material introduced to the fermenter sludge. This explanation seems quite probable as

these fatty acids completely disappeared within four days of fermentation.

Figure 2 shows the DGGE patterns of nucleic acids amplified with archaea-specific primers. Results prove a distinct dominance of *Methanothermobacter thermoautotrophicus* and *Methanothermobacter wolfei* at t = 0. These two members of *Methanothermobacter* as well as all other organisms, which will be discussed within the present paper, exactly matched the reference lines of the respective pure culture (as far as available) and/or were identified by sequencing. The dominance of *Methanothermobacter* sp. confirms previous investigations of the thermophilic fermenter where the inoculum was taken from for the present investigation, and where *M. wolfei* has been proven to be the dominant methanogenic organism [9]. Both species of *Methanothermobacter* are efficient hydrogenotrophic organism (following reaction 1, shown in Table 1) and so this efficient pathway of methanogenesis started after a very short lag phase of not more than two days. This was also proven by a very sharp decrease in the concentration of H_2, which fell beneath the detection limit (0.005%) again within one week. However, conditions for the initial dominant species seemed to become unfavorable as these organisms completely disappeared till t = 4 and another hydrogenotrophic organism, *Methanoculleus thermophilus*, became increasingly abundant. A very sharp increase in the abundance of *M. thermophilus* was also proven by dHPLC analyses, which again proved to be an efficient fingerprint method for investigating post-PCR mixtures of nucleic acids [22].

Figure 3 shows the course of the abundance of *M. thermophilus* during the whole fermentation. The DNA from the peaks was gathered after HPLC analysis, sequenced and proven to derive from *M. thermophilus* (100% identity). Obviously, this methanogen could rapidly respond to the harsh initial conditions and to a certain extent better handle the steadily increasing concentrations of VFAs accompanied with the decreasing pH. Besides that, the hydrogenotrophic methanogenesis has once again turned out to be more efficient, not only under standard but also under *in situ* conditions (Reaction 1 vs. 3, Table 1) as proven in an earlier investigation [26].

Quantitative analyses proved the total DNA content to be maximal at t = 0, probably because of the above mentioned input of eukaryotic cell material and thus nucleic acids. Contrary, the numbers of total archaea, which we could prove to be equivalent to methanogens in this environment, were minimal at t = 0 and distinctly increased till t = 4 (Figure 4).

Stagnation

After t = 4 the number of methanogens remained constant till t = 18 (Figure 4) and thus mark a stagnation phase of approximately 14 days, during which nearly no further methane was produced. Concentrations of CH_4, CO_2 and H_2 remained constant at 30%, 50% and <0.005%, respectively, and also the pH remained unchanged at a low level of about 6.7 (Figure 1A). The only parameters which changed within this phase were the concentrations of acetic and butyric acid (Figure 1B). Whereas the first one steadily increased during this whole phase the latter one was characterized by a distinct degradation starting at t = 14. This degradation of butyric acid was nearly the only sign of microbial activity within this phase of fermentation. As the degradation of butyric acid usually follows the reaction 2, shown in Table 1, this degradation should additionally account for the increasing concentration of acetic acid (Figure 1B). It is important to notice that this reaction is endergonic under standard conditions but becomes slightly exergonic under *in situ* conditions ($52°C$ and real pH and concentrations of gases and VFAs) [26]. Furthermore, there is a syntrophic connection with acetate-degrading organisms

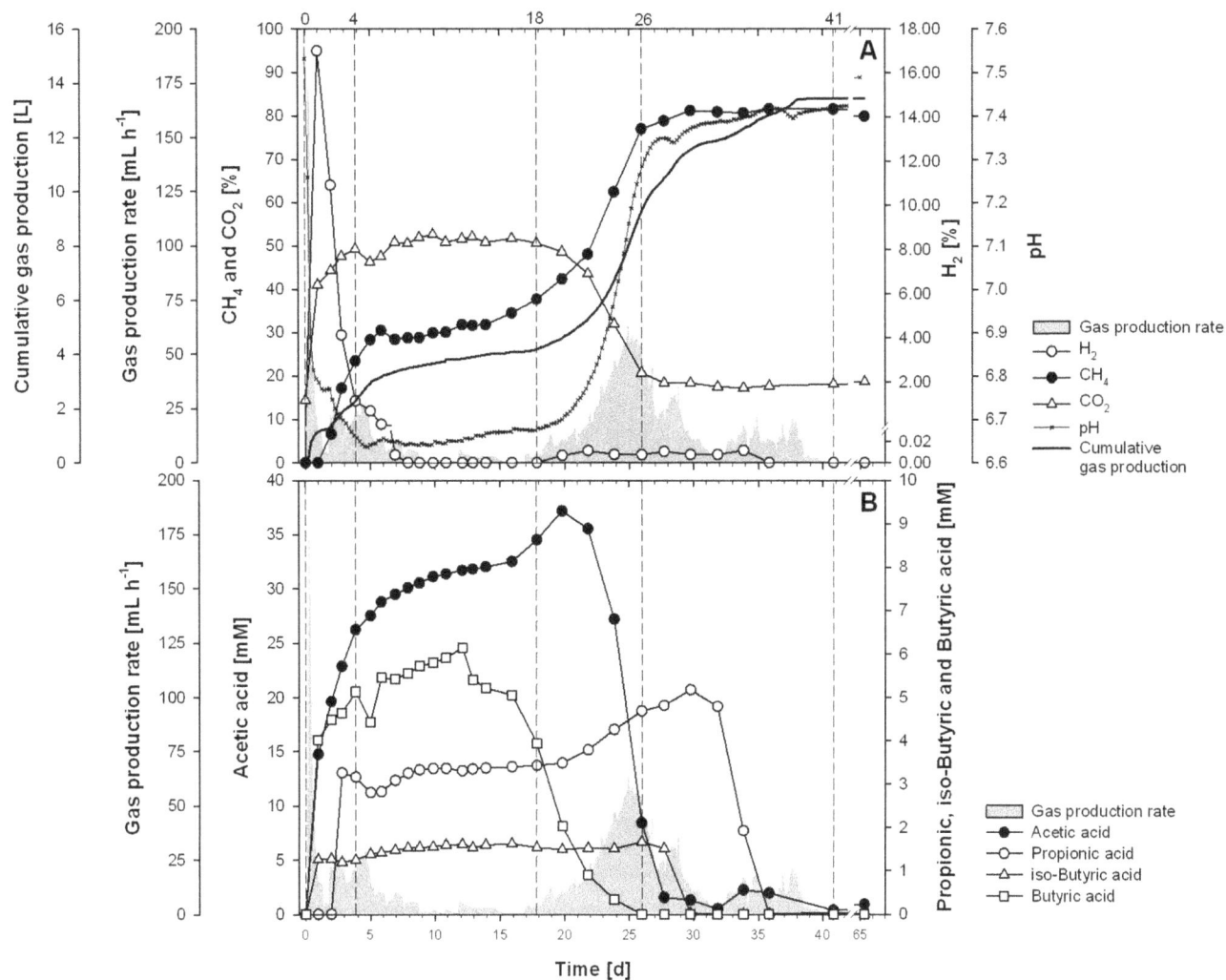

Figure 1. Fermenter performance. pH-values, qualitative and quantitative properties of biogas (A) and concentrations of VFAs (B) during the fermentation. Gas production rate (grey background) is given in A and B to ease the comparison. Dashed lines at t = 0, 4, 18, 26, 41 outline the samples which were additionally investigated by molecular approaches.

(e.g. reaction 3, Table 1) so that the sum of the reactions (Reaction 4, Table 1) becomes exergonic, both under standard and even more under realistic conditions. And indeed, the acetic acid oxidation resulted in a distinct production of methane at the end of the stationary phase (Figure 1). When concentration of acetic acids exceeds about 32 mM the oxidation of propionic acid is hampered leading to a second appearance of H_2 which might favor hydrogenotrophic methanogens.

Microbial analyses proved that *M. wolfei* and *M. thermoautotrophicus* have completely disappeared, that the dominance of *Methanoculleus* sp. steadily decreased and that *Methanosarcina thermophila* very slightly appeared in several strains during this stagnation phase (Figure 2). Obviously, there is a connection between the bad fermenter performance and low gas production during the stagnation phase on the one hand, and the lack of an acetoclastic methanogenic organism, being able to efficiently use the high amounts of acetic acid on the other hand. However, the growth rate of *M. thermophila* was obviously quite low or somehow suppressed, which resulted in a long lag phase of acetoclastic methane production. Although *Methanosarcina* sp. was first detectable at t = 4, it took this strain a long time, till t = 26, until it

became dominant. The DGGE band quantification did not only point to the dominance of *M. thermophila* but also to a greater archaeal diversity at t = 26 as several weak and unidentified bands appeared.

Obviously, *M. thermophila* is a very robust acetotrophic methanogen and it was the only one in our investigation which was able to handle high concentrations of acetic acid (35 mM) and the corresponding low pH values, confirming results from the literature [27,28]. The distinct drop in pH seems to be the reason for the break in CH_4 production, which only *Methanosarcina* sp. was able to resolve. In former investigations always using the very same inoculum from the 750 000 L large-scale fermenter, sometimes a break with a hampered gas production occurred and sometimes the second phase of gas production was directly connected to the first hydrolytic phase [7,12]. The reason for this different and hardly predictable behavior is not clear yet but we assume that it corresponds with the presence or lack of *Methanosarcina* sp. Slight differences in buffer capacities of the media resulting in different extents of the pH decreases and thus different growth rates of *Methanosarcina* sp. might be a possible explanation. Another possibility, however not likely, might be that *Methanosarcina* sp.

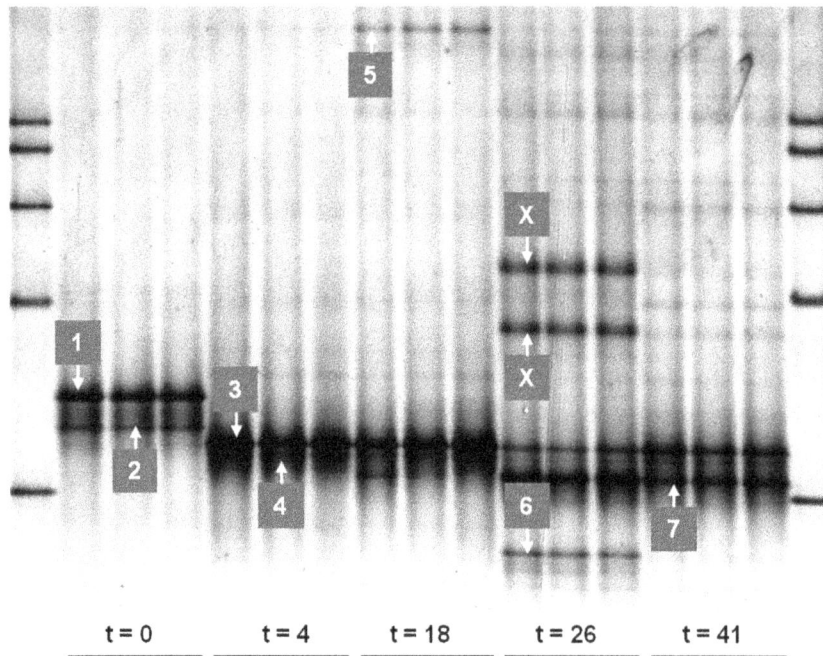

Figure 2. Archaeal DGGE. DGGE of archaeal PCR-products out of samples taken at day = 0, 4, 18, 26 and 41. Assignment of different bands: 1 *Methanothermobacter thermoautotrophicus*, 2 *Methanothermobacter wolfei*, 3 *Methanoculleus thermophilus*, 4 *Methanosarcina* sp., 5 *Thermoplasma* sp., 6 *Methanosarcina thermophila*, 7 *Methanosarcina thermophila*.

did not grow on acetic acid in the original fermenter sludge and thus had to adapt its metabolism towards the acetoclastic instead of methylo- or hydrogenotrophic pathway leading to a lag-phase. [29] observed that the pregrowth conditions for *Methanosarcina* spp., which define the pathway for methanogenesis, had a significant impact on the occurrence and duration of the lag-phases. Also a very recent investigation proved that bioaugmentation of enriched inocula with *Methanosarcina* sp. led to an improved start-up of digestions suffering from high acetic acid loads [30].

Results from PLFA analyses point to an increase of fatty acids typical for gram positive bacteria and confirm the absence of eukaryotic cells during the stagnation phase (data not shown). These latter results clearly proved that anaerobic fungi like *Neocallimastix* sp., which are sometimes discussed to be engaged in anaerobic digestion [31,32] should not play any role during the investigated fermentation.

Second phase of efficient methane production

As mentioned above, at t = 14 the degradation of butyric acid started but it was not before the rapid degradation of acetic acid, started at t = 19, that the second increase in gas production occurred (Figure 1B). Within a few days the concentration of CH_4 increased to the final concentration of about 80%, whereas the content of CO_2 decreased to about 15%. Finally the cumulative gas production reached about 14 L standing for about 850 mL gas per gram of carbon, which is a remarkable result compared with gas yields, known from literature [33,34]. Interestingly, H_2 became detectable again at the end of the stagnation phase. It should be derived from the degradation of VFAs (see Table 1) and might have promoted the growth of hydrogenotrophic methanogens. As several bands appeared in the DGGE, in all three independent DNA-extractions, and because the number of operons per organisms (probably not more than three) should be constant for a single organism, we think that the bands represent different species of *Methanosarcina* sp. or at least different strains of M. *thermophila*. It is important to keep in mind that *Methanosarcina* sp. is

Table 1. Thermodynamic properties of selected reactions at standard and at *in situ* conditions*).

Reaction	Standard conditions (kJ reaction^{-1})	*In situ* conditions*) (kJ reaction^{-1})
(1) $4H_2+HCO_3^-+H^+\rightarrow CH_4+3H_2O$	−135.5	−43.4
(2) Butyric acid$^-+2H_2O\rightarrow 2$ acetic acid$^-+H^++2H_2$	48.2	−8.3
(3) Acetic acid$^-+H_2O\rightarrow CH_4+HCO_3^-$	−31.0	−20.7
Reaction (1) plus two times reaction (2) resulting in		
(4) Butyric acid$^-+4H_2O\rightarrow 2CH_4+2HCO_3^-+H^++2H_2$	−13.8	−49.7

*) 52°C and real concentrations of gases and VFAs according to [26].

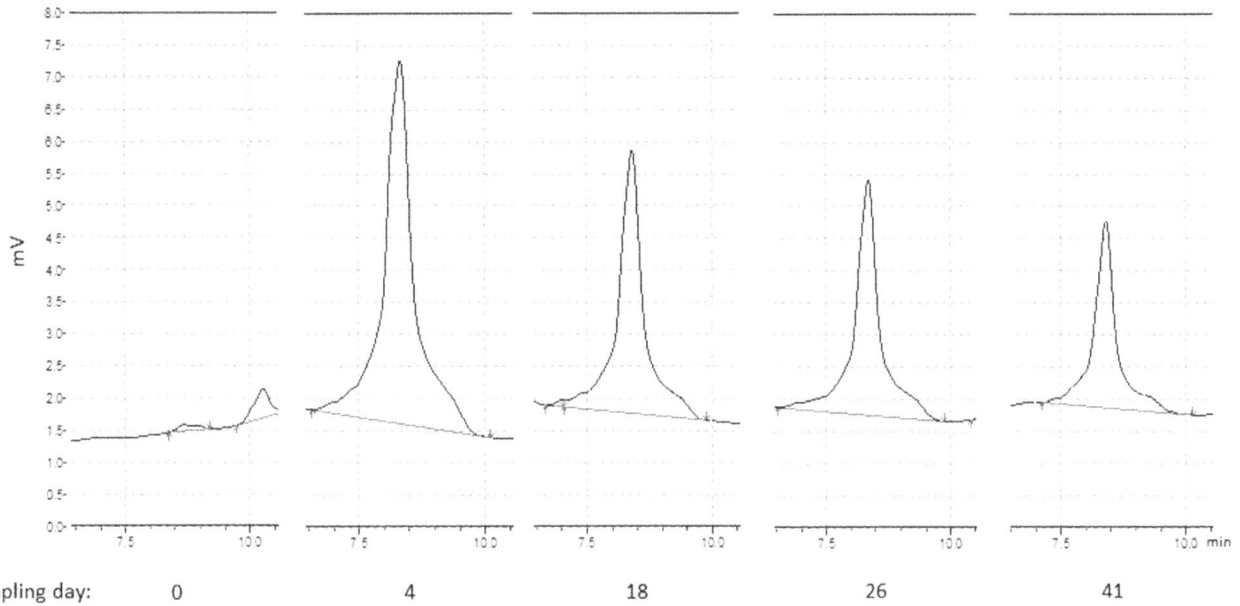

Figure 3. dHPLC of *Methanoculleus* **sp.** dHPLC signals [mV] of *Methanoculleus* sp. within PCR-products of different samples taken at day = 0, 4, 18, 26 and 41.

a very versatile methanogen with respect to its substrates because it is able to use all four known methanogenic pathways, which are the hydrogenotrophic, acetoclastic, methylotrophic and the methyl reduction way [11,35,36]. Altogether, our results as well as the referred literature, emphasize the potential of *Methanosarcina* sp. as the central key player under high organic loads or deteriorated conditions, as it was the case during this second phase of gas production. Although within complete different habitats – an abandoned coal mine and in a rice field – [37] and [35] could also

prove Methanosarcinales to govern CH_4 formation by utilizing acetic acid rather than H_2.

Accompanying the distinct degradation of acetic acid, the pH rose again and reached a level of about 7.4. Obviously this was favorable for a greater variety of methanogens, apart from *M. thermophila*. The Shannon index calculated on the basis of DGGE-data proved the highest archaeal diversity at t = 26. Besides *M. thermophila* and *M. thermophilus* a further organism could be identified by all the methods applied, namely *Thermoplasma* sp. or at least some closely-related archaeon representing a non

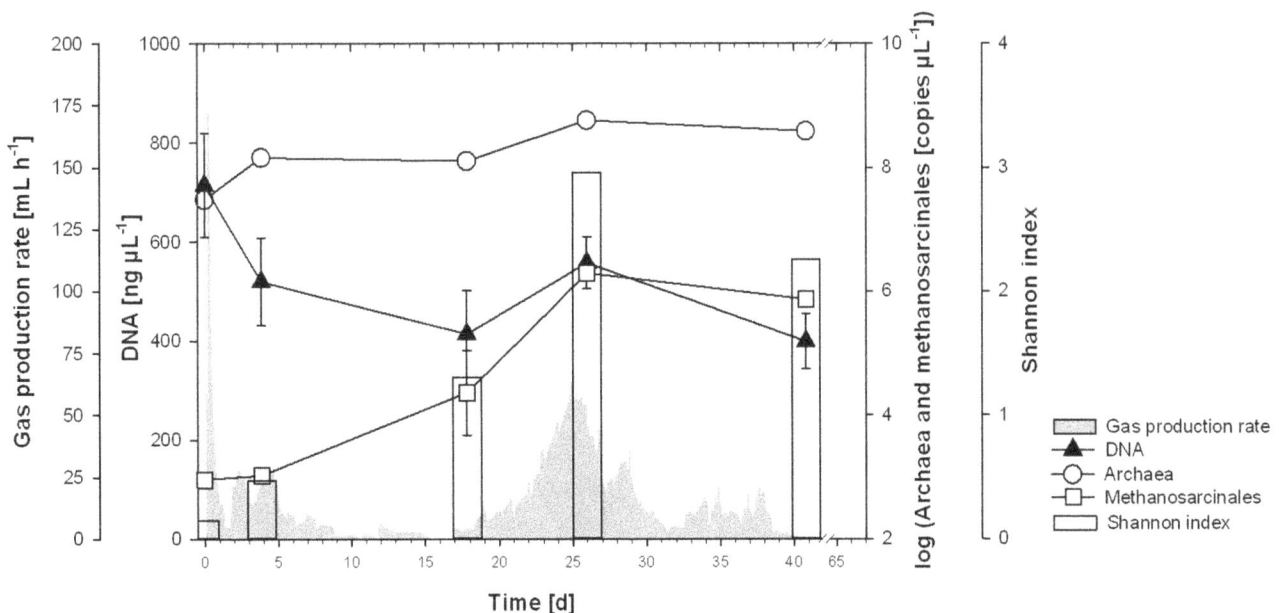

Figure 4. qPCR and archaeal diversity. Content of DNA, copy numbers of Archaea and Methanosarcinales determined via qPCR and Shannon index (bars) on the basis of archaeal DGGE bands before the background of gas production (see Figure 1).

methanogenic organism, which is usually known for its extremophilic way of living [38]. However, despite the occurrence of *Thermoplasma* sp. and despite the high archaeal diversity, *M. thermophila* remained the dominant organism, and also the gas production rate reached its optimum at this time (Figure 1). At t = 26 the concentration of DNA increased again and the number of archaea (determined via qPCR) reached its maximum (Figure 4).

Final phase

At the end of the fermentation the gas production ceased and the concentrations of VFAs, CH_4 and CO_2 were at a constant level. All other parameters describing abundance and activities of the engaged microorganisms distinctly decreased and reached final minima.

The distinct succession from a hydrolytic to a hydrogeno- and acetotrophic community was well documented by DGGE and dHPLC and confirmed by qPCR measurements as well as sequencing data. PLFA analyses in contrast seemed to be of limited evidence in anaerobic systems due to the uncertainties in assignment of specific fatty acids to microbial groups. However, within the present investigation we could prove that there were only very few key players engaged in the investigated digestion, i.e. *Methanothermobacter thermoautotrophicus* and *M. wolfei* at the beginning, *Methanoculleus thermophilus* during the intermediate and to minor quantities in the second phase of high gas production, and *Methanosarcina thermophila* most dominant during the second phase of high gas production.

Members of the genera *Methanosarcina* and *Methanosaeta* are the only methanogens able to degrade acetic acid. While *Methanosarcina* sp. usually dominates at high acetic acid concentrations because of its high conversion rates and low affinity, *Methanosaeta* sp. dominate at opposite conditions due to its high affinity but low conversion rates [10,11,30]. In our investigation threshold values for acetoclastic methanogenesis were found to be around 0.6 mM, which corresponds to findings of [39] who determined for *M. barkeri* and *M. mazei* 1.2 and 0.4 mM, respectively, whereas distinct lower values (0.07 mM) were calculated for *Methanosaeta* sp.. Additionally, in an early, anyway excellent work, kinetics of *Methanosarcina* sp. MSTA-1 was investigated [38]. Under optimum temperature and pH conditions Km for acetate kinase and threshold values for acetate were 10.7 and 0.7 mM respectively, thus again confirming our data. Besides, *Methanosarcina* sp. was shown to have a wide pH-range optimum for growth, and slightly acidic conditions even seem to induce increased growth rates [40].

Overall, *Methanosarcina thermophila* seems to be the most important methanogen in the investigated environment, as it was able to terminate a stagnation phase, most probably caused by a decreased pH and accumulated acetic acid. Thus, besides the inoculation with *Methanosarcina* sp., an adaptation of fermenter conditions towards properties favorable for this organism might be a promising possibility to skip phases of low gas production and to optimize CH_4 yields during anaerobic fermentation. Nevertheless, further research is required, also with respect to potential for up-scaling and applicability.

Author Contributions

Conceived and designed the experiments: PI. Performed the experiments: CR AOW TS PL. Analyzed the data: PI AOW PL. Contributed reagents/materials/analysis tools: PI. Wrote the paper: PI.

References

1. Weiland P (2010) Biogas production: Current state and perspectives. Appl Microbiol Biotechnol 85: 849–860.

2. Illmer P, Gstraunthaler G (2009) Effect of Seasonal changes in Quantities of biowaste on Full scale anaerobic digester performance. Waste Management 29: 162–167.

3. Illmer P, Schwarzenauer T, Malin C, Wagner AO, Miller LM, et al. (2009) Process parameters within a 750,000 litre anaerobic digester during a year of disturbed fermenter performance. Waste Manag 29: 1838–1843.

4. Supaphol S, Jenkins SN, Intomo P, Waite IS, O'Donnell AG (2011) Microbial community dynamics in mesophilic anaerobic co-digestion of mixed waste. Bioresource Technol 102: 4021–4027.

5. Chandra R, Takeuchi H, Hasegawa T (2012) Methane production from lignocellulosic agricultural crop waste: A review in context to second generation of biofuel production. Renewable and Sustainable Energy Reviews 16: 1462–1476.

6. Esposito G, Frunzo L, Giordano A, Liotta F, Panico A, et al. (2012) Anaerobic co-digestion of organic wastes. Reviews in Environmental Science and Biotechnology 11: 325–341.

7. Lins P, Malin C, Wagner AO, Illmer P (2010) Reduction of accumulated volatile fatty acids by an acetate-degrading enrichment culture. FEMS Microbiol Ecol 71: 469–478.

8. Wagner AO, Gstraunthaler G, Illmer P (2010) Utilisation of single added fatty acids by consortia of digester sludge in batch culture. Waste Manag 30: 1822–1827.

9. Malin C, Illmer P (2008) Ability of DNA-content and DGGE analysis to reflect the performance condition of an anaerobic biowaste fermenter. Microbial Res 163: 503–511.

10. Demirel B, Scherer P (2008) The roles of acetotrophic and hydrogenotrophic methanogens during anaerobic conversion of biomass to methane: A review. Reviews in Environmental Science and Biotechnology 7: 173–190.

11. De Vrieze J, Hennebel T, Boon N, Verstraete W (2012) *Methanosarcina*: The rediscovered methanogen for heavy duty biomethanation. Bioresource Technol 112: 1–9.

12. Lins P, Reitschuler C, Illmer P (2012) Development and evaluation of inocula combating high acetate concentrations during the start-up of an anaerobic digestion. Bioresource Technol 110: 167–173.

13. Wagner AO, Hohlbrugger P, Lins P, Illmer P (2012) Effects of different nitrogen sources on the biogas production - a lab-scale investigation. Microbial Res 167: 630–636.

14. Yu Y, Lee C, Hwang S (2004) Group-specific primer and probe sets to detect methanogenic communities using quantitative real-time polymerase chain reaction. Biotechnology and Bioengineering 89: 670–679.

15. Muyzer G, DeWaal EC, Uitterlinden AG (1993) Profiling of complex microbial populations by denaturing gradient gel electrophoresis of polymerase chain reaction-amplified genes coding for 16S rRNA. Applied and Environmental Microbiology 59: 695–700.

16. Heuer H, Krsek M, Baker P, Smalla K, Wellington EMH (1997) Analysis of actinomycete communities by specific amplification of genes encoding 16S rRNA and gel-electrophoretic separation in denaturing gradients. Applied and Environmental Microbiology 63: 3233–3241.

17. Großkopf R, Janssen PH, Liesack W (1998) Diversity and structure of the methanogenic community in anoxic rice paddy soil microcosms as examined by cultivation and direct 16S rRNA gene sequence retrieval. Applied and Environmental Microbiology 64: 960–969.

18. Watanabe T, Asakawa S, Nakamura A, Nagaoka K, Kimura M (2004) DGGE method for analyzing 16S rDNA of methanogenic archaeal community in paddy field soil. FEMS Microbiol Lett 232: 153–163.

19. Sekiguchi H, Tomioka N, Nakahara T, Uchiyama H (2001) A single band does not always represent single bacterial strains in denaturing gradient gel electrophoresis analysis. Biotechnology Letters 23: 1205–1208.

20. Reitschuler C, Illmer P (2012) Evaluation of primers used for the investigation of methanogenic and methanotrophic communities and the potential for improvement of commonly used qPCR applications. World Journal of Microbiology and Biotechnology, in press DOI 10.1007/s11274-013-1450-x.

21. Pereyra LP, Hiibel SR, Prieto Riquelme MV, Reardon KF, Pruden A (2010) Detection and quantification of functional genes of cellulose- degrading, fermentative, and sulfate-reducing bacteria and methanogenic archaea. Applied and Environmental Microbiology 76: 2192–2202.

22. Wagner AO, Malin C, Illmer P (2009) Application of Denaturing High-Performance Liquid Chromatography in Microbial Ecology: Fermentor Sludge, Compost, and Soil Community Profiling. Applied and Environmental Microbiology 75: 956–964.

23. White DC, Davis WM, Nickels JS, King JD, Bobbie RJ (1979) Determination of the sedimentary microbial biomass by extractible lipid phosphate. Oecologia 40: 51–62.

24. Sönnichsen M, Müller BW (1999) A rapid and quantitative method for total fatty acid analysis of fungi and other biological samples. Lipids 34: 1347–1349.

25. Schwarzenauer T, Lins P, Reitschuler C, Illmer P (2012) The use of FAME analyses to discriminate between different strains of *Geotrichum klebahnii* with different viabilities. World J Microbiol Biotechnol 28: 755–759.

26. Lins P, Illmer P (2009) Thermodynamische Aspekte der anaeroben Vergärung. Müll und Abfall 12: 604–608.

27. Lü F, Hao L, Guan D, Qi Y, Shao L, et al. (2013) Synergetic stress of acids and ammonium on the shift in the methanogenic pathways during thermophilic anaerobic digestion of organics. Water Res 47: 2297–2306.

28. Hao L, Lü F, Li L, Wu Q, Shao L, et al. (2013) Self-adaption of methane-producing communities to pH disturbance at different acetate concentrations by shifting pathways and population interaction. Bioresource Technol 140: 319–327.

29. Ferguson TJ, Mah RA (1983) Effect of H$_2$-CO$_2$ on methanogenesis from acetate or methanol in *Methanosarcina* sp. Applied and Environmental Microbiology 46: 348–355.

30. Lins P, Reitschuler C, Illmer P (2014) *Methanosarcina* spp., the key to relieve the start-up of a thermophilic anaerobic digestion suffering from high acetic acid loads. Bioresource Technol 152: 347–354.

31. Lockhart RJ, Van Dyke MI, Beadle IR, Humphreys P, McCarthy AJ (2006) Molecular biological detection of anaerobic gut fungi (*Neocallimastigales*) from landfill sites. Applied and Environmental Microbiology 72: 5659–5661.

32. Mountfort DO, Orpin CG (1994) Anaerobic fungi: biology, ecology and function. New York: Marcel Dekker, Inc.

33. Chae KJ, Jang A, Yim SK, Kim IS (2008) The effects of digestion temperature and temperature shock on the biogas yields from the mesophilic anaerobic digestion of swine manure. Bioresource Technol 99: 1–6.

34. Nielsen HB, Heiske S (2012) Anaerobic digestion of macroalgae: methane potentials, pre-treatment, inhbition and co-digestion. Water Sci Tech 64: 1723–1729.

35. Goevert D, Conrad R (2009) Effect of substrate concentration on carbon isotope fractionation during acetoclastic methanogenesis by *Methanosarcina barkeri* and *M.acetivorans* and in rice field soil. Applied and Environmental Microbiology 75: 2605–2612.

36. Kulkarni G, Kridelbaugh DM, Guss AM, Metcalf WW (2009) Hydrogen is a preferred intermediate in the energy-conserving electron transport chain of *Methanosarcina barkeri*. PNAS 106: 15915–15920.

37. Beckmann S, Lueders T, Krüger M, von Netzer F, Engelen B, et al. (2011) Acetogens and acetoclastic *Methanosarcinales* govern methane formation in abandoned coal mines. Applied and Environmental Microbiology 77: 3749–3756.

38. Shimada H, Nemoto N, Shida Y, Oshima T, Yamagishi A (2008) Effects of pH and temperature on the composition of polar lipids in *Thermoplasma acidophilum* HO-62. J Bacteriol 190: 5404–5411.

39. Westermann P, Ahring BK, Mah RA (1989) Threshold acetate concentrations for acetate catabolism by aceticlastic methanogenic bacteria. Applied and Environmental Microbiology 55: 514–515.

40. Clarens M, Moletta R (1990) Kinetic studies of acetate fermentation by *Methanosarcina* sp. MSTA-1. Appl Microbiol Biotechnol 33: 239–244.

16

MetaBinG: Using GPUs to Accelerate Metagenomic Sequence Classification

Peng Jia[2,3,4], Liming Xuan[4,5], Lei Liu[2,3,4]*, Chaochun Wei[1,4]*

1 Department of Bioinformatics and Biostatistics, School of Life Sciences and Biotechnology, Shanghai Jiao Tong University, Shanghai, China, 2 Key Laboratory of Systems Biology, Shanghai Institutes for Biological Sciences, Chinese Academy of Sciences, Shanghai, China, 3 Graduate School of the Chinese Academy of Sciences, Shanghai, China, 4 Shanghai Center for Bioinformation Technology, Shanghai, China, 5 Department of Biochemistry and Molecular Biology, School of Bioengineering, East China University of Science and Technology, Shanghai, China

Abstract

Metagenomic sequence classification is a procedure to assign sequences to their source genomes. It is one of the important steps for metagenomic sequence data analysis. Although many methods exist, classification of high-throughput metagenomic sequence data in a limited time is still a challenge. We present here an ultra-fast metagenomic sequence classification system (MetaBinG) using graphic processing units (GPUs). The accuracy of MetaBinG is comparable to the best existing systems and it can classify a million of 454 reads within five minutes, which is more than 2 orders of magnitude faster than existing systems. MetaBinG is publicly available at http://cbb.sjtu.edu.cn/~ccwei/pub/software/MetaBinG/MetaBinG.php.

Editor: Jonathan H. Badger, J. Craig Venter Institute, United States of America

Funding: This work was supported by grants from the National High-Tech R&D Program (863) (2009AA02Z310, 009AA02Z306), the National Natural Science Foundation of China (60970050), the Shanghai Pujiang Program (09PJ1407900), and the Science and Technology Innovation Program of the Basic Science Foundation of Shanghai (08JC1416700). The funders had no role in study design, data collection and analysis, decision to publish, or preparation of the manuscript.

Competing Interests: The authors have declared that no competing interests exist.

* E-mail: leiliu@sibs.ac.cn (LL); ccwei@sjtu.edu.cn (CW)

Introduction

The culture-independent metagenomics methods try to sequence all genetic materials recovered directly from an environment. It has the potential to provide a global view of a microbial community [1]. However, one of the challenging tasks is to assign these raw reads or assembled contigs into classes according to the evolutionary distances among their source genomes. This process is called metagenomic sequence classification.

There are two major types of computational methods for metagenomic sequence classification: alignment-based and composition-based. Alignment-based methods can determine that a sequence is from an organism only if the source genome or a genome with similar sequence has been sequenced. When the source genome is fully sequenced, alignment-based methods are accurate in general. However, it is difficult for alignment-based methods to do classification when the sequences of the source genomes or closely-related genomes are not available. Unfortunately, this is the case for many metagenomes. It is a significant limitation for alignment-based methods. Composition-based methods, on the other hand, are less accurate but are able to assign every read to a source bin, which can be one or more species, genera or other taxonomy ranks.

Advantages of the next-generation sequencing (NGS) technologies such as the high throughput and low cost make them more and more attractive for metagenome sequencing. Among NGS platforms, the 454 sequencers provide the longest reads (up to 400 bps in average), and they can generate more than half a million reads in just one run [2]. However, NGS technologies make classification more challenging by providing a large amount of shorter reads than a traditional sequencing platform does. These fast growing numbers of metagenomic sequences from NGS platforms put efficient and reliable classification systems in high demand.

There are many existing metagenomic sequence classification systems, such as Phymm and PhymmBL [3]. Phymm uses interpolated Markov models (IMMs) to classify short reads, and has obtained pretty good sensitivity and specificity on its own test dataset. However, the computational cost of Phymm is very expensive and it can be a problem when the size of a dataset to be classified is huge. For example, it may take Phymm 100 hours or more to classify a single run of 454 sequencing data (see Results part). PhmmBL added alignment-based method to Phymm, and achieved better accuracy. PhymmBL is about 50% slower than Phymm (data not shown). Recent updated version of PhymmBL [4] can run multiple jobs simultaneously in a multi-processor computer. PhymmBL can also run on multiple machines. But this parallelization of PhymmBL requires extra splitting and merging steps for each list of input reads. For other similar systems with webservers, such as CAMERA [5], MG-RAST [6,7], the time to classify a run of 454 sequencing data varies from hours to weeks [8]. This can be a serious problem when the sample size increases.

Graphic processing units (GPUs) were originally designed to accelerate graphic display only. In the past few years, GPUs have evolved to GPGPUs (general purpose GPUs), which can do general purpose scientific and engineering computing. In many cases, programs implemented on GPUs can run significantly faster than on multi-core CPU-based systems since a GPU may have hundreds of cores. With the success of a parallel programming model called CUDA for GPUs, programming on GPUs for general scientific computing becomes much easier than before [9]. Therefore, using GPUs is becoming very attractive for researchers who need to boost the performance of their applications in a wide range of scientific areas, including bioinformatics [10]. However, it is not straightforward to apply GPUs to a new research area. A GPU-based version of BLAST has been developed, and it is nearly four times faster than the CPU-based version [11].

In metagenomics, "metagenomic sequence classification" is sometimes distinguished from "metagenomic sequence binning, " which refers to the grouping of a dataset into subgroups but the subgroups remain unlabeled [3]. Obviously, a metagenomic sequence classification system can also be used as a binning system by treating all distinct subgroups without considering their names or labels.

In this paper, we present a fast metagenomic sequence classification system (MetaBinG) using the power of GPUs. MetaBinG is able to classify accurately a single run of NGS-based metagenome shotgun sequence data in minutes instead of hours or days in a single desktop workstation.

Results and Discussion

In order to compare the performance of MetaBinG and Phymm, 1212 fully sequenced bacterial genomes were downloaded from the NCBI FTP site (ftp://ftp.ncbi.nih.gov/genomes/Bacteria/) on 14 December 2010. With NCBI taxonomy information, 390 genomes were removed to guarantee that every genus has at least two genomes. The remaining 822 genomes belong to 133 genera (534 species). The species in each genus were assigned to training and test groups. All genomes of one species were assigned into a same set, either the training set or the test set. In the end, we generated a training set of 468 genomes (288 species, 133 genera) and a test set of 354 genomes (246 species, 133 genera) (The complete list of training and test genomes is available as Table S1). This simulated the situation that the source organisms of short sequences were not present in the reference database. At least at species level, there was no overlap between the training and test sets. Ten different sequence lengths were tested. For each sequence length, 10 sequences were randomly sampled from every chromosome or plasmids of the 354 test genomes. Therefore, there were 6,640 reads for each sequence length. Then the test datasets were classified using MetaBinG trained from the 468 training genomes with K = 5 (We observed that 5th-order Markov model was enough to get accurate results).

Although better results could be achieved with higher order of Markov models, there may be not sufficient data to train the models when K increases. For example, in a 4-million-bp genome, 4^K is larger than 4 million when K is bigger than 11. Therefore, there must be insufficient training data for models if K is bigger than 11. In addition, there will be 4 times more computing time when K increases one since the complexity of MetaBinG is $O(N' * 4^K)$, where N' is the number of genomes used for training. Therefore, the best result may be obtained by a relatively low order of Markov model (Table 1). The speed of classification is also an important aspect. Considering all these issues, we finally chose 5th-order Markov models for metagenomic sequence classification.

Both MetaBinG and Phymm V3.2 perform well at high ranks (Table 2). Although the accuracy of MetaBinG is lower than Phymm in most cases, the differences between MetaBinG and Phymm decrease when the read length increases. MetaBinG performs classification at all taxonomy levels. The accuracy results at the lower ranks (from phylum to genus) were reported in Table

Table 1. Impact of the order (K) of Markov models in MetaBinG.

Sequence Length (bps)	K = 3	K = 4	K = 5	K = 6	K = 7
100	1/47.5	2/49.6	4/50.6	15/50.6	**55/50.7**
200	2/56.7	2/58.9	5/60.8	**16/61.7**	55/61.6
300	2/62.2	3/64.9	**6/67.7**	16/67.6	57/66.9
400	3/66.3	4/69.7	**6/71.6**	17/71.6	57/71.4
500	3/69.2	4/72.7	**8/74.5**	18/74.5	59/73.3
600	4/71.9	5/75.3	8/77.2	**19/78.0**	59/74.9
700	5/74.5	6/77.8	9/79.2	**20/79.3**	61/76.9
800	6/75.4	7/78.4	**10/80.3**	22/79.9	62/77.2
900	8/76.6	9/79.2	**12/80.8**	22/80.6	63/78.4
1000	8/78.5	10/81.5	13/82.4	**25/82.7**	65/78.9

The impact of the order of Markov models (K) in MetaBinG has been tested. The K values various from 3 to 7. The sequence data sets are the same as in Table 2. Ten different sequence lengths from 100 bps to 1000 bps have been used for testing. Each sequence length contains 6,640 sequences. Each column is for a K value, which is the order of a Markov model. The total computing time (in seconds) and accuracy was measured as in Table 2. Each cell contains the total computing time and the accuracy separated by a "/".
For each sequence length, the best performance is in a bold font. K is set to 5 by default in MetaBinG.

S2 and S3 for Phymm and MetaBinG respectively. The results show that the MetaBinG is at least 140-fold faster than Phymm with a comparable or slightly worse accuracy (Table 2). When applied to real data, the high-order Markov models can be pre-computed from all the 1212 genomes for both MetaBinG and Phymm. It took MetaBinG about 3.5 hours to build 5th-order Markov Models for 1212 genomes with a single thread version training program, while Phymm spent 30 hours to build IMMs for the 1212 genomes. Unlike Phymm, MetaBinG package contains the pre-built 5th-order Markov Models so that users could use MetaBinG directly without any training steps. In addition, we provide an interface to add new training genomes to the pre-built Markov Models.

MetaBinG has been tested on a real dataset, a biogas reactor dataset containing 616,072 454 reads with an average length of 230 bps [12]. Using all the 1212 training genomes, MetaBinG spent 248 seconds and Phymm spent 4 days 5 hours 57 minutes and 56 seconds to classify the biogas reactor dataset, which means MetaBinG is almost 1500-fold faster than Phymm when dealing with real high throughput sequencing data. However, the microbial community recovered by MetaBinG and Phymm is quite similar (Figure 1). In practice, multiple instances of Phymm can be run in a computer with multiple cores. Therefore, the actual speedup may be much lower. For example, the speedup for the biogas dataset analysis will be 188 (= 1500/8) if 8 instances of Phymm are run simultaneously in one computer with 8 cores. However, it is safe to say that MetaBinG is 2 orders of magnitude faster than Phymm.

It may seem inefficient for sequences with lengths shorter than 4^K bases. However, k-th order Markov model provides a uniformed representation (4^{k+1} dimension vectors) for genomes in the reference database, which makes it very simple to parallelize the computing in a GPU. In practice, this computing is done by matrix multiplication. The matrix multiplication functions are from CUBLAS library, which is a Basic Linear Algebra Subprograms (BLAS) library ported to CUDA (Compute Unified

Table 2. Comparison of Phymm and MetaBinG.

Sequence Length (bps)	Phymm		MetaBinG		Speedup
	Accuracy (%)	Time (s)	Accuracy (%)	Time (s)	
100	53.62	573	50.61	4	143
200	64.21	880	60.82	5	176
300	70.71	1262	67.66	6	210
400	73.36	1652	71.56	6	275
500	76.02	1949	74.48	8	244
600	78.47	2330	77.24	8	291
700	79.89	2632	79.21	9	292
800	81.86	3006	80.25	10	301
900	82.40	3403	80.77	12	284
1000	84.18	3795	82.35	13	292

Ten different sequence lengths from 100 bps to 1000 bps have been used for testing. Each sequence length contains 6,640 sequences. The accuracy and total computing time (in seconds) for 6,640 sequences is listed in the table. Accuracy is measured at phylum level. The last column in the table shows the speedup of MetaBinG compared to Phymm. Both Phymm and MetaBinG were tested in the same Linux machine with 2 Intel Xeon E5520 processors (8 cores in total), 16 GB RAM and one NVDIA Tesla C1060 GPU card (240 cores). Default parameters were used for Phymm. The same input sequences and reference databases were used for both MetaBinG and Phymm. The accuracy is defined by dividing the number of correctly predicted sequences by the total number of test sequences since both methods assign every sequence to a source genome. The time measured here included all overhead except the creating of reference databases.

Device Architecture) [13]. CUBLAS supports high density of parallelization, and the parallelization is managed by CUDA. In addition, matrix multiplication functions in CUBLAS are optimized for parallel computing in GPUs. Compared to a single threaded CPU version of the same algorithm, the speedup can even be larger than the actual number of cores in GPUs.

In order to check the impact of GPUs on the speed of MetaBinG, MetaBinG and its CPU version were compared. A naive single-threaded version of MetaBinG was implemented without using the BLAS library. K was set to its default value 5. For the same 6,640 input sequences as in Table 2, MetaBinG was about 200 to 500 times faster than its CPU version for sequences with lengths from 100 bps to 1000 bps when all 1212 genomes were included in the reference database (data not shown). In the test for the biogas dataset with more than half million sequences, the speedup could go up to about 600 times. A parallel CPU version of MetaBinG was also implemented using BLAS library (http://www.netlib.org, last updated on Jan. 20th, 2011). For the biogas dataset, the GPU version was about 25 times faster than the parallel CPU version (data not shown). Therefore, the speedup of MetaBinG is partially from BLAS library, and partially from GPUs. Meanwhile, the speedup of using GPUs is related to the size of inputs including the number of sequences and the size of reference genome database. The bigger input data sets, the higher speedup it can achieve.

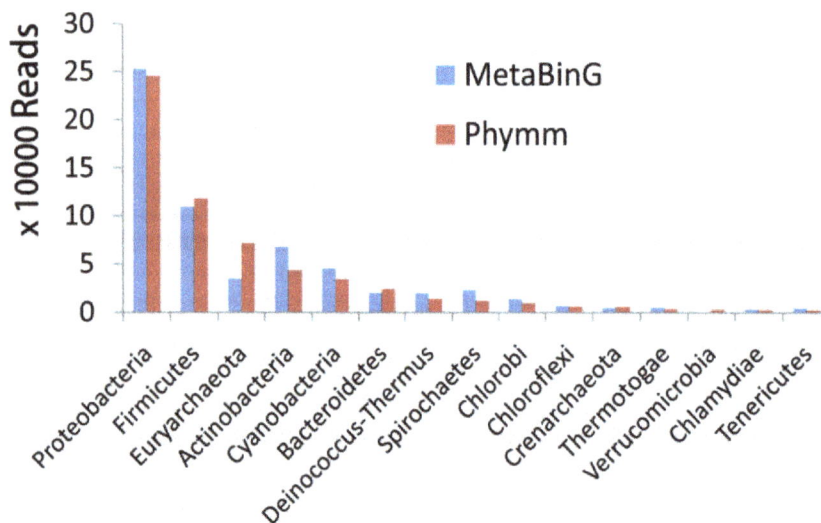

Figure 1. Biogas metagenome recovered by MetaBinG and Phymm. The 616,072 454 reads contained in the biogas metagenome dataset have been classified using MetaBinG and Phymm. The classification accuracy was measured at phylum level. The histogram shows only the top 15 phylum from the metagenomes recovered by Phymm. In general, the results recovered from MetaBinG and Phymm are similar except some small differences in *Euryarchaeota* and *Actinobacteria*. Among the top 15 phyla generated by Phymm, 14 was in the list of top 15 produced by MetaBinG. The relative ranks for these phyla generated by different methods varies at most by a value of two. MetaBinG is almost 1500-fold faster than Phymm.

In general, MetaBinG is an ultra-fast metagenomic sequence classification system for high-throughput sequence data. We demonstrated that MetaBinG could provide competitive results for sequences with long lengths in a speed 2 orders of magnitude faster. Due to the progress of sequencing technologies, the throughput gets higher and the reads get longer. The demand for a fast tool to analyze a huge amount of metagenomic sequences is constantly increasing. Therefore, MetaBinG can be a useful tool for the metagenomic classification.

Latest version of PhymmBL can produce a confidence score for an input sequence at each taxonomy rank, which is very convenient for users to assess the reliability of the classification. MetaBinG has not implemented this though it will be a welcome feature to come in the near future.

MetaBinG is publicly available at http://cbb.sjtu.edu.cn/~ccwei/pub/software/MetaBinG/MetaBinG.php. MetaBinG contains a pre-built 5th-order Markov Model for each of the existing 1212 genomes, so users do not need to train the models any more. The file size of current version of the full package is about 17 MB. MetaBinG has been tested on 64-bit Linux OS. One CUDA device is needed and it should be installed appropriately before running MetaBinG. In addition, users can add new Markov models (or genomes to the training set) using the addref.pl script in the package. In the near future, MetaBinG may become more accurate with more genomes available. This ultra-fast tool can be useful for a wide range of related research communities.

Methods

A kth-order Markov models is used in MetaBinG. A state in the Markov model is defined as an oligonucleotide of length k, and each state connects to 4 other states. The previous state shares k-1 bases with the next state. Therefore, there are 4^{k+1} transitions in total. A genomic sequence under the kth-order Markov model can be viewed as a sequence of state-transitions. The transition probabilities can be calculated for each genome in the training data set according to its Markov model as following:

$$kMM_{i,mn} = P_i(O_m|O_n) = \frac{F_i(O_m|O_n)}{F_i(O_m)} \quad (1)$$

where O_m and O_n are oligonucleotides of length k, $P(O_m \mid O_n)$ represents the transition probability from O_m to O_n, $F(O_m \mid O_n)$ represents observed count of transitions from O_m to O_n in a genomic sequence i and $F(O_m)$ is the observed count of O_m. A 4^{k+1} diemension vector is created to represent each genome in the training set. In practice, the minus logarithm value of each transition probability is saved.

A short sequence of length l can be considered as l-k transitions and a score S_i, which represents the distance between the short sequence and a genome i, can be computed as following:

$$S_i = - \sum_{j=0}^{l-k-1} ln(p_i(O_j|O_{j+1})) \quad (2)$$

where O_j and O_{j+1} are two oligonucleotides of length k, and $P(O_j|O_{j+1})$ is the transition probability from O_j to O_{j+1} observed in the i-th genome. When the transition from O_j to O_{j+1} does not exist in the i-th genome, the logarithm value of the transition probability will be set to a constant (default is 10). The high-order Markov models can be pre-computed from genomes in the training dataset. For each sequence, a genome in the database with the minimum score is selected as the source genome. At the end, each

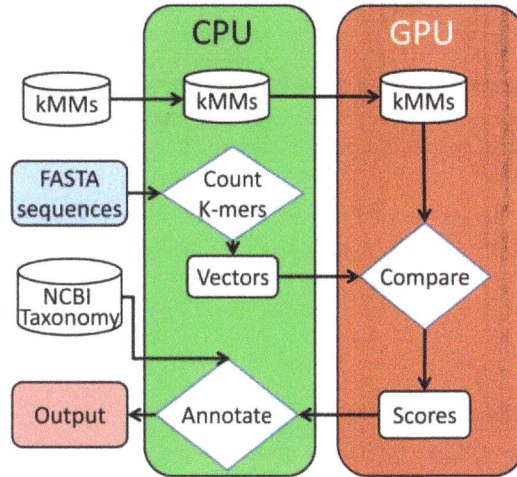

Figure 2. The system design of MetaBinG. First, the pre-built kth-order Markov Models (kMMs) are loaded to the GPU memory. Second, a CPU transforms input FASTA sequences into vectors of k-mer frequencies, which are then transferred to the GPU memory. Comparison of vectors against pre-built Markov models is done in the GPUs. The minimum scores are then output to the CPU, and the input sequence will be annotated with the NCBI taxonomy information in the CPU.

test sequence will be annotated with the taxonomy information of its source genome.

The algorithm complexity is determined by the number of genomes in the database and the order of Markov Models. It can be defined as follows

$$T(k,N') = O(N' * 4^k) \quad (3)$$

where k represents the length of oligonucleotides and N' stands for the number of genomes used for training.

In practice, the score S_i in equation (2) is calculated by matrix multiplication. First, the transitions generated from each genome in the reference database are converted into a 4^{k+1} diemension vector. Then, a matrix can be created from all vectors generated from genomes in the reference database. These can be prebuilt. For each short metagenomic sequence, the transitions generated from it are converted into a 4^{k+1} dimension vector as well. Then, the scores are computed by matrix multiplication, which is done by calling the SGEMM() function of CUBLAS library. At the end, the best score is picked and the associated genome is selected as its source genome. These are done by GPUs and the taxonomy information about the source genomes is printed out by CPUs.

MetaBinG is implemented in C with CUBLAS library. The system design of MetaBinG is shown in Figure 2. It has been tested in a Linux machine with 2 Intel Xeon E5520 CPUs (8 CPU cores), 16 GB memory and one NVDIA Tesla C1060 GPU card (240 cores). NVIDIA CUDA compiler driver nvcc release 3.0, V0.2.1221 with options "-L/usr/local/cuda/lib64 -lcudart -lcublas" was used to compile the GPU version source code.

Supporting Information

Table S1 The Complete list of training and test genomes. We downloaded 1212 fully sequenced bacterial genomes from the NCBI FTP site (ftp://ftp.ncbi.nih.gov/

genomes/Bacteria/) on 14 Dec 2010. Using the NCBI taxonomy, 390 genomes were removed to guarantee that every genus has at least two genomes. The remaining 822 genomes were assigned to training and test groups. Genomes from a species were assigned to one and only one set, either the training set or the test set. In the end, we generated 468 training genomes and 354 test genomes.

Table S2 Accuracy of Phymm at different ranks. The accuracy of Phymm was reported at different ranks. The data sets, the software and parameters are all the same as in Table 2.

Table S3 Accuracy of MetaBinG at different ranks. The accuracy of MetaBinG was reported at different ranks. The data sets, the software and parameters are all the same as in Table 2.

Author Contributions

Conceived and designed the experiments: CW PJ. Performed the experiments: PJ LX. Analyzed the data: PJ LX. Contributed reagents/materials/analysis tools: CW PJ. Wrote the paper: CW PJ. Revised the manuscript: CW PJ LL. Developed and implemented the algorithm, collected datasets, and tested the system: PJ. Implemented and tested the parallel CPU (parallel) version of MetaBinG: LX. Conceived and designed the research work: PJ. Conceived, designed and directed the research project: CW.

References

1. Hugenholtz P, Goebel BM, Pace NR (1998) Impact of culture-independent studies on the emerging phylogenetic view of bacterial diversity. Journal of Bacteriology 180: 4765–4774.
2. Shendure J, Ji H (2008) Next-generation DNA sequencing. Nat Biotechnol 26: 1135–1145.
3. Brady A, Salzberg SL (2009) Phymm and PhymmBL: metagenomic phylogenetic classification with interpolated Markov models. Nat Methods 6: 673–676.
4. Brady A, Salzberg S (2011) PhymmBL expanded: confidence scores, custom databases, parallelization and more. Nat Methods 8: 367.
5. Seshadri R, Kravitz SA, Smarr L, Gilna P, Frazier M (2007) CAMERA: a community resource for metagenomics. PLoS Biol 5: e75.
6. Glass EM, Wilkening J, Wilke A, Antonopoulos D, Meyer F (2010) Using the metagenomics RAST server (MG-RAST) for analyzing shotgun metagenomes. Cold Spring Harb Protoc 2010: pdb prot5368.
7. Meyer F, Paarmann D, D'Souza M, Olson R, Glass EM, et al. (2008) The metagenomics RAST server - a public resource for the automatic phylogenetic and functional analysis of metagenomes. BMC Bioinformatics 9: 386.

8. Rosen GL, Reichenberger ER, Rosenfeld AM (2011) NBC: the Naive Bayes Classification tool webserver for taxonomic classification of metagenomic reads. Bioinformatics 27: 127–129.
9. NVIDIA (2011) What is GPU Computing. Available: http://www.nvidia.com/object/GPU_Computing.html. Accessed 2011 Sept 1.
10. Dematte L, Prandi D (2010) GPU computing for systems biology. Brief Bioinform 11: 323–333.
11. Vouzis PD, Sahinidis NV (2011) GPU-BLAST: using graphics processors to accelerate protein sequence alignment. Bioinformatics 27: 182–188.
12. Schluter A, Bekel T, Diaz NN, Dondrup M, Eichenlaub R, et al. (2008) The metagenome of a biogas-producing microbial community of a production-scale biogas plant fermenter analysed by the 454-pyrosequencing technology. J Biotechnol 136: 77–90.
13. NVIDIA (2011) CUBLAS. Available: http://developer.nvidia.com/cuBLAS. Accessed 2011 Sept 1.

17

Effects of Temperature and Carbon-Nitrogen (C/N) Ratio on the Performance of Anaerobic Co-Digestion of Dairy Manure, Chicken Manure and Rice Straw: Focusing on Ammonia Inhibition

Xiaojiao Wang[1]*, Xingang Lu[2], Fang Li[1], Gaihe Yang[1]

1 College of Agronomy, Northwest A&F University, Yangling, Shaanxi, People's Republic of China, **2** School of Chemical Engineering, Northwest University, Xian, Shaanxi, People's Republic of China

Abstract

Anaerobic digestion is a promising alternative to disposal organic waste and co-digestion of mixed organic wastes has recently attracted more interest. This study investigated the effects of temperature and carbon-nitrogen (C/N) ratio on the performance of anaerobic co-digestion of dairy manure (DM), chicken manure (CM) and rice straw (RS). We found that increased temperature improved the methane potential, but the rate was reduced from mesophilic (30~40°C) to thermophilic conditions (50~60°C), due to the accumulation of ammonium nitrogen and free ammonia and the occurrence of ammonia inhibition. Significant ammonia inhibition was observed with a C/N ratio of 15 at 35°C and at a C/N ratio of 20 at 55°C. The increase of C/N ratios reduced the negative effects of ammonia and maximum methane potentials were achieved with C/N ratios of 25 and 30 at 35°C and 55°C, respectively. When temperature increased, an increase was required in the feed C/N ratio, in order to reduce the risk of ammonia inhibition. Our results revealed an interactive effect between temperature and C/N on digestion performance.

Editor: Wenjun Li, National Center for Biotechnology Information (NCBI), United States of America

Funding: This work was supported by science and technology support projects 'the biological technology integration and demonstration of high yield biogas digestion from the mix ingredients' (2011 BAD15B03) from Ministry of Science and Technology Department of the People's Republic of China and Research Fund for the Doctoral Program of Higher Education of Northwest A & F University, China(2013BSJJ057). The funders had no role in study design, data collection and analysis, decision to publish, or preparation of the manuscript.

Competing Interests: The authors have declared that no competing interests exist.

* E-mail: w-xj@nwsuaf.edu.cn

Introduction

Anaerobic digestion is an effective way of converting agricultural waste into biogas that can be used to generate energy, which is especially efficient in rural western China. In the past decade, this technology has received great attention in both scientific research and practice. However, the efficiency of anaerobic digestion may be limited by inadequate amount and diversity of waste from a single resource, which is insufficient for large-scale digesters, as well as the drawbacks of using single substrates, such as improper carbon-nitrogen (C/N) ratios, low pH of the substrate itself, poor buffering capacity, and high concentrations of ammonia [1,2,3]. Therefore, co-digestion of mixture substrates for biogas production has recently attracted more interest.

Co-digestion of various biosolid wastes, a process that utilizes the nutrients and bacterial diversity in those wastes to optimize the digestion process, is an attractive approach for improving the efficiency of biotransformation [4]. A primary advantage of co-digestion is that it could efficiently balance feedstock carbon and nitrogen and a balanced C/N ratio of feedstock is likely to improve methane production. An early study conducted by Wu *et al.* revealed that swine manure co-digested with corn stalks at a C/N ratio of 20 obtained increased cumulative biogas production up to 11-fold and increased cumulative net methane volume up to 16-fold, when compared to swine manure digested alone [5]. Recent study by Wang *et al.* also suggested that co-digestion of dairy manure, chicken manure and wheat straw, had better digestion performance with stable pH and low concentrations of total ammonium nitrogen (TAN) and free ammonia (FA) at adjusted C/N ratios of 25 and 30 [6]. Similar observations were also reported by Hills for dairy manure, demonstrating that the greatest methane production was achieved when the C/N ratio was adjusted to 25 using glucose [7]. By optimizing the substrate C/N ratio, co-digestion of wastes of different C/N characteristics can greatly enhance the efficiency of biogas digestion.

Although many studies indicated that the optimal C/N ratios in methane fermentation were 25~30 [8,9,10], the depletion of carbon and nitrogen could be affected by operating conditions, such as temperature, resulting in the occurrence of inhibitory effects. It has been reported that the high FA concentration could inhibit thermophilic more seriously than mesophilic digestion [11,12,13]. A decrease in operating temperature from 60°C to 37°C in anaerobic digesters with a high ammonia concentration provided relief from FA inhibition, leading to increase in biogas yield [14,15]. FA concentration under mesophilic digestion is already inhibitory in the range of 80~150 mg L^{-1} at a pH of 7.5 [16,17,18]. However, under thermophilic conditions, when the concentration of FA was increased to 620 mg L^{-1} in the ammonia

Table 1. Chemical characteristics of raw materials used in this study.

Substrate	[a] TS content/%	VS content/%	pH	Total carbon/g kg^{-1}VS	Total Kjeldahl nitrogen/g kg^{-1}VS	C/N
DM	15.8±0.34	81.5±1.41	7.26±0.03	65.8±1.19	2.96±0.05	22.2±0.22
CM	29.9±0.67	65.3±1.26	6.93±0.11	58.6±1.77	6.11±0.08	9.6±0.16
RS	89.2±1.59	92.3±1.34	-	328±5.67	6.34±0.11	51.7±1.62

[a] ±shows the standard error.

toxicity test, a gradual decrease of 21% was observed in biogas [19]. Another study also indicated that thermophilic flora tolerated at least twice as much FA compared to mesophilic flora [20]. Because the concentrations of TAN and FA originally depend on the content of organic nitrogen in the reactor and on C/N ratios, the indicator of substrate carbon and nitrogen content may also interact with temperature and that interaction results in different concentrations of ammonia and FA, as well as inhibitory effects.

Base on previous studies mentioned above, there are interactive effects between temperature and ammonia in the digestion process and the digestion efficiency is dramatically affected by the temperature and C/N ratio. Thus, to investigate this interaction, we first examined the effect of a series of temperatures on the mixtures of certain ratios of C/N (25), and secondly, compared the digestion performance of mixtures with a series of C/N ratios by adjusting the proportions of each substrate, dairy manure (DM), chicken manure (CM) and rice straw (RS) under mesophilic and thermophilic conditions.

Materials and Methods

Substrate characteristics

DM and CM were collected from a livestock farm located in Yangling, China. RS was obtained from a local villager. Before being put into the reactor, the air-dried RS was cut into pieces (2~3 cm). The substrates were individually homogenized and subsequently stored at 4°C for further use. The chemical characterization of each substrate tested in this study is shown in Table 1. All samples were collected and tested in triplicate, and the averages of the three measurements are presented.

Ethics statement

The collections of DM and CM were permitted by livestock farms belonging to 'Besun' group in Yangling, China. The RS was provided voluntarily by a local villager in Qishan, Baoji, China. The inoculum was obtained from a household biogas digester in a biogas demonstration village named Cuixigou in Yangling and the collection was permitted by the hosts. The all experimental procedures conformed to the regulations established by the Ethics Committee of the Research Center of Recycle Agricultural Engineering and Technology of Shaanxi Province, China.

Experimental design and set-up

Experiment 1: Three mixture sets were investigated in this experiment: set A (DM+ RS), set B (CM+RS), and set C (DM+ CM+RS). For set A and set B, the C/N ratio was 25, achieved by adjusting the DM/RS or CM/RS ratio. For set C, based on a DM/CM ratio of 1:1, multi-component substrates were prepared by adding RS to the DM-CM mixtures in order to adjust the C/N ratio to 25. The proportions of all substrates in each mixture were in a volatile solid (VS) state. The operation temperatures were 20, 30, 40 (mesophilic), 50, and 60°C (thermophilic), respectively.

Experiment 2: For all mixture sets, RS was added into the DM-CM mixtures with a VS ratio of 1:1, in order to adjust the C/N ratio to selected levels. C/N ratios of 15, 20, 25, 30 and 35 were selected in tests at a temperature of 35°C, but ratios of 20, 25, 30, 35 and 40 were selected in tests at a temperature of 55°C.

The initial VS ratio of substrate to inoculum was kept at 1:2 for all experimental setups. Each reactor had a 1 L capacity and contained 600 mL of total liquid, including 200 mL of inoculum and mixed substrate of 15gVS/L. The inoculum used for digestion at 20, 30, 35 and 40°C was digested cattle manure, taken from a lab-scale reactor operated at 35°C with a hydraulic

Table 2. Effects of temperature on pH value in anaerobic co-digestion with a C/N ratio of 25.

Temperature (°C)	DM+RS		CM+RS		DM+CM+RS	
	[a] Average	Final	Average	Final	Average	Final
20	6.12±0.13	6.64±0.14	5.42±0.11	5.01±0.07	5.92±0.10	6.58±0.11
30	6.89±0.15	7.12±0.16	6.42±0.03	6.92±0.02	7.11±0.11	7.35±0.14
40	7.21±0.09	7.44±0.13	7.19±0.08	7.38±0.12	7.48±0.13	7.67±0.11
50	7.58±0.11	7.61±0.13	7.66±0.12	7.79±0.07	7.56±0.12	7.74±0.03
60	7.69±0.15	7.88±0.15	7.82±0.10	8.11±0.11	7.72±0.16	7.92±0.13
[b]LSD$_{0.05}$= 0.47 [c]LSD$_{0.05}$=0.61						

[a] ±shows the standard error
[b]LSD value at the 5% level based on all average values from three mixture sets at all operation temperatures
[c]LSD value at the 5% level based on all final values from three mixture sets at all operation temperatures

Table 3. Effects of temperature on total ammonia content in anaerobic co-digestion with a C/N ratio of 25.

Temperature (°C)	DM+RS		CM+RS		DM+CM+RS	
	[a] Average	Final	Average	Final	Average	Final
20	182.3±3.9	229.3±4.8	495.2±7.2	532±9.2	477±8.3	521±1.88
30	260.2±5.8	518.5±8.5	552.5±2.0	674±14.3	531±10.2	778±12.8
40	421.2±7.5	772.7±13.4	768.4±12.6	995±22.2	737±10.7	921±17.7
50	541.5±9.4	968.6±18.7	938.3±14.8	1261±18.3	869±3.1	1116±16.3
60	593.6±11.4	1052.4±18.8	951.8±12.5	1201±4.8	906±14.9	1256±20.6

^{b}LSD$_{0.05}$ = 52.6 ^{c}LSD$_{0.05}$ = 81.9

a +shows the standard error
b LSD value at the 5% level based on all average values from three mixture sets at all operation temperatures
c LSD value at the 5% level based on all final values from three mixture sets at all operation temperatures.

retention time (HRT) of 15 days. Additionally, digestion at 50, 55 and 60°C was inoculated with digested cattle manure from the lab-scale reactor operated at 55°C with a HRT of 15 days. A control with only inoculum was used to determine biogas production due to endogenous respiration. Each treatment was performed in triplicate. All reactors were tightly closed with rubber septa and screw caps. The headspace of each reactor was flushed with nitrogen gas for about 3 min to assure anaerobic conditions prior to starting the digestion tests. To provide mixing of the reactor contents, all reactors were shaken manually for about 1 min, once a day prior to measurement of biogas volume.

Analytical techniques

Total solids, VS, pH, total Kjeldahl nitrogen (TKN), and total ammonium nitrogen (TAN) analysis were performed according to APHA Standard Methods [21]. Total organic carbon was determined by the method described by Cuetos et al. [22]. For all treatments, FA concentration was calculated in accordance with Hansen et al. [23]. The volume of biogas was measured by displacement of water. Methane content in the produced biogas was analyzed with a fast methane analyzer (Model DLGA-1000, Infrared Analyzer, Dafang, Beijing, China). The C/N ratio was determined by dividing the total organic carbon content by the total nitrogen content, according to the following equation.

$$C/N = \frac{W1 \times C1 + W2 \times C2 + W3 \times C3}{W1 \times N1 + W2 \times N2 + W3 \times N3}$$

Where W1, W2 and W3 were the VS weight in a single substrate in the mixture, C1, C2 and C3 were the organic carbon content (g kg^{-1}VS) in each substrate and N1, N2 and N3 were the nitrogen content (g kg^{-1}VS) in each substrate.

Results

Effects of temperature on the performance of anaerobic co-digestion based on experiment 1

Increased temperature resulted in pH increases in all three mixtures (Table 2). The pH values in digesters at 20°C, with average values of 6.12, 5.42 and 5.92 in the mixtures of DM+RS, CM+RS and DM+CM+RS, respectively, were far lower than those under other temperatures. From 30 to 60°C, the average pH values were in the range of 6.42 ~7.82.

A linear correlation between TAN and temperature (20 – 60°C) was observed and the highest TAN value was 1,261 mg L^{-1} in the mixture of CM+RS at 50°C (Table 3). The relationship between FA (Y, mg L^{-1}) and temperature (T, °C) was evaluated by the following equations: Y = 0.0302e$^{1.82T}$ in the mixture of DM+RS, Y = 0.0216e$^{2.0T}$ in the mixture of CM+RS and Y = 0.101e$^{1.65T}$ in the mixture of DM+CM+RS. On average, the mixture of DM+CM+RS had significantly higher TAN and FA concentrations than the mixture of DM+RS, but was lower than the mixture of CM+RS (Tables 3 and 4).

With the increase of temperature, methane potential continuously increased, but the increasing rate was lower under thermophilic than under mesophilic conditions (Fig. 1). The mixture of DM+CM+RS had a little higher methane potential than the mixtures of DM+RS and CM+RS.

Table 4. Effects of temperature on free ammonia content in in anaerobic co-digestion with a C/N ratio of 25.

Temperature (°C)	DM+RS		CM+RS		DM+CM+RS	
	[a]Average	Final	Average	Final	Average	Final
20	0.1±0.004	0.4±0.009	0.1±0.002	0.9±0.01	0.2±0.001	0.8±0.01
30	1.6±0.04	5.4±0.08	1.2±0.03	4.4±0.06	5.4±0.09	13.8±0.1
40	12.7±0.5	31.4±0.6	17.9±0.8	35.5±1.6	32.8±1.1	63.7±2.1
50	65.4±2.2	101.2±4.8	108.6±3.7	189.2±4.3	82.0±1.8	153.3±3.2
60	142.5±3.6	324.5±4.9	240.7±5.2	479.2±3.2	192.6±3.2	376.7±8.2
[b]$LSD_{0.05}$ = 23.2 [c]$LSD_{0.05}$ = 45.6						

[a] ± shows the standard error
[b] LSD value at the 5% level based on all average values from three mixture sets at all operation temperatures
[c] LSD value at the 5% level based on all final values from three mixture sets at all operation temperatures

Effects of C/N ratio on the performance of anaerobic co-digestion based on experiment 2

The pH value and the concentrations of TAN and FA were significantly influenced by C/N ratios at 35°C. For digesters with C/N ratios of 15 and 20, the pH values were higher than 7.0 during the whole digestion process, and the final pH values reached to 8.09 and 7.68, respectively (Fig. 2A).The average pH value was as low as 6.67 when the C/N ratio increased to 35. C/N ratios of 25 and 30 resulted in average pH values of 7.12 and 7.02, respectively. In addition, the contents of TAN and FA decreased with increased C/N ratios (Fig. 2B and C). Low C/N ratios of 15 and 20 resulted in TAN and FA concentrations as high as 2610, 2258 mg L^{-1} and 314, 108 mg L^{-1}, respectively. Treatments with C/N ratios of 25, 30 and 35 resulted in low and stable TAN and FA during the anaerobic process. The average concentrations of TAN were 985, 739 and 568 mg L^{-1} when C/N ratios were of 25, 30, and 35, respectively and the average concentrations of FA were 9.1, 7.5 and 2.2 mg L^{-1} when C/N ratios were of 25, 30, and 35, respectively.

Under 55°C, pH values were between 7.0 and 7.92 in treatments with C/N ratios of 20 and 25. Stable pH values around 7.0 were observed when C/N ratios were of 30 and 35. When the C/N ratio was increased to 35, the pH value was lower, at around 6.2 (Fig. 3A). The concentrations of TAN in treatments with C/N ratios of 20 and 25 increased up to 1500 mg L^{-1} by day 10 and reached peaks as high as 2415 and 1932 mg L^{-1},

respectively (Fig. 3B). FA increased continuously in digestion with final concentrations of 461 and 235 mg L^{-1} when C/N ratios were of 20 and 25. For C/N ratios between 30 and 40, TAN and FA concentrations were in the range of 430~1426 mg L^{-1} and 2~131 mg L^{-1}, respectively (Fig. 3B and C).

Methane potential increased first and then decreased with increases of C/N ratios. The highest methane potential was observed with a C/N ratio of 25 at 35°C with 272 mL g^{-1}VS and with a C/N ratio of 30 at 50°C with 286 mL g^{-1}VS, respectively (Fig. 4). The quadratic models for methane potential in terms of the C/N ratio as a variable were significant and the equations at 35°C (1) and 55°C (2) were expressed as follows:

$$Y = -0.8475X^2 + 45.36X - 345.3, R^2 = 0.9652 \qquad (1)$$

$$Y = -1.16X^2 + 71.16X - 781.4, R^2 = 0.8922 \qquad (2)$$

Where Y was methane potential and X was the C/N ratio. The optimum conditions for maximum methane potential were calculated as a C/N of 26.76 at 35°C and a C/N ratio of 30.67 at 55°C, respectively. Accordingly, the highest methane potential was estimated as 265.7 and 309.9 mL g^{-1} VS.

Discussion

According to the study by Calli *et al.*, ammonia inhibition occurs in the range of 1500~3000 mg L^{-1} TAN when the pH value is over 7.4 [24]. Then, TAN concentrations of three mixtures were in a safe range below 1261 mg L^{-1} at temperatures between 20 and 60°C (Table 3). Compared with ammonium nitrogen, FA has been suggested as the active component causing ammonia inhibition, since it is freely membrane-permeable [25]. It has been reported that a range between 80 and 150 mg L^{-1} FA was inhibitory for methanogens [16,26]. In our study, FA concentrations were in this range at 50°C and far higher than 150 mg L^{-1} at 60°C (Table 4), indicating the occurrence of ammonia inhibition. Based on experiment 1, temperature obviously played a greater role in methane production in the range of 20 ~40°C than in the range of 40 ~60°C. Methane potentials in three mixtures were an average of 2.49 times higher at 40 than 20°C, but only 1.20 times higher at 60 than at 40°C. And no significant difference was found in methane potential between 50 and 60°C.

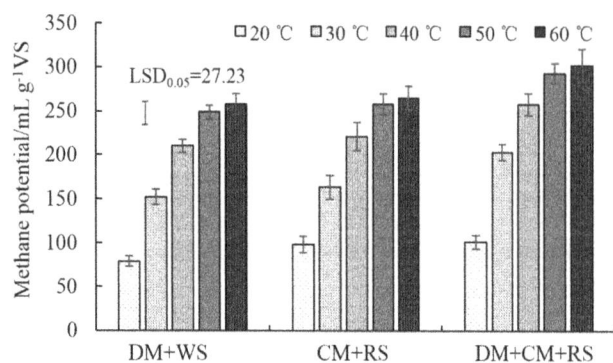

Figure 1. Effects of temperature on methane potential in mixtures with a C/N ratio of 25. Values are presented as the mean ±standard error of three replicates (n = 3). Vertical bars represent LSD at the 5% level.

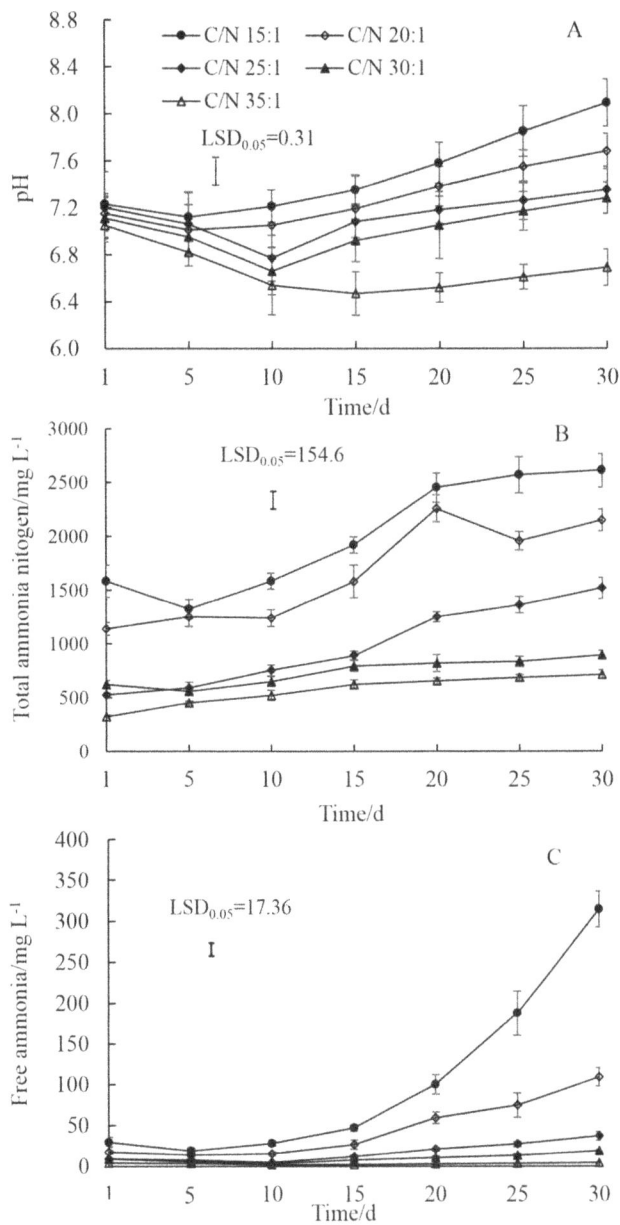

Figure 2. Changes of pH, total ammonium nitrogen, and free ammonia with different C/N ratios in the mixture of dairy manure (DM), chicken manure (CM), and rice straw (RS) in anaerobic co-digestion at 35°C. Values are presented as the mean ±standard error of three replicates (n = 3). Vertical bars represent LSD at the 5% level.

Figure 3. Changes of pH, total ammonium nitrogen, and free ammonia with different C/N ratios in the mixture of dairy manure (DM), chicken manure (CM), and rice straw (RS) IN anaerobic co-digestion at 55°C. Values are presented as the mean ±standard error of three replicates (n = 3). Values are presented as the mean ±standard error of three replicates (n = 3). Vertical bars represent LSD at the 5% level.

These results also suggest the existence of an inhibitory effect by ammonia under thermophilic conditions. However, in the anaerobic digestion of organic wastes, it has been reported that methane production was inhibited up to 50% by 220 mg L^{-1} FA at 37°C and by 690 mg L^{-1} FA at 55°C [20]. That is, thermophilic flora tolerated at least twice as much FA as compared to mesophilic flora. The higher methane potential under thermophilic conditions suggested that increased ammonia did not completely inhibit the digestion process and did not offset the advantage of increased temperature in thermodynamics and kinetics, which might result from proper C/N ratios of mixture substrates.

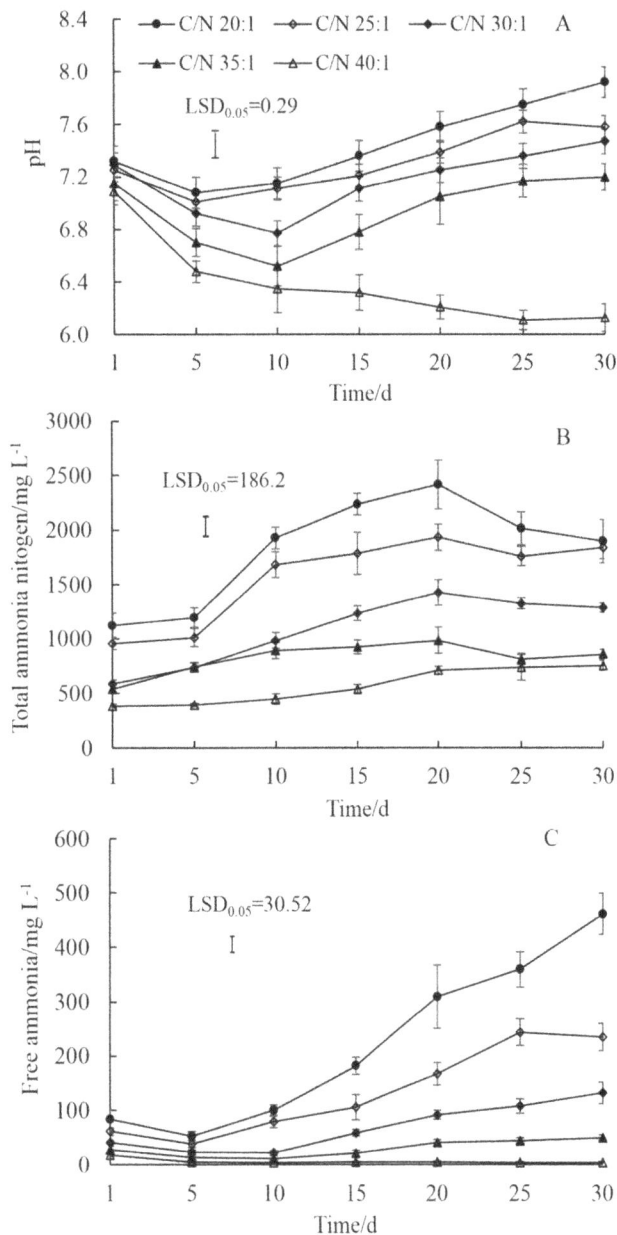

Due to the potential role of the C/N ratio in regulating the inhibitory effects of ammonia, digestions with different C/N ratios were tested in experiment 2 under mesophilic and thermophilic conditions to further obtain optimal C/N ratios with less ammonia inhibition. We found that the mixture of DM+CM+RS had better digestion performance in methane potential than the mixtures of DM+RS and CM+RS (Fig. 1), which might be due to the increased buffering capacity and the synergistic effect, which was inconsistent with the result reported by Wang *et al.* [6]. The mixture of DM+CM+RS was then selected for follow-up studies.

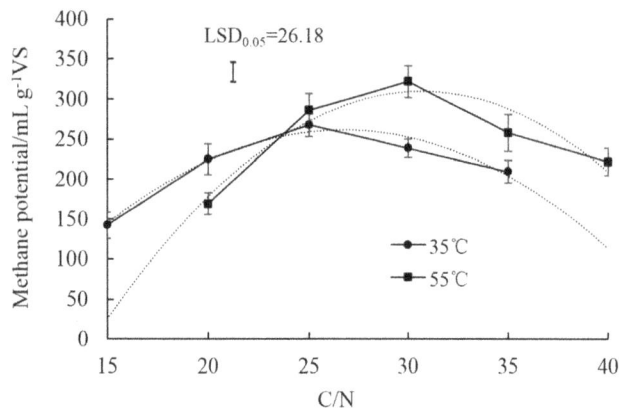

Figure 4. Changes of methane potential with different C/N ratios in the mixture of dairy manure (DM), chicken manure (CM), and rice straw (RS) in anaerobic co-digestion at 35°C and 55°C. The dotted lines were fitting curves for both temperatures. Values are presented as the mean ±standard error of three replicates (n = 3). Vertical bars represent LSD at the 5% level.

Substrates that have low C/N ratios contain relatively high concentrations of ammonia, exceeding concentrations necessary for microbial growth, and probably inhibiting anaerobic digestion [3,23]. TAN concentrations were as high as 2500 mg L^{-1} and FA increased up to final concentrations of 314 and 461 mg L^{-1}, when the C/N ratio was of 15 at 35°C and was 20 at 55°C, respectively (Figs. 2 and 3). Therefore, methane potential was reduced down to 142 mLg^{-1}VS at 35°C and 169 mLg^{-1}VS at 55°C, accounting for just 53.0% and 52.5%, compared with their maximum values (Fig. 4). Under both temperatures, with the increase of C/N ratios, TAN and FA concentrations decreased. For example, the average FA concentrations at 35°C were reduced 56.5, 83.7, 90.7 and 97.1% from a C/N ratio of 15 to 20, 25, 30 and 35, respectively. Previous reports suggested that using a feedstock C/N ratio from 27 to 32 promotes steady digester operation at optimum ammonia nitrogen levels and feedstock with a C/N ratio of 32 producing a lower concentration of ammonia nitrogen and FA [25,27]. Thus, the digestion system was sensitive to the feed C/N ratio and a higher C/N ratio reduced the protein solubilization rate and hence produced lower TAN and FA concentration within the system, which was found to be advantageous.

Ammonia inhibition under mesophilic and thermophilic conditions has been compared in previous studies. It has been observed that an increase in temperature resulted in a reduction of the biogas yield, due to the increased inhibition of FA under higher temperature [14,15,28]. In our study, ammonia inhibition occurred with a C/N ratio of 20 at 55°C, whereas a C/N ratio of 15 experienced inhibition at 35°C, suggesting that higher temperature improved the degradation efficiency of organic nitrogen to ammonia nitrogen. However, when C/N ratios were higher than 25, methane potential at 55°C was higher than at 35°C (Fig. 4), indicating higher C/N ratios reduced the risk of ammonia inhibition under thermophilic conditions. Moreover, the optimal C/N ratios were obtained at 26.76 and 30.67 under mesophilic and thermophilic conditions, respectively, by optimizing the quadratic models between methane potential and C/N ratio. These results showed that ammonia inhibition occurring under thermophilic conditions might be avoided by optimizing the C/N ratio in co-digestion of different substrates. However, a very high C/N ratio promotes the growth of methanogen populations that are able to meet their protein requirements and will, therefore, no longer react with the remaining carbon content of the substrate, resulting in a low production of gas.

Conclusions

This study demonstrated an interactive effect between C/N ratio and temperature on the performance of anaerobic co-digestion of dairy manure, chicken manure and rice straw. Our results suggest that increased temperature from mesophilic to thermophilic conditions resulted in ammonia inhibition, however, this kind of inhibition could be reduced or avoided by increasing the C/N ratio of mixed feedstock to an appropriate level. In anaerobic co-digestion of DM, CM and RS, the optimal C/N level was 26.76 at 35°C and 30.67 at 55°C. Adjusting the proportions of mixture substrates in anaerobic co-digestion to obtain suitable feed characteristics, such as the C/N ratio, pH and nutrients, is an effective way to achieve desired digestion performance.

Author Contributions

Conceived and designed the experiments: XJW XGL GHY. Performed the experiments: XJW FL. Analyzed the data: XJW XGL GHY. Wrote the paper: XJW.

References

1. Banks C, Humphreys P (1998) The anaerobic treatment of a ligno-cellulosic substrate offering little natural pH buffering capacity. Water Sci Technol 38:29–35.
2. Zhang T, Liu LL, Song ZL, Ren GX, Feng ZY, et al.(2013) Biogas Production by Co-Digestion of Goat Manure with Three Crop Residues. PLoS ONE 8(6): e66845.
3. Procházka J, Dolejš P, Máca J, Dohányos M (2012) Stability and inhibition of anaerobic processes caused by insufficiency or excess of ammonia nitrogen. Appl Microbiol Biotechnol 93:439–447.
4. Wang XJ, Yang GH, Feng YZ, Ren GX, Han XH (2012) Optimizing feeding composition and carbon-nitrogen ratios for improved methane yield during anaerobic co-digestion of dairy chicken manure and wheat straw. Bioresour Technol 120:78–83.
5. Wu X, Yao WY, Zhu J, Miller C (2010) Biogas and CH(4) productivity by co-digesting swine manure with three crop residues as an external carbon source. Bioresour Technol 101: 4042–4047.
6. Wang XJ, Yang GH, Li F, Feng YZ, Ren GX, et al.(2013) Evaluation of two statistical methods for optimizing the feeding composition in anaerobic co-digestion: Mixture design and central composite design. Bioresour Technol 131:172–178.
7. Hills DJ (1979) Effects of carbon: nitrogen ratio on anaerobic digestion of dairy manure. Agr Wastes 1:267–278.
8. Kayhanian M, Tchobanoglous G (1992) Computation of C/N ratios for various organic fractions. Biocycle 33:58–60.
9. Marchaim U, Krause C (1993) Propionic to acetic-acid ratios in overloaded anaerobic- digestion. Bioresour Technol 43:195–203.
10. Yen HW, Brune DE (2007) Anaerobic co-digestion of algal sludge and waste paper to produce methane. Bioresour Technol 98:130–134.
11. Zeeman G, Wiegant W, Koster-Treffers M, Lettinga G (1985) The influence of the total-ammonia concentration on the thermophilic digestion of cow manure. Agr Wastes 14:19–35.
12. Wiegant W, Zeeman G (1986) The mechanism of ammonia inhibition in the thermophilic digestion of livestock wastes. Agr Wastes 16:243–253.
13. Angelidaki I, Ahring B (1993) Thermophilic anaerobic digestion of livestock waste: the effect of ammonia. Appl Microbiol Biot 38:560–564.
14. Angelidaki I, Ahring B (1994) Anaerobic thermophilic digestion of manure at different ammonia loads: effect of temperature. Water Res 28:727–731.
15. Hansen KH, Angelidaki I, Ahring BK (1999) Improving thermophilic anaerobic digestion of swine manure. Water Res 33:1805–1810.
16. Braun R, Huber P, Meyrath J (1981) Ammonia toxicity in liquid piggery manure digestion. Biotechnol Lett 3:159–164.
17. Kroeker E, Schulte D, Sparling A, Lapp H (1979) Anaerobic treatment process stability. J Water Pollut Control Fed 718–727.

18. Siles JA, Martin MA, Chica AF, Martin A (2010) Anaerobic co-digestion of glycerol and wastewater derived from biodiesel manufacturing. Bioresour Technol 101:6315– 6321.
19. Gallert C, Winter J (1997) Mesophilic and thermophilic anaerobic digestion of source- sorted organic wastes: effect of ammonia on glucose degradation and methane production. Appl Microbiol Biot 48:405–410.
20. PHA (1995) Standard methods for the examination of water and wastewater. Washington. DC, American Public Health Association.
21. Cuetos MJ, Fernandez C, Gomez X, Moran A (2011) Anaerobic co-digestion of swine manure with energy crop residues. Biotechnol Bioprocess Eng 16:1044– 1052.
22. Hansen KH, Angelidaki I, Ahring BK (1998) Anaerobic digestion of swine manure: Inhibition by ammonia. Water Res 32:5–12.
23. Calli B, Mertoglu B, Inanc B, Yenigun O (2005) Effects of high free ammonia concentrations on the performances of anaerobic bioreactors. Process Biochem 40:1285–1292.
24. Kayhanian M (1999) Ammonia inhibition in high-solids biogasification: An overview and practical solutions. Environ Technol 20:355–365.
25. Ahring BK, Angelidaki I, Johansen K (1992) Anaerobic treatment of manure together with industrial-waste. Water Sci Technol 25:311–318.
26. Koster I, Lettinga G (1988) Anaerobic digestion at extreme ammonia concentrations. Biol Waste 25:51–59.
27. Zeshan Karthikeyan OP, Visvanathan C (2012) Effect of C/N ratio and ammonia-N accumulation in a pilot-scale thermophilic dry anaerobic digester. Bioresour Technol 113:294–302.
28. Garcia ML, Angenent LT (2009) Interaction between temperature and ammonia in mesophilic digesters for animal waste treatment. Water Res 43:2373–2382.

18

Selection and Characterization of Biofuel-Producing Environmental Bacteria Isolated from Vegetable Oil-Rich Wastes

Almudena Escobar-Niño[1,2]**, Carlos Luna**[3]**, Diego Luna**[3]**, Ana T. Marcos**[1]**, David Cánovas**[1]***,
Encarnación Mellado[2]*

1 Department of Genetics, Faculty of Biology, University of Seville, Seville, Spain, **2** Department of Microbiology and Parasitology, Faculty of Pharmacy, University of Seville, Seville, Spain, **3** Department of Organic Chemistry, University of Córdoba, Córdoba, Spain

Abstract

Fossil fuels are consumed so rapidly that it is expected that the planet resources will be soon exhausted. Therefore, it is imperative to develop alternative and inexpensive new technologies to produce sustainable fuels, for example biodiesel. In addition to hydrolytic and esterification reactions, lipases are capable of performing transesterification reactions useful for the production of biodiesel. However selection of the lipases capable of performing transesterification reactions is not easy and consequently very few biodiesel producing lipases are currently available. In this work we first isolated 1,016 lipolytic microorganisms by a qualitative plate assay. In a second step, lipolytic bacteria were analyzed using a colorimetric assay to detect the transesterification activity. Thirty of the initial lipolytic strains were selected for further characterization. Phylogenetic analysis revealed that 23 of the bacterial isolates were Gram negative and 7 were Gram positive, belonging to different clades. Biofuel production was analyzed and quantified by gas chromatography and revealed that 5 of the isolates produced biofuel with yields higher than 80% at benchtop scale. Chemical and viscosity analysis of the produced biofuel revealed that it differed from biodiesel. This bacterial-derived biofuel does not require any further downstream processing and it can be used directly in engines. The freeze-dried bacterial culture supernatants could be used at least five times for biofuel production without diminishing their activity. Therefore, these 5 isolates represent excellent candidates for testing biofuel production at industrial scale.

Editor: Ligia O Martins, Universidade Nova de Lisboa, Portugal

Funding: This work was supported by grants from Junta de Andalucía (P08-RNM-03515, P11-CVI-7427 MO and P11-TEP-7723), as well as Spanish Ministry of Economy and Competitiveness (Project ENE 2011-27017) and FEDER funds. The funders had no role in study design, data collection and analysis, decision to publish, or preparation of the manuscript.

Competing Interests: The authors have declared that no competing interests exist.

* Email: davidc@us.es (DC); emellado@us.es (EM)

Introduction

With a growing world population, fossil fuels are currently consumed too rapidly. Thus, it is expected that we will deplete these non-renewable resources from the planet in a relatively short period of time. The increasing demand for fossil fuels has additional consequences, such as the concomitant increased prices of crude oil, the environmental concerns about the pollution due to crude oil derivatives and the global greenhouse effects, which altogether are triggering the exploration of novel alternative sources of fuels. Therefore, it is imperative to develop alternative, sustainable and inexpensive fuels. Biodiesel can be such a fuel with the appropriate technologies. Triglycerides (oils and fats) can not be directly used in the available diesel engines due to the high viscosity and the acidic composition of these lipids, the formation of free fatty acids resulting in gum formation by oxidation and polymerization, the carbon deposition and the thickening of the lubricant. Thus, the use of vegetable oils as alternative sources of fuels requires their processing to reach a viscosity and a volatility similar to crude oil derived fuels, a fact that will allow them to be directly used in the current configuration of diesel engines. These reasons are driving the development of new vegetable oil derivatives displaying properties similar to those of diesel fuels. Nowadays the most promising and accepted vegetable oil derivative is biodiesel [1,2].

Biodiesel is a renewable source of energy considered to be carbon neutral because the carbon delivered during its combustion was fixed only a few years before from the atmosphere. In addition, its use in internal combustion engine does not produce sulfur oxide and minimizes three fold the formation of soot particulates in comparison with petrochemical diesel [3]. Biodiesel is a mixture of mono-alkyl esters that can be obtained from vegetable oils, animal fats, waste cooking oils, greases and algae. Three processing techniques are mainly used to catalyze the conversion of vegetable oils into ready-to-use biofuels: pyrolysis, microemulsification and transesterification. The most popular method is the transesterification (ethanolysis or methanolysis) of vegetable oils to produce biodiesel. The transesterification process can be performed using an alkali catalyst, an acid catalyst, a biocatalyst or heterogeneous catalysts, however only the alkaline process is currently carried out at industrial scale, because it is cost effective and highly efficient [4]. In the alkaline process sodium hydroxide (NaOH) or potassium hydroxide (KOH) is used as a catalyst along with methanol or ethanol. NaOH or KOH react with the alcohol to give an alcoxy group. Then the alcoxy moiety reacts with any triglyceride to form the corresponding methyl- or ethyl-esters and glycerol. After the synthesis, the production of biodiesel requires a series of downstream processes that include repeated washings to reach sufficient purity. These downstream operations are usually hampered with problems, such as the separation of catalyst and unreacted methanol from biodiesel, the risk of free fatty acid or water contamination and soap formation during the transesterification reaction [1,2].

The production of biodiesel using a biocatalyst eliminates the disadvantages of the chemical alkaline process by obtaining a product of very high purity with less or no downstream operations. In particular, to avoid the need to separate glycerol from biodiesel, an alternative strategy based on incorporating some of the glycerol derivatives in the final product has been considered. In this way, a biofuel formed by a mix of glycerol derivatives and fatty acid methyl esters (FAME) or fatty acid ethyl esters (FAEE) not only prevents the need to separate the glycerol from the biodiesel, but also increases the yield of the process as the total number of carbons involved in the reaction is maintained in the final product [5,6]. Some examples of methodologies considering this option are based on the transesterification reaction of triglycerides (TG) with dimethyl carbonate [7], methyl acetate [8] or ethyl acetate [9] resulting in a mixture of three molecules of FAME or FAEE and one of glycerol carbonate or glycerol triacetate. Another example is Ecodiesel-100, a patented biofuel obtained using 1,3-selective partial ethanolysis of triglycerides with a porcine pancreatic lipase. Ecodiesel-100 is a mixture of FAEE and monoglycerides (MG) [5,6]. Despite of the advantages of Ecodiesel-100, the main drawback is the high cost of the purified enzyme. Therefore, there is a need to solve this issue before the process can be implemented at the industrial scale. The enzymatic production of biofuel is possible using extracellular or intracellular lipases.

Lipases (triacylglycerol acylhydrolases EC 3.1.1.3) are very versatile biocatalysts: they are stable in organic solvents; they do not require any cofactor; they have broad substrate specificity; and they show high enantioselectivity. The number of cloned lipases has increased since the 1980s, in part because of this versatility, but also because of the increased number of characterized lipolytic microorganisms. Several enzymatic activities have been described for lipases, such as hydrolysis, transesterification and esterification. However, the hydrolytic activity of lipases has been studied in detail in contrast to the transesterification activity. It is hypothesized that the transesterification activity of lipases is similar to the serine-protease catalytic activity, because both enzymes share the same catalytic triad Ser–His–Asp/Glu [10].

Extensive and persistent screening for new microorganisms and their lipolytic enzymes can open novel alternatives for synthetic processes and consequently, novel possibilities to contribute to solve environmental problems. The aim of this study is to find and select lipolytic microorganisms showing transesterification activity, which can be used to produce a cheap enzymatic extract, avoiding laborious downstream processing. Employing this enzymatic extract able to produce biofuel could reduce the cost of enzyme production. Therefore the selected microorganisms have the potential to be used in white biotechnology for the industrial production of a yet-to-be commercialized biofuel.

Materials and Methods

Site description and sample collection

Four samples were collected from different locations in an oil mill in Ecija (Sevilla, Spain) at the end of the harvesting season (January 2011). Pool 1 (AE1B) and pool 2 (AE2B) were sampled from ponds containing liquid wastes from the oil mill. Alpechín (AEA) was obtained from the solid wastes of the oil mill and the last sample (AEDH) was obtained from the wastes of the olive harvests containing mainly olive tree leftovers (i.e. dehydrated leaves and branches).

Samples were taken in sterile plastic 50 ml tubes and stored in the dark at 4°C until they were processed. The pH, electric conductivity (C.E.) and humidity of soil were measured as previously reported [11]. The extraction of bioavailable Ca, Mg, Na, K, Fe, Cu, Mn and Zn was performed as previously reported [12,13]. Determination of metals was performed using atomic absorption spectrometry in a spectrophotometer UNICAM (Thermo). Determination of total C, N and S was performed in a LECO's CNS-2000 elemental autoanalyzer. Extraction and determination of bioavailable P was performed as previously reported [14,15]. Determination of organic matter was done by calcination (UNE-EN 13039 standard) and the determination of total carbonates following the calcimetry protocol of Bernard [11].

Isolation of microorganisms showing lipolytic activity (hydrolysis)

One gr of soil sample was resuspended in 3 ml of sterile saline solution (ClNa 0.85% w/v). This suspension was diluted with saline solution (ClNa 0.85% w/v) to obtain single colonies and plated on a battery of solid media containing 0.5% tributyrine. The media used were LB (Luria Bertani medium), Potato Dextrose Broth (PDB) (Difco), 2% Agar (w/v) in distilled water, 9K A (1 L 9K solution, 1% (w/v) glucose and 1% (w/v) yeast extract), 9K G (1 L 9K solution, 0.5% (w/v) glucose, 0.5% (w/v) yeast extract, and 1% (w/v) malt extract) or 9K Gamp (1 L 9K solution, 1% (w/v) glucose, 0.5% (w/v) yeast extract, 1% (w/v) malt extract and 20 μg/ml ampicillin). The 9K solution was prepared as described in Silverman [16] without $FeSO_4 \cdot 7H_2O$. To obtain solid media 2% agar was added to the media. The plates were incubated at 30°C for 4–7 days. Those microorganisms displaying a clearing zone around the colony were grown in plates containing non-selecting LB (bacteria), PDB (fungi) or YPD (yeast) media, and then the lipolytic activity was confirmed by growing the isolates again on the corresponding media containing 0.5% tributyrine.

Screening for lipolytic microorganisms showing transesterification activity

For screening microorganisms displaying transesterification activity we developed a simple colorimetric method based on a previously reported one [17] with some modifications. The method consists on the transesterification of *para*-nitrophenyl palmitate (*p*-NPP) with ethanol in the absence of water to release the yellow colored compound *para*-nitrophenol (*p*-NP), which can be subsequently detected by using a spectrophotometer. A water-free environment promotes the transesterification reaction over the water-dependent hidrolisis reaction of *p*-NPP. A 5 ml overnight culture of each positive lipolytic bacterial isolate was employed to inoculate flasks containing 25 ml of LB media plus 2% tributyrine at a final OD_{600} of 0.1. These cultures were grown for 3 days at 30°C and 200 rpm. After this time cells were pelleted by centrifugation and 1.8 ml of the supernatants were freeze-dried for 24 hours. The freeze-dried supernatant was mixed with 1 ml of 10 mM *p*-NPP (in n-hexane) and 60 μl of absolute ethanol. The mixture was made in 2 ml safe-lock eppendorf tubes and incubated with shaking at 200 rpm in a rotator shaker at 37°C for 16 hours. A negative control was prepared by using a mixture of absolute ethanol and p-NPP. As a control of the hydrolytic activity of lipases we used a mixture of p-NPP and freeze-dried supernatant (without ethanol). After 16 hours of reaction the lipase was allowed to decant at the bottom of the tube for 10 minutes, then 25 µl of supernatant was mixed with 1 ml of 0.05 M NaOH in a 1.5 ml eppendorf tube. The p-NP produced during the transesterification reaction was extracted by the aqueous alkaline phase, transferred to a 1 ml cuvette. and the absorbance was quantified at 410 nm using a Beckman DU 640 spectrophotometer.

Production of biodiesel at benchtop scale

The production of biofuel was analyzed by gas chromatography as previously reported [18]. A 5 ml overnight culture of each positive lipolytic bacterial isolate was employed to inoculate a flask containing 25 ml of LB media plus 2% tributyrine at a final OD_{600} of 0.1. Cultures were grown for 3 days at 30°C and 200 rpm. The whole volume of the supernatants was freeze-dried and all the extract was used directly in a transesterification reaction (nonprocessed supernatant). To eliminate salts and other components of the media and tributyrine from the supernatants they were concentrated in dialysis bags (12 KDa, Sigma) using polyethylene glycol (average Mw 8,000, Sigma) at 4°C overnight, then dialyzed in 0.05 M potassium phosphate buffer (pH 7.6) four times. The resulting concentrated and dialyzed supernatant was finally freeze-dried and used in a transesterification reaction (processed supernatant).

The transesterification reaction was performed with continous shaking at 37°C for 24 hours and contained 6 ml sunflower oil, 1.75 ml absolute ethanol, 0.05 ml NaOH 10N and the entire processed or non-processed supernatant coming from 25 ml of culture. A negative control was prepared by using a mixture of 6 ml sunflower oil, 1.75 ml absolute ethanol and 0.05 ml NaOH 10N. The ethyl esters and glycerides produced in the transesterification reaction were analyzed using gas chromatography (GC) as described below.

To study the kinetics of the reaction, samples were taken at the indicated time points, and analyzed by GC. To quantify the biofuel production after the re-utilization of the bacterial extracts, the reaction mix was centrifuged, the supernatant discarded and the pellet was employed to repeat the transesterification reaction and the GC analysis. Both analyses were performed using only processed supernatant.

Chemical analysis of the biofuel

The method used integrates two official methods for the detection of esters (UNE EN ISO 14103) and glycerides (UNE EN ISO 14105), using cetane (n-hexadecane) as an internal standard to quantify the contents of glycerol, ethyl esters and glycerides (mono-, di- and triglycerides) as previously described [5]. Briefly, a gas chromatograph Varian 430 GG fitted with a capillary column HT5, 0.1 µm (25 m×0.32 mm, SGE, Supelco) with a flame ionization detector (FID) and *splitless* injection was used. The transesterification reaction product (12.5 µl) was mixed with 4 ml of a 1:1 (v/v) ethanol/dichloromethane mixture that contained the internal standard (cetane), and 0.5 µl of the prepared sample was employed for the analysis.

The results were expressed as relative quantities of the corresponding fatty acid ethyl esters (FAEE) and some monoglycerides (MG) of lower retention time (RT<25 minutes) and the sum of the quantities of the other MG (RT>25 minutes) and the

diglycerides (DG). The yield refers to the relative amount (%) of FAEE + MG (with lower RT) produced. The conversion includes the total amount (%) of triglyceride transformed into FAEE, MG and DG. Blank reactions containing only the mixture of oil, ethanol and NaOH were performed. Conversion of the starting oil material was below 15% under these experimental conditions. Data shown are the average of at least two independent experiments.

Determination of the viscosity

The viscosity was determined in a capillary viscometer Oswald Proton Cannon-Fenske Routine Viscometer 33200, size 2150. The method used was based on determining the time needed for a given volume of fluid to pass between two points marked in the instrument. It correlates with the speed reduction suffered by the liquid flow as a result of internal friction of its molecules, depending on their viscosity. From the flow time, t, in seconds, the kinematic viscosity (v, centistokes, cSt) can be obtained from the equation: $v \times t = C$, where C is the constant calibration of the measuring system in cSt·s, which is given by the manufacturer (0.10698 mm^2 s^{-1}, at 40°C) and t the flow time in seconds. The kinematic viscosity also represents the ratio between the dynamic viscosity and the density (ρ, $v = \eta/\rho$).

Analysis of hydrolytic activities (amylase, protease, DNAse, pullulanase and xylanase)

Amylase activity was screened on LB solid medium supplemented with 0.2% (w/v) soluble starch, 0.5% (w/v) peptone and 0.3% (w/v) meat extract. After 7 days of incubation at 30°C the plates were flooded with 0.3% (w/v) I_2–0.6% (v/v) KI solution. Hydrolysis of starch results in a clearing zone around the colonies [19].

The presence of protease activity was determined in LB solid medium supplemented with 2% (w/v) skim milk. The appearance of zones of precipitation of paracasein around the colonies after 3 days of incubation at 30°C indicated the presence of proteolytic activity [19].

DNAse activity was analyzed by growing bacteria on DNAse test agar plates containing 4.2% (w/v) Agar DNA. The plates were incubated for 7 days at 30°C and then flooded with 1N HCl solution. A clearing zone around the colonies indicated DNase activity [20].

The pullulanase and xylanase activities were detected by screening for zones of blue halos produced due to the hydrolysis and solubilization of the AZCL-pullulano and AZCL-xylano, respectively [21].

Isolation of DNA and 16S rRNA gene sequence analysis

The bacterial DNAs were isolated following standard protocols [22]. The total DNA isolated was used as the template for the amplification of the 16S rRNA by PCR using the universal primers designed for Bacteria 16F27 (5′-AGAGTTT-GATCMTGGCTCAG-3′) and 16R1488 (5′-CGGTTACCTTGTTAGGACTTCACC-3′) [23]. The program used for the amplification was: one cycle of 95°C for 5 minutes; 25 cycles of 94°C for 1 minute, 50°C for 1 minute and 72°C for 2 minutes; and a final extension cycle for 10 minutes at 72°C. Partial 16S rRNA gene sequences (c. 650 bp corresponding to positions 39 to 689 of the 16S rRNA gene from *Escherichia coli*) were obtained and aligned to the most similar 16S rRNA gene sequence in the GenBank database using the BLASTn algorithm. Using the most similar sequences found in the GenBank database a multiple sequence alignment of the DNA sequences was

Table 1. Chemical and physico-chemical analysis of the oil mill samples.

Parameter	Samples[a]			
	AE1B	AE2B	AEA	AEDH
C total (%)	16.24	22.490	66.410	22.970
N total (%)	0.894	0.616	1.169	0.819
S Total (%)	0.05594	N.D	0.100	0.052
K (%)	0.144	0.110	0.456	0.269
Na (mg/Kg)	61.234	130.597	168.019	235.668
Mg (%)	0.018	0.030	0.022	0.059
Ca (%)	1.602	1.500	0.074	1.861
P (mg/Kg)	36.755	43.541	280.390	168.848
Mn (mg/Kg)	63.528	95.033	12.015	125.528
Fe (mg/Kg)	599.663	651.608	139.812	151.260
Cu (mg/Kg)	73.277	49.632	27.195	140.325
Zn (mg/Kg)	19.163	14.253	29.750	11.955
pH (1/5)	6.837	6.55	4.68	6.27
E.C. 1/5 (mS/cm)[b]	1.68	2.62	3.89	2.14
Humidity (%)	5.460	4.690	5.819	5.179
O.M. (%)[c]	30.709	40.186	95.801	40.782
CaCO$_3$ total (%)	9.499	8.808	1.262	6.314

[a]All the results are referred to the dried mass of the samples.
[b]E.C.: Electric Conductivity.
[c]O.M.: Organic Matter.

constructed using the ClustalW software [24]. The phylogenetic trees were obtained using the MEGA4 software [25] with the neighbor-joining method [26]. The data set was bootstrapped 500 times to ensure reliability of each branch. The evolutionary distances were computed using the maximum composite likelihood method [27] and are shown in number of base substitutions per site.

Nucleotide sequence accession numbers

The nucleotide sequences reported in this work have been deposited in the GenBank database under accession numbers KC880159 to KC880188.

Results and Discussion

Chemical composition of the collected samples

One of the aims of this work was to find microorganisms capable of producing biofuel from vegetable oil. Therefore, locations rich in vegetable oil were selected to screen for microorganisms harboring lipase activity against vegetable oil. In particular, these samples were collected from an olive oil mill located in Ecija, Southern Spain.

The chemical and physico-chemical characteristics of the collected samples are shown in Table 1. The analysis of the samples revealed that the most abundant metal element in all samples was Fe, although in the sample AEDH there were only minor differences between the Fe, Mn, Cu and Zn content. The rest of elements analyzed were C>Ca>N>K>S>Mg>Na>P (from most to less abundant). The exceptions were sample AE2B where S was no detected, sample AEA where P was more abundant than Na, and AEDH where Mg was more abundant than S. The pH of all samples was around 6.5, except in AEA (*alpechín*), which showed pH 4.7. In most of the samples the organic material accounted for 30–40% except for the AEA, which was the richest one with a 95.8%.

Table 2. Distribution of types of lipolytic microorganisms isolated from the oil mill samples.

Sample	Bacteria	Yeast	Fungi	Total
AE1B	61	21	174	256
AE2B	166	208	10	392
AEDH	57	190	1	248
AEA	7	70	51	128
Total	291	489	236	**1016**

Table 3. Classification and selection of bacteria showing transesterification activity according to the *para*-nitrophenylpalmitate test.

Absorbance 410 nm	Number of strains	Positive/Negative
>1	30	Positive
1–0.8	28	Positive
<0.8	233	Negative

First step of the screening: selection of microorganisms showing lipolytic activity (hydrolysis)

As a first step of the screening process, we assayed for the hydrolytic activity of the microorganisms. This is an easy, quick and cheap test to perform on solid media based on the visual inspection of plates containing the lipid substrate tributyrine for microorganisms showing a clearing zone around the colony edges. To obtain single colonies and select lipolytic microorganisms, samples were diluted in sterile saline solution and plated on LB, PDB, water-agar, K9 A, K9 G or K9 Gamp solid media supplemented with 0.5% tributyrine. During the first round of selection, 1,016 colonies (fungi, yeasts and bacteria) were identified to produce a clearing zone of hydrolysis (Table 2). Most microorganisms were obtained from media K9 A, K9 G and K9 Gamp (68% fungi, 65% bacteria, 65% yeasts). We decided to study only the lipolytic bacteria and thus, we selected 291 bacteria for further studies. These strains were first inoculated on non-selective media and re-checked for hydrolytic activity on LB supplemented with 0.5% tributyrine. Sample AEA yielded the poorest number of lipolytic bacteria. The AEA sample was obtained from *alpechín*, which is the toxic wastewater resulting after pressing the olives during the production of oil. The analysis of this sample showed that it was acid and rich in organic matter. This result is in agreement with previous reports describing that *alpechín* is slightly acid and rich in soluble organic compounds

[28,29]. Although AEA was the sample containing the highest organic material content, usually lignin accounts for ca. 50% of the total organic material in this type of samples. Lignin is a recalcitrant polymer that may also affect microbial activity and survival negatively. In fact, the phytotoxic and antimicrobial effects of phenols, organic acids and fatty acids that are usually present in this wastewater *alpechín* have been previously reported [30]. This may explain the low number of lipolytic microorganisms found in this sample.

Second step of the screening: Analysis of the transesterification activity of lipolytic microorganisms

We have employed a method based on the transesterification of *p*-nitrophenolpalmitate (*p*-NPP) with ethanol (or other alcohol) to release the yellow-colored *p*-nitrophenol (*p*-NP) as described in material and methods. In order to select only extracellular enzymes, this second step of the screening was performed using freeze-dried supernatant of the bacteria selected in the first step of the screening.

During the screening procedure, a diverse set of control assays was carried out in the absence of ethanol or lipolytic bacteria. The *p*-NP released under these control conditions was considered to result from the hydrolysis or a non-enzymatic reaction of *p*-NPP and it was employed as the threshold to determine the existence of transesterification activity. The maximum absorbance value

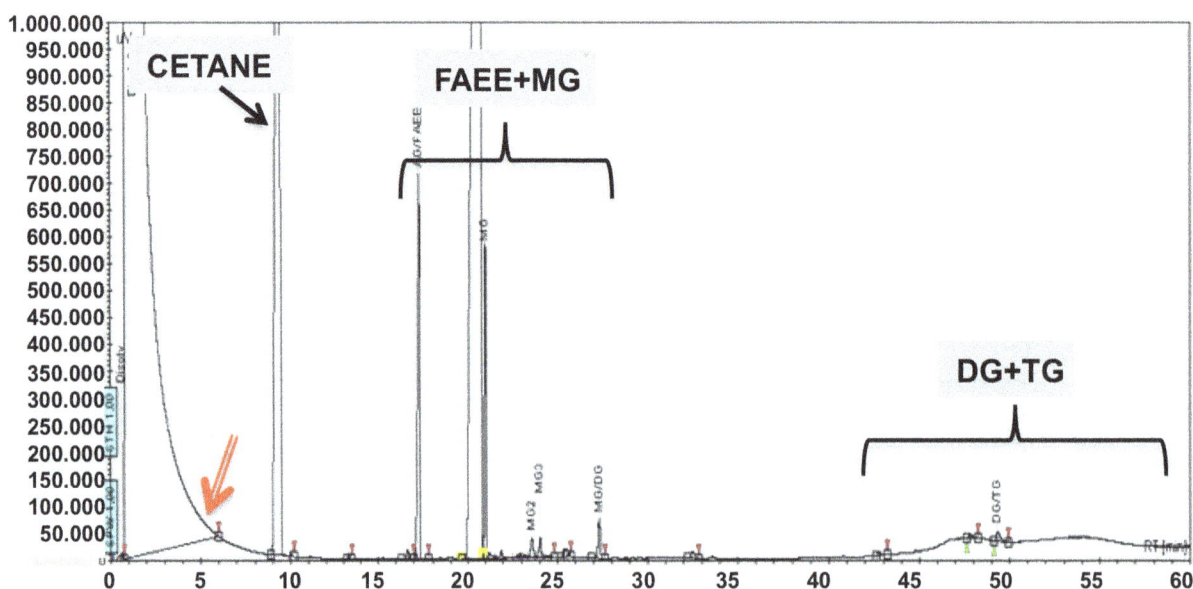

Figure 1. Chromatogram obtained by GC of the reaction mix after the transesterification of sunflower oil and ethanol performed by the processed supernatant of strain AE2B 122. Cetane was used as an internal standard. FAEE: Fatty Acid Ethyl Ester; MG: MonoGlyceride; DG: DiGlyceride; TG: TriGlyceride. The double-lined arrow indicates the retention time expected for glycerol as previously reported [5,6].

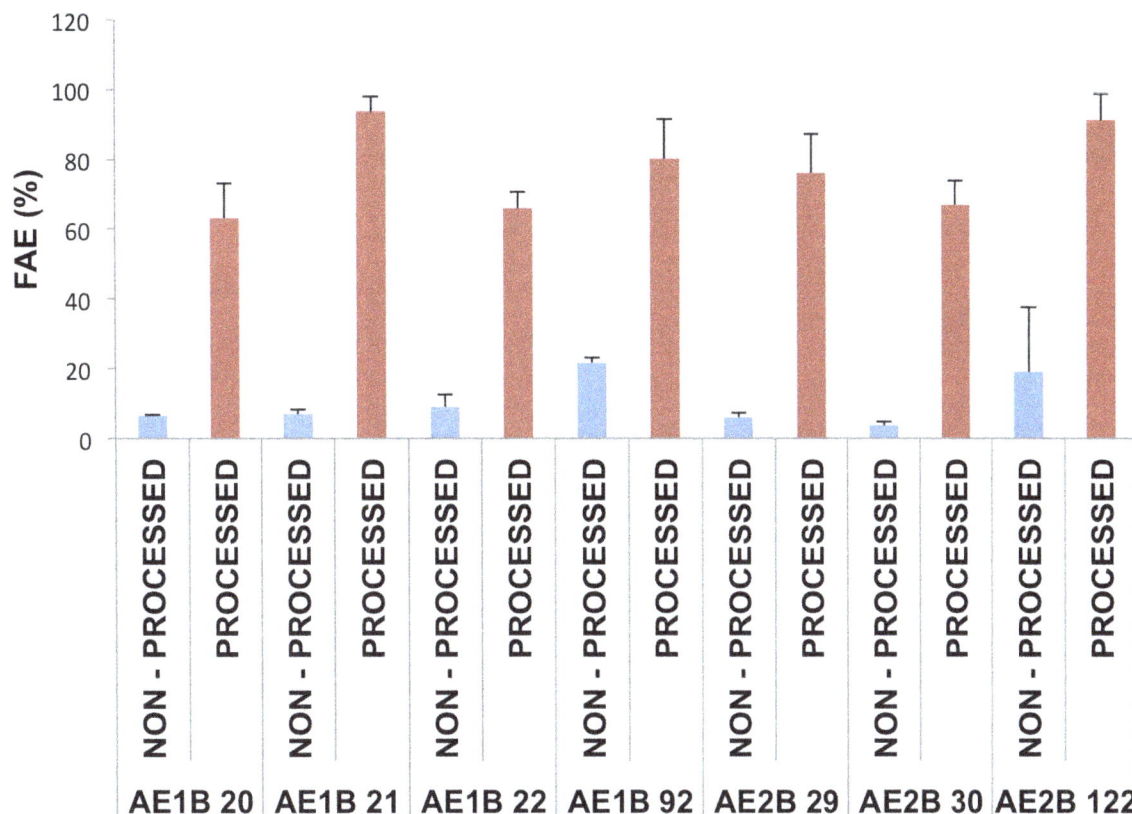

Figure 2. Comparison of biofuel production using processed or non-processed supernatants. The bacterial supernatants were processed as described in Materials and Methods or not processed. Both processed and non-processed supernatants were freeze-dried and employed for transesterification reaction with sunflower oil and ethanol. The data show the percentage of FAE (FAEE+MG) produced after 24 hours of reaction using 7 randomly selected bacterial isolates. Data are the average of at least 2 independent experiments.

obtained in the control reactions was 0.8. The transesterification reaction with all the strains selected during the first step of the screening was performed in parallel. All strains showing absorbance values over this threshold of 0.8 were considered positive and values below the threshold were considered background levels (Table 3). According to the above criterion, 58 strains were considered positive. Out of the 58 positive strains, the top 30 bacteria, i.e. those showing absorbance values over 1.0, were selected for further analysis and quantification of biofuel production.

During this step we selected only for bacteria showing extracellular transesterification activity because it is described that at industrial level the highest yields of biodiesel production are obtained with extracellular lipases rather than with whole cells [1,3]. In fact, whole cell transesterification procedures produce low substrate conversion yields due to the toxicity of the solvent to the host cells, and also due to the low mass transfer rate of high molecular weight substrates (oil) from the solvent phase to the whole cell biocatalyst [31]. Another noticeable difference when the transesterification reaction is performed with whole cells is the reaction time. Novozym 435 gave yields of 87% after 3.5 h of reaction in a fed-batch operation. However, to obtain yields of 80–90% of biodiesel using whole cells, it was required to perform the reaction process for 70 h [32].

Third step of the screening: Selection of the bacterial supernatants producing biofuel at benchtop scale

The aim of the third step of the screening was to identify the best performing bacteria in biofuel production at benchtop scale. To optimize the screening protocol, we first grew the 30 selected bacterial strains as described in Materials and Methods. In a first attempt the supernatants (non-processed) were freeze-dried to obtain a fine dried powder that can be used directly for the production of biofuel. The transesterification reaction to produce biofuel was performed at basic pH with sunflower oil, ethanol and the powdered supernatant. After completion of the reaction, the biofuel/lipid/fatty acid mixture was analyzed by GC. The chromatograms show clear different retention times for the final (FAEE + MG), and the initial (TG) and intermediate (DG) products (Fig. 1). The yield obtained with the bacteria initially tested was lower than 22% in all cases (Fig. 2). In order to try to increase the yield, the supernatants of the cultures were processed as described in materials and methods, and the samples were then freeze-dried and assayed for biofuel production. We observed that processing of the bacterial supernatants drastically increased the performance up to 17.8 fold in comparison to the non-processed samples (Fig. 2). Therefore, we found out that an easy and cheap process of concentration and dialysis could increase the yield of biofuel production. Once the protocol was optimized, we proceeded to quantify biofuel production in the 30 selected bacterial strains. The complete results of the reactions using processed supernatants are shown in Table 4. Since the composition of biodiesel is a mix of FAEEs and the composition of biofuel

Table 4. Transesterification reaction of sunflower oil by the bacterial extracts determined by GC analysis.

Sample	FAE %[a]	DG %	TG %	Conversion %
Sunflower oil	3.4	29.9	66.7	33.3
Negative Control	13.8	41.9	44.3	55.7
AE1B 20	63.1	6.7	30.2	69.8
AE1B 21	93.7	6.3	0.0	100
AE1B 22	65.9	2.5	31.7	68.4
AE1B 26	80.1	12.9	7.1	92.9
AE1B 27	57.9	4.3	37.7	62.3
AE1B 28	58.7	3.1	38.2	61.8
AE1B 35	67.4	2.5	30.2	69.8
AE1B 89	72.2	11.6	16.2	83.8
AE1B 90	58.6	2.7	38.7	61.3
AE1B 92	80.1	7.5	12.4	87.6
AE2B 29	76.0	7.8	16.2	83.8
AE2B 30	66.8	5.1	28.1	71.9
AE2B 85	41.2	13.8	45.1	54.9
AE2B 120	35.1	15.1	50.1	50.2
AE2B 122	91.16	1.75	6.8	93.2
AE2B 130	71.8	8.1	20.1	79.9
AE2B 131	61,27	9,55	14,59	85,41
AE2B 133	34.8	19.5	45.7	54.3
AE2B 134	31.7	2.1	66.3	33.7
AE2B 199	25.0	3.3	71.7	28.3
AE2B 222	77.5	3.2	12.9	87.1
AE2B 232	48.1	0.7	51.2	48.8
AE2B 250	46.4	10.7	42.9	57.1
AE2B 259	30.7	8.3	61.0	39.0
AE2B 261	82.6	7.8	9.6	90.4
AE2B 263	49.3	10.5	40.2	59.8
AE2B 264	24.1	2.6	73.3	26.7
AE2B 332	54.0	0.7	45.4	54.7
AE2B 340	51.4	1.0	47.6	52.4
AEDH 145	61.0	4.0	35.0	65.0

The percentage of FAE shows the yield of biofuel production in the reaction mix. Conversion shows the percentage of TG metabolized during the reaction.
[a]FAE, Fatty Acid Esters (fatty acid ethyl esters + monoglycerides) (Biofuel). DG, Diglycerides. TG, Triglycerides. Conversion, percentage of TG converted into FAE+DG.

is a mix of FAEs, the most important value to be considered is the percentage of FAE, which are a mix of two moles of FAEE and one mol of MG. The analysis of the data revealed that the supernatants of 5 of the bacterial strains were capable of producing over 80% FAE, 9 bacterial supernatants produced between 60–80% FAE, and the production of the other 16 isolates was below 60% FAE. Eight bacterial strains were capable of processing over 80% of the TG in the reaction mixture (conversion of TG into DG plus FAE).

During the screening for the best biofuel-producing strains, we noticed that glycerol did not appear at the expected retention time (5 min). This suggests that TGs were converted into ethyl esters (FAEEs) and MG, which co-elute under our experimental conditions (Fig. 1). This was previously observed in transesterification reactions performed using porcine pancreatic lipase [6]. From an industrial perspective, the biofuel produced by these bacteria would not require downstream processing to separate the glycerol contaminant because glycerol was integrated in the final mix as MG. This could be due to a 1,3 selective ethanolysis of the sunflower oil by the enzyme, that means a partial ethanolysis of the oil that produce 1 mol of MG for each 2 moles of FAEE [6]. The chemical analysis also suggests that this methodology is not suitable to produce conventional biodiesel according to EN 14214, but rather the new type of biofuel, which incorporates glycerol as previously described [5,6].

Phylogenetic analyses of the selected biofuel producing bacteria

In order to assign the bacterial biofuel producers to specific phylogenetic groups, the sequence of the 16S rRNA gene was determined. Phylogenetic reconstruction performed with different methods was consistent, and consequently only the tree obtained with Neighbor-Joining for the evolutionary history and Maximum Composite Likelihood for evolutionary distances is shown (Fig. 3).

A

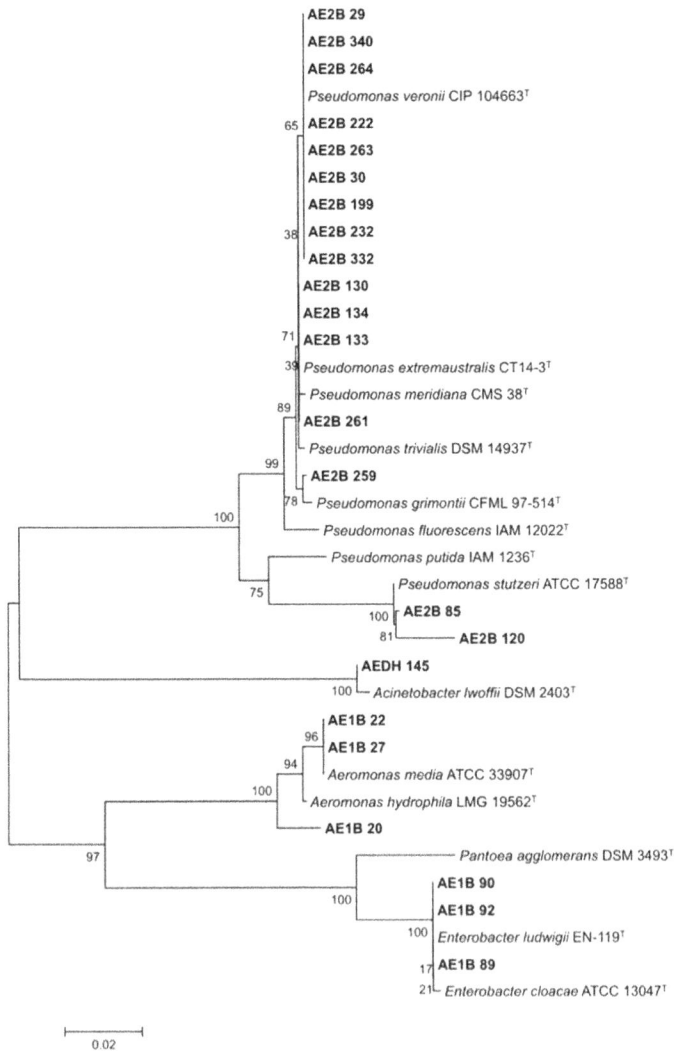

B

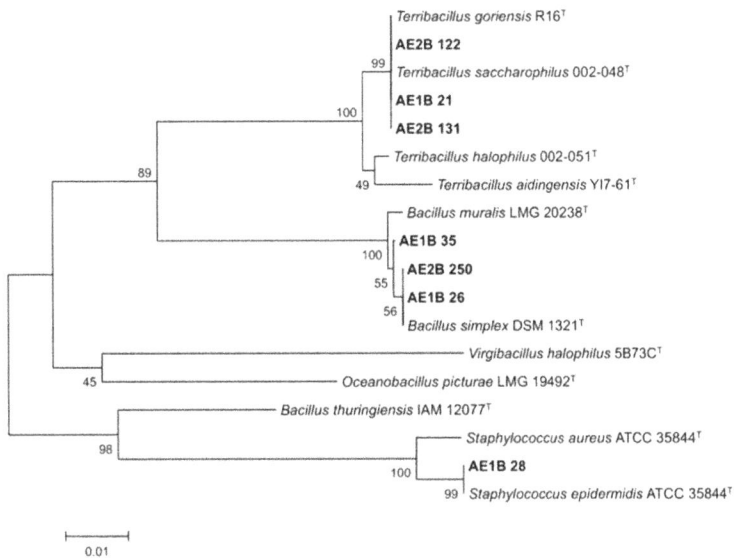

Figure 3. Evolutionary relationships of the selected strains. Phylogenetic trees were inferred from the 16S rRNA sequences of the 23 Gram negative (A) and the 7 Gram positive (B) bacteria with Neighbor-Joining clustering. The distances were calculated using Maximum Composite Likelihood. The bacterial strains isolated in this work are indicated in bold. 16S rRNA gene sequences from the isolates correspond to 650 bp. Bar represents a 2% (A) or 1% (B) of sequence difference.

The strains isolated were very diverse with representatives of both Gram positive and Gram negative bacteria. This analysis together with the high number of eukaryotic microorganisms (yeasts and fungi) also isolated during this work suggests that there is a high diversity of living organisms in the samples of the olive-mill wastes. Table 5 shows the closest relative to each of the isolates according to the results obtained after the phylogenetic analysis.

Most of the lipolytic isolates were obtained from the sample AE2B (19 isolates): 16 of them were related to the genus *Pseudomonas*, 2 of them related to the genus *Terribacillus*, and one isolate related to the genus *Bacillus*. Eleven lipolytic isolates were obtained from the sample AE1B, which were related to the genera *Enterobacter* (3 isolates), *Aeromonas* (3 isolates), *Bacillus* (2 isolates) and *Terribacillus* (one isolate). Only one isolate was obtained from the sample AEDH. It was related to the genus *Acinetobacter*. Interestingly, no lipolytic isolates were selected from the sample AEA (Table 5).

The isolates that showed the highest yield in the production of biofuel (selectivity over 60%) were 14 strains belonging to the genera *Pseudomonas*, *Acinetobacter*, *Enterobacter*, *Bacillus*, and *Terribacillus* (Table 4 and 5). Their closest relatives were *Pseudomonas veronii* [33] (AE2B 30, AE2B 29, AE2B 222), *Pseudomonas extremaustralis* [34] (AE2B 130, AE2B 261), *Acinetobacter lwoffii* [35] (AEDH 145), *Enterobacter ludwigii* [36] (AE1B 89 and AE1B 92), *Bacillus simplex* [37] and *B. muralis* [38] (AE1B 26 and AE1B 35), *Aeromonas hydrophila* and *A. media* (AE1B 22 and AE1B 20), and *Terribacillus goriensis* [39] and *Terribacillus saccharophilus* [40] (AE1B 21 and AE2B 122). Several of the selected isolates belong to genera that have been already reported to express transesterification activity, e.g. *Pseudomonas*, *Enterobacter* and *Bacillus*. Other microorganisms that have been reported to have enzymes with transesterification activity are *P. fluorescens*, *P. cepacia*, *Rhizomucor miehei*,

Rhizopus oryzae, *Candida rugosa*, *Thermomyces lanuginosus*, *Candida antarctica*, *Chromobacterium viscosum*, *Burkholderia cepacia*, *E. aerogenes*, *Mucor miehei*, *Penicillium expansum* and *B. subtilis* [31]. It is interesting to note that the strains AE2B 122 and AE1B 21 showing the best performance in the production of biofuel under our experimental conditions belong to the genus *Terribacillus*. To the best of our knowledge there are no reports on the employment of members of the genus *Terribacillus* for transesterification reactions, and therefore, they constitute novel producers of biofuel synthesizing enzymes.

Viscosity measurements

The high similarity of the chromatography retention values obtained between different MG and FAEE compounds suggests that the rheological properties of these compound mixes were similar. This is an important factor for future applications as biofuel in diesel engines. Therefore, we measured the viscosity of the bacteria-derived biofuel, as a critical parameter for employing the biofuel. Figure 4 shows the viscosity of the supernatants of the top 5 biofuel-producing bacteria. All bacterial supernatants selected for biofuel production reduced the viscosity of the reaction product at least three times compared to the viscosity of the sunflower oil. Processing the bacterial supernatants as described in materials and methods further reduced the viscosity of the reaction product (Fig. 4), providing another evidence of the advantage of employing processed supernatants over non-processed supernatants.

Kinetics of biodiesel production and re-utilization of bacterial supernatants

All previous experiments of biofuel production were performed for 24 h. Three strains were randomly selected to test whether the reaction times could be shorten under our experimental condi-

Table 5. The closest relative of each isolate is indicated based on the phylogenetic reconstruction shown in Figure 3.

Sample	Closest relative	Accession number	% Similarity
AE2B 29 AE2B 30 AE2B 199 AE2B 222 AE2B 232 AE2B 263 AE2B 264 AE2B 332 AE2B 340	*Pseudomonas veronii*	AF064460	99–100
AE2B 130 AE2B 133 AE2B 134 AE2B 261	*Pseudomonas extremaustralis*	AJ583501	99–100
AE2B 259	*Pseudomonas grimontii* CFML 97-514T; AF268029	AF268029	99
AE2B 85 AE2B 120	*Pseudomonas stutzeri*	AF094748	98–100
AEDH 145	*Acinetobacter lwoffii*	X81665	99
AE1B 89 AE1B 90 AE1B 92	*Enterobacter ludwigii*	AJ853891	99–100
AE1B 20	*Aeromonas hydrophila* strain: LMG 19562; AJ508765	AJ508765	99
AE1B 22 AE1B 27	*Aeromonas media*	X60410	99–100
AE1B 26 AE1B 250	*Bacillus simplex*	AJ439078	99–100
AE1B 35	*Bacillus muralis*	AJ316309	99
AE1B 28	*Staphylococcus epidermidis* ATCC 14990; D83363	D83363	99
AE1B 21 AE2B 122 AE2B 131	*Terribacillus goriensis*[a] *T. saccharophilus*[a]	DQ519571 AB243845	99–100

Samples AE1B and AE2B were obtained from ponds containing liquid wastes from the oil mill. Sample AEDH was obtained from the wastes of the olive harvests containing mainly olive tree leftovers (leaves and branches).
[a]Strains AE1B21, AE2B122 and AE2B131 showed the same % of similarity to both *T. goriensis* and *T. saccharophilus*.

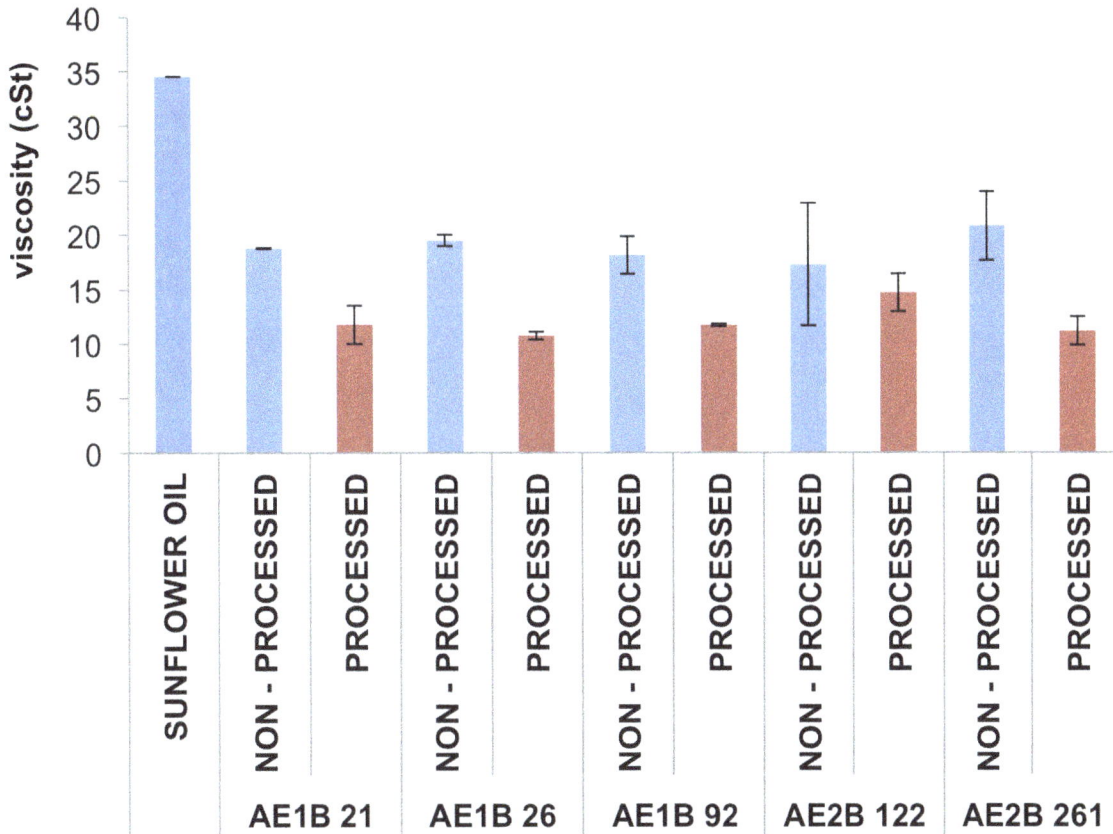

Figure 4. Viscosity of the sunflower oil, and the transesterification product using processed and non-processed supernatants of selected bacteria. Biofuel (FAEE+MG) was obtained by transesterification reaction of sunflower oil and ethanol with processed or non-processed supernatants of the best 5 biofuel-producing bacterial strains. Data are the average of at least two independent experiments.

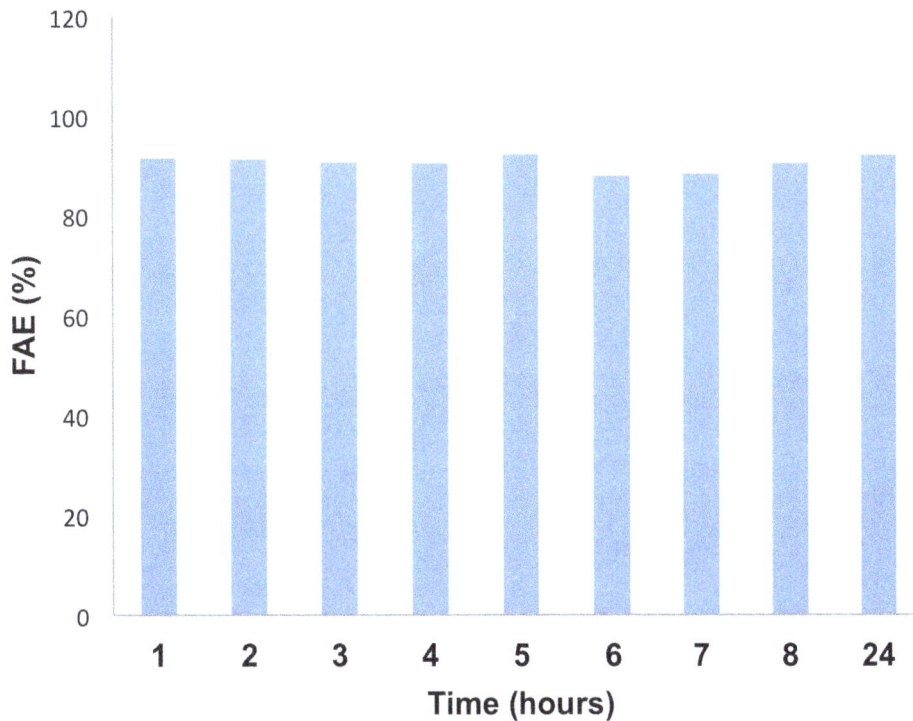

Figure 5. Kinetic analysis of the production of biofuel. The transesterification reaction of sunflower oil and ethanol by processed supernatants was set up, and samples were withdrawn and analyzed by GC at the indicated time points. The vexperiment was performed with 3 different bacterial strains. The data show one representative experiment of the kinetic analysis performed with *Enterobacter* sp. AE1B 92 supernatants.

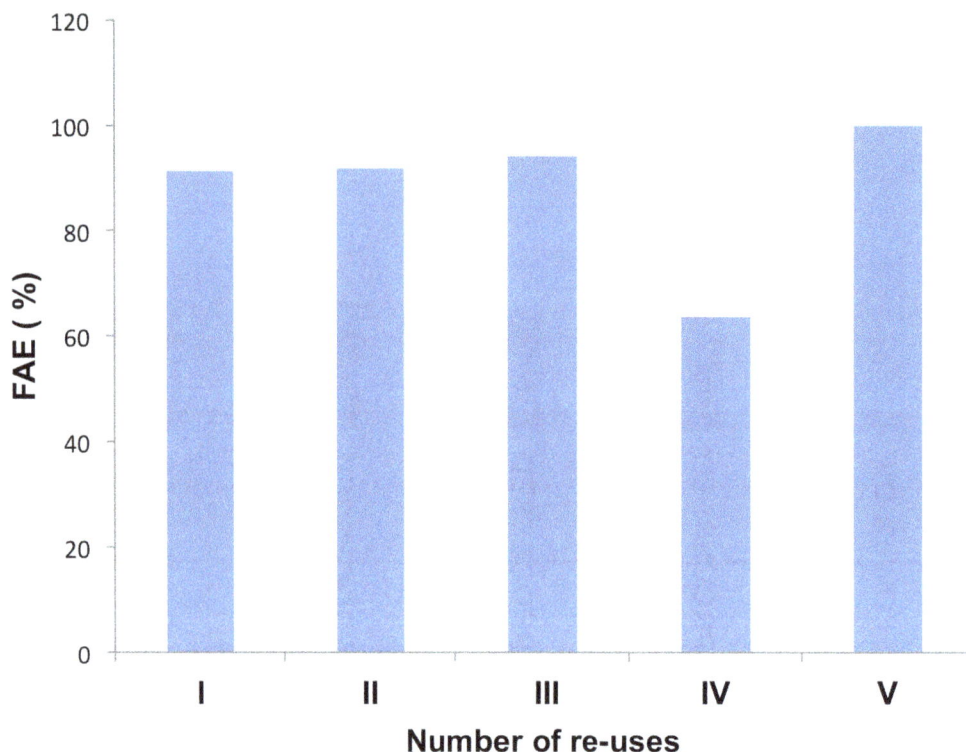

Figure 6. Re-utilization of bacterial supernatants for the production of biofuel. The transesterification reaction of sunflower oil and ethanol was performed with processed supernatants of the bacterial strains. After completion of the reaction, the mix was centrifuged, and the biofuel was analyzed by GC. The pellet was employed for another transesterificaction reaction. The reaction was repeated 5 times with the same extracts. The experiment was performed with 5 different bacterial strains. The data show one representative experiment performed with *Enterobacter* sp. AE1B 92 supernatant.

tions. The reaction was performed with processed supernatants as described above for 24 h and samples were withdrawn at the indicated time points to analyze the chemical composition of the reaction mixture. Figure 5 shows a representative example of a kinetic experiment performed with strain *Enterobacter* sp. AE1B 92. After one hour of reaction, the bacterial supernatant was capable of producing over 90% of FAE. There was slight variation in the reaction composition over the 24 h period of incubation. Similar results were found with two other bacterial strains (data not shown).

An economical factor to be considered during the industrial process affects the number of times that a supernatant can be employed during the biofuel production. Therefore, we randomly selected five bacterial strains to study the number of times that they can be employed. In all cases, the supernatants could be employed at least five times without losing their activity significantly (Fig. 6 shows a representative experiment of the 5 strains).

Characterization of the hydrolytic activities of the selected biofuel-producing bacteria

To further characterize the biofuel-producing bacteria, we analyzed the hydrolytic activities of the 30 strains. All the bacterial strains were tested in LB plates containing the corresponding substrates. The amylase activity was the most abundant hydrolytic activity found in these isolates, as 11 isolates showed amylase activity. Ten bacteria showed DNAse activity, 9 showed protease activity and 5 showed pullulanase activity. However, none of them showed xylanase activity (Table S1 in File S1). Three of the isolates (*Aeromonas* sp. AE1B 27, *Aeromonas* sp. AE1B 22 and *Aeromonas*

sp. AE1B 20) showed all the hydrolytic activities (except xylanase). There were 12 bacteria that did not show any hydrolytic activity under our assay conditions, which accounts for 40% of the isolates.

Conclusion

In this work we describe the selection and isolation of microorganisms able to perform a transesterification reaction. We have employed a three-step screening to isolate lipolytic microorganisms able to produce a biofuel, consisting of a mix of FAEE and MG with lower viscosity than sunflower oil. Five of the bacterial strains isolated were capable of producing biofuel with yields higher than 80%. The best two performing strains belong to the genus *Terribacillus*, which has not been reported as biofuel producer so far. These two strains can be of great interest at industrial scale. In addition, here we report an easy and cheap method for processing the bacterial supernatant, which is capable of producing biofuel efficiently without further downstream processing or the need for expensive protein purification.

Supporting Information

File S1   Characterization of the hydrolytic activities of the selected biofuel-producing bacteria and Table S1.

Acknowledgments

We are grateful to Gabriel Gutiérrez Pozo for help with the phylogenetic analysis and to the oil mill "Coperativa Agropecuaria Industrial" (CAPI) located in Écija (Sevilla) for allowing us to sample at its facilities.

Author Contributions

Conceived and designed the experiments: AEN DL ATM DC EM. Performed the experiments: AEN CL DL. Analyzed the data: AEN CL DL DC EM. Contributed reagents/materials/analysis tools: DL DC EM. Contributed to the writing of the manuscript: AEN DL DC EM.

References

1. Robles-Medina A, Gonzalez-Moreno PA, Esteban-Cerdan L, Molina-Grima E (2009) Biocatalysis: towards ever greener biodiesel production. Biotechnol Adv 27: 398–408.
2. Meher LC, Vidya Sagar D, Naik SN (2006) Technical aspects of biodiesel production by transesterification. Energy Rev 10: 248–268.
3. Ranganathan SV, Narasimhan SL, Muthukumar K (2008) An overview of enzymatic production of biodiesel. Bioresour Technol 99: 3975–3981.
4. Zhang Y, Dube MA, McLean DD, Kates M (2003) Biodiesel production from waste cooking oil: 1. Process design and technological assessment. Bioresour Technol 89: 1–16.
5. Verdugo C, Luque R, Luna D, Hidalgo JM, Posadillo A, et al. (2010) A comprehensive study of reaction parameters in the enzymatic production of novel biofuels integrating glycerol into their composition. Bioresour Technol 101: 6657–6662.
6. Luna C, Sancho E, Luna D, Caballero v, Calero J, et al. (2013) Biofuel that keeps glycerol as monoglyceride by 1,3-Selective ethanolysis with Pig Pancreatic Lipase covalently Immobilized on AlPO4 support. Energies: 3879–3900.
7. Ilham Z, Saka S (2010) Two-step supercritical dimethyl carbonate method for biodiesel production from Jatropha curcas oil. Bioresour Technol 101: 2735–2740.
8. Ognjanovic N, Bezbradica D, Knezevic-Jugovic Z (2009) Enzymatic conversion of sunflower oil to biodiesel in a solvent-free system: process optimization and the immobilized system stability. Bioresour Technol 100: 5146–5154.
9. Modi MK, Reddy JR, Rao BV, Prasad RB (2007) Lipase-mediated conversion of vegetable oils into biodiesel using ethyl acetate as acyl acceptor. Bioresour Technol 98: 1260–1264.
10. Ribeiro BD, de Castro AM, Coelho MA, Freire DM (2011) Production and use of lipases in bioenergy: a review from the feedstocks to biodiesel production. Enzyme Res 2011: 615803.
11. Porta Casanellas J, López Acevedo M, Rodríguez Ochoa R (1986) Técnicas y Experimentos en Edafología: Col.legi Oficial d'Enginyer Agrònoms. 282 p.
12. Sumner ME, Miller WP (1996) Chemical methods. In: Sparks DL, editor. Methods of soil Analysis Part 3 1ed. Madison, USA: Soil Science Society of America Inc. pp.1201–1229.
13. Lindsay WL, Norwell WA (1978) Development of DTPA soil for zinc, iron, manganese and copper. Soil Sci Soc Am J: 421–462.
14. Olsen SR, Cole CV, Watanabe FS, Dean LA (1954) Estimation of available phosphorous in soil by extraction with sodium bicarbonate. Circular (United States. Dept. of Agriculture): U.S. Dept. of Agriculture. 19 p.
15. Murphy J, Riley JP (1962) A modified single solution method for the determination of phosphate in natural waters. Anal Chim 7: 31–36.
16. Silverman MP, Lundgren DG (1959) Studies on the chemoautotrophic iron bacterium Ferrobacillus ferrooxidans. I. An improved medium and a harvesting procedure for securing high cell yields. J Bacteriol 77: 642–647.
17. Teng Y, Xu Y (2007) A modified para-nitrophenyl palmitate assay for lipase synthetic activity determination in organic solvent. Anal Biochem 363: 297–299.
18. Verdugo C, Luque R, Luna D, Hidalgo JM, Posadillo A, et al. (2010) A comprehensive study of reaction parameters in the enzymatic production of novel biofuels integrating glycerol into their composition. Bioresour Technol: 6657–6662.
19. Cowan ST, Steel KJ (2004) Manual for the Identification of Medical Bacteria. Cambridge: Cambridge University Press. 352 p.
20. Jeffries CD, Holtman DF, Guse DG (1957) Rapid method for determining the activity of microorganisms on nucleic acids. J Bacteriol 73: 590–591.
21. Morgan FJ, Adams KR, Priest FG (1979) A culture method for the detection of pullulan-degrading enzymes in bacteria and its application to the genus Bacillus. J Appl Bacteriol 46: 291–294.
22. Green MR, Sambrook J (2012) Molecular Cloning, A Laboratory manual; fourth, editor. Cold Spring Harbor, New York: Cold Spring Harbor Laboratory Press. 2028 p.
23. Moreno ML, Garcia MT, Ventosa A, Mellado E (2009) Characterization of Salicola sp. IC10, a lipase- and protease-producing extreme halophile. FEMS Microbiol Ecol 68: 59–71.
24. Thompson JD, Gibson TJ, Plewniak F, Jeanmougin F, Higgins DG (1997) The CLUSTAL_X windows interface: flexible strategies for multiple sequence alignment aided by quality analysis tools. Nucleic Acids Res 25: 4876–4882.
25. Tamura K, Dudley J, Nei M, Kumar S (2007) MEGA4: Molecular Evolutionary Genetics Analysis (MEGA) software version 4.0. Mol Biol Evol 24: 1596–1599.
26. Saitou N, Nei M (1987) The neighbor-joining method: a new method for reconstructing phylogenetic trees. Mol Biol Evol 4: 406–425.
27. Tamura K, Nei M, Kumar S (2004) Prospects for inferring very large phylogenies by using the neighbor-joining method. Proc Natl Acad Sci U S A 101: 11030–11035.
28. de la Fuente C, Clemente R, Bernal MP (2008) Changes in metal speciation and pH in olive processing waste and sulphur-treated contaminated soil. Ecotoxicol Environ Saf 70: 207–215.
29. Alburquerque JA, Gonzalvez J, Garcia D, Cegarra J (2006) Measuring detoxification and maturity in compost made from "alperujo", the solid by-product of extracting olive oil by the two-phase centrifugation system. Chemosphere 64: 470–477.
30. Alburquerque JA, Gonzalvez J, Garcia D, Cegarra J (2004) Agrochemical characterisation of "alperujo", a solid by-product of the two-phase centrifugation method for olive oil extraction. Bioresour Technol 91: 195–200.
31. Bajaj A, Lohan P, Prabhat NJ, Mehrotra R (2010) Biodiesel production through lipase catalyzed transesterification: An overview. J Mol Catal B: Enzym 62: 9–13.
32. Ribeiro BD, de Castro AM, Coelho MA, Freire DM (2011) Production and use of lipases in bioenergy: a review from the feedstocks to biodiesel production. Enzyme Res 2011: 1–16.
33. Elomari M, Coroler L, Hoste B, Gillis M, Izard D, et al. (1996) DNA relatedness among Pseudomonas strains isolated from natural mineral waters and proposal of Pseudomonas veronii sp. nov. Int J Syst Bacteriol 46: 1138–1144.
34. Lopez NI, Pettinari MJ, Stackebrandt E, Tribelli PM, Potter M, et al. (2009) Pseudomonas extremaustralis sp. nov., a Poly(3-hydroxybutyrate) Producer Isolated from an Antarctic Environment. Curr Microbiol 59: 514–519.
35. Brisou J, Prevot AR (1954) Studies on bacterial taxonomy. X. The revision of species under Acromobacter group. Ann Inst Pasteur (Paris) 86: 722–728.
36. Hoffmann H, Stindl S, Stumpf A, Mehlen A, Monget D, et al. (2005) Description of Enterobacter ludwigii sp. nov., a novel Enterobacter species of clinical relevance. Syst Appl Microbiol 28: 206–212.
37. Meyer A, Gottheil O (1901) "Botanische beschreibung einiger bodenbakterien.". Zentralbl Bakteriol Parasitenkd Infektionskr Hyg Abt II 7: 680–691.
38. Heyrman J, Logan NA, Rodriguez-Diaz M, Scheldeman P, Lebbe L, et al. (2005) Study of mural painting isolates, leading to the transfer of 'Bacillus maroccanus' and 'Bacillus carotarum' to Bacillus simplex, emended description of Bacillus simplex, re-examination of the strains previously attributed to 'Bacillus macroides' and description of Bacillus muralis sp. nov. Int J Syst Evol Microbiol 55: 119–131.
39. Krishnamurthi S, Chakrabarti T (2008) Proposal for transfer of Pelagibacillus goriensis Kim, et al. 2007 to the genus Terribacillus as Terribacillus goriensis comb. nov. Int J Syst Evol Microbiol 58: 2287–2291.
40. An S-Y, Asahara M, Goto K, Kasai H, Yokota A (2007) Terribacillus saccharophilus gen. nov., sp. nov. and Terribacillus halophilus sp. nov., spore-forming bacteria isolated from field soil in Japan. Int J Syst Evol Microbiol 57: 51–55.

19

Lipid Profile Remodeling in Response to Nitrogen Deprivation in the Microalgae *Chlorella* sp. (Trebouxiophyceae) and *Nannochloropsis* sp. (Eustigmatophyceae)

Gregory J. O. Martin[1]⑨, David R. A. Hill[1]⑨, Ian L. D. Olmstead[1], Amanda Bergamin[1], Melanie J. Shears[2,3], Daniel A. Dias[2], Sandra E. Kentish[1], Peter J. Scales[1], Cyrille Y. Botté[3¶]*, Damien L. Callahan[2,4¶]

1 Department of Chemical and Biomolecular Engineering, The University of Melbourne, Parkville, Victoria, Australia, **2** Metabolomics Australia, The School of Botany, The University of Melbourne, Parkville, Victoria, Australia, **3** Apicolipid Group, Laboratoire Adaption et Pathogenie des Microorganismes UMR5163, CNRS, University of Grenoble I, La Tronche, France, **4** Centre for Chemistry and Biotechnology, School of Life and Environmental Sciences, Deakin University, Burwood, Victoria, Australia

Abstract

Many species of microalgae produce greatly enhanced amounts of triacylglycerides (TAGs), the key product for biodiesel production, in response to specific environmental stresses. Improvement of TAG production by microalgae through optimization of growth regimes is of great interest. This relies on understanding microalgal lipid metabolism in relation to stress response in particular the deprivation of nutrients that can induce enhanced TAG synthesis. In this study, a detailed investigation of changes in lipid composition in *Chlorella* sp. and *Nannochloropsis* sp. in response to nitrogen deprivation (N-deprivation) was performed to provide novel mechanistic insights into the lipidome during stress. As expected, an increase in TAGs and an overall decrease in polar lipids were observed. However, while most membrane lipid classes (phosphoglycerolipids and glycolipids) were found to decrease, the non-nitrogen containing phosphatidylglycerol levels increased considerably in both algae from initially low levels. Of particular significance, it was observed that the acyl composition of TAGs in *Nannochloropsis* sp. remain relatively constant, whereas *Chlorella* sp. showed greater variability following N-deprivation. In both algae the overall fatty acid profiles of the polar lipid classes were largely unaffected by N-deprivation, suggesting a specific FA profile for each compartment is maintained to enable continued function despite considerable reductions in the amount of these lipids. The changes observed in the overall fatty acid profile were due primarily to the decrease in proportion of polar lipids to TAGs. This study provides the most detailed lipidomic information on two different microalgae with utility in biodiesel production and nutraceutical industries and proposes the mechanisms for this rearrangement. This research also highlights the usefulness of the latest MS-based approaches for microalgae lipid research.

Editor: Howard Riezman, University of Geneva, Switzerland

Funding: The study was funded by The University of Melbourne: Interdisciplinary Seed Funding, Victorian Node of Metabolomics Australia (Bioplatforms Australia Ltd, Federal Government grants), Australian Commonwealth Government (Second Generation Biofuels Research and Development Grant Program), Bio Fuels Ltd (Victoria Australia), Agence Nationale pour la Recherche (ApicoLipid Project), and Atip-Avenir-Finovi (ApicoLipid project). The funders had no role in study design, data collection and analysis, decision to publish, or preparation of the manuscript.

Competing Interests: The funders, including Bioplatforms Australia Ltd and Bio Fuels Ltd had no role in study design, data collection and analysis, decision to publish, or preparation of the manuscript.

* Email: cyrille.botte@ujf-grenoble.fr

⑨ These authors contributed equally to this work.

¶ These authors are joint senior authors on this work.

Introduction

Increasing and unstable oil prices, diminishing fossil fuel reserves, and increasing atmospheric CO_2 levels have revitalized interest in microalgae as a renewable, lipid-rich feedstock for biofuel production [1–5]. Microalgae are attractive as feedstock for biofuel production primarily due to their ability to synthesize and accumulate high levels of TAGs in specialized lipid bodies located in the cytoplasm and to a lesser extent the chloroplast [6]. The fatty acids (FA) associated with the TAGs can be trans-esterified into fatty acid methyl esters (FAMEs) to produce biodiesel. The TAGs are the preferred source of FAs for biodiesel production as

they can be converted to biodiesel using conventional methods [7], however, these acyl chains (used here interchangeably with FA) are mainly present as building blocks for the algal membrane lipids. These include major plastid and photosynthesis-related glycolipids such as monogalactosyldiacylglycerol (MGDG), digalactosyldiacylglycerol (DGDG) and sulfoquinovosyldiacylglycerol (SQDG); phosphoglycerolipids such as phosphatidylglycerol (PG), phosphatidylcholine (PC), phosphatidylethanolamine (PE), phosphatidylserine (PS) and phosphatidylinositol (PI), together with sphingolipids and other neutral lipids such as diacylglycerol (DAG). As the physical properties and quality of the biodiesel are directly dependent on the nature and chemical structures of the recovered FAs [8,9], the relative amounts of each lipid class and their FA profiles largely determine the suitability of a given feedstock for biodiesel production.

The lipid composition of plant and algal cells are in a constant state of flux, with the relative amounts and localization of lipid species able to change in response to environmental conditions. The effect of changing conditions on total FA profile in microalgae has been reported, with studies showing that the FA profile can be affected by parameters such as the availability of phosphorus [10], nitrogen [11–13]; silicon [14]; temperature [15,16]; salinity [17]; light level [18]; light/dark cycle [19,20]; and growth phase of the culture [21,22]. Changing environmental conditions have also been shown to affect the relative proportion of each lipid class in microalgae. Critically, certain types of stress, and in particular as N-deprivation, are known to increase the production of TAGs [23]. N-deprivation is seen as a practical means of enhancing biodiesel production from microalgae, and an increasingly detailed understanding of the underlying metabolism is required for its optimal implementation.

Our understanding of the regulation of lipid metabolism in photosynthetic organisms has been developing steadily over recent decades. Lipid metabolism in plants and algae involves a combination of *de novo* synthesis pathways, recycling pathways and lipid trafficking between sub-cellular compartments [24,25]. These pathways must be tightly regulated to achieve lipid homeostasis and ensure the composition of each sub-cellular membrane is maintained or altered as appropriate for the environmental conditions. The available data on microalgae suggest that enhanced TAG production under N-deprivation results, at least in part, from the recycling of glycolipids combined with *de novo* synthesis pathways [25–28]. However, the precise nature of the perturbations to the lipidome that occur in microalgae under N-deprivation are yet to be determined.

To better understand the changes that occur in microalgae in response to N-deprivation, it is essential to obtain accurate and quantitative information about lipid composition. The identification and quantification of individual lipid species in biological samples has been greatly facilitated by recent advances in LC-MS and GC-MS [29]. These techniques have been successfully employed for the analysis of plant lipids [30,31], including under normal versus physiologically-stressed conditions [32], and enabled the characterization of MGDG and DGDG in photosynthetic marine protists [33,34], and SQDG [35], betaine lipids [16] and TAGs [14,36] in various species of algae.

In this work, we use LC-MS and GC-MS to obtain detailed lipidomic information for two model microalgal species from the chromophyte (golden-brown) and chlorophyte (green) lineages under N-replete and N-deplete conditions. These data provide important new information on the effect of N-deprivation on FA profiles across several key lipid classes, and in particular within TAGs, which is discussed in the context of current knowledge of the metabolic responses of microalgae to nutrient stress.

Materials and Methods

2.1 Algal strains

Cultures of Chlorella sp. and Nannochloropsis sp. were maintained in 25 mL clear plastic tissue culture flasks at 17°C in a light:dark cycle of 12:12 hr with a photon flux density of 48–55 $\mu E.m^{-2}.s^{-1}$ in a marine medium as previously described [37]. For the experimental work, 1.2 L of aerated cultures of the two algal species were grown in aerated 2 L Schott bottles at 17°C with a photon flux density of 60–70 $\mu E.m^{-2}.s^{-1}$. The aeration

provided both a source of carbon and agitation for the algal cells in culture. Nitrogen-replete (control) cultures were established with 5 mM NO_3^-, at this concentration nitrogen was not growth-limiting within the time of the experiment. N-deplete cultures were established in fresh medium which contained 0.5 mM NO_3^- for both species. These cultures were allowed to grow for 7–10 days, then harvested by continuous centrifugation (5000 g) and immediate lipid extraction.

2.2 Lipid extraction

Lipids were extracted using a modified Bligh and Dyer [38] procedure as previously described [37]. The extracts were dried under nitrogen, weighed and stored under nitrogen at −20°C until used for the LC-MS analysis.

2.3 LC-MS

The dried algal lipid extracts were re-suspended in butanol/methanol (1:1, v/v) containing 10 mM butylated hydoxy toluene (BHT) for analysis by LC-MS. The BHT was added to improve the oxidative stability of extracts. An Agilent 1200 series LC system equipped with a vacuum degasser, binary pump, temperature controlled autosampler and column oven was used for chromatography. Lipids were fractionated using an Ascentis Express 50 mm×2.1 mm, 2.7 μm particle size RP amide column (Supelco-SigmaAldrich) which was maintained at 35°C. Lipids were eluted using a binary mobile phase gradient at a flow rate of 0.2 mL·min^{-1}. Mobile phase (A) comprised water/methanol/tetrahydrofuran (50:20:30, v/v/v) with 10 mM ammonium formate, mobile phase (B) comprised water/methanol/tetrahydrofuran (5:20:75, v/v/v) with 10 mM ammonium formate. The gradient started at 100% A and linearly decreased to 0% A over 10 min with a one min hold at 0% A, then re-equilibration at 100% A for 4 minutes. Lipids were measured using an Agilent 6410b electrospray ionization-triple quadrupole (ESI-QQQ)-MS.

Multiple reaction monitoring (MRM) lists for phospholipids were developed for quantification by initial untargeted scans using the precursor and neutral loss scan functions. The following lipids were analysed using precursor ion scanning in positive ion mode: PC (precursors of m/z 184.1), cholesterol esters (m/z 369.4), PG (m/z 189) and in negative ion mode: PI (m/z 241), phosphatidic acid PA (m/z 153) and SQDG (m/z 225). Positive ion neutral loss scanning was used to identify PE (neutral loss of 141 u), MGDG (neutral loss of 179 u), DGDG (neutral loss of 341 u) and in negative ion neutral loss mode: PS (neutral loss of 87 u). The neutral lipids were analysed in single ion monitoring mode (SIM). The SIM targets for mono-, di- and triacylglycerols were created using all combinations of the identified FAs from the total GC-MS data [37]. A maximum of 100 MRMs or SIMs (5 ms dwell times) were analysed per time segment providing approximately 12–16 data points across a chromatographic peak. Optimized parameters for capillary, fragmentor, and collision voltages were 4,000, 60–160, and 0–60 V, respectively (details in Table S1).

Single ion monitoring was used to quantify neutral lipids due to the different response factors which arise from fragmentation of DAGs and TAGs. The predominant product ions arising from MS/MS of DAGs and TAGs are from the neutral loss of fatty acids. The product ions produced from TAG's with three different fatty acids do not necessarily produce a 1:1:1 ratio from the neutral loss of each fatty acid. Also, the fragmentation of a TAG with the same fatty acids in all three SN positions also produces a higher response factor when compared with a TAG with a mix of fatty acids. This means that a single MRM cannot be selected for TAGs and DAGs when using a single representative standard, in this case triolein (TAG18:1/18:1/18:1) and diolein (DAG18:1/18:1) and

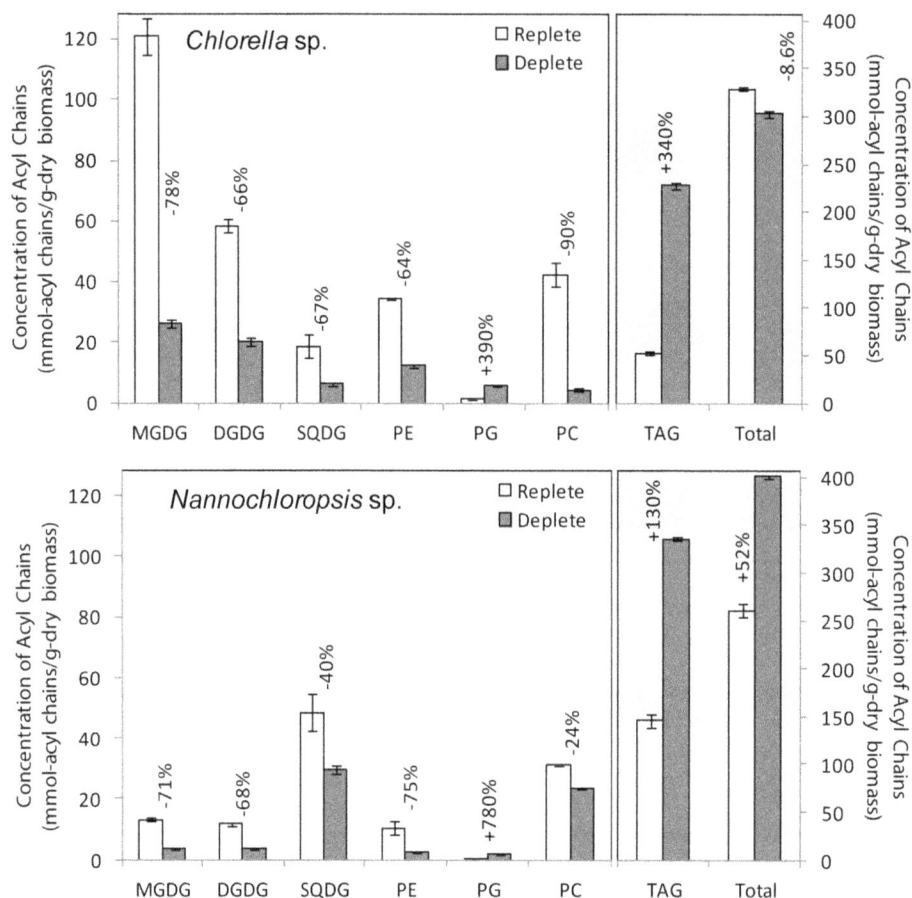

Figure 1. Total acyl chain concentrations within different lipid classes recovered from nitrogen replete and deplete *Chlorella* **sp. and** *Nannochloropsis* **sp.** The TAGs and total lipids read from right hand axis, all other lipid classes read from the left hand axis. Error bars represent the standard deviation of triplicate experiments.

therefore single ion monitoring was used for DAGs and TAGs. This acquisition approach measures the abundance of intact lipids species.

The data were processed using Agilent Mass Hunter Qualitative (for scan data) and Quantitative (for MRM and SIM data) analysis software. A combined 50 μM external calibration standard containing deuterated phospholipids PC(16:0/18:1)d31 (5 isotopomers), PE(16:0/18:1)d31, PG(16:0/18:1)d31 (3-isotopomers present), PA(16:0/18:1)d31, PS(16:0/18:1)d31, PI(16:0/18:1)d31 and glycolipids MGDG(34:6/36:6), DGDG(34:6/36:6), SQDG(34:1) was prepared. A second combined external standard (100 μM) of mono, di- and triacylglycerol standard (MAG-18:1; DAG-18:1/18:1; TAG-18:1/18:1/18:1; Nu-Chek Prep >98%) was also prepared. A single point calibration was used for quantification for each lipid against the standard from the same class and the final concentrations calculated using the extracted lipid mass for each algal extract. Where more than one isotopomer or lipid was present (e.g. PE the sum of the areas for the multiple peaks in the standard was used for quantification. Detected lipid species were annotated as follows; lipid class (sum of carbon atoms in the two fatty acid chains:sum of double bonds in the fatty acid chains).

Results and Discussion

The LC-MS data provided quantitative distributions of key lipid classes on the basis of total acyl chain concentration in *Chlorella* sp. and *Nannochloropsis* sp. under N-replete and N-deplete conditions (Figure 1) and profiles of the combined acyl content of each lipid class (Figures 2 and 3). The LC-MS approach used here provided the combined mass of the acyl chains attached to a particular head group (see Table S2 for complete data set). As it is possible for lipids with multiple acyl chains to have different combinations of FAs sharing the same overall mass, the individual acyl content was not determined directly. While an MS/MS analysis on each lipid could determine the exact FA composition this was not practical for the extensive number of lipids examined in this work. The assignment of fatty acid compositions for the glycolipids was ambiguous therefore MS/MS of the high abundant MGDG and DGDG lipids was carried out to confirm the FA composition (Figure S1) For the other classes the LC-MS data was combined with previously obtained GC-MS data for these samples [37], allowing the most probable combination of individual fatty acids to each lipid species to be ascertained (Figures 4 and 5, and Table S3). The compatibility and consistency of the two data sets was cross-checked by comparing the FA profile of the overall lipid content, TAGs (neutral lipids), and polar lipids (glycolipids and phospholipids) as determined by the two different methods (Table S4). Considering the fundamen-

Chlorella sp. Replete
TAG — Number of Double Bonds

Number of Acyl Carbons	0	1	2	3	4	5	6	7	8	9	10	11	12
42													
44	0.01	0.07			0.01								
46	0.01	0.01	0.01	0.01				0.01					
48	0.01	0.01	0.07	0.17	0.09	0.06	0.07	0.01					
50	0.01	0.08	0.06	0.35	0.36	0.41	0.44	0.17	0.02	0.03			
52		0.04	0.13	0.18	1.00	0.79	0.37	0.31	0.02	0.04	0.04	0.04	
54	0.23		0.01	0.04	0.07	0.33	0.03	0.03	0.03		0.02	0.04	0.03
56				0.01	0.14	0.01	0.01	0.01	0.01	0.01		0.01	
58		0.01	0.01								0.15	0.77	
60					0.05								
62		0.01	0.01										
64			0.01	0.03	0.05	0.10		0.05	0.02				0.01
66	0.03	0.02		0.01	0.07	0.05	0.20	0.20	0.24	0.27	0.05	0.01	0.02

Chlorella sp. Deplete
TAG — Number of Double Bonds

Number of Acyl Carbons	0	1	2	3	4	5	6	7	8	9	10	11	12
42													
44													
46		0.01	0.01	0.01									
48	0.01	0.01	0.05	0.16	0.06	0.02	0.03						
50	0.10	0.08	0.05	0.33	0.39	0.27	0.40	0.10		0.01			
52	0.01	0.21	0.19	0.16	1.00	0.60	0.38	0.41	0.01	0.01	0.02	0.03	
54	0.01	0.01	0.11	0.12	0.12	0.62	0.35	0.17	0.01	0.01	0.01	0.03	0.02
56								0.01					
58													
60													
62													
64				0.01	0.02	0.05		0.01	0.02				
66	0.01	0.01			0.03	0.03	0.11	0.09	0.11	0.10	0.14	0.06	

Nannochloropsis sp. Replete
TAG — Number of Double Bonds

Number of Acyl Carbons	0	1	2	3	4	5	6	7	8	9	10	11	12
42													
44	0.01	0.05	0.02										
46	0.03	0.23	0.12	0.02									
48	0.16	1.00	0.80	0.29	0.06			0.01					
50	0.03	0.04	0.18	0.11	0.04	0.03		0.01	0.01	0.04			
52		0.01	0.01	0.02	0.09	0.06	0.12			0.01			
54		0.01	0.03		0.02	0.02	0.01						
56								0.01					
58													
60		0.02		0.03	0.01			0.01					
62		0.01	0.01	0.01	0.06	0.05		0.02					
64			0.06	0.22	0.18	0.12	0.07	0.01		0.02			
66			0.02	0.01	0.01	0.03	0.02	0.02		0.01			0.01

Nannochloropsis sp. deplete
TAG — Number of Double Bonds

Number of Acyl Carbons	0	1	2	3	4	5	6	7	8	9	10	11	12
42													
44		0.06	0.01										
46	0.02	0.25	0.09	0.01									
48	0.16	1.00	0.69	0.14	0.02			0.01					
50	0.05	0.06	0.39	0.11	0.05	0.03	0.02	0.01		0.02			
52		0.01	0.03	0.01	0.07	0.05	0.15						
54			0.02		0.01	0.02	0.02						
56								0.01		0.01	0.01		
58													
60			0.01		0.03	0.01							
62		0.01		0.01	0.04	0.03		0.01		0.02			
64		0.01		0.05	0.13	0.10	0.04	0.06	0.01			0.01	0.01
66			0.02	0.01			0.02	0.02	0.01				0.01

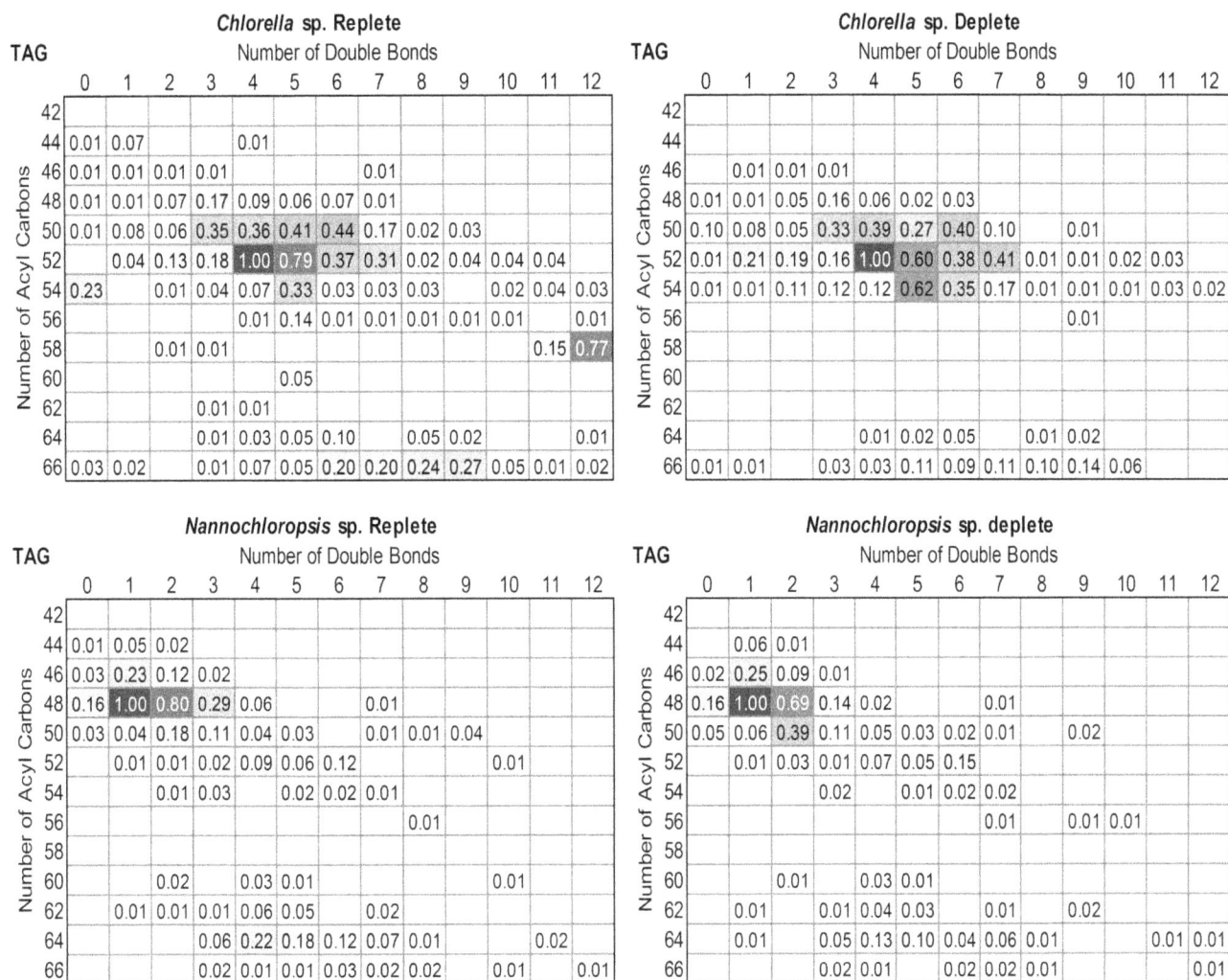

Figure 2. Heat mapped profile of combined acyl content (3 acyl chains) of TAG in nitrogen replete and nitrogen deplete _Chlorella_ sp. and _Nannochloropsis_ sp. The assigned values represent the relative proportion of each species normalised to the most prevalent species in each (e.g. 52:4 for nitrogen replete _Chlorella_ sp.). Values less than 0.01 are not presented.

tal differences between these two methods, the two data sets were in good agreement.

3.1 Quantitative distribution of lipid classes before and after N-deprivation

In agreement with previous studies [12,28,37,39–41] N-deprivation led to a considerable accumulation of TAGs in both species (Figure 1), increasing by 340% in _Chlorella_ sp. and 130% in _Nannochloropsis_ sp. Despite the differences in percentage increase, however, after N-deprivation the total amount of TAGs was still higher in _Nannochloropsis_ sp. at 336 ± 2 mmol$_{acyl\ chains}$/g$_{dry\ biomass}$ compared to 228 ± 3 mmol$_{acyl\ chains}$/g$_{dry\ biomass}$ in _Chlorella_ sp.

In both species, the increase in TAG content was associated with a decrease in total phosphoglycerolipid content, on a dry biomass basis (Figure 1). The amount of phosphoglycerolipid by weight decreased from 78 ± 4 to 23 ± 1 mmol$_{acyl\ chains}$/g$_{dry\ biomass}$ in _Chlorella_ sp. and from 42 ± 2 to 28 ± 1 mmol$_{acyl\ chains}$/g$_{dry\ biomass}$ in _Nannochloropsis_ sp. Analysis of individual phosphoglycerolipid classes revealed that following N-deprivation, PE decreased by 64–75% and PC decreased by 24–90%, while the

levels of PS, PI and PA remained below the threshold of detection. As both PE and PC contain nitrogen in their head group, these decreases most likely reflect reduced capacity to synthesize these species following N-deprivation. By contrast, N-deprivation resulted in elevated levels of PG in both species, increasing by 390% in _Chlorella_ sp. and 780% in _Nannochloropsis_ sp. Despite these large percentage increases, however, the PG content by weight remained relatively low in both species.

The increase in TAGs following N-deprivation was similarly associated with a decrease in total glycolipid content. The amount of glycolipid by weight decreased from 200 ± 12 to 52 ± 3 mmol$_{acyl}$ $_{chains}$/g$_{dry\ biomass}$ in _Chlorella_ sp. and from 73 ± 7 to 37 ± 2 mmol$_{acyl\ chains}$/g$_{dry\ biomass}$ in _Nannochloropsis_ sp. Analysis of individual glycolipid classes revealed that N-deprivation led to decreases in MGDG and DGDG of 66–78% and a decrease SQDG of 40–67%. Interestingly, the most abundant glycolipid species differed between the two algae. In _Chlorella_ sp., MGDG was the most abundant of the chloroplast glycolipids, consistent with observations in a variety of photosynthetic organisms [42]. In _Nannochloropsis_ sp., SQDG was the most abundant chloroplast

Chlorella sp. Replete

MGDG — Number of Double Bonds

Acyl Carbons	0	1	2	3	4	5	6
32				0.01		0.06	0.01
34		0.01	0.07	0.21	0.52	1.00	
36				0.02	0.03	0.02	
38							
40							

DGDG — Number of Double Bonds

Acyl Carbons	0	1	2	3	4	5	6
32				0.03	0.02	0.03	
34		0.03	0.03	0.11	0.20	0.50	1.00
36					0.05	0.09	0.04
38							
40							

SQDG — Number of Double Bonds

Acyl Carbons	0	1	2	3	4	5	6
32	0.17	1.00	0.05	0.01			
34		0.01	0.02	0.13			
36				0.01	0.01	0.01	
38							
40							

PE — Number of Double Bonds

Acyl Carbons	0	1	2	3	4	5	6
32		0.01	0.05	0.16			
34		0.09	1.00	0.14	0.02	0.03	0.02
36			0.02	0.09	0.27	0.15	0.03
38							
40							

PG — Number of Double Bonds

Acyl Carbons	0	1	2	3	4	5	6
32	0.40	0.28	0.01	0.01			
34	0.06	0.69	1.00	0.73	0.72		
36		0.01	0.06	0.01			
38							
40							

PC — Number of Double Bonds

Acyl Carbons	0	1	2	3	4	5	6
32		0.01	0.02	0.14	0.01	0.01	
34		0.03	0.23	0.30	0.10	0.15	0.15
36			0.07	0.20	0.41	1.00	0.69
38				0.01	0.01	0.01	0.01
40							

Chlorella sp. Deplete

MGDG — Number of Double Bonds

Acyl Carbons	0	1	2	3	4	5	6
32						0.01	
34			0.02	0.07	0.25	0.26	1.00
36						0.01	0.01
38							
40							

DGDG — Number of Double Bonds

Acyl Carbons	0	1	2	3	4	5	6
32					0.01		
34	0.01	0.13	0.05	0.20	0.41	0.34	1.00
36					0.01	0.01	0.02
38							
40							

SQDG — Number of Double Bonds

Acyl Carbons	0	1	2	3	4	5	6
32	0.19	1.00	0.01	0.02			
34				0.06			
36							
38							
40							

PE — Number of Double Bonds

Acyl Carbons	0	1	2	3	4	5	6
32		0.01	0.02	0.06			
34		0.08	1.00	0.06	0.01	0.01	0.01
36			0.04	0.09	0.25	0.06	
38							
40							

PG — Number of Double Bonds

Acyl Carbons	0	1	2	3	4	5	6
32	0.16	0.08	0.01	0.01			
34	0.02	0.23	0.72	1.00	0.80	0.01	
36			0.02	0.01			
38							
40							

PC — Number of Double Bonds

Acyl Carbons	0	1	2	3	4	5	6
32		0.01	0.02	0.12	0.01		
34		0.02	0.30	0.24	0.11	0.19	0.13
36			0.09	0.28	0.92	1.00	0.37
38							
40							

Nannochloropsis sp. Replete

MGDG — Number of Double Bonds

Acyl Carbons	0	1	2	3	4	5	6
32			0.00	0.01	0.01	0.15	
34			0.02	0.01	0.12	1.00	0.02
36					0.05	0.30	0.32
38							
40							

DGDG — Number of Double Bonds

Acyl Carbons	0	1	2	3	4	5	6
32	0.02	0.29	0.18	0.03	0.01	0.02	
34			0.02	0.02	0.05	0.40	0.01
36				0.01	0.07	0.53	1.00
38							
40							

SQDG — Number of Double Bonds

Acyl Carbons	0	1	2	3	4	5	6
32	0.04						
34		0.01	0.10	1.00	0.01	0.01	0.02
36				0.02	0.01	0.07	
38							
40							

PE — Number of Double Bonds

Acyl Carbons	0	1	2	3	4	5	6
32		0.07	0.12	0.03			
34			0.05	0.05	0.02	0.01	
36			0.02	0.02	0.12	1.00	0.87
38				0.02	0.06	0.08	
40	0.01					0.01	0.16

PG — Number of Double Bonds

Acyl Carbons	0	1	2	3	4	5	6
32	0.01	0.36	0.01				
34	0.01	0.04	0.02	0.01			
36		0.01	0.06	0.01	0.07	0.76	1.00
38						0.01	
40							

PC — Number of Double Bonds

Acyl Carbons	0	1	2	3	4	5	6
32		0.08	0.30	0.18	0.06	0.01	
34		0.09	0.30	0.27	0.34	0.14	0.02
36			0.03	0.08	0.25	0.55	1.00
38					0.01	0.03	0.09
40							

Nannochloropsis sp. Deplete

MGDG — Number of Double Bonds

Acyl Carbons	0	1	2	3	4	5	6
32					0.01	0.14	
34					0.13	1.00	0.01
36			0.01		0.04	0.32	0.58
38							
40							

DGDG — Number of Double Bonds

Acyl Carbons	0	1	2	3	4	5	6
32	0.02	0.17	0.02			0.01	
34					0.03	0.23	
36					0.04	0.35	1.00
38							
40							

SQDG — Number of Double Bonds

Acyl Carbons	0	1	2	3	4	5	6
32	0.04						
34		0.02	0.09	1.00			
36						0.01	0.02
38							
40							

PE — Number of Double Bonds

Acyl Carbons	0	1	2	3	4	5	6
32		0.01	0.05	0.02			
34			0.02	0.04	0.01		
36			0.04	0.02	0.02	0.52	1.00
38					0.01	0.04	0.04
40							0.05

PG — Number of Double Bonds

Acyl Carbons	0	1	2	3	4	5	6
32	0.01	0.53	0.01				
34	0.01	0.01	0.01				
36			0.01	0.00	0.10	1.00	0.82
38							
40							

PC — Number of Double Bonds

Acyl Carbons	0	1	2	3	4	5	6
32		0.17	0.28	0.10	0.02		
34		0.24	0.77	1.00	0.94	0.05	
36			0.04	0.08	0.31	0.04	0.55
38					0.01	0.04	0.16
40							

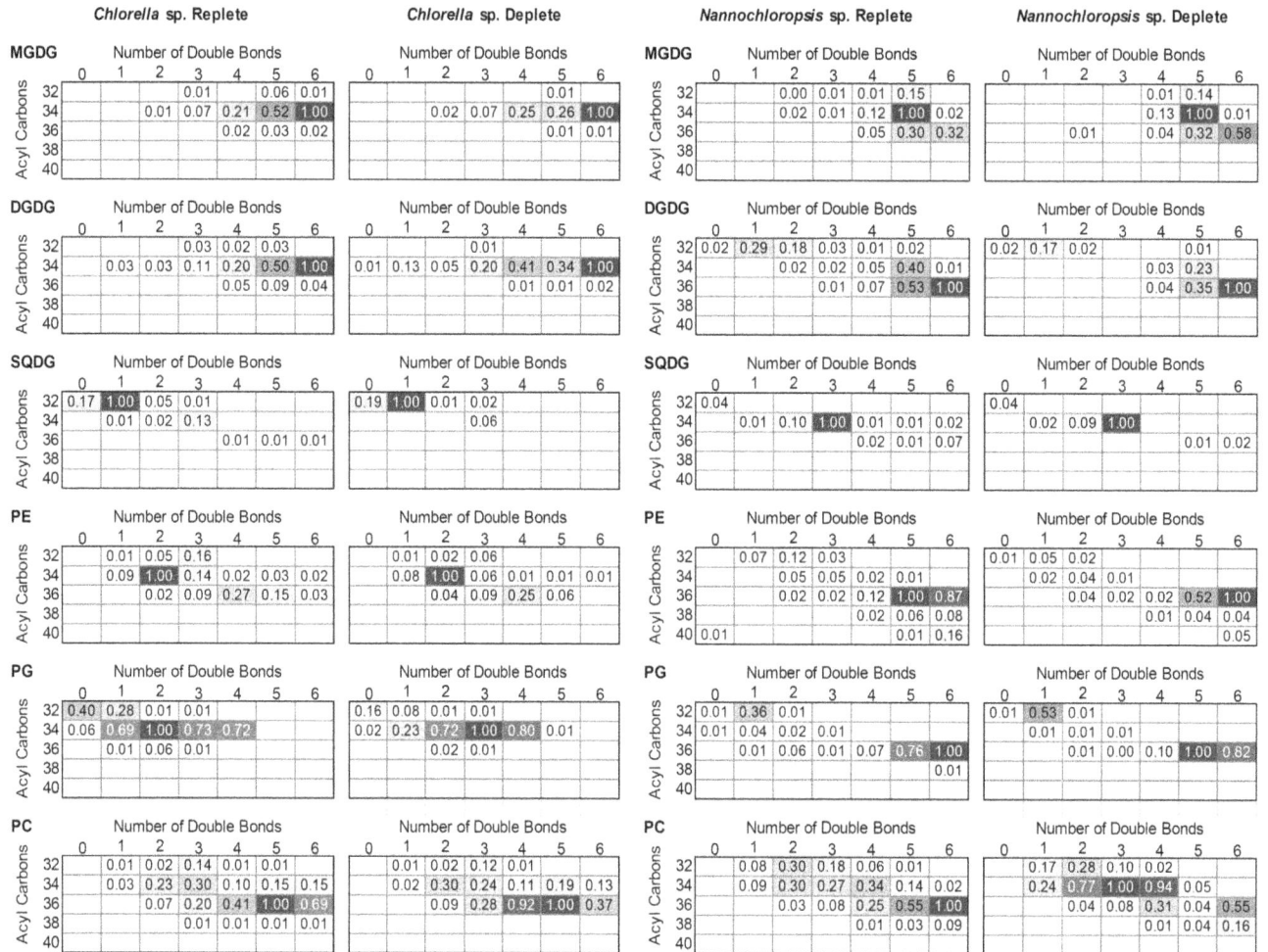

Figure 3. Heat mapped profiles of combined acyl content (2 acyl chains) of glyco- and phosphoglycerolipids in nitrogen replete and nitrogen deplete *Chlorella* sp. and *Nannochloropsis* sp. The assigned values represent the relative proportion of each species normalised to the most prevalent species in each. Values less than 0.01 are not presented.

glycolipid, a feature shared with only a few photosynthetic bacteria and algae.

Data from studies in other photosynthetic organisms suggest the increase in PG and the decrease in chloroplast glycolipids are likely to be related. Phosphatidylglycerol is the only phosphoglycerolipid that is found in the photosynthetic membranes of plastids [42], where it is necessary for the functions of photosystems I and II and the light harvesting complex II [43]. Photosynthesis and chloroplast development are also known to be dependent on both MGDG and DGDG [33,44], and SQDG [43]. The significant increase in PG with N-deprivation may therefore be to compensate for the loss of chloroplast glycolipids in an attempt to maintain proper photosynthetic activity. Alternatively, as N-deprivation has been shown to produce algal cells with smaller chloroplasts containing fewer thylakoids (photosynthetic membranes), less pigment and a disruption to photosynthetic capacity along with a reduction in the level of proteins involved in the photosynthetic electron transport chain [28,45], the decrease in glycolipids observed could also be due to a reduction in chloroplast size or thylakoid number. Another contributing factor to the observed decrease in membrane lipids following N-deprivation is the increase in average cell size that also results from generation of large TAGs-containing lipid bodies (or lipid droplets). If the data

were obtained on a per cell basis rather than on a dry biomass basis as presented here, the apparent reduction in membrane lipids would be less and the increase in TAGs would be greater.

3.2 Comparison of acyl content of different lipid classes

The LC-MS was used to identify and quantify individual lipid species of a particular mass within each of the major lipid classes (Figures 2 and 3). The most prominent TAG species in *Nannochloropsis* sp. under both N-replete and N-deplete conditions were 48:1 and 48:2 (Figure 2), which likely consist of 16:0 and 16:1 acyl chains (Table S3). The TAG profile in *Nannochloropsis* sp. shows only a few changes in response to N-deprivation. The most notable change is the increase in the levels of 50:2 (likely comprising 18:1, 16:1 and 16:0). As the *de novo* production of these FA species has previously been observed with nutrient deprivation e.g. [46], the increase in the 50:2 species suggests a similar response occurred here upon N-deprivation. There was also a notable decrease in the 64:4 and 64:5 species in *Nannochloropsis* sp. following N-depletion (Figure 2), indicating long polyunsaturated FA species (e.g. 20:4, 20:5) did not accumulate in the TAGs. These long FA chains and more particularly C20:5 (eicosapentaenoic acid or EPA) are usually found in the chloroplast galactolipids MGDG and DGDG

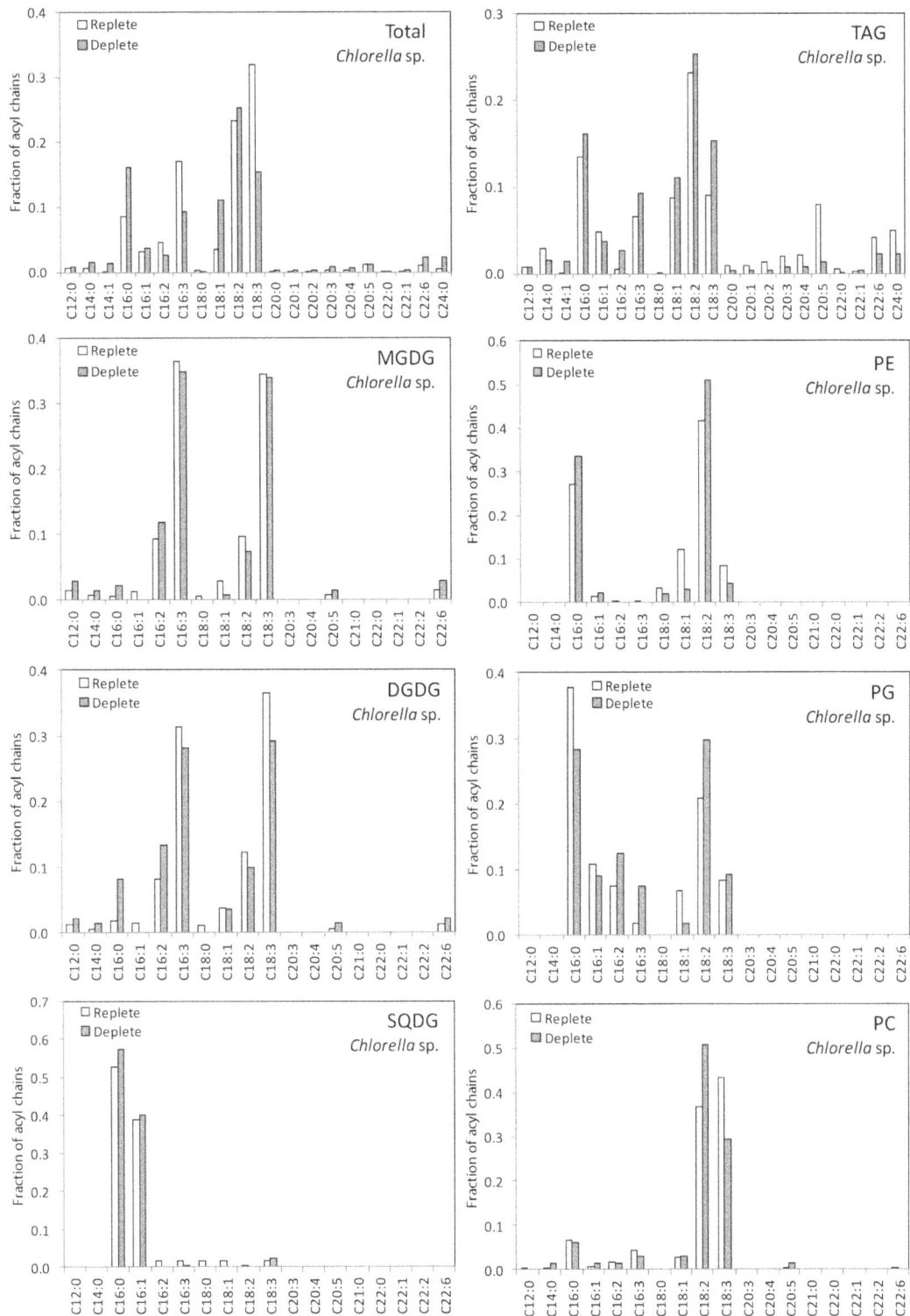

Figure 4. Estimated acyl chain composition as a function of lipid class based on combined LC-MS and lipid class fractionated GC-MS data [37] for nitrogen replete and deplete *Chlorella* sp.

(Figure 5). Here EPA was found be globally decreased in PG, SQDG and PC but was found in higher abundance in DGDG and PE under nitrogen deprivation. These observations suggest that EPA could be exported outside the chloroplast via DGDG or free FA to be eventually re-utilised for purposes other than TAG synthesis. Since EPA is a good source of reducing power, it could be recycled by the cell as an easy energy source during nitrogen-stress. Future experiments could confirm this by measuring the activity of beta oxidation and/or reducing power stocks under stress. There was a slight increase in 48:1 relative to 48:2 following

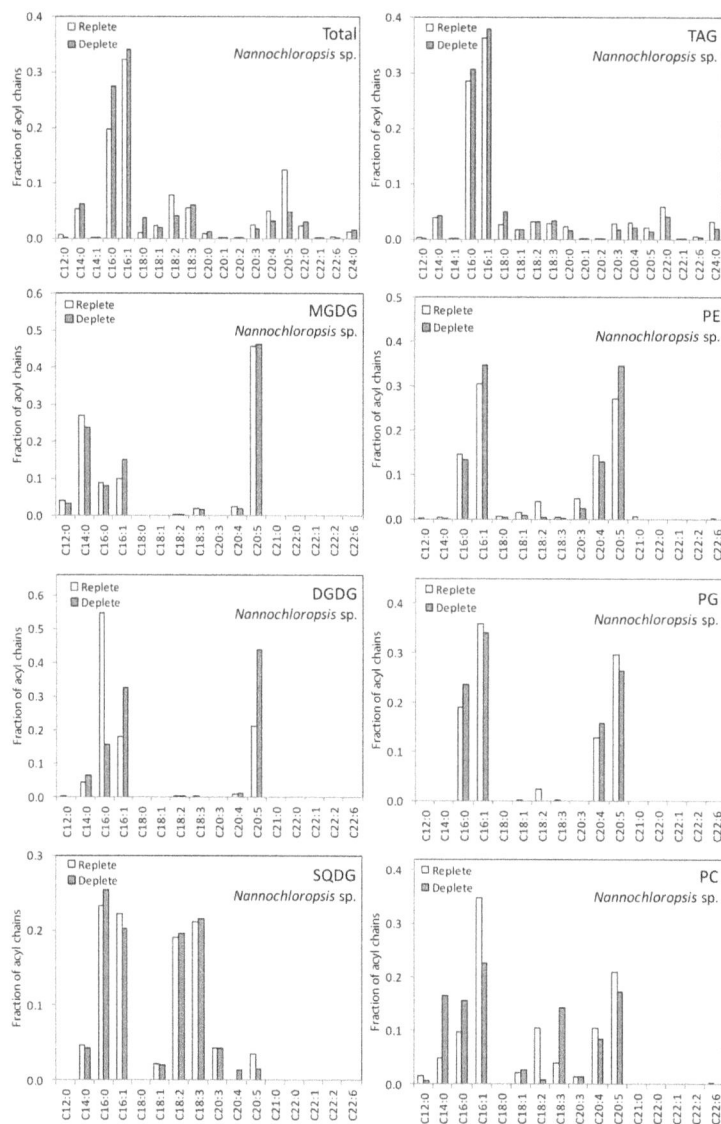

Figure 5. Estimated acyl chain composition as a function of lipid class based on combined LC-MS and lipid class fractionated GC-MS data [37] for nitrogen replete and deplete *Nannochloropsis* sp.

N-depletion (Figure 2), reflecting the increase in 16:0 compared to 16:1 observed in GC-MS analysis (Figure 5). This suggests that newly synthesized FAs could be directly incorporated into TAGs prior to their desaturation, potentially as an energy saving mechanism.

The profile of TAGs in *Chlorella* sp. was more diverse than in *Nannochloropsis* sp. The most abundant species were 50:4/5, 52:3/4/5/6, 54:0/3, 58:12 and 66:5–10 (Figure 2, Table S2). In comparison to *Nannochloropsis* sp., the TAG content of *Chlorella* sp. showed more variability upon N-deprivation, indicating that more FA exchange and trafficking could occur during the stress (Figure 2). Of significance was the near disappearance of 58:12 (comprising 20:5, 20:5 and 18:2). This disappearance was associated with an increase in TAG species possessing a single 20:5 in combination with shorter and more saturated FAs such as 52:7 (20:5/16:1/16:1) and 54:5 (20:5/18:0/16:0). There was also a clear increase of 54:2–7 species, suggesting TAGs were enriched in C18:0/C18:1/C18:2 (Figure 2, Table S4). The stable levels of 20:5 in TAGs under N-deprivation suggest that 20:5 was being

recycled by the cell. Usually, EPA (20:5) is associated with chloroplast lipids as observed here (Figure 4), although not as their major components as detected in *Nannochloropsis* sp. (Figure 5). Furthermore, since none of lipid classes of *Chlorella* sp. showed any significant decrease in EPA, it could indicate that this FA species is mainly recycled directly from existing TAGs and replaced by other FA species in the TAG pool during nitrogen deprivation. The reason for recycling or specific replacement of EPA remains to be elucidated. A noticeable change in TAG content is the increase of C16:3, C18:3 and C18:2. C18:3 and C16:3 are typically found in chloroplast membranes (Figure 4). Both species were found to significantly decrease in DGDG and at a lesser extent, in MGDG, both major chloroplasts lipids species. A possible explanation for the C16:3/C18:3-containing TAG population is the direct recycling of chloroplast galactolipids. The cell could eventually compensate its loss of chloroplast fatty acid by increasing *de novo* fatty acid production under stress conditions, as suggested by (i) an increase of mid-long fatty acid chains C16:0/C18:0/C18:1/C18:2 in MGDG, DGDG (and SQDG), (ii) an

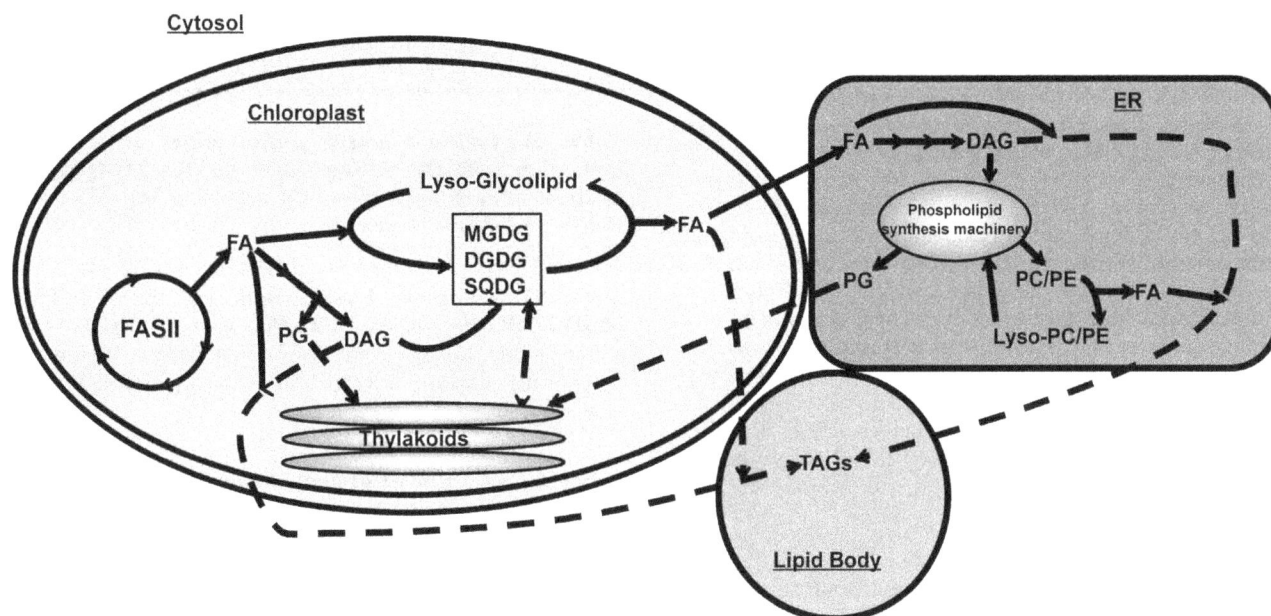

Figure 6. Proposed pathways for TAG synthesis during N-starvation.

increase of C16:3/C18:3 in less abundant chloroplast classes PG and SQDG and (iii) a global increase of thylakoids/photosystems-interacting lipid class PG (Figure 5).There was also a decrease of very long chain-containing TAGs (66:6–9), without a significant decrease of 22:6 or 20:4 in other lipid species (Figure 4) indicating these TAGs were likely utilized for other purposes (such as beta-oxidation) rather than being remodelled. Finally, levels of 58:1–3 were also higher in N-deprived conditions (16:0, 18:1 and 18:2), again likely representing an increase in *de novo* FA synthesis in TAGs. The 16:0 and 18:1 are normally the newly synthesized FA species, so their increase in TAGs suggests that on top of recycling "mature" lipids for TAG generation, the cell is also actively using its *de novo* synthesis pathways, and potentially recycling lipids for other purposes.

Other changes could be detected in glycerophospholipids during N-deprivation, each specific to FA abundances/utilisation depending on the microalgal species. Indeed, the PC, PE and PG of *Chlorella* sp. were all subjected to an increase in C18:2 content, whilst containing less C18:3 (with the exception of PG for the latter). This increase in C18:2 was also detected in the TAG profile (Figure 4) concomitant with the increase of C18:3. On the other hand, the glycerophospholipids of *Nannochloropsis* sp. showed different behaviour. Here the main changes occurred at the level of C14:0, C16:0/1 and C20:4/5 with shorter and less unsaturated chains found to increase in PC, PE and PG, whilst EPA was reduced in PC and PG.

One key difference in the overall FA profiles of the two species (compare Figures 4 and 5) is that *Chlorella* sp. has a preference for the production of the prokaryotic- and eukaryotic-generated C16:0 and C18:3 typically found in chloroplast glycolipids in higher plants whereas *Nannochloropsis* sp. produces more EPA (20:5). The significance of this distribution is unclear, but follows a general rule that chromist (golden-brown) algae (including *Nannochloropsis*) produce EPA while chlorophyte (green) algae (including *Chlorella*), if any, produce DHA.

3.3 Interpretation of biosynthetic rearrangements due to nitrogen deprivation

Our data shows N-deprivation leads to a considerable increase in TAG and PG content and a concomitant decrease in polar lipid content in both species of microalgae. Despite these changes, however, the FA profiles within each lipid class remained relatively constant. As previously mentioned, phosphoglycerolipids are important structural components of extra-plastidial membrane. The maintenance of their FA profiles therefore shows the composition of extra-plastidial membranes is not significantly altered by N-deprivation. Conversely, the glycolipids which are key components of chloroplast membranes are required to support photosynthesis and although their FA and species profiles were also largely maintained, the increase in PG content suggests that this alone was insufficient for optimal photosynthetic activity.

We therefore propose that N-depletion triggers major remodelling of intracellular lipid pools in both *Chlorella* sp. and *Nannochloropsis* sp. in order to boost TAG synthesis in lipid bodies whilst maintaining lipid homeostasis in other compartments (Figure 6). Even though the two species responded differently, the major trends appear to be conserved. In both *Chlorella* sp. and *Nannochloropsis* sp., the chloroplast glycolipids appear to be broken down as previously proposed [26–28,45], that are then recycled for TAG synthesis. *Nannochloropsis spp.* seemed to incorporate shorter and more saturated FA (C16:0, C16:1) from neosynthesized galactolipids, whereas *Chlorella spp.* used a combination of C18:2 and C18:3 species, likely originating from mature galactolipids, and likely neosynthesized C18:0/C18:1. Interestingly, both species contained EPA (C20:5) in TAGs or galactolipids prior to nitrogen stress but this FA specie did not increase and/or decreased in TAGs species and other glycerolipids upon N-deprivation, suggesting it was either metabolized (e.g. by beta-oxidation) or modified for a yet to be determined reason.

An enzyme capable of cleaving FAs from MGDG has been identified in *Chlamydomonas reinhardtii* [45] providing a possible mechanism for recycling of this glycolipid. Based on our data, we would suggest that similar enzymes also exist for the liberation of

FAs from DGDG and SQDG. To compensate for the loss of the chloroplast glycolipids, both algae appear to up-regulate PG synthesis. In both cases, the significant increases in PG levels were concomitant to the loss of SQDG, the other acidic lipid class found in the plastid. Thus, FAs derived from the breakdown of SQDG could be used in TAG synthesis while PG replaced SQDG in the maintenance of chloroplast function and structural integrity. Indeed, an increase in PG has been observed in SQDG-deficient mutant and sulphur-starved *Chlamydomonas*, suggesting the replacement of SQDG with PG may be common to both stress responses [47]. At the same time, additional FAs for TAG synthesis appear to be generated by the chloroplast *de novo* FA synthesis pathway [48]. These FAs may be exported to the lipid body directly, or may transit indirectly after first being incorporated into DAG in the endoplasmic reticulum (ER) or chloroplast (Figure 6). Finally, the breakdown and recycling of existing phosphoglycerolipids in the ER is likely to provide yet another source of FAs for TAG synthesis during N-deprivation.

Conclusions

This work provides the most detailed analysis of the microalgal lipidome after nitrogen deprivation using new LC-MS approach. These data showed the expected increase in the concentrations of TAGs with a concomitant decrease in the levels of the chloroplast glycolipids, MGDG, DGDG and SQDG, as well as two major nitrogen containing phosphoglycerolipids, PC and PE after N-deprivation. In contrast, PG levels were highly increased in both models (from very low starting levels), a previously unnoticed effect of N-deprivation. Some modifications of FAs derived from the breakdown of existing glycolipids and phosphoglycerolipids in *Nannochloropsis spp.* and long FA chains from TAGs in *Chlorella* sp. occurs during nitrogen deprivation. This finding has great implications for the nutraceutical and biodiesel industry. It also shows that major remodelling of the intracellular lipid pools in both species occurs in order to boost TAG synthesis whilst attempting to maintain lipid homeostasis in other essential compartments such as the chloroplast.

The detailed quantitative information provided by the application of LC-MS illustrates the utility of this technique for studying lipid biochemistry and metabolism in microalgae and for the development of biotechnological applications such as biodiesel production. Liquid chromatography-MS provides the most comprehensive coverage of lipids with minimal sample handling. This enables investigations to be performed using small scale cultures, facilitating more extensive investigations across a range of growth variables and species.

Supporting Information

Figure S1 Example LC-MS/MS spectra of high abundant glycolipids in *Nannochloropsis* sp. and *Chlorella* sp. Fatty acid assignments are included on the spectra. All

fragment ions listed show the fatty acid chain length and degree of saturation for an ion of the following structure: [MAG(FA)-H₂O+ H⁺]⁺. Note: Gal = neutral loss of galactose.

Table S1 Collision energy, fragmentor voltage and ionisation polarity settings used for the LC-QQQ-MS analysis of each lipid class. The ESI source settings were the same across all lipid classes.

Table S2 Complete LC-MS data set for the TAG, MGDG, DGDG, SQDG, PE, PG, and PC lipids. Data represent the average and standard deviation of triplicate experiments and are presented in units of mmol-lipid/g-dry biomass.

Table S3 Assignment of acyl chains to individual lipid species determined by manual matching of LC-MS and GC-MS data (from Olmstead et al. 2013). Acyl chains were assigned to: LC-MS TAG data based on GC-MS data of the neutral lipid SPE fraction; LC-MS data of the combined pool of MGDG, DGDG and SQDG based on GC-MS data of the glycolipid SPE fraction; LC-MS data of the combined pool of PE, PG, PC based on GC-MS data of the phospholipid SPE fraction.

Table S4 Heat map comparison of relative acyl chain abundance in the total lipids, TAGs, and polar lipids (glycol- plus phospholipids) of nutrient replete and deplete *Chlorella* sp. and *Nannochloropsis* sp. as determined by direct GC-MS analysis and inferred from LC data cross checked with GC results. The assigned values represent the fraction of the total acyl in each lipid group represented by a particular fatty acid species. The GC-MS data for C16:2 and C16:3 acyl groups were quantified assuming consistent detector responses with the C16:1 peaks, as no standards were available.

Acknowledgments

The authors would like to thank Prof. R. Wetherbee, School of Botany, The University of Melbourne for the use of their Phytoplankton Laboratory and algal culture facility. We would also like to acknowledge the significant contributions made by the reviewers.

Author Contributions

Conceived and designed the experiments: GJOM DRAH CYB DLC. Performed the experiments: GJOM DRAH CYB ILDO AB MJS DD SEK PJS DLC. Analyzed the data: GJOM DRAH CYB DLC. Contributed reagents/materials/analysis tools: DLC GJOM. Wrote the paper: GJOM DRAH CYB MJS DLC.

References

1. Chisti Y (2007) Biodiesel from microalgae. Biotechnol Adv 25: 294–306.
2. Gouveia L, Oliveira AC (2009) Microalgae as a raw material for biofuels production. J Ind Microbiol Biot 36: 269–274.
3. Hu Q, Sommerfield M, Jarvis E, Ghirardi M, Posewitz M, et al. (2008) Microalgal triacylglycerols as feedstocks for biofuel production: perspectives and advances. Plant J 54: 621–639.
4. Rodolfi L, Chini Zittelli G, Bassi N, Padovani G, Biondi N, et al. (2009) Microalgae for oil: strain selection, induction of lipid synthesis and outdoor mass cultivation in a low-cost photobioreactor. Biotechnol Bioeng 102: 100–112.
5. Williams PLlB, Laurens LML (2010) Microalgae as biodiesel and biomass feedstocks: review and analysis of the biochemistry, energetics and economics. Energ Environ Sci 3: 554–590.
6. Volkman JK, Jeffrey SW, Nichols PD, Rogers GI, Garland CD (1989) Fatty acid and lipid composition of 10 species of microalgae used in mariculture. J Exp Mar Biol Ecol 128: 219–240.
7. Olmstead ILD, Kentish SE, Scales PJ, Martin GJO (2013) Low solvent, low temperature method for extracting biodiesel lipids from concentrated microalgal biomass. Biores Technol 148: 615–619.
8. Stansell GR, Gray VM, Sym SD (2012) Microalgal fatty acid composition: implications for biodiesel quality. J Appl Phycol 24: 791–801.
9. Zendejas FJ, Benke PI, Lane PD, Simmons BA, Lane TW (2012) Characterization of the acylglycerols and resulting biodiesel derived from vegetable oil and microalgae (Thalassiosira pseudonana and Phaeodactylum tricornutum). Biotechnol Bioeng 109: 1146–1154.

10. Siron R, Giusti G, Berland B (1989) Changes in the fatty acid composition of *Phaeodactylum tricornutum* and *Dunaliella tertiolecta* during growth and under phosphorus deficiency Mar Ecol-Prog Ser 55: 95–100.

11. Breuer G, Lamers PP, Martens DE, Draaisma RB, Wijffels RH (2012) The impact of nitrogen starvation on the dynamics of triacylglycerol accumulation in nine microalgae strains. Biores Technol 124: 217–226.

12. Recht L, Zarka A, Boussiba S (2012) Patterns of carbohydrate and fatty acid changes under nitrogen starvation in the microalgae *Haematococcus pluvialis* and *Nannochloropsis* sp. Appl Microbiol Biot 94: 1495–1503.

13. Tornabene TG, Holzer G, Lien S, Burris N (1983) Lipid composition of the nitrogen starved green alga *Neochloris oleoabundans*. Enzyme Microb Technol 5: 435–440.

14. Yu E, Zendejas F, Lane P, Gaucher S, Simmons B, et al. (2009) Triacylglycerol accumulation and profiling in the model diatoms *Thalassiosira pseudonana* and *Phaeodactylum tricornutum* (Baccilariophyceae) during starvation. J Appl Phycol 21: 669–681.

15. Converti A, Casazza AA, Ortiz EY, Perego P, Del Borghi M (2009) Effect of temperature and nitrogen concentration on the growth and lipid content of *Nannochloropsis oculata* and *Chlorella vulgaris* for biodiesel production. Chem Eng Process 48: 1146–1151.

16. Roche SA, Leblond JD (2010) Betaine lipids in chlorarachniophytes. Phycol Res 58: 298–305.

17. Takagi M, Karseno, Yoshida T (2006) Effect of salt concentration on intracellular accumulation of lipids and triacylglyceride in marine microalgae *Dunaliella* cells. J Biosci Bioeng 101: 223–226.

18. Guihéneuf F, Mimouni V, Ulmann L, Tremblin G (2009) Combined effects of irradiance level and carbon source on fatty acid and lipid class composition in the microalga *Pavlova lutheri* commonly used in mariculture. J Exp Mar Biol Ecol 369: 136–143.

19. Sicko-Goad L, Simmons MS, Lazinsky D, Hall J (1988) Effect of light cycle on diatom fatty acid composition and quantitative morphology. J Phycol 24: 1–7.

20. Sukenik A, Carmeli Y (1990) Lipid synthesis and fatty acid composition in *Nannochloropsis* sp. (Eustigmatophyceae) grown in a light-dark cycle. J Phycol 26: 463–469.

21. Dunstan GA, Volkman JK, Barrett SM, Garland CD (1993) Changes in the lipid composition and maximisation of the polyunsaturated fatty acid content of three microalgae grown in mass culture. J Appl Phycol 5: 71–83.

22. Roncarati A, Meluzzi A, Acciarri S, Tallarico N, Meloti P (2004) Fatty acid composition of different microalgae strains (*Nannochloropsis* sp., *Nannochloropsis oculata* (Droop) Hibberd, *Nannochloris atomus* Butcher and *Isochrysis* sp.) according to the culture phase and the carbon dioxide concentration. J World Aquacult Soc 35: 401–411.

23. Sharma KK, Schuhmann H, Schenk PM (2012) High lipid induction in microalgae for biodiesel production. Energies 5: 1532–1553.

24. Benning C (2009) Mechanisms of lipid transport involved in organelle biogenesis in plant cells. Annu Rev Cell Dev Biol 25: 71–91.

25. Joyard J, Ferro M, Masselon C, Seigneurin-Berny D, Salvi D, et al. (2010) Chloroplast proteomics highlights the subcellular compartmentation of lipid metabolism. Prog Lipid Res 49: 128–158.

26. Boyle NR, Page MD, Liu B, Blaby IK, Casero D, et al. (2012) Three acyltransferases and nitrogen-responsive regulator are implicated in nitrogen starvation-induced triacylglycerol accumulation in *Chlamydomonas*. J Biol Chem 287: 15811–15825.

27. Sanjaya, Miller R, Durrett TP, Kosma DK, Lydic TA, et al. (2013) Altered lipid composition and enhanced nutritional value of *Arabidopsis* leaves following introduction of an algal diacylglycerol acyltransferase 2. Plant Cell 25: 677–693.

28. Simionato D, Block MA, La Rocca N, Jouhet J, Marechal E, et al. (2013) The response of *Nannochloropsis gaditana* to nitrogen starvation includes *de novo* biosynthesis of triacylglycerols, a decrease of chloroplast galactolipids, and reorganization of the photosynthetic apparatus. Eukaryot Cell 12: 665–676.

29. Cui Z, Thomas MJ (2009) Phospholipid profiling by tandem mass spectrometry. J Chromatogr B 877: 2709–2715.

30. Samarakoon T, Shiva S, Lowe K, Tamura P, Roth M, et al. (2012) *Arabidopsis thaliana* Membrane Lipid Molecular Species and Their Mass Spectral Analysis. In: Normanly J, editor. High-Throughput Phenotyping in Plants: Humana Press. pp. 179–268.

31. Shiva S, Vu HS, Roth MR, Zhou Z, Marepally SR, et al. (2013) Lipidomic analysis of plant membrane lipids by direct infusion tandem mass spectrometry. Methods Mol Biol 1009: 79–91.

32. Welti R, Li W, Li M, Sang Y, Biesiada H, et al. (2002) Profiling Membrane Lipids in Plant Stress Responses: role of phospholipase Dα in freezing-induced lipid changes in *Arabisopsis*. J Biol Chem 277: 31994–32002.

33. Botté CY, Yamaryo-Botté Y, Janouškovec J, Rupasinghe T, Keeling PJ, et al. (2011) Identification of Plant-like Galactolipids in *Chromera velia*, a Photosynthetic Relative of Malaria Parasites. J Biol Chem 286: 29893–29903.

34. Gray CG, Lasiter AD, Li C, Leblond JD (2009) Mono- and digalactosyldiacylglycerol composition of dinoflagellates. I. Peridinin-containing taxa. Eur-J Phycol 44: 191–197.

35. Keusgen M, Curtis J, Thibault P, Walter J, Windust A, et al. (1997) Sulfoquinovosyl diacylglycerols from the alga *Heterosigma carterae*. Lipids 32: 1101–1112.

36. MacDougall K, McNichol J, McGinn P, O'Leary SB, Melanson J (2011) Triacylglycerol profiling of microalgae strains for biofuel feedstock by liquid chromatography–high-resolution mass spectrometry. Anal Bioanal Chem 401: 2609–2616.

37. Olmstead ILD, Hill DRA, Dias DA, Jayasinghe NS, Callahan DL, et al. (2013) A quantitative analysis of microalgal lipids for optimization of biodiesel and omega-3 production. Biotechnol Bioeng 110: 2096–2104.

38. Bligh EG, Dyer WJ (1959) A rapid method of total lipid extraction and purification. Can J Biochem Physiol 37: 911–917.

39. Gong Y, Guo X, Wan X, Liang Z, Jiang M (2013) Triacylglycerol accumulation and change in fatty acid content of four marine oleaginous microalgae under nutrient limitation at different culture ages. J Basic Microbiol 53: 29–36.

40. Pal D, Khozin-Goldberg I, Cohen Z, Boussiba S (2011) The effect of light, salinity, and nitrogen availability on lipid production by *Nannochloropsis* sp. Appl Microbiol Biotechnol 90: 1429–1441.

41. Roleda MY, Slocombe SP, Leakey RJG, Day JG, Bell EM, et al. (2013) Effects of temperature and nutrient regimes on biomass and lipid production by six oleaginous microalgae in batch culture employing a two-phase cultivation strategy. Bioresour Technol 129: 439–449.

42. Jouhet J, Maréchal E, Block MA (2007) Glycerolipid transfer for the building of membranes in plant cells. Prog Lipid Res 46: 37–55.

43. Härtel H, Essigmann B, Lokstein H, Hoffmann-Benning S, Peters-Kottig M, et al. (1998) The phospholipid-deficient pho1 mutant of *Arabidopsis thaliana* is affected in the organization, but not in the light acclimation, of the thylakoid membrane. Biochim Biophys Acta 1415: 205–218.

44. Kobayashi K, Narise T, Sonoike K, Hashimoto H, Sato N, et al. (2013) Role of galactolipid biosynthesis in coordinated development of photosynthetic complexes and thylakoid membranes during chloroplast biogenesis in *Arabidopsis*. Plant J 73: 250–261.

45. Li X, Moellering ER, Liu B, Johnny C, Fedewa M, et al. (2012) A Galactoglycerolipid Lipase Is Required for Triacylglycerol Accumulation and Survival Following Nitrogen Deprivation in *Chlamydomonas reinhardtii*. Plant Cell 24: 4670–4686.

46. Arisz SA, van Himbergen JAJ, Musgrave A, van den Ende H, Munnik T (2000) Polar glycerolipids of *Chlamydomonas moewusii*. Phytochemistry 53: 265–270.

47. Sugimoto K, Midorikawa T, Tsuzuki M, Sato N (2008) Upregulation of PG synthesis on sulfur-starvation for PS I in *Chlamydomonas*. Biochem Biophys Res Commun 369: 660–665.

48. Fan J, Andre C, Xu C (2011) A chloroplast pathway for the de novo biosynthesis of triacylglycerol in *Chlamydomonas reinhardtii*. FEBS Lett 585: 1985–1991.

20

Effect of Inlet and Outlet Flow Conditions on Natural Gas Parameters in Supersonic Separation Process

Yan Yang[1], Chuang Wen[1]*, Shuli Wang[1], Yuqing Feng[2]

1 Jiangsu Key Laboratory of Oil-Gas Storage and Transportation Technology, Changzhou University, Changzhou, Jiangsu Province, China, **2** Computational Informatics, Commonwealth Scientific and Industrial Research Organization, Melbourne, The State of Victoria, Australia

Abstract

A supersonic separator has been introduced to remove water vapour from natural gas. The mechanisms of the upstream and downstream influences are not well understood for various flow conditions from the wellhead and the back pipelines. We used a computational model to investigate the effect of the inlet and outlet flow conditions on the supersonic separation process. We found that the shock wave was sensitive to the inlet or back pressure compared to the inlet temperature. The shock position shifted forward with a higher inlet or back pressure. It indicated that an increasing inlet pressure declined the pressure recovery capacity. Furthermore, the shock wave moved out of the diffuser when the ratio of the back pressure to the inlet one was greater than 0.75, in which the state of the low pressure and temperature was destroyed, resulting in the re-evaporation of the condensed liquids. Natural gas would be the subsonic flows in the whole supersonic separator, if the mass flow rate was less than the design value, and it could not reach the low pressure and temperature for the condensation and separation of the water vapor. These results suggested a guidance mechanism for natural gas supersonic separation in various flow conditions.

Editor: Gongnan Xie, Northwestern Polytechnical University, China

Funding: This work was supported in part by the National Natural Science Foundation of China (No. 51176015) and Jiangsu Key Laboratory of Oil–Gas Storage and Transportation Technology (No. SCZ1211200004/001). The funders had no role in study design, data collection and analysis, decision to publish, or preparation of the manuscript.

Competing Interests: The authors have declared that no competing interests exist.

* Email: wenchuang2008@gmail.com

Introduction

As the global economy rises, the demand for energy supply is increasing continuously in the last two decades. Natural gas plays a significant strategic role in the energy supply [1]. Natural gas is gaseous mixture, primarily composed of methane, ethane, propane and butane, with some heavier alkanes, carbon dioxide, hydrogen sulfide, nitrogen and a small amount of water vapor [2]. The presence of water vapor in natural gas increases the risk of the formation of gas hydrates with line plugging due to hydrate deposition on the pipe walls, results in corrosion combined with acid gases including carbon dioxide and hydrogen sulfide, and reduces the delivery capacity of the pipelines because of the collection of free water [3]. Consequently, the water vapor must be removed from natural gas early on.

At present, many conventional techniques are employed for the natural gas separation, such as absorption, adsorption, refrigeration, membranes and so on. A supersonic separator, as a novel technique, has been introduced to natural gas processing from the beginning of this century [4–6]. In essence, the supersonic separation technique causes refrigeration like the Joule-Thompson effect and Turbine expansion, both of which induce a low temperature for the condensation of water vapor. The supersonic separator mainly consists of a Laval nozzle, a swirl device and a diffuser.

Malyshkina [7,8] obtained the distribution of gas dynamic parameters through a supersonic separator with a computational method, and a procedure was developed to predict the separation capability of water vapor and higher hydrocarbons from natural gas by using a supersonic separator determined by the initial parameters. Karimi and Abdi [9] studied the flow fields of natural gas in a Laval nozzle of 0.12 m long. But the working fluid was assumed to be a supercritical flow. The geometric construction and flow conditions are quite different from the actual flow states of natural gas in a supersonic separator for dehydration. Jiang et al. [10] employed the corrected Internally Consistent Classical Theory and Gyarmathy theory to modelling the nucleation and droplet growth of natural gas in the supersonic separation process. A supersonic separator was compared to a Joule-Thomson valve with TEG and the results demonstrated the high economic performance and natural gas liquids recovery of a supersonic separator [11]. The generalized radial basis function artificial neural networks were used to optimize the geometry of a supersonic separator [12]. Rajaee Shooshtari and Shahsavand developed a new theoretical approach based on mass transfer rates to calculate the liquid droplet growth in supersonic conditions for binary mixtures [13]. In our preliminary studies, a central body was incorporated in a supersonic separator with a swirling device composed of vanes and an ellipsoid [14]. The effects of swirls on natural gas flow in supersonic separators were computationally simulated with the Reynolds stress model [15]. The particle separation characteristic in a supersonic separator was calculated using the discrete particle method [16].

The mechanisms of the upstream and downstream influences are not well understood for various flow conditions from the wellhead and the back pipelines. The purpose of this study is to

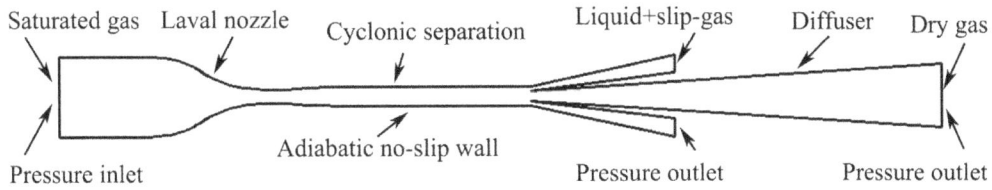

Figure 1. Schematic diagram of a supersonic separator.

investigate the effects of the operating parameters on natural gas supersonic separation process, including the back pressure, inlet mass flow rates, inlet pressures and inlet temperatures. The Redlich-Kwong real gas model is employed to calculate the gas thermal properties in high pressure and low temperatures in our simulation.

Governing equations

Natural gas can be accelerated to supersonic velocities with a Laval nozzle in a supersonic separator and, accordingly, low pressure and temperature conditions are achieved for water vapor condensation. The fluid structure of natural gas flows can be described by the conservation equations of mass, momentum and energy. To close the partial differential equations, the Shear Stress Transport (SST) [17] turbulence model was used in our simulation to solve the supersonic gas flows.

The mass equation of gas phase (continuity equation) is described as:

$$\frac{\partial}{\partial x_i}(\rho u_i) = 0 \tag{1}$$

where ρ and u are the gas density and velocity, respectively.

The conservation of momentum for gas phase can be written as follows:

$$\frac{\partial}{\partial x_j}(\rho u_i u_j + p\delta_{ij} - \tau_{ji}) = 0 \tag{2}$$

where p is the gas pressure; τ_{ij} is the viscous stress; δ_{ij} is the Kronecker delta.

The energy equation for gas phase is expressed as Eq. 3.

$$\frac{\partial}{\partial x_j}(\rho u_j E + u_j p + q_j - u_i \tau_{ij}) = 0 \tag{3}$$

where E is the total energy; q_j is the heat flux; t is the time.

The turbulent kinetic energy and the specific dissipation rate equations in SST model are as follows [17,18]:

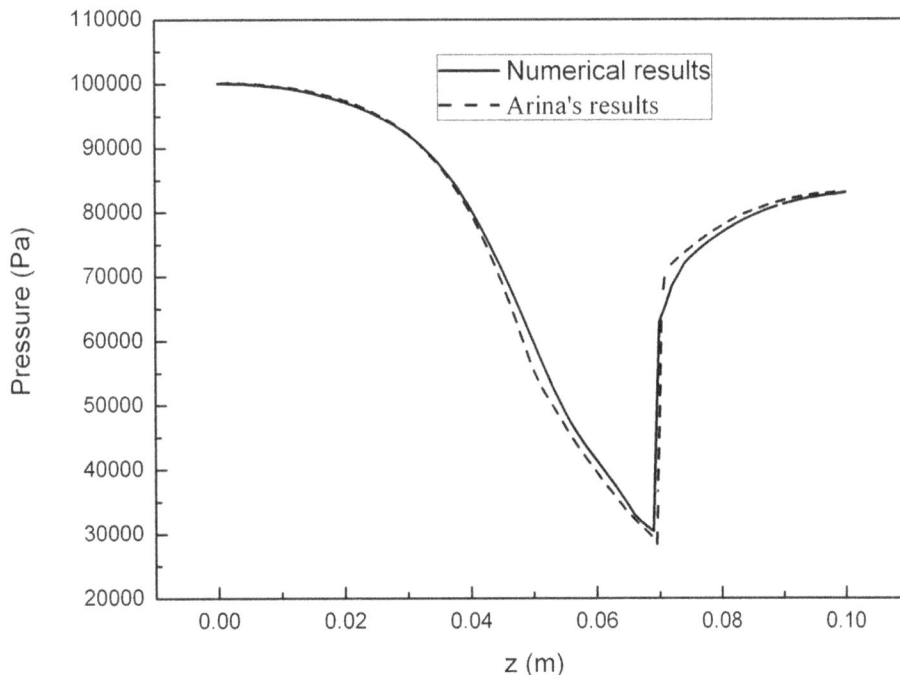

Figure 2. Pressure profile for nozzle flow.

Table 1. Mole composition of natural gas.

Natural gas composition	Mole fraction (%)
CH_4	91.36
C_2H_6	3.63
C_3H_8	1.44
i-C_4H_{10}	0.26
n-C_4H_{10}	0.46
i-C_5H_{12}	0.17
n-C_5H_{12}	0.16
H_2O	0.03
CO_2	0.45
N_2	2.04

$$\frac{\partial}{\partial x_i}(\rho k u_i) = \frac{\partial}{\partial x_j}\left(\Gamma_k \frac{\partial k}{\partial x_j}\right) + \bar{G}_k - Y_k + S_k \tag{4}$$

$$\frac{\partial}{\partial x_j}(\rho \omega u_j) = \frac{\partial}{\partial x_j}\left(\Gamma_\omega \frac{\partial \omega}{\partial x_j}\right) + G_\omega - Y_\omega + D_\omega + S_\omega \tag{5}$$

where k is the turbulent kinetic energy, ω is the specific dissipation rate. Γ_k and Γ_ω represent the effective diffusivity of k and ω, respectively. $\bar{G}_k$ represents the generation of turbulence kinetic energy due to mean velocity gradients. G_ω represents the generation of the specific dissipation rate, ω. Y_k and Y_ω represent the dissipation of k and ω due to turbulence. D_ω represents the cross-diffusion term. S_k and S_ω are user-defined source terms.

An equation of state must be developed to calculate the physical property of fluids in supersonic flows. In this simulation, the Redlich-Kwong real gas equation of state model [19] was employed to predict gas dynamic parameters, described in Eq. (6).

$$p = \frac{RT}{V_m - b} - \frac{a}{\sqrt{T} V_m (V_m + b)} \tag{6}$$

where p is the gas pressure, R is the gas constant, T is temperature, V_m is the molar volume (V/n), a is a constant that corrects for attractive potential of molecules, and b is a constant that corrects for volume.

The constants a and b are different depending on which gas is being analyzed. They can be calculated from the critical point data of the gas:

$$a = \frac{0.4275 R^2 T_c^{2.5}}{p_c} \tag{7}$$

$$b = \frac{0.08664 R T_c}{p_c} \tag{8}$$

where T_c and p_c are the temperature and pressure at the critical point, respectively.

For the multi-component mixtures, such as natural gas, mixing laws are utilized to calculate the parameters a and b. The Van Der Waals mixing rules [20,21] were applied to obtain the parameters for the mixtures from those pure components. The mathematical expressions of this mixing rule can be written,

$$a = \sum_{i=1}^{n} \sum_{j=1}^{n} x_i x_j \sqrt{a_i a_j}(1 - k_{ij}) \tag{9}$$

$$b = \sum_{i=1}^{n} x_i b_i \tag{10}$$

where x is molar fraction; n is the total number of the gas components; k_{ij} is the binary interaction parameter between components i and j.

Mathematical modelling

Computational domain and boundary conditions

A Laval nozzle is a key part of a supersonic separator to generate supersonic flows for the condensation and separation of natural gas. Thus, the nozzle needs to be designed specifically, as shown in Figure 1. The cubic polynomial equation was employed to calculate the converging contour of the nozzle, as shown in Eq. (11), while the Foelsch's analytical calculation method was used to design the diverging part of the nozzle [22]. This design of the converging part will accelerate the gas flow uniformly to achieve the sound speed in the throat area. The critical cross-section area is 0.0002378 m^2. The nozzle entrance and exit areas are 0.007854 m^2 and 0.0004460 m^2, respectively. In addition, a straight tube with the length of 100 mm was connected to the nozzle upstream and diffuser downstream, respectively.

$$\begin{cases} \dfrac{D - D_{cr}}{D_1 - D_{cr}} = 1 - \dfrac{1}{X_m^2}\left(\dfrac{x}{L}\right)^3 & \left(\dfrac{x}{L} \le X_m\right) \\ \dfrac{D - D_{cr}}{D_1 - D_{cr}} = \dfrac{1}{(1 - X_m)^2}\left(1 - \dfrac{x}{L}\right)^3 & \left(\dfrac{x}{L} > X_m\right) \end{cases} \tag{11}$$

where D_1, D_{cr} and L are the inlet diameter, the throat diameter and the convergent length, respectively. $X_m = 0.45$. x is the distance between arbitrary cross section and the inlet, and D is the convergent diameter at arbitrary cross section of x.

A structured grid was generated for the supersonic separator while a finer grid scheme in the boundary layer was employed in Laval nozzle and supersonic channel. The grid independence was

Table 2. Initial conditions for back pressure simulation.

Cases	Inlet pressure (bar)	Inlet temperature (K)	Back pressure (bar)
1	100	300	85
2	100	300	80
3	100	300	75
4	100	300	62

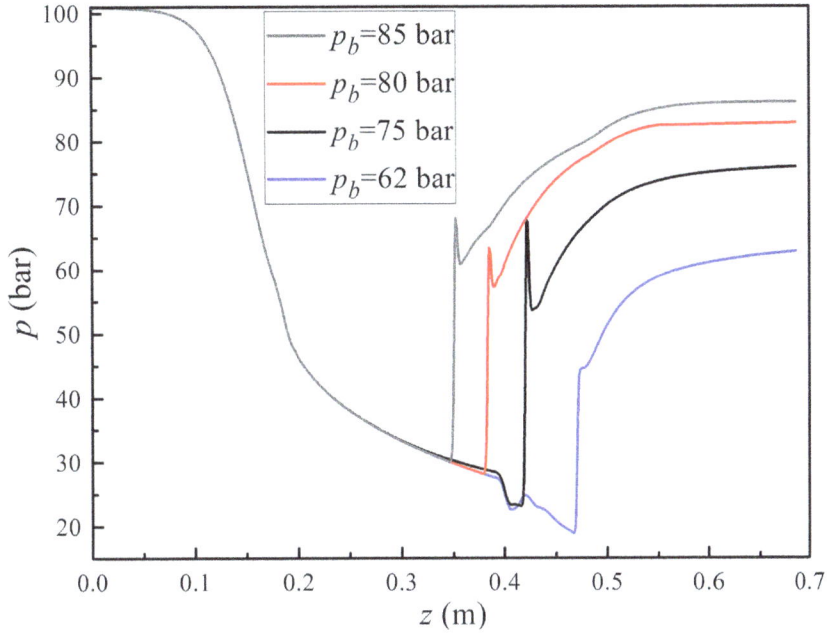

(a) Static pressure

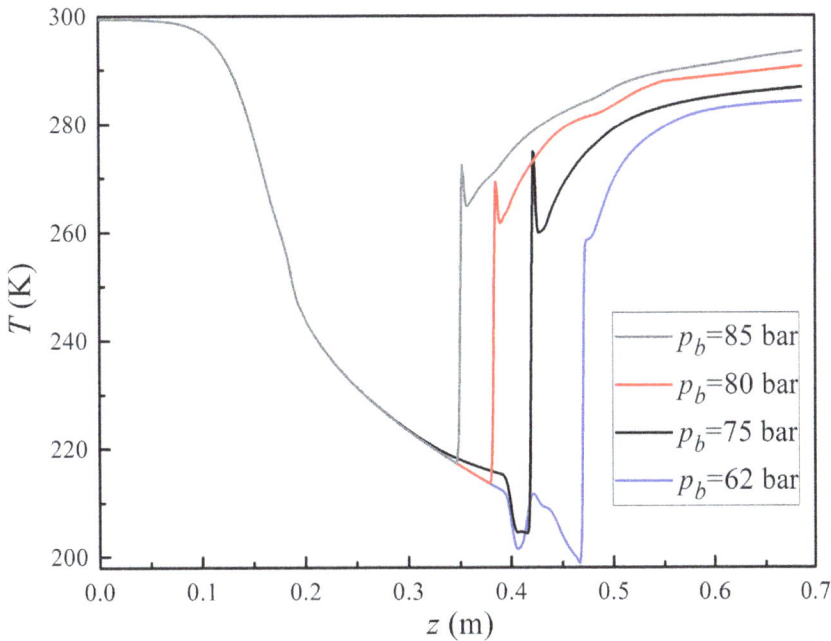

(b) Static temperature

Figure 3. Effect of back pressure on natural gas dynamic parameters.

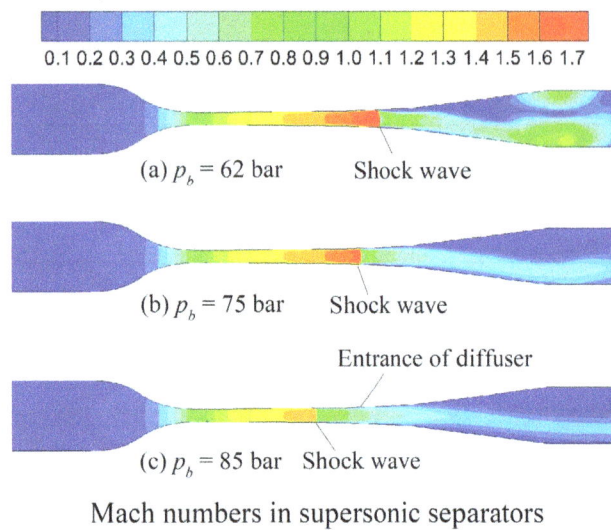

(a) p_b = 62 bar   Shock wave

(b) p_b = 75 bar   Shock wave

Entrance of diffuser

(c) p_b = 85 bar   Shock wave

Mach numbers in supersonic separators

Figure 4. Mach numbers in supersonic separators with various back pressures.

tested before we carried out the simulation. Boundary conditions played a significant role in a numerical simulation. In our case related to a supersonic separator, the pressure boundary conditions were assigned for the inlet and outlet of the supersonic separator, respectively, according to the flow characteristics of the supersonic compressible fluid,. No-slip and adiabatic boundary conditions were specified for the walls. The turbulent kinetic energy and turbulent dissipation rate were employed as the turbulence parameters.

Computational methods

The finite volume methods were used to discretize the partial differential equations of the supersonic gas flows. The pressure based implicit solver was employed to solve the governing equations. The SIMPLE algorithm [23] was applied to couple the velocity field and pressure. The standard pressure scheme was adopted to interpolate the pressure values on the surface of the control volume. The second-order upwind scheme was used for other variables, such as density, momentum, turbulence kinetic energy, turbulence dissipation rate.

Validation

For the validation of our computational methods in supersonic flows, it was validated with Arina's results before we applied it to

our designed supersonic separator [24,25]. Figure 2 depicts the pressure profiles in a Laval nozzle with the numerical results and Arina's work. It could be seen that the same flow behavior was obtained and the shock wave position was accurately captured by our simulation method. Therefore, the numerical results agree with Arina's results well. It was demonstrated that our developed model could be used in the prediction of the supersonic flow for natural gas dehydration.

Results and Discussion

Effect of back pressure

The flow characteristics of natural gas were numerically simulated in the supersonic separation process. The multi-components gas mixture in Baimiao gas well of Zhongyuan Oil Field was selected for the calculation. The composition of natural gas in mole fraction is shown in Table 1.

The incoming flow parameters are fixed when we study the effect of the back pressures on the supersonic separation process. The detailed initial conditions for the back pressure simulation are shown in Table 2. Figure 3 presents the static pressure and static temperature profiles along the flow direction in the conditions of different back pressures. The shock wave position moves into the nozzle from the diffuser with the rise of the back pressure. The shock wave will stay in the diffuser while the back pressure is about less than 75 bar with the inlet pressure of about 100 bar. If the back pressure increases to 80 bar, the shock wave will move into the supersonic channel across the diffuser entrance. The pressure and temperature profiles exhibit several fluctuations close to the shock wave and away from it. This is induced by the interaction between the boundary layer separation and the shock boundary layer.

Figure 4 depicts the contours of gas Mach numbers in the supersonic separators with various back pressures. It clearly shows the obvious differences of the shock wave position with the increasing back pressure. In this simulation case, the shock wave even goes into the nozzle diverging part when the back pressure reaches 85 bar. In this condition, the shock wave will destroy the state of the low pressure and temperature, resulting in the re-evaporation of the condensed liquids to decline the separation efficiency of the supersonic separators.

Effect of inlet mass flow rate

A Laval nozzle is a key part in a supersonic separator, and the critical area at the nozzle throat determines the gas mass flow rate through this device. The detailed initial conditions for inlet mass flow rate simulation are shown in Table 3. Figure 5 describes the gas dynamic parameters with various inlet mass flow rate, namely, including the gas Mach number, the static pressure and static

Table 3. Initial conditions for inlet mass flow rate simulation.

Cases	Inlet mass flow rate (kg/s)	Inlet temperature (K)	Back pressure (bar)
1	1.343	300	85
2	2.687	300	85
3	3.493	300	85
4	3.896	300	85
5	4.000	300	85

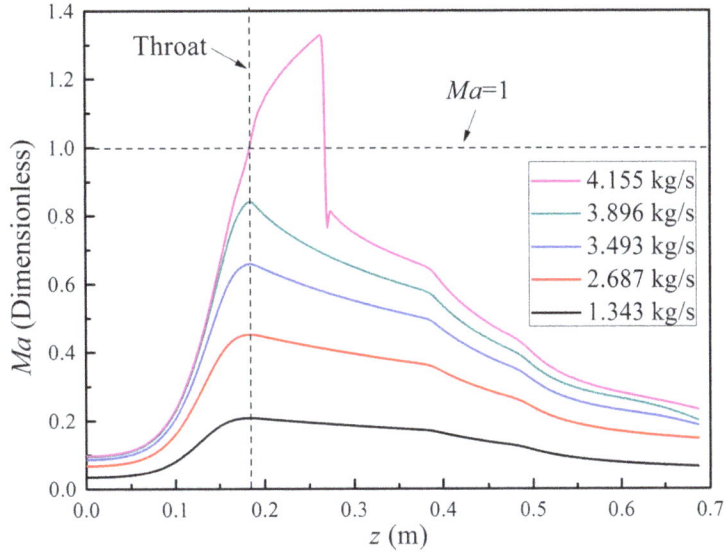

(a) Mach number

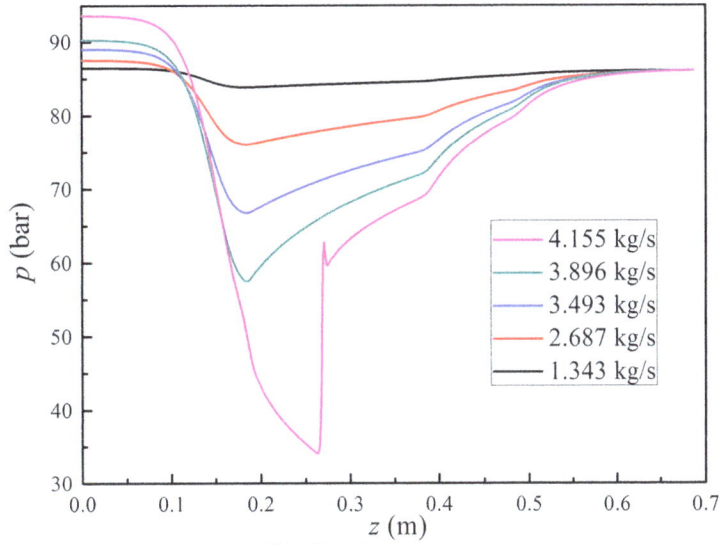

(b) Static pressure

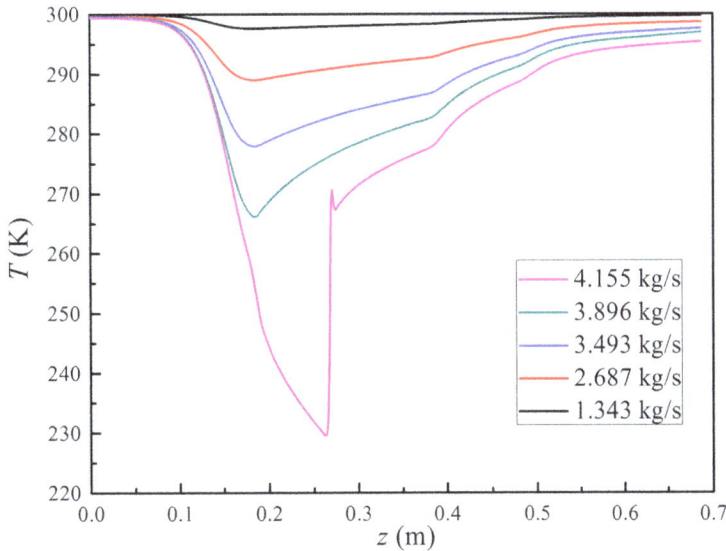

(c) Static temperature

Figure 5. Effect of inlet mass flow rate on natural gas dynamic parameters.

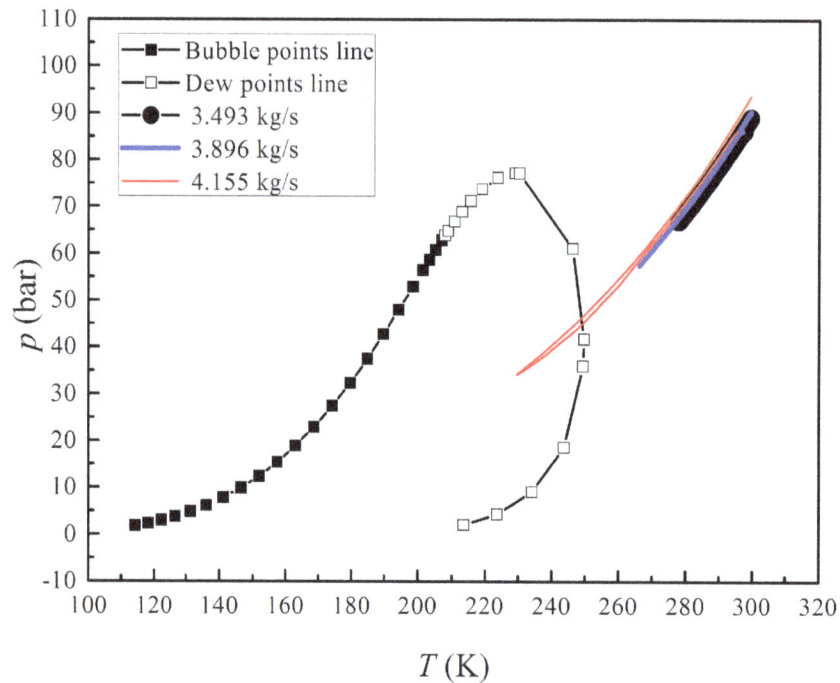

Figure 6. Phase envelope and pressure–temperature relationships with various inlet mass flow rates.

temperature. If the mass flow rate is less than the design value, the gas velocity at the nozzle throat is less than the critical value, although the converging part speeds up the gas flows. Because of the Mach number at the throat is less than unity, the gas velocity declines in the diverging part of the Laval nozzle. In this situation, the maximum velocity is obtained at the nozzle throat. That is, natural gas is the subsonic flows in the whole supersonic separator, which cannot reach the low pressure and temperature for the condensation of the water vapor. The gas Mach number rises with the increase of the inlet gas mass flow rate, resulting in the decline of the static pressure and temperature. When the inlet gas flow rate reaches the design value, the choked flow conditions will be achieved. In our simulation cases, the critical flow condition is obtained when the inlet gas mass flow rate is about 4.155 kg/s. In this condition, the natural gas flow continues to expand in the diverging part of the Laval nozzle, and the maximum Mach number is around 1.33.

Figure 6 depicts the phase envelope curve and the pressure–temperature (P-T) profiles with various inlet mass flow rates. We can see that P-T profile doesn't reach the phase envelope curve because of the high pressure and temperature in the supersonic separator, when the inlet gas mass flow rate is smaller than the design value. Therefore, the water vapor can hardly be removed from natural gas when the inlet mass flow rate is less than the designed rate.

Effect of inlet pressure

The inlet temperature is fixed and the back pressure is set to be the 85% of the inlet one, when we studied the effect of the inlet pressures on the gas dynamic parameters. The detailed initial conditions for inlet pressure simulation are shown in Table 4. The gas mass flow rate in a supersonic separator increases with the rises of the inlet pressure. It indicates that the processing capacity of a supersonic separator can be improved by increasing the inlet pressure in natural gas processing. Figure 7 presents the gas static

Table 4. Initial conditions for inlet pressure simulation.

Cases	Inlet pressure (bar)	Inlet temperature (K)	Back pressure (bar)
1	50	300	42.5
2	100	300	85
3	200	300	170
4	300	300	255

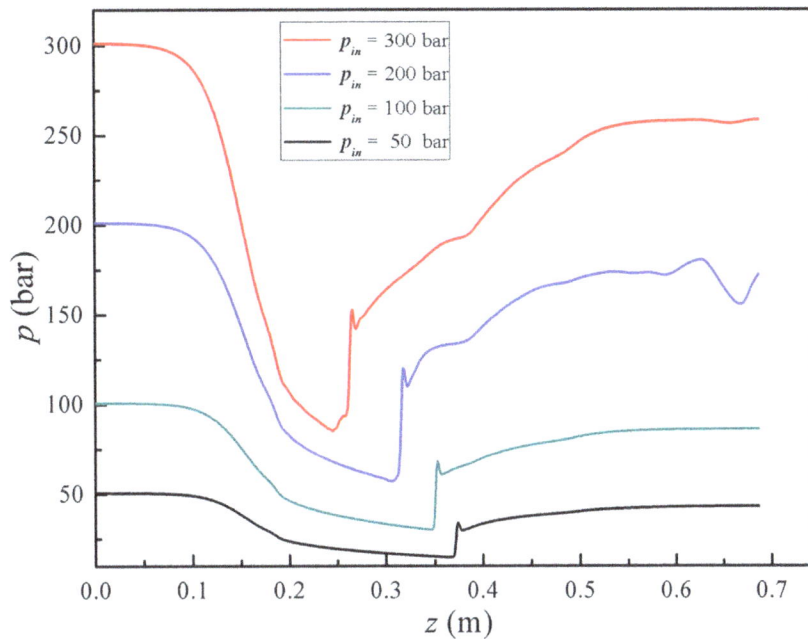

(a) Static pressure

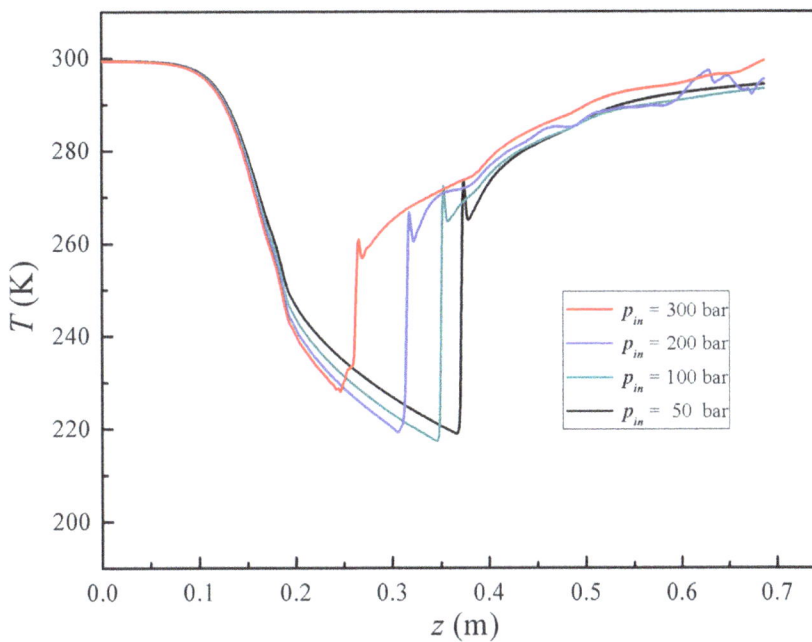

(b) Static temperature

Figure 7. Effect of inlet pressure on natural gas dynamic parameters.

pressure and temperature profiles along the designed supersonic separator. The shock wave position shifts forward to the nozzle with a higher inlet pressure. For example, the shock position stayed at z = 0.370 m at an inlet pressure of about 50 bar. However, the shock location goes to the upstream of the nozzle divergent part, at z = 0.263 m, when the inlet increases to 300 bar. That is, the shock wave position shifts forward by a distance of

about 97 mm. This numerical simulation indicates that the pressure recovery capacity of the supersonic separator will decline in a higher inlet pressure.

Figure 8 depicts the phase envelope curve and the pressure–temperature (P-T) profiles with various inlet pressures. The P-T profile goes into the gas-liquid two phase zone, although the inlet pressure is changed, when the inlet pressure is lower than 100 Bar.

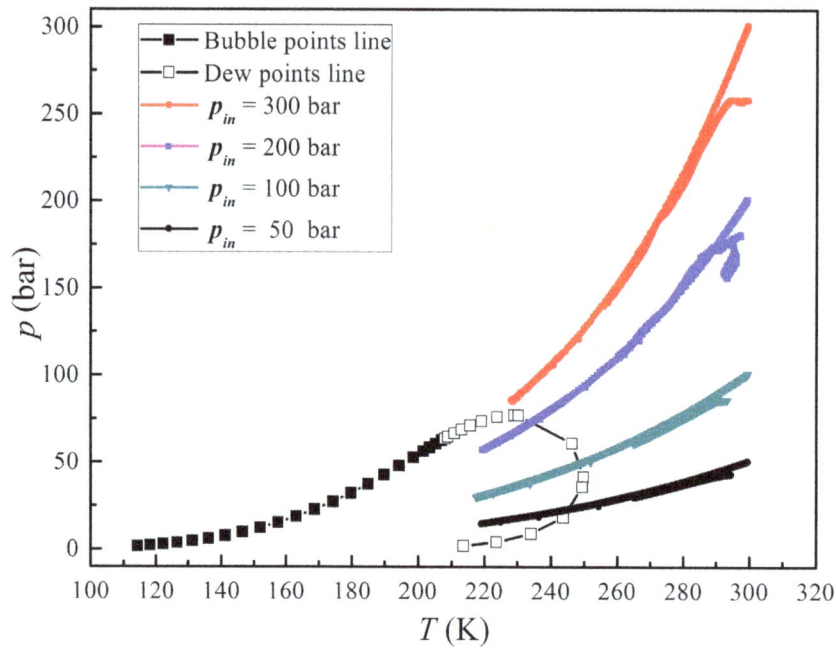

Figure 8. Phase envelope and pressure–temperature relationships with various inlet pressures.

In these conditions the static pressure and temperature is low enough for the condensation of the water vapor in natural gas. But if the inlet pressure exceeds 200 bar, the natural gas flow will present a supercritical fluid in the supersonic separator, which is not suitable for the gas dehydration. Therefore, we suggest that the maximum inlet pressure should be around 100 bar for natural gas dehydration using a supersonic separator.

Effect of inlet temperature

The inlet and back pressure are fixed to study the influence of the inlet temperature. The detailed initial conditions for inlet temperature simulation are shown in Table 5. The gas mass flow rate decreases with the rise of the inlet temperature in the supersonic separator. It indicates that the processing capacity of a supersonic separator can be improved by decreasing the inlet temperature in natural gas processing. It can be seen in Figure 9 that the shock position moves backward from nozzle to diffuser with the increase of the inlet temperature. However, the shock position moves just by a distance of about 5 mm with the increase of the inlet temperature from 10°C to 70°C, which is the normal

temperature in natural gas processing. Hence, we can neglect the effect of the inlet temperature on the shock wave position in the supersonic separator. Figure 10 shows that the P-T profile goes further into the gas-liquid two phase zone with the decline of the inlet temperature. This is because the lower inlet temperature will cause a lower static temperature in the Laval nozzle, when the pressure ratio is fixed in the supersonic separator.

Conclusion

The gas dynamic parameters in a supersonic separator were simulated using the Shear Stress Transport (SST) turbulence model and Redlich–Kwong real gas model. The effect of the inlet and outlet flow conditions on the gas dynamic parameters was analyzed in the supersonic separation process, especially on the shock wave position. The gas flow cannot be choked in the supersonic separator, when the inlet mass flow rate is less than the designed one. It results in a high pressure and temperature inside the device and the water vapor cannot be removed from natural gas. The shock wave position shifts forward to the nozzle with a

Table 5. Initial conditions for inlet temperature simulation.

Cases	Inlet pressure (bar)	Inlet temperature (K)	Back pressure (bar)
1	100	283	85
2	100	303	85
3	100	323	85
4	100	343	85

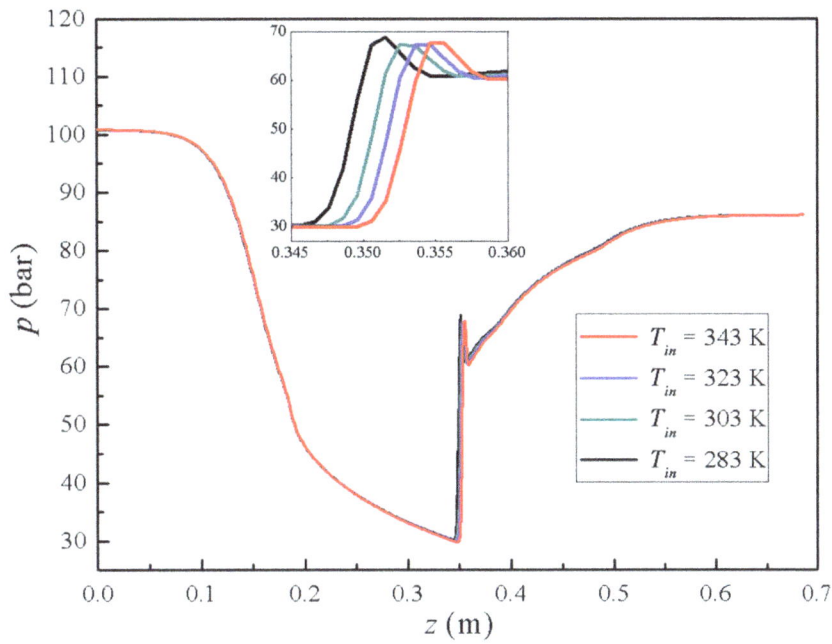

(a) Static pressure

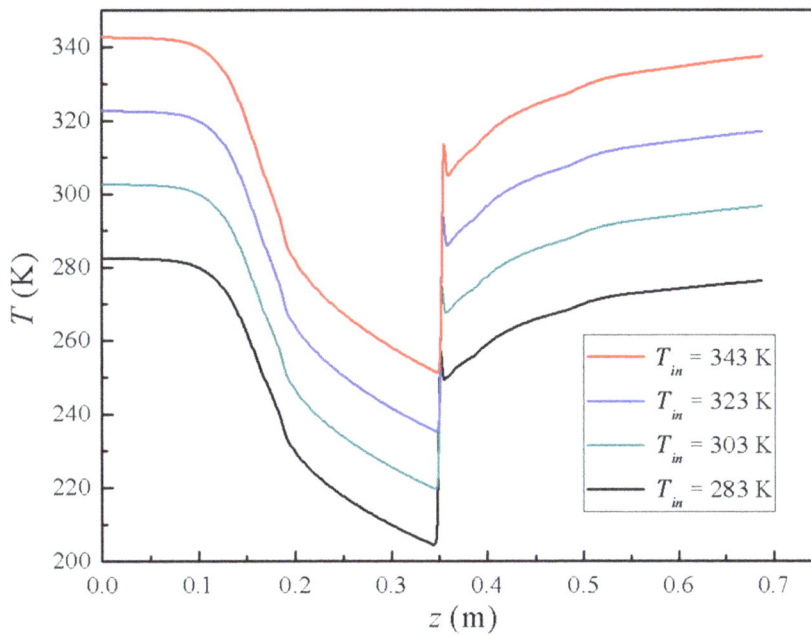

(b) Static temperature

Figure 9. Effect of inlet temperature on natural gas dynamic parameters.

higher inlet pressure. The effect of the inlet temperature on the shock wave position can be neglected when the inlet temperature increases from 10°C to 70°C. The increasing back pressure induces the shock wave position to move forward from the diffuser to Laval nozzle. The shock wave moves into the supersonic channel or Laval nozzle when the back pressure is about more than 75 bar with the inlet pressure of about 100 bar. The shock

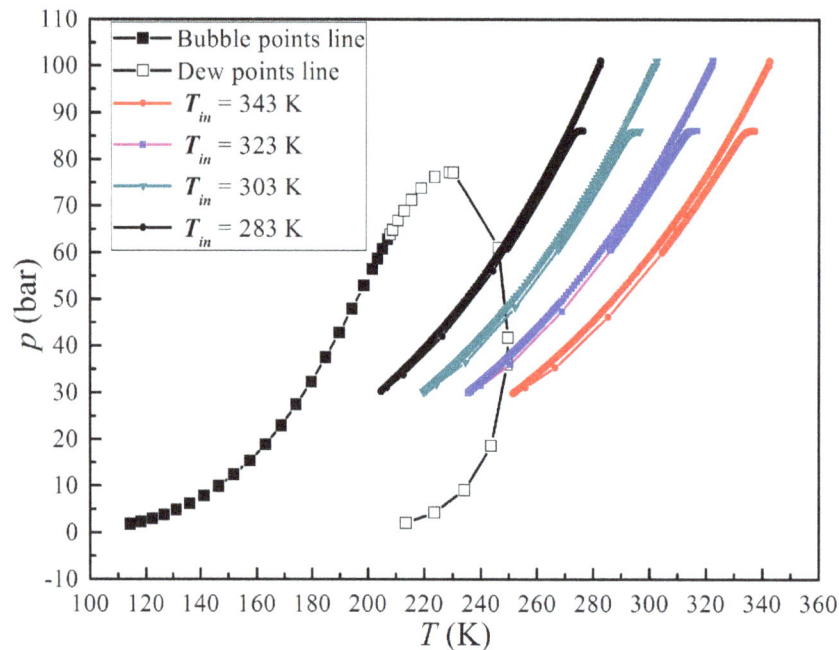

Figure 10. Phase envelope and pressure–temperature relationships with various inlet temperatures.

wave will destroy the state of the low pressure and temperature, resulting in the re-evaporation of the condensed liquids.

Acknowledgments

This work was supported in part by the National Natural Science Foundation of China (No. 51176015) and Jiangsu Key Laboratory of Oil–Gas Storage and Transportation Technology (No. SCZ1211200004/001).

Author Contributions

Conceived and designed the experiments: CW YY. Performed the experiments: CW YY. Analyzed the data: CW YY. Contributed reagents/materials/analysis tools: CW YY. Wrote the paper: CW YY. Revised the manuscript: CW YY SW YF.

References

1. Soldo B (2012) Forecasting natural gas consumption. Appl Energy 92: 26–37.
2. Economides MJ, Wood DA (2009) The state of natural gas. J Nat Gas Sci Eng 1: 1–13.
3. Mokhatab S, Poe WA, Speight JG (2006) Handbook of natural gas transmission and processing, Gulf Professional Publishing, Burlington, MA, USA.
4. Okimoto D, Brouwer J (2002) Supersonic gas conditioning. World Oil 223: 89–91.
5. Alferov VI, Baguiro LA, Dmitriev L, Feygin V, Imaev S, et al. (2005) Supersonic nozzle efficiently separates natural gas components. Oil Gas J 103: 53–58.
6. Wen C, Cao X, Yang Y, Zhang J (2011) Swirling effects on the performance of supersonic separators for natural gas separation. Chem Eng Technol 34(9): 1575–1580.
7. Malyshkina MM (2008) The structure of gas dynamic flow in a supersonic separator of natural gas. High Temperature 46: 69–76.
8. Malyshkina MM (2010) The procedure for investigation of the efficiency of purification of natural gases in a supersonic separator. High Temperature 48: 244–250.
9. Karimi A, Abdi MA (2009) Selective dehydration of high-pressure natural gas using supersonic nozzles. Chem Eng Process 48: 560–568.
10. Jiang D, Eri Q, Wang C, Tang L (2011) A fast and efficient numerical-simulation method for suprsonic gas processing. SPE projects, facilities & construction 6(2): 58–64.
11. Machado PB, Monteiro JGM, Medeiros JL, Epsom HD, Araujo OQF (2012) Supersonic separation in onshore natural gas dew point plant. J Nat Gas Sci Eng 6: 43–49.
12. Mahmoodzadeh Vaziri B, Shahsavand A (2013) Analysis of supersonic separators geometry using generalized radial basis function (GRBF) artificial neural networks. J Nat Gas Sci Eng 13: 30–41.
13. Rajaee Shooshtari SH, Shahsavand A (2013) Reliable prediction of condensation rates for purification of natural gas via supersonic separators. Sep Purif Technol 116: 458–470.
14. Wen C, Cao X, Yang Y (2011) Swirling flow of natural gas in supersonic separators. Chem Eng Process 50(7): 644–649.
15. Wen C, Cao X, Yang Y, Li W (2012) Numerical simulation of natural gas flows in diffusers for supersonic separators. Energy 37: 195–200.
16. Wen C, Cao X, Yang Y, Zhang J (2012) Evaluation of natural gas dehydration in supersonic swirling separators applying the discrete particle method. Adv Powder Technol 23: 228–233.
17. Menter FR (1994) Two-equation eddy-viscosity turbulence models for engineering applications. AIAA Journal 32(8): 1598–1605.
18. ANSYS Fluent User Manual, 2011, ANSYS INC.
19. Kwak TY, Mansoori GA (1986) Van der waals mixing rules for cubic equations of state. applications for supercritical fluid extraction modeling, Chem Eng Sci 41: 1303–1309.
20. Benmekki EH, Kwak TY, Mansoori GA (1987) Supercritical fluids, American Chemical Society, Washington.
21. Redlich O, Kwong JNS (1949) On the thermodynamics of solutions. V. an equation of state. fugacities of gaseous solutions. Chem Rev 44: 233–244.
22. Foelsch K (1949) The analytical design of an axially symmetric Laval nozzle for a parallel and uniform jet. J Aero Sci 16: 161–166.
23. Patankar SV, Spalding DB (1972) A calculation procedure for heat, mass and momentum transfer in three-dimensional parabolic flows. Int J Heat Mass Transfer 1: 1787–1806.
24. Yang Y, Wen C, Wang S, Feng Y (2014) Theoretical and numerical analysis on pressure recovery of supersonic separators for natural gas dehydration. Appl Energy 132: 248–253.
25. Arina R (2004) Numerical simulation of near-critical fluids. Appl Numer Math 51: 409–426.

21

The Functional Potential of Microbial Communities in Hydraulic Fracturing Source Water and Produced Water from Natural Gas Extraction Characterized by Metagenomic Sequencing

Arvind Murali Mohan[1,2]**, Kyle J. Bibby**[1,3,4]**, Daniel Lipus**[1,3]**, Richard W. Hammack**[1]**, Kelvin B. Gregory**[1,2]*

1 National Energy Technology Laboratory, Pittsburgh, Pennsylvania, United States of America, **2** Department of Civil and Environmental Engineering, Carnegie Mellon University, Pittsburgh, Pennsylvania, United States of America, **3** Department of Civil and Environmental Engineering, University of Pittsburgh, Pittsburgh, Pennsylvania, United States of America, **4** Department of Computational and Systems Biology, University of Pittsburgh Medical School, Pittsburgh, Pennsylvania, United States of America

Abstract

Microbial activity in produced water from hydraulic fracturing operations can lead to undesired environmental impacts and increase gas production costs. However, the metabolic profile of these microbial communities is not well understood. Here, for the first time, we present results from a shotgun metagenome of microbial communities in both hydraulic fracturing source water and wastewater produced by hydraulic fracturing. Taxonomic analyses showed an increase in anaerobic/facultative anaerobic classes related to *Clostridia*, *Gammaproteobacteria*, *Bacteroidia* and *Epsilonproteobacteria* in produced water as compared to predominantly aerobic *Alphaproteobacteria* in the fracturing source water. The metabolic profile revealed a relative increase in genes responsible for carbohydrate metabolism, respiration, sporulation and dormancy, iron acquisition and metabolism, stress response and sulfur metabolism in the produced water samples. These results suggest that microbial communities in produced water have an increased genetic ability to handle stress, which has significant implications for produced water management, such as disinfection.

Editor: Robert J. Forster, Agriculture and Agri-Food Canada, Canada

Funding: Work was supported by the National Energy Technology Laboratory's Regional University Alliance (NETL-RUA), a collaborative initiative of the NETL, this technical effort was performed under the RES contract DE-FE0004000 (http://www.netl.doe.gov/). The funders had no role in study design, data collection and analysis, decision to publish, or preparation of the manuscript.

Competing Interests: The authors have declared that no competing interests exist.

* Email: kelvin@cmu.edu

Introduction

High-volume hydraulic fracturing operations for natural gas development from deep shale produce millions of gallons of wastewater over the lifetime of the well [1], [2], [3], commonly termed as 'produced water'. This produced water contains elevated concentrations of salts, metals, hydrocarbons and radioactive elements [3], [4], [5], [6], [7]. Microbial communities in produced water can utilize hydrocarbons as sources of carbon and energy [8] and transform redox labile salts and metals. This can give rise to significant water management challenges [9] and increased production costs [10], [11]. For instance, sulfidogenic and acid producing bacteria can cause corrosion of metal infrastructure, souring of natural gas, and reduced formation permeability [10], [11], [12], [13].

Deleterious microbial activity is commonly controlled with biocides at significant cost to the driller. However, despite biocide use, microbial activity is prevalent in produced water. Previous studies have shown that biocide effectiveness may be limited by high salt concentrations, organic compounds, and long residence times in the subsurface [14], [15], [16]. Other studies have shown that microbial communities in produced water are distinct from those in the injected fracturing fluid, and correlate well with changes in geochemical and environmental conditions [5], [15], [17]. This implies that the common practice of recycling produced water for subsequent hydraulic fracturing may introduce adapted populations into the formation [5].

Over the past decade molecular ecology surveys based on the 16S rRNA gene have increased our knowledge about the taxonomic composition of microbial communities in reservoir environments [5], [15], [17], [18], [19], [20], [21], [22]. However, these studies offer limited insights on the metabolic capabilities of the microbial community, as they rely on taxonomic inference based on 16S rRNA gene similarity to previously isolated microorganisms. As an example of the limitations of using previously isolated microorganisms to infer metabolic capability, the 'core genome' of the well-studied *Escherichia coli* is typically less than 50% of the genes in the genome, and <30% of the *E. coli* pan-genome [23]. On the other hand, shotgun metagenomic surveys enable access to complete genetic information within microbial genomes from uncultured, mixed consortia [24], [25], [26]. These surveys have provided significant insights on the functional potential of microorganisms in diverse environments such as marine samples [25], corals [27], activated sludge [28], permafrost [29], hydrocarbon and sandstone reservoirs [30], [31], and swine gut [32]. Despite the importance of microbial activity in produced water brines from hydraulic fracturing operations, the functional potential of associated microbial communities has not yet been studied. In this study, the metagenome of fracturing source water and produced water at two different time points from a Marcellus Shale natural gas well in Westmoreland County, PA was generated using Illumina MiSeq technology. The microbial ecology from 16S rRNA surveys and chemical composition of these samples has been described in a previous publication [5]. Sequences from each sample were assembled into contiguous sequences (contigs) and analyzed for taxonomic affiliations and functional potential of the microbial communities.

Materials and Methods

Sampling

Samples of hydraulic fracturing source water, and produced water on days 1 and 9 were collected from a horizontally drilled Marcellus Shale natural gas well in Westmoreland County, Pennsylvania, U.S.A in October 2011. The source water used for fracturing was a mix of fresh reservoir water (~80%) and produced water (~20%) from previous fracturing operations. Fracturing additives amended to the source water included proppant (silica sand), scale inhibitor (ammonium chloride), biocide (mixture of tributyl tetradecyl phosphonium chloride, methanol and proprietary chemicals), hydrochloric acid, gel (paraffinic solvent), breaker (sodium persulfate) and friction reducer (hydrotreated petroleum distillate). Details regarding the sampling procedure and chemical additives used in the fracturing process are described elsewhere [5]. The aqueous geochemical characteristics of these samples were described previously [5] (Table S1).

DNA extraction, library preparation and Illumina sequencing

Unfiltered water samples were centrifuged at 6,000 g for 30 min in an Avanti J-E centrifuge (Beckman Coulter, Brea, CA) to pellet cells. DNA was extracted from 0.25 g of cell pellet using MO BIO power soil DNA isolation kit (MO BIO, Carlsbad, CA) according to the manufacturer's instructions. DNA was prepared using Nextera XT DNA sample preparation kit (Illumina, San Diego, CA) according to manifacturer's instructions at Genewiz (South Plainfield, NJ). DNA for sequencing was quantified using qPCR prior to clustering, and sequenced using the Illumina MiSeq (Illumina, San Diego, CA) with a 2×250 PE configuration at Genewiz, NJ. Sequencing demultiplexing was performed on the Illumina MiSeq instrument using sample-specific barcodes.

Bioinformatic analyses

The raw unpaired sequences were checked for sequencing tags and adapters using the predict function implemented within the TagCleaner program [33]. No sequencing tags or adapters were identified. Sequences were then subjected to quality control using the FastX toolkit within the Galaxy platform [34] with a minimum length 100 and minimum quality score 20. The velvet assembler [35] was used to assemble sequences that passed quality control into contiguous sequences. The assembly parameters were empirically optimized for the dataset prior to assembly (Table S2); the dataset was processed using a kmer length of 77. Generated contigs >500 bp in length were uploaded to the MG-RAST server [36] with associated metadata files for taxonomic affiliations and functional annotations. Sequence similarity searches in MG-RAST was performed using the BLAT tool [37]. The metagenomes from fracturing source water, day 1 produced water, and day 9 produced water are available in the MG-RAST server [36] under accession nos. 4525703.3, 4525704.3 and 4525705.3, respectively. Taxonmic assignments of selected funcional categories from MG-RAST were excecuted in MGTAXA [38], [39], on the Galaxy bioinformatics workbench [40], [34], using default parameters and taxonomy as defined by the NCBI taxonomic tree. Data is for contig abundance and does not reflect read mapping.

As an additional assembly-independent analysis, sequence data was mapped against reference genomes downloaded from NCBI (Table S3) with CLC Genomics Workbench (Version 6.5.1, CLC Bio, Aarhus, Denmark) [41] using default parameters and no masking. Reference genomes were selected based upon taxonomic observations in MG-RAST annotation and a previous microbial ecology investigation [5]. Prior to mapping, sequencing data was trimmed to a minimum length of 100 bp and minimum quality score of 20. Furthremore, sequences for the sulfite reductase subunits A and B (dsrA/dsrB) (Table S4a) and the suflur metabolism gene adenylyl sulfate reductase subunit A (apsA) (Table S4b) were downloaded from NCBI and mapped against the trimmed sequencing data using CLC Genomics Workbench (Version 6.5.1, CLC Bio, Arhus, Denmark).

Results and Discussion

A total of 10 002, 17 055 and 16 661 contigs from the fracturing source water, produced water day 1 and day 9 samples, respectively, were uploaded to MG-RAST for downstream analyses. All uploaded contigs passed MG-RAST quality control and de-replication filters. The metagenomics sequence statistics are summarized in Table 1.

Taxonomic composition

Taxonomic affiliations were assigned to contigs with predicted proteins and rRNA genes based on comparison with the M5NR database. Alpha diversity (predicted phylotypes) for the fracturing source water, produced water day 1 and day 9 samples were 90, 79 and 88, respectively (Figure S1). Rarefaction curves for each of the samples were asymptotic suggesting that the majority of taxonomic diversity was recovered from the samples (Figure S1). Alpha diversity values and rarefaction curves were obtained using the MG-RAST tool.

Bacteria constituted the dominant domain (97–99% of the total community) in all samples. However, a shift in bacterial community composition was detected between the samples at the class and order levels (Figure 1, 2). Contigs affiliated to the class *Alphaproteobacteria* constituted the majority of the community in the fracturing source water (81%) and produced water day 1 (67%) samples (Figure 1). Within *Alphaproteobacteria*, the dominant order detected was *Rhodobacterales* (68–88% of the *Alphaproteobacteria*; 55–59% of the total community) in both the source water and produced water day 1 samples (Figure 2). The relative abundance of *Alphaproteobacteria* decreased to <2% of the community in the produced water day 9 sample. Previous

Table 1. Metagenomic sequence statistics of fracturing source water (SW), produced water day 1 (PW day 1) and produced water day 9 (PW day 9).

	SW	Pw day 1	PW day 9
Total base pair (bp) count	7,939,565 bp	18,254,354 bp	15,253,129 bp
No. of Contigs	10,002	17,055	16,661
Mean length of Contigs	793±809 bp	1,070±1,195 bp	915±651 bp
% GC content in Contigs	59±8%	55±13%	43±9%
% Contigs containing predicted proteins with known functions	83%	93.1%	80.8%
% Contigs containing predicted proteins with unknown functions	16.6%	6.6%	18.9%
% Contigs containing rRNA genes	0.4%	0.3%	0.3%
Identified protein features	9,919	20,687	16,982
Identified functional categories	8,041	16,948	13, 570

qPCR analysis of these samples suggests that that the total bacterial population remained constant at 10^6–10^7 copies of 16S RNA gene/ml [5].

An increase in the number of contigs associated with the class *Clostridia* was observed in the produced water day 1 sample (17%) as compared to the fracturing source water (1%). However, the relative abundance of *Clostridia* decreased to 3% in the produced water day 9 sample. The majority of the *Clostridia* in the produced water day 1 sample were affiliated to the order *Thermoanaerobacterales* (94% of *Clostridia*; 16% of the total community) (Figure 2). *Gammaproteobacteria* sequences constituted a minor fraction (6%) of the total community in the fracturing source water and produced water day 1 samples but increased in relative abundance to constitute the dominant class (52%) in the produced water day 9 sample. Within the *Gammaproteobacteria* of the produced water day 9 sample, dominant orders included *Vibrionales* (67% of *Gammaproteobacteria*) and *Alteromonadales* (23% of *Gammaproteobacteria*) (Figure 2). The day 9 samples also

showed an increase in relative abundance of *Epsilonproteobacteria* (16%) and *Bacteroidia* (10%) classes as compared to the other samples (<2% of the total community). The major bacterial phyla, classes and orders identified in this study were consistent with previous 16S rRNA gene based clone library and pyrosequencing surveys of these samples (Figure S2) [5]. These results indicate a shift towards facultative anaerobic/anaerobic and halophilic communities in the produced water samples as compared to a predominantly aerobic community in the fracturing source water. At the class level, in each of the samples less that 3% of the total sequences did not affiliate to any taxonomic group.

A minor fraction of the total community was represented by contigs affiliated to *Archaea* (0.1–0.4%), *Viruses* (0.3–1%) and *Eukaryota* (0.4–1.4%) domains. These domains were not analyzed for in the previous 16S rRNA gene survey of these samples [5], and were not considered in more detailed functional classification of the metagenomes.

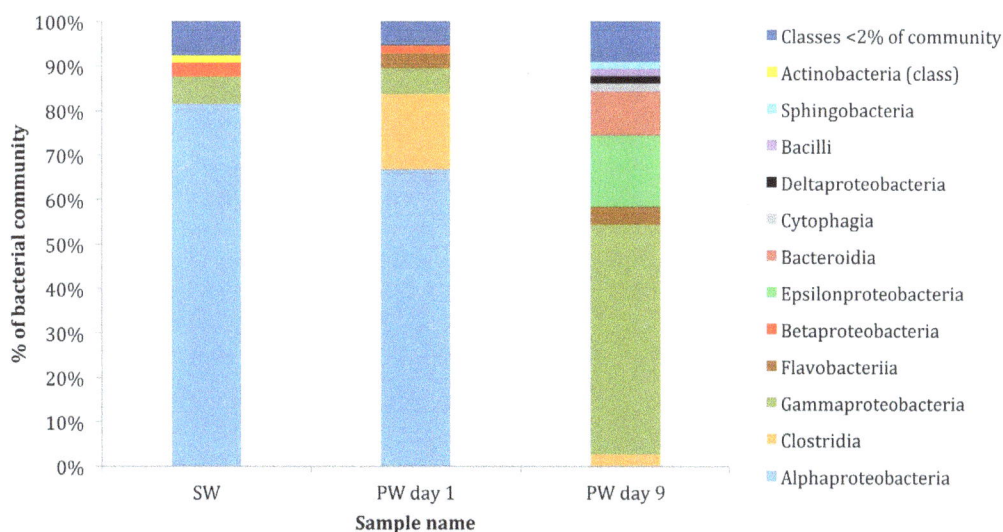

Figure 1. Class level affiliations assigned to contigs with predicted proteins and rRNA genes in source water (SW), produced water day 1 (PW day 1) and produced water day 9 (PW day 9). Total community includes *Bacteria, Archaea, Viruses* and *Eukaryota*.

Order (Class)	SW	PW day 1	PW day 9
Rhodobacterales (Alphaproteobacteria)	>40-60%	>40-60%	
Caulobacterales (Alphaproteobacteria)	>10-20%		
Rhizobiales (Alphaproteobacteria)	2-5%		
Sphingomonadales (Alphaproteobacteria)			
Burkholderiales (Betaproteobacteria)			
Bacteroidales (Bacteroidia)			
Thermoanaerobacterales (Clostridia)		>10-20%	
Clostridiales (Clostridia)			
Flavobacteriales (Flavobacteria)			
Alteromonadales (Gammaproteobacteria)			>10-20%
Vibrionales (Gammaproteobacteria)			>30-40%
Enterobacteriales (Gammaproteobacteria)			
Cytophagales (Cytophagia)			
Campylobacterales (Epsilonproteobacteria)			>5-10%

Color code

2-5%	
>5-10%	
>10-20%	
>20-30%	
>30-40%	
>40-60%	

Figure 2. Order level affiliations assigned to contigs with predicted proteins and rRNA genes in source water (SW), produced water day 1 (PW day 1) and produced water day 9 (PW day 9). Total community includes *Bacteria, Archaea, Viruses* and *Eukaryota*. Only orders representing >2% of the total community are shown in the figure.

Mapping results

Metagenomic reads were mapped against a diverse set of reference genomes to confirm MG-RAST taxonomic results and only reference genomes with good mapping results are discussed in this section. Reference genome mapping results confirmed taxonomic MG-RAST contig analysis. The best mapping results

Coverage of Reference Genomes	SW	PW day 1	PW day 9
Dinoroseobacter shibae	0.36	0.39	0.20
Thermoanaerobacter sp.	0.01	0.86	0.05
Thermoanaerobacter pseudethanolicus	0.02	0.87	0.05
Ruegeria pomeroyi	0.40	0.43	0.23
Thermoanaerobacter tengcongensis	0.01	0.45	0.03
Roseobacter denitrificans	0.30	0.33	0.15
Flavobacterium psychrophilum	0.18	0.27	0.05
Arcobacter butzleri	0.31	0.52	0.47
Arcobacter nitrofigilis	0.23	0.50	0.39
Marinobacter hydrocarbonoclasticus	0.14	0.19	0.83
Bacteroides fragilis	0.02	0.06	0.13
Sulfospirllium deleyianum	0.05	0.07	0.08
Sulfurimonas denitrificans	0.15	0.13	0.12
Parabacteroides distasonis	0.02	0.07	0.10
Phenylobacterium zucineum	0.33	0.27	0.08
Jannaschia sp.	0.25	0.27	0.12
Rhodobacter sphaeroides	0.38	0.42	0.21
Hyphomonas neptunium	0.17	0.22	0.06
Flavobacterium johnsoniae	0.09	0.14	0.03
Rhodobacter capsulatus	0.34	0.37	0.18
Marinobacter adhaerens	0.14	0.17	0.68
Clostridium difficile	0.02	0.06	0.06
Roseovarius sp.	0.79	0.79	0.51
Roseovarius nubinhibens	0.47	0.50	0.28
Vibrio campbelli	0.06	0.25	0.51
Thermoanaerobacter mathranii	0.01	0.88	0.05

Color code

>0-0.1	
>0.1-0.2	
>0.2-0.3	
>0.3-0.4	
>0.4-0.5	
>0.5-0.6	
>0.6-0.7	
>0.7-0.8	
>0.8-0.9	

Figure 3. Fraction of genome coverage for source water (SW), produced water day 1 (PW day 1) and produced water day 9 (PW day 9) samples. Reads were mapped against reference genomes using CLC Genomic workbench version 6.5.1 using default parameters. Shown are fractions of reads mapped against each reference genome included in the analysis for all three samples.

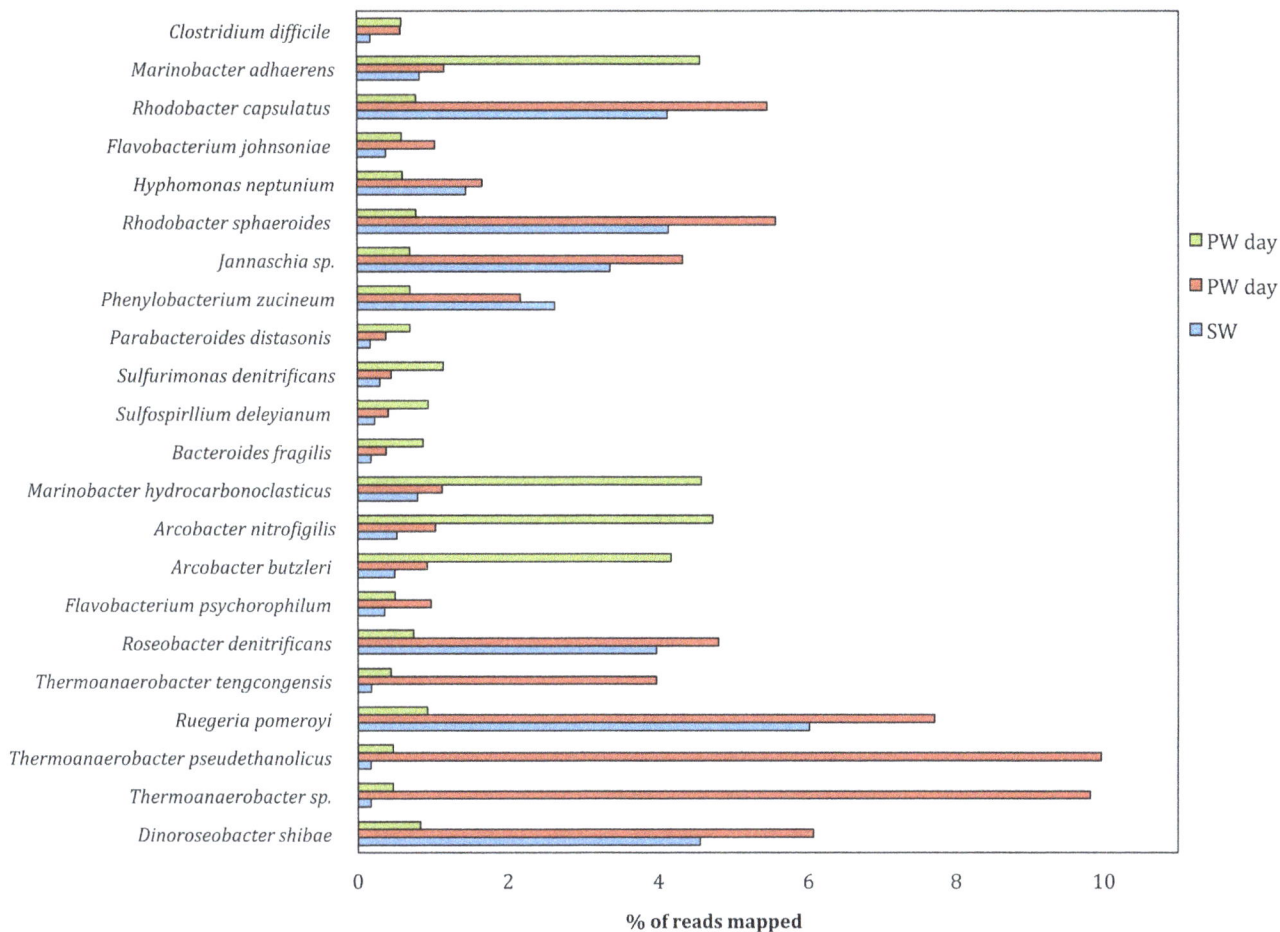

Figure 4. Read distribution for source water (SW), produced water day 1 (PW day 1) and produced water day 9 (PW day 9) samples. Reads were mapped against reference genomes using CLC Genomic workbench version 6.5.1 using default parameters. Shown are percentages of reads mapped against each reference genome included in the analysis for all three samples.

for source water were obtained when sequences were mapped against reference genomes of *Alphaproteabacteria*, specifically of the order *Rhodobacterales* (Figures 3, 4). Similarly, produced water day 1 sample mapping results suggest that it was dominated by bacteria of the orders *Rhodobacterales* and *Thermoanaerobacterales* (Figures 3, 4). A distinct shift in bacterial community was observed between produced water day 1 samples and produced water day 9 samples based on mapping results. Best mapping results for produced water day 9 samples were obtained for reference genomes in the order *Campylobacterales* and *Alteromondales* further supporting the MG-RAST results (Figures 3, 4). Produced water samples demonstrated a distinctive signature with reads mapping best to few select reference genomes, while source water sample reads were distributed more evenly throughout all included reference genomes. For four reference genomes (*Thermoanaerobacter sp.* X514, *Thermoanaerobacter pseudethanolicus*, *Thermoanaerobacter mathranii* in produced water day 1 samples and *Marinobacter hydrocarbonoclasticus* DSM 7299 in produced water day 9 sample) more than 80% coverage was achieved suggesting that these species could play important roles in the microbial community of the representative sample (Figure 3). Highest observed reference genome coverage for source water sample sequences were 79% for *Roseovarius* sp. 217, 40% for *Ruegeria pomeroyi* and 38% for *Rhodobacter sphaeroides* (Fig-

ure 3). For produced water day 1 samples, about 10% of all trimmed sequencing reads mapped against the three *Thermanaerobacter* genomes included in the analysis and 8–13% of reads mapped successfully against *Roseovarius sp.* 217 and *Roseovarius nubinhibens* genomes (Figure 4). 7.7% of produced water day 1 reads mapped against the *Ruegeria pomeroyi* genome (Figure 4). 4–6% of reads for produced water day 9 samples mapped against two different *Marinobacter* and *Arcobacter* reference genomes and one *Vibrio* reference genome (Figure 4). Almost 16% of all reads from source water samples mapped against *Roseovarius sp.* 217 and approximately 4–6% of reads for source water sample mapped against each *Dinoroseobacter shibae*, *Ruegeria pomeroyi*, *Rhodobacter sphaeroides* and *Rhodobacter capsulatus* genomes (Figure 4). All mapping results are summarized in Table S3. The high number of reads form source water and produced water day 1 samples mapping against *Roseovarius* species is in agreement with previous 16S rRNA gene sequencing [5], implying the *Roseovarius* species might be of importance in these waters. *Roseovarius sp.* was previously identified in natural gas brines from the Marcellus shale and its potential implications are discussed elsewhere [9].

The goal of this analysis was to provide an independent confirmation of MG-RAST results. Mapping results depend on the reference genomes selected and these reference genomes might

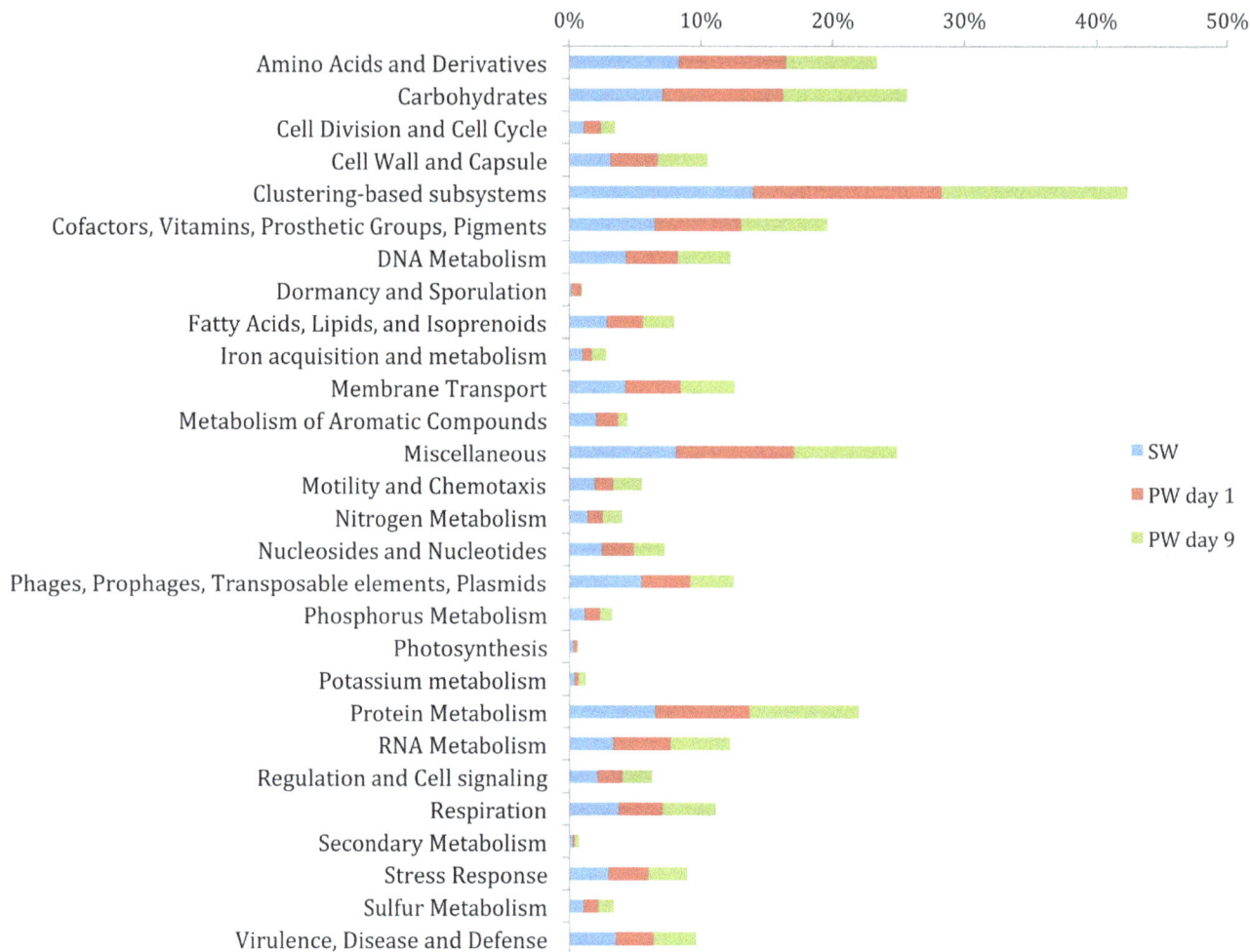

Figure 5. Actual abundance of contigs belonging to Level 1 functional categories in source water (SW), produced water day 1 (PW day 1) and produced water day 9 (PW day 9). Functional annotations were assigned based on the Subsystems database.

not be the same isolates found in the environment. While reference genomes for uncultured microorganisms from oil/gas environments are limited, the positive results achieved by this mapping analysis confirm the initial taxonomic assessment.

Sulfur metabolism gene mapping results

Very few reads in all three samples were successfully mapped against the sulfur metabolism genes dsrA and dsrB. 7 reads of produced water day 1 sample and 55 reads of produced water day 9 sample were successfully mapped against the dsrA/dsrB gene of *Desulfovibrio desulfuricans* with a coverage of 28% and 78% respectively (Table S4a). In addition 10 reads of produced water day 9 sample were successfully mapped against the dsrA/dsrB gene of *Desulfotignum balticum* with a coverage of 19% (Table S4a). For aspA genes, the produced water day 9 sample showed best results with 16, 11, 9 and 6 reads successfully mapped against aspA genes of *Desulfovirbio alaskensis*, *Desulfococcus mulitvorans*, *Desulfotignum balcticum* and *Desulfobacterium autotorphicum* with a coverage of 94%, 46%, 33% and 31% respectively (Table S4b). Very few source water and produced water day 1 reads were mapped successfully against the aspA genes included in the analysis (Table S4b). These results suggest that sulfur metabolism could play a more important role in produced water day 9 sample due the higher abundance of genes associated with sulfur

metabolism. Organisms that can metabolize sulfur compounds to sulfide are of interest in oil and gas environments because of their potential role in infrastructure corrosion, gas souring, worker safety as well as environmental health concerns.

Functional classification of metagenomes

The SEED subsystems database [42], was used to predict the metabolic potential of fracturing source water and produced water samples. Level 1 indicates the broadest set of functional categories to which sequences are assigned, and Level 2 refers to more specific functional assignments within Level 1 categories. The abundance of contigs designated to Level 1 functional categories is illustrated in Figure 5. The metabolic potential (based on Level 1 and Level 2 functional categories) between the samples was compared in a normalized manner (Figure 6, 7) to account for differences in community structure, size of the library, gene content between samples and to effectively compare low abundance functional categories [43]. Read normalization was performed within the MG-RAST analysis pipeline, in accordance with standards for metagenomic analysis.

The five most abundant Level 1 functional categories in all three samples were found to be clustering-based subsystems (e.g. genes where functional coupling is evident but function is unknown;~14%), carbohydrate metabolism (7–9%), amino acids

Level 1 functional categories	SW	PW day 1	PW day 9
Amino Acids and Derivatives			
Carbohydrates			
Cell Division and Cell Cycle			
Cell Wall and Capsule			
Clustering-based subsystems			
Cofactors, Vitamins, Prosthetic Groups, Pigments			
DNA Metabolism			
Dormancy and Sporulation			
Fatty Acids, Lipids, and Isoprenoids			
Iron acquisition and metabolism			
Membrane Transport			
Metabolism of Aromatic Compounds			
Miscellaneous			
Motility and Chemotaxis			
Nitrogen Metabolism			
Nucleosides and Nucleotides			
Phages, Prophages, Transposable elements, Plasmids			
Phosphorus Metabolism			
Photosynthesis			
Potassium metabolism			
Protein Metabolism			
RNA Metabolism			
Regulation and Cell signaling			
Respiration			
Secondary Metabolism			
Stress Response			
Sulfur Metabolism			
Virulence, Disease and Defense			

Color code:

>0-0.1	
>0.1-0.2	
>0.2-0.3	
>0.3-0.4	
>0.4-0.5	
>0.5-0.6	
>0.6-0.7	
>0.7-0.8	
>0.8-0.9	
>0.9-1	

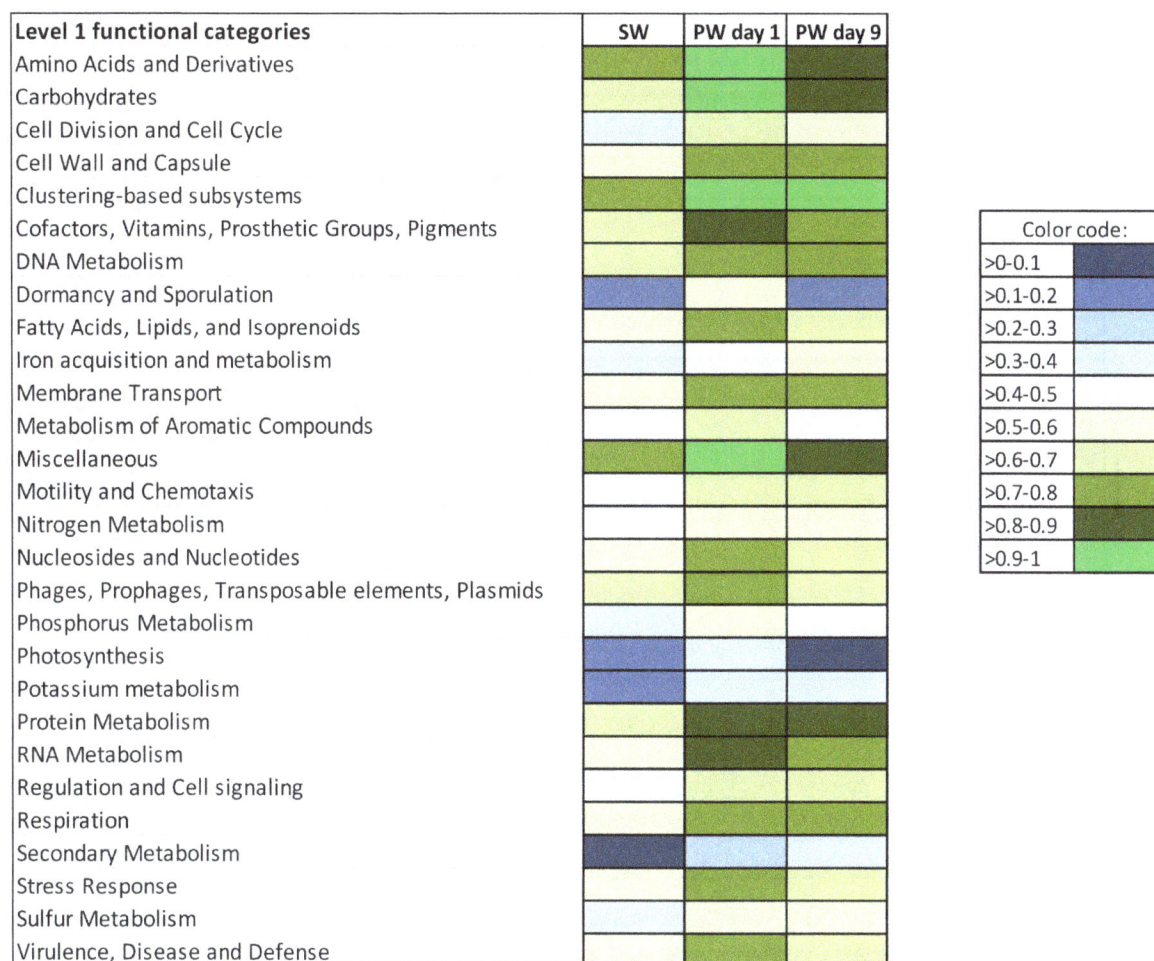

Figure 6. Normalized abundance (values of 0–1) of contigs belonging to Level 1 functional categories in source water (SW), produced water day 1 (PW day 1) and produced water day 9 (PW day 9). Functional annotations were assigned based on the Subsystems database.

and derivatives (7–8%), miscellaneous (eg: genes associated with iron sulfur cluster assembly and Niacine-Choline transport and metabolism; 8–9%), protein metabolism (6–8%), suggesting the dominant role of these functional categories in all samples (Figure 5). These functional categories were similarly identified as dominant in previous studies of soil [44], [45], marine samples [24],[46], activated sludge [24], freshwater [24] and hypersaline environments [24]. Normalization of gene abundance data shows a relative increase in each of the above functional categories in the produced water samples as compared to the fracturing source water (Figure 6) implying that core systems necessary for survival are enriched in the produced water community.

While comparison of gene abundance affiliated with the dominant broad Level 1 categories suggests similar functional profiles across samples, analysis of more specific Level 2 functional categories shows sample specific differences in metabolic capabilities (Figure 7). Differences in metabolic potential indicate a selective pressure exerted in the subsurface for microbes with particular metabolic capabilities. For instance, within the Level 1 carbohydrate metabolism category, sequences related to Level 2 functional categories such as mono-, di-, oligo- and polysaccharides, and aminosugar metabolism were present in higher relative abundance in the produced water samples (Figure 7). This finding correlates well with the expected higher content of carbohydrates

in produced water samples [5]. Carbohydrates and polysaccharide compounds added during hydraulic fracturing can serve as carbon and energy sources for microbial activity [8]. Within the Level 1 protein metabolism category, sequences affiliated with the Level 2 selenoprotein category were detected only in the produced water samples (Figure 7). One possible explanation is the role of selenoproteins in combating oxidative stress [47], which may arise from elevated concentrations of organic or inorganic dissolved constituents in produced water [48]. Results showed that *Rhodobacterales* were the dominant population involved in oxidative stress response in source water and produced water day 1 samples (Figure 8). However, *Alteromonadales* and *Vibrionales* were the dominant orders involved in oxidative stress response in produced water day 9 sample (Figure 8). Within the Level 1 clustering subsystem, genes affiliated with the Level 2 carbohydrate metabolism show a relative increase in the produced water samples as compared to fracturing source water (Figure 7). An increase in the relative abundance of genes related to carbohydrate metabolism in produced water compared to fracturing source water suggests the potential for utilization of hydrocarbons added either as fracturing fluid amendments or those derived from the shale formation and an overall shift to a more heterotrophic microbial community.

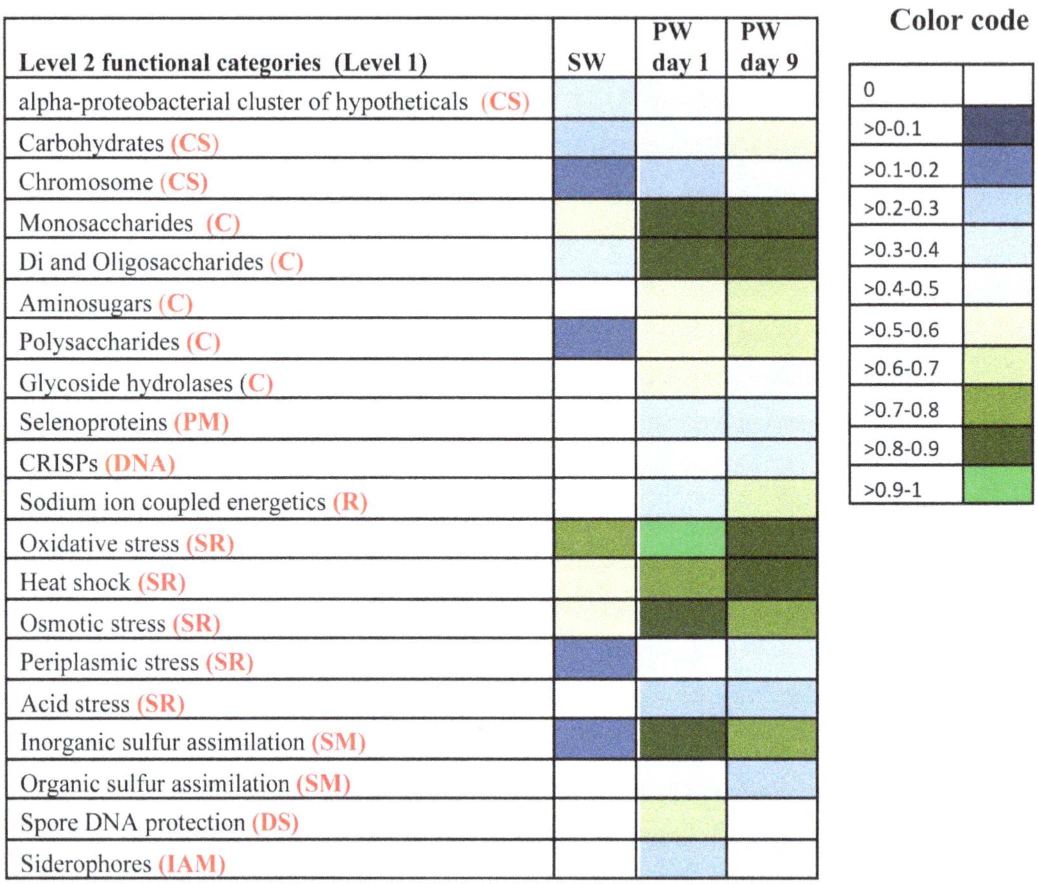

Level 2 functional categories (Level 1)	SW	PW day 1	PW day 9
alpha-proteobacterial cluster of hypotheticals (CS)			
Carbohydrates (CS)			
Chromosome (CS)			
Monosaccharides (C)			
Di and Oligosaccharides (C)			
Aminosugars (C)			
Polysaccharides (C)			
Glycoside hydrolases (C)			
Selenoproteins (PM)			
CRISPs (DNA)			
Sodium ion coupled energetics (R)			
Oxidative stress (SR)			
Heat shock (SR)			
Osmotic stress (SR)			
Periplasmic stress (SR)			
Acid stress (SR)			
Inorganic sulfur assimilation (SM)			
Organic sulfur assimilation (SM)			
Spore DNA protection (DS)			
Siderophores (IAM)			

Color code

0	
>0-0.1	
>0.1-0.2	
>0.2-0.3	
>0.3-0.4	
>0.4-0.5	
>0.5-0.6	
>0.6-0.7	
>0.7-0.8	
>0.8-0.9	
>0.9-1	

Figure 7. Normalized abundance (values of 0–1) of contigs belonging to selected Level 2 functional categories within associated Level 1 categories in source water (SW), produced water day 1 (PW day 1) and produced water day 9 (PW day 9). Functional annotations were assigned based on the Subsystems database. The affiliations of Level 2 categories to Level 1 categories are coded as follows CS- Clustering based subsystems; C- Carbohydrates; PM- Protein metabolism; DNA- DNA metabolism; R- Respiration; SR- Stress response; SM- Sulfur metabolism; DS- Dormancy and sporulation; IAM- Iron acquisition and metabolism.

Less abundant Level 1 functional categories showing an increase in normalized abundance in produced water samples (Figure 6) included genes affiliated with stress response (3%), respiration (3–4%), iron acquisition and metabolism (1%), sulfur metabolism (1%), and dormancy and sporulation (0.2–1%). Analysis of Level 2 functional categories within these Level 1 domains identified differences in metabolic potential between these samples (Figure 7). Within the Level 1 stress response domain, produced water samples showed a greater relative abundance of sequences affiliated with Level 2 categories such as acid stress, heat shock, periplasmic stress and osmotic stress (Figure 7). The increase in the relative abundance of these genes suggests a response to external stress experienced by the produced water microbial community. Results suggest that produced water day 1 population involved in osmotic stress response was dominated by the order *Rhodobacterales* and produced water day 9 population involved osmotic stress response was dominated by the orders *Vibrionales* and *Alteromonadales* (Figure 9). Subsurface stresses can include increased subsurface temperatures (>40°C) [49], addition of HCl and biocides to fracturing fluid, and higher concentrations of dissolved salts (Table S1) [5]. Within the Level 1 respiration category, sequences affiliated to the Level 2 category of sodium ion coupled energetics were undetected in fracturing source water (Na+ 2.9 g/L) but increased in relative abundance with time in produced water samples (Na+ concentrations in PW day 1 and day 9 were

13.9 and 43 g/L) (Figure 7). This suggests that the produced water microbial community could use sodium ion coupled energetics for their energy needs, consistent with previous observations in saline environments [50]. In the Level 1 domain of sulfur metabolism, the relative abundance of genes affiliated with Level 2 functional categories of inorganic and organic sulfur assimilation increased in produced water samples as compared to fracturing source water (Figure 7). Genes recovered from produced water day 1 show that populations involved in sulfur metabolism were dominated by the orders *Rhodobacterales* and *Thermoanaerobacterales* (Figure 10). However, sulfur metabolism in produced water day 9 samples was dominated by the orders *Vibrionales* and *Bacteroidales* (Figure 10). Within the Level 1 domain of iron metabolism, sequences affiliated with siderophores, undetected in the fracturing source water, increased with time in produced water samples (Figure 7). Siderophores are strong chelators of ferric iron secreted and are utilized by bacteria for iron metabolism [51]. Relative increase in siderophore affiliated genes correlates with an increase in total iron concentrations with time in produced water (4.2–81.6 mg/L) (Table S1). Within the Level 1 dormancy and sporulation category, high relative abundance of Level 2 spore DNA protection related sequences in produced water day 1 sample (Figure 7) suggests the potential for long term dormancy of cells through DNA protection [52]. BLAT analysis [37] showed that these genes were similar to those present in *Thermoanaerobacter*, a

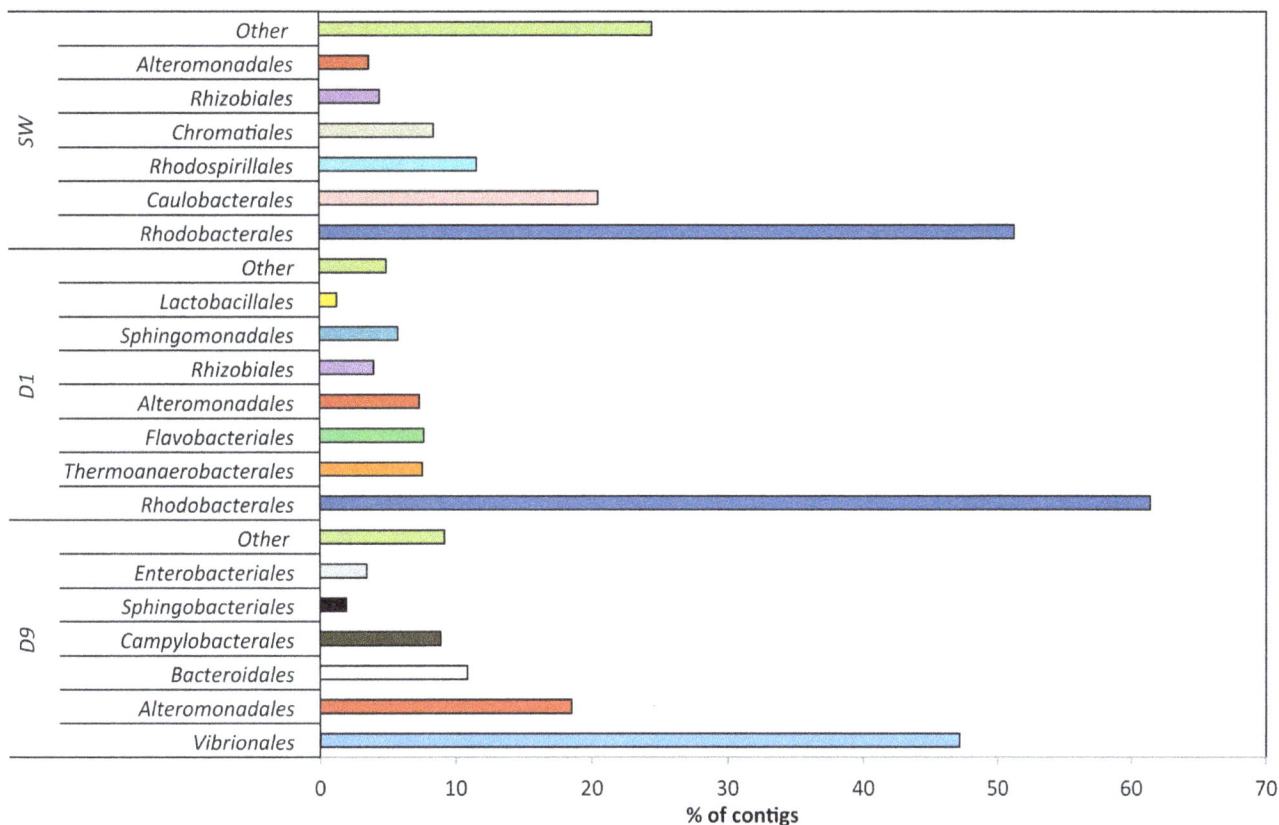

Figure 8. Taxonomic classification of oxidative stress contigs for each analyzed water sample as assigned by MGTAXA. SW- Source water; D1- Produced water day 1; D9- Produced water day 9. Only the top six bacterial orders to which most contigs were assigned to are shown in the figure. The less abundant bacterial orders are grouped as "other".

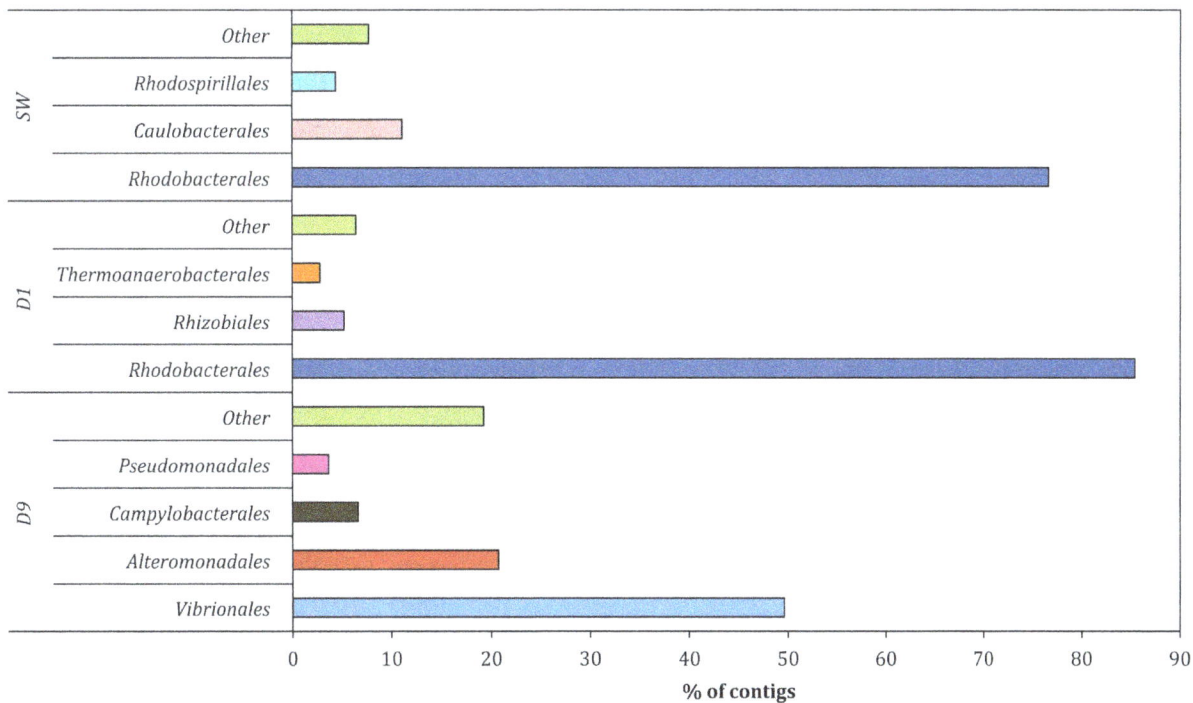

Figure 9. Taxonomic classification of osmotic stress contigs for each analyzed water sample as assigned by MGTAXA. SW- Source water; D1- Produced water day 1; D9- Produced water day 9. Only the top four bacterial orders to which most contigs were assigned to are shown in the figure. The less abundant bacterial orders are grouped as "other".

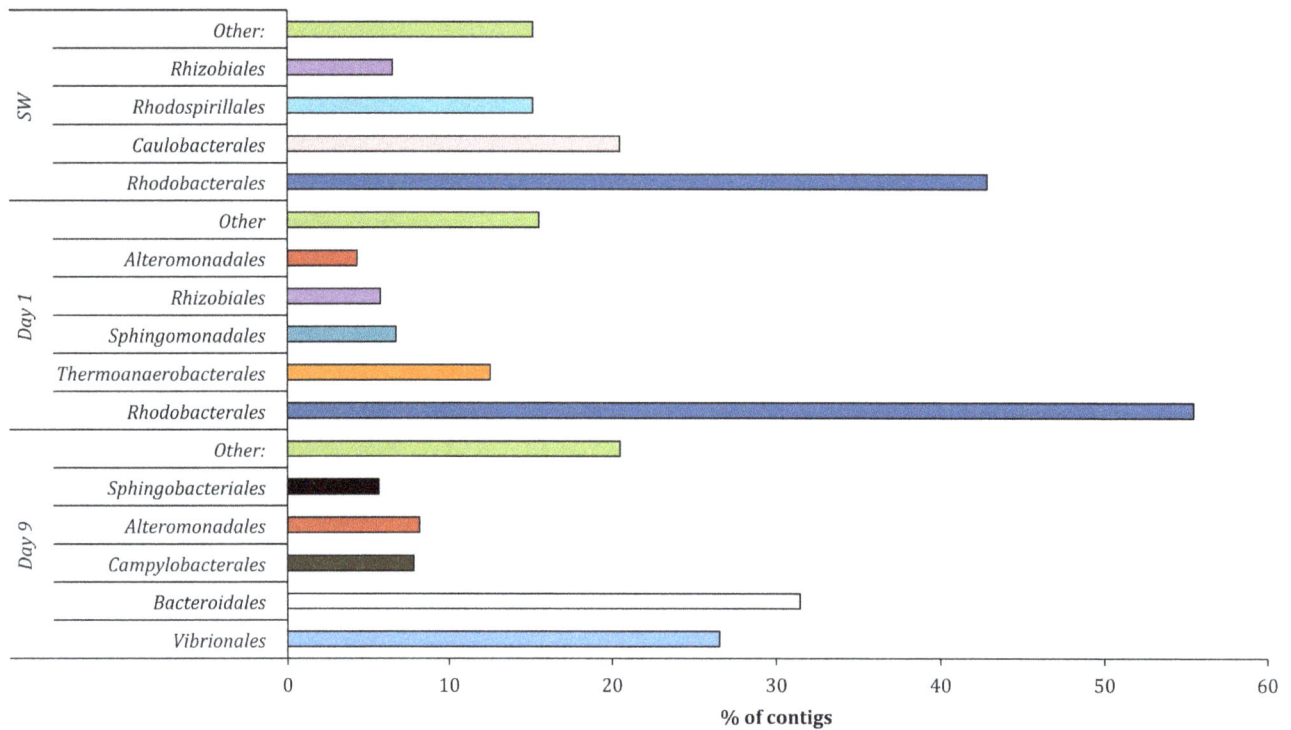

Figure 10. Taxonomic classification of sulfur metabolism contigs for each analyzed water sample as assigned by MGTAXA. SW-Source water; D1- Produced water day 1; D9- Produced water day 9. Only the top five bacterial orders to which most contigs were assigned to are shown in the figure. The less abundant bacterial orders are grouped as "other".

bacterial order that constituted 16% of the total community in this sample (Figure 2). An increase in the relative abundance of spore forming bacteria and genes affiliated with sporulation and dormancy is an important consideration in biocide application, and may provide an explanation for the previously observed limited efficacy of biocides [5].

Concluding Remarks. This study is the first shotgun metagenomic analysis of produced water from hydraulic fracturing for natural gas production and provides novel insights on taxonomic and functional potential of this pertinent yet unexplored environment. Taxonomic analysis showed that *Bacteria* constituted the dominant (>98%) domain in both fracturing source water and produced water samples. Results demonstrated the emergence of distinct bacterial classes and orders in the produced water samples and fracturing source water samples. These bacterial taxa were consistent with results from a previous 16S rRNA gene based survey of these samples [5]. The metabolic profile showed both a relative increase and functional changes in genes responsible for carbohydrate metabolism, respiration, sporulation and dormancy, iron acquisition and metabolism, stress response and sulfur metabolism in the produced water samples as compared to the fracturing source water sample. These results suggest that the microbial community is responsive to changes in hydrocarbon content, induced stresses such as increase in temperature, addition of biocides, and an increase in concentration of dissolved salts such as iron and sulfur. The detection of genes affiliated with sodium ion coupled energetics exclusively in the produced water samples suggests the use of sodium ion based energetics by microorganisms in these sodium rich environments. Understanding the evolving metabolic capabilities of microbial communities in produced water will help the

industry and its regulators improve environmental and economic sustainability of oil and gas extraction through more informed water management decisions.

Supporting Information

Figure S1 Plot of refraction curves with associated Alpha diversity in fracturing source water (SW), produced water day 1 (PW day 1) and produced water day 9 (PW day 9).

Figure S2 Sequences affiliated to major bacterial phyla in source water, Produced water day 1 and Produced water day 9 using 16S rRNA gene pyrosequencing and metagenomics.

Table S1 Chemical composition of source water and produced water (PW) samples days 1, 9 and 187.

Table S2 Assembly optimization statistics. Velvet 1.2.08 was used to optimize assembly of Source Water derived sequences.

Table S3 Mapping results for source water, produced water day 1 and produced water day 9 sequencing data against selected bacteria species reference genomes. Mapping analysis was performed using CLC Genomics Workbench version 6.5.1 with default parameters.

Table S4 Mapping results, (A), for source water, produced water day 1 and produced water day 9 sequencing data against the genome sequences of the dsrA/dsrB gene of selected microbial organisms. Mapping analysis was performed using CLC Genomics Workbench version 6.5.1 with default parameters. (B) Mapping results for source water, produced water day 1 and produced water day 9 sequencing data against the genome sequences of the apsA gene of selected microbial organisms. Mapping analysis was performed using CLC Genomics Workbench version 6.5.1 with default parameters.

Author Contributions

Conceived and designed the experiments: AMM KBG KJB RWH. Performed the experiments: AMM. Analyzed the data: AMM KJB DL. Contributed reagents/materials/analysis tools: KBG KJB. Wrote the paper: AMM KJB DL RWH KBG.

References

1. Veil JA (2010), Water Management Technologies Used by Marcellus Shale Gas Producers, ANL/EVS/R-10/3, prepared by Environmental Science Division, Argonne National Laboratory for the U.S. Department of Energy, Office of Fossil Energy, National Energy Technology Laboratory, July ANL/EVS/R-10/3.

2. Arthur JD, Bohm B, Coughlin B, Layne M, Cornue D (2009) Evaluating the Environmental Implications of Hydraulic Fracturing in Shale Gas Reservoirs, *SPE 121038*, In: SPE Americas Environmental and Safety Conference. San Antonio, TX, March 23–25.

3. Gregory KB, Vidic RD, Dzombak DA (2011) Water Management Challenges Associated with the Production of Shale Gas by Hydraulic Fracturing. Elements 7: 181–186.

4. Barbot E, Vidic N, Gregory KB, Vidic RD (2013) Spatial and Temporal Correlation of Water Quality Parameters of Produced Waters from Devonian-Age Shale following Hydraulic Fracturing. Environ Sci Technol 47: 2562–2569.

5. Murali Mohan A, Hartsock A, Bibby K, Hammack RW, Vidic RD, et al. (2013) Microbial Community Changes in Hydraulic Fracturing Fluids and Produced Water from Shale Gas Extraction. Environ Sci Technol 47(22): 13141–13150.

6. Soeder DJ, Kappel WM (2009) USGS Fact Sheet 2009–3032.

7. Hill D, Lombardi T, Martin J (2004) Fractured Shale Gas Potential In New York. Northeastern Geol. Environ Sci 26: 57–78.

8. Moore SL, Cripps CM (2010). Bacterial Survival in Fractured Shale Gas Wells of the Horn River Basin (CSUG/SPE 137010). CSUG pp. 1–14.

9. Murali Mohan A, Hartsock A, Hammack RW, Vidic RD, Gregory KB (2013) Microbial Communities in Flowback Water Impoundments from Hydraulic Fracturing for Recovery of Shale Gas. FEMS Microbiology Ecology 86(3): 567–580.

10. Kermani M, Harrop D (1996) The impact of corrosion on oil and gas industry. SPE Production Facilities 11: 186–190.

11. Little BJ, Lee JS (2007) Microbiologically influenced corrosion.Wiley and Sons Inc., Hoboken, NJ.

12. Fichter JK, Johnson K, French K, Oden R (2008) Use of Microbiocides in Barnett Shale Gas Well Fracturing Fluids to Control Bacteria Related Problems (Paper No. 08658). In NACE International Corrosion Conference and Expo pp. 1–14.

13. Roberge PR (2000) Handbook of Corrosion Engineering. McGraw-Hill, New York.

14. Struchtemeyer CG, Morrison MD, Elshahed MS (2012) A critical assessment of the efficacy of biocides used during the hydraulic fracturing process in shale natural gas wells. International Biodeterioration & Biodegradation 71: 15–21.

15. Struchtemeyer CG, Elshahed MS (2012) Bacterial communities associated with hydraulic fracturing fluids in thermogenic natural gas wells in North Central Texas, USA. FEMS Microbiology Ecology 81: 13–25.

16. Williams TM, Mcginley HR (2010) Deactivation of Industrial Water Treatment Biocides (Paper No. 10049). In NACE International Corrosion Conference and Expo pp. 1–15.

17. Davis JP, Struchtemeyer CG, Elshahed MS (2012) Bacterial communities associated with production facilities of two newly drilled thermogenic natural gas wells in the Barnett Shale (Texas, USA). Microbial Ecology 64: 942–954.

18. Dahle H, Garshol F, Madsen M, Birkeland NK (2008) Microbial community structure analysis of produced water from a high-temperature North Sea oilfield. Antonie van Leeuwenhoek 93: 37–49.

19. Pham VD, Hnatow LL, Zhang S, Fallon RD, Jackson SC, et al. (2009) Characterizing microbial diversity in production water from an Alaskan mesothermic petroleum reservoir with two independent molecular methods. Environmental Microbiology 11: 176–187.

20. Grabowski A, Nercessian O, Fayolle F, Blanchet D, Jeanthon C (2005) Microbial diversity in production waters of a low-temperature biodegraded oil reservoir. FEMS Microbiology Ecology 54: 427–443.

21. van der Kraan GM, Bruining J, van Loosdrecht MCM, Muyzer G (2010) Microbial diversityofan oil water processing site and its associated oil field: the possible role of microorganisms as information carriers from oil-associated environments. FEMS Microbiology Ecology 71: 428–443.

22. Gittel A, Sorensen KB, Skovhus TL, Ingvorsen K, Schramm A (2009) Prokaryotic community structure and sulfate reducer activity in water from high-temperature oil reservoirs with and without nitrate treatment. Appl Environ Microbiol 75: 7086–7096.

23. Hendrickson H (2009) Order and disorder during *Escherichia coli* divergence. PLoS genetics, 5, e1000335. Available: http://www.plosgenetics.org/article/info%3Adoi%2F10.1371%2Fjournal.pgen.1000335

24. Dinsdale EA, Edwards RA, Hall D, Angly F, Breitbart M et al. (2008) Functional metagenomic profiling of nine biomes. Nature 452: 629–632.

25. DeLong EF, Preston CM, Mincer T, Rich V, Hallam SJ, et al. (2006) Community genomics among stratified microbial assemblages in the ocean's interior. Science 311: 496–503.

26. Tringe SG, Rubin EM (2005) Metagenomics: DNA sequencing of environmental samples. Nature reviews. Genetics 6: 805–814.

27. Wegley L, Edwards R, Beltran Rodriguez-Brito1 HL, Rohwer F (2007) Metagenomic analysis of the microbial community associated with the coral *Porites astreoides*. Environmental Microbiology 9: 2707–2719.

28. Yu K., Zhang T (2012) Metagenomic and metatranscriptomic analysis of microbial community structure and gene expression of activated sludge. PloS one 7, e38183. Available: http://www.plosone.org/article/info%3Adoi%2F10.1371%2Fjournal.pone.0038183

29. Yergeau E, Hogues H, Whyte LG, Greer CW (2010) The functional potential of high Arctic permafrost revealed by metagenomic sequencing, qPCR and microarray analyses. The ISME journal 4: 1206–1214.

30. Dongshan AN, Caffrey SM, Soh J, Agrawal A, Brown D, et al. (2013) Metagenomics of Hydrocarbon Resource Environments Indicates Aerobic Taxa and Genes to be Unexpectedly Common. Environ Sci Technol 47: 10708–10717.

31. Dong Y, Kumar CG, Chia N, Kim PJ, Miller P, et al. (2013) *Halomonas sulfidaeris*-dominated microbial community inhabits a 1.8 km-deep subsurface Cambrian Sandstone reservoir. Environmental Microbiology 16(6): 1695–1708.

32. Lamendella R, Domingo JWS, Ghosh S, Martinson J, Oerther DB (2011) Comparative fecal metagenomics unveils unique functional capacity of the swine gut. BMC Microbiology 11:103. Available: http://www.biomedcentral.com/1471-2180/11/103

33. Schmieder R, Lim YW, Rohwer F, Edwards R (2010) TagCleaner: Identification and removal of tag sequences from genomic and metagenomic datasets. BMC Bioinformatics 11:341. Available: http://www.ncbi.nlm.nih.gov/pmc/articles/PMC2910026/

34. Goecks J, Nekrutenko A, Taylor J, The Galaxy Team (2010) Galaxy: A comprehensive approach for supporting accessible, reproducible, and transparent computational research in the life sciences. Genome biology 11:R86. Available: http://genomebiology.com/2010/11/8/R86

35. Zerbino DR (2010) Using the Velvet de novo assembler for short-read sequencing technologies. Curr Protoc Bioinformatics 31:11.5.1–11.5.12. Available: http://www.ncbi.nlm.nih.gov/pmc/articles/PMC2952100/

36. Meyer F, Paarmann D, D'Souza M, Olson R, Glass EM, et al. (2008) The metagenomics RAST server - a public resource for the automatic phylogenetic and functional analysis of metagenomes. *BMC Bioinformatics* 2008, 9: 386. Available: http://www.biomedcentral.com/1471-2105/9/386

37. Kent WJ (2002) BLAT—The BLAST-Like Alignment Tool. Genome Research 12: 656–664. Available: http://www.ncbi.nlm.nih.gov/pmc/articles/PMC187518/

38. Tovchigrechko A, Sul SJ, MGTAXA- A free software for taxonomic classification of metagenomic sequences with machine learning techniques. Available: http://andreyto.github.io/mgtaxa/

39. Brady A, Salzberg SL (2009) Classification with interpolated markov models. Nature Methods 6: 673–676.

40. Giardine B, Riemer C, Hardison C, Burhans R, Elnitski L (2005) Galaxy: A platform for interactive large scale genome analysis. Genome Research 15: 1451–1455. Available: http://www.ncbi.nlm.nih.gov/pmc/articles/PMC1240089/

41. CLC Genomics Workbench, "Version 6.5.1", CLC bio A/S Science Park Aarhus Finlandsgade, 10–12. Available: http://www.clcbio.com/products/clc-genomics-workbench/

42. Overbeek R, Begley T, Butler RM, Choudhuri JV, Chuang HY, et al. (2005) The subsystems approach to genome annotation and its use in the project to annotate 1000 genomes. Nucleic acids research 33: 5691–5702.

43. Shi Y, Tyson GW, Eppley JM, DeLong EF (2011) Integrated metatranscriptomic and metagenomic analyses of stratified microbial assemblages in the open ocean. The ISME journal 5: 999–1013.

44. Delmont TO, Prestat E, Keegan KP, Faubladier M, Robe P, et al. (2012) Structure, fluctuation and magnitude of a natural grassland soil metagenome. The ISME Journal 6 (9): 1677–1687.

45. Urich T, Lanzén A, Qi J, Huson DH, Schleper C, et al. (2008) Simultaneous assessment of soil microbial community structure and function through analysis of the meta-transcriptome. PloS one 3, e2527. Available: http://www.plosone.org/article/info%3Adoi%2F10.1371%2Fjournal.pone.0002527
46. Gilbert JA, Field D, Huang Y, Edwards R, Li W, et al. (2008) Detection of large numbers of novel sequences in the metatranscriptomes of complex marine microbial communities. PloS one 3, e3042. Available: http://www.plosone.org/article/info%3Adoi%2F10.1371%2Fjournal.pone.0003042
47. Lu J, Holmgren A (2009) Selenoproteins. The Journal of biological chemistry 284: 723–727.
48. Valavanidis A, Vlahogianni T, Dassenakis M, Scoullos M (2006) Molecular biomarkers of oxidative stress in aquatic organisms in relation to toxic environmental pollutants. Ecotoxicol Environ Saf 64: 178–189.
49. Driscoll FG (1986) Groundwater and Wells. Johnson Filtration Inc.: St Paul, MN.
50. Kogure K (1998) Bioenergetics of marine bacteria. Current Opinion in Biotechnology 9: 278–282.
51. Sandy M, Butler A (2010) Microbial Iron Acquisition: Marine and Terrestrial Siderophores. Chem Rev 109: 4580–4595.
52. Setlow P (1992) Mini Review: I Will Survive: Protecting and Repairing Spore DNA. Journal of Bacteriology 174: 2737–2741.

22

Effect of Increasing Total Solids Contents on Anaerobic Digestion of Food Waste under Mesophilic Conditions: Performance and Microbial Characteristics Analysis

Jing Yi[ꝯ], **Bin Dong**[ꝯ], **Jingwei Jin, Xiaohu Dai***

National Engineering Research Center for Urban Pollution Control, College of Environmental Science and Engineering, Tongji University, Shanghai, People's Republic of China

Abstract

The total solids content of feedstocks affects the performances of anaerobic digestion and the change of total solids content will lead the change of microbial morphology in systems. In order to increase the efficiency of anaerobic digestion, it is necessary to understand the role of the total solids content on the behavior of the microbial communities involved in anaerobic digestion of organic matter from wet to dry technology. The performances of mesophilic anaerobic digestion of food waste with different total solids contents from 5% to 20% were compared and the microbial communities in reactors were investigated using 454 pyrosequencing technology. Three stable anaerobic digestion processes were achieved for food waste biodegradation and methane generation. Better performances mainly including volatile solids reduction and methane yield were obtained in the reactors with higher total solids content. Pyrosequencing results revealed significant shifts in bacterial community with increasing total solids contents. The proportion of phylum *Chloroflexi* decreased obviously with increasing total solids contents while other functional bacteria showed increasing trend. *Methanosarcina* absolutely dominated in archaeal communities in three reactors and the relative abundance of this group showed increasing trend with increasing total solids contents. These results revealed the effects of the total solids content on the performance parameters and the behavior of the microbial communities involved in the anaerobic digestion of food waste from wet to dry technologies.

Editor: Dwayne Elias, Oak Ridge National Laboratory, United States of America

Funding: This research has been supported financially by National Key Technologies R&D Program of China (2010BAC67B04) and the key projects of National Water Pollution Control and Management of China (2011ZX07316-004). The funders had no role in study design, data collection and analysis, decision to publish, or preparation of the manuscript.

Competing Interests: The authors have declared that no competing interests exist.

* Email: yijing4321@163.com

[ꝯ] These authors contributed equally to this work.

Introduction

Food waste (FW), usually from residential, commercial establishments, institutional and industrial sources, is generated at an ever-increasing rate (higher than 10% every year) with the rapid population growth and rising living standards in China [1]. It seems to be a good idea to reuse this favorable feedstock for energy recovery and municipal solid waste (MSW) reduction because FW contains high moisture and biodegradable organics and accounts for 40–50% of the weight of MSW. Anaerobic digestion (AD) is the most attractive and cost-effective technology for treating sorted organic fraction of MSW, especially food wastes [2]. Various AD processes have been widely developed in many countries for the treatment of FW.

So far, three main types of AD technologies have been developed according to the total solids (TS) content of feedstocks: conventional wet ($\leqq$ 10% TS), semi-dry (10–20% TS) and modern dry ($\geq$ 20% TS) processes. Dry anaerobic digestion, so called "high-solids" technology, has become attractive and was applied widely because it requires smaller reactor volume, lower energy

requirements for heating, less material handling, and so on [2–4]. The TS content of solid waste influences anaerobic digestion performance, especially biogas and methane production efficiency [5]. Previous reports have investigated that role of TS content on AD performance in order to determine conditions for optimum gas production. Abbassi-Guendouz et al., showed that the total methane production decreased with TS contents increasing from 10% to 25% in batch anaerobic digestion of cardboard under mesophilic conditions [6]. The results obtained by Duan et al., showed that high-solids system could reach much higher volumetric methane production rate compared with low-solids system at the same solid retention time (SRT) in mesophilic anaerobic reactors treating sewage sludge [3]. Forster-Carneiro et al., showed that the biogas and methane production decreased with the total solids contents increasing from 20% to 30% in dry batch anaerobic digestion of food waste [2].

Anaerobic digestion is a multi-stage biochemical process in which the complex organic materials undergo hydrolysis, acidogenesis, and methanogenesis in series and each metabolic stage is functioned by different types of microorganisms [4]. They are present in a mixed culture but differ in their nutritional and pH requirement, growth kinetics, and their ability to tolerate environment stresses [7]. Characterization of microbial community structures in anaerobic digesters has been attractive from the point of review of engineering because understanding of microbial behavior can provides valuable information to optimize fermentation process to favor efficient breakdown of wastes [8]. However, the available literature is mainly about performance and corresponding the structure and dynamic of microbial community in either thermophilic or mesophilic anaerobic digestion of food waste, or only simply about performance comparisons. The AD performances at steady state and the comprehensive characterizations of microbial community in anaerobic digestion of FW with different TS contents (wet, semi-dry and dry) were not compared in parallel. In order to increase the efficiency of anaerobic digestion of FW, it is necessary to understand the role of the TS contents on the behavior of the microbial community structure involved in the anaerobic digestion of degradation from wet to dry technology.

Recently, various molecular microbial ecology tools have been applied in numerous studies to analyze microbial communities in different anaerobic digesters and their influences on the efficiency and stability of AD processes [9]. Pyrosequencing, as a next generation sequencing technology, has gained increasing attention as a novel tool for studying the microbial diversity [4]. Recently, this technology has been widely and successfully used to characterize the microbial community structures in various environmental samples, such as source waster [10], membrane filtration systems [11], soil [12]. Meanwhile, the microbial community structures were compared by this technology in anaerobic digestion of food waste at different organic loading rates (OLRs) [4].

Hence, the aim of this study was to conduct a comprehensive comparison of the microbial community structure using 454 high throughput pyrosequencing technology and related these microbial findings to their respective performances of mesophilic anaerobic digesters treating FW with different TS contents ranging from 5%–20%. It was expected that the reported work herein will reveal the role of the TS content on the behavior of the microbial community structure to increasing TS contents and hence to effective guide high solids anaerobic digestion of FW and to optimize the operational conditions for high anaerobic digestion efficiency.

Materials and Methods

Substrates and inoculums

FW used in this study was collected every 30 days from a dining room at Tongji University in Shanghai. After removing bones, shells, and other indigestible materials, the FW was finely smashed using an electrical crusher and sufficient mixed and stored at 4°C. The TS of the FW ranged from 26% to 28% (w/w) and volatile solid (VS) accounted for 92%–95% of TS. The mesophilic seed sludge was obtained from a full-scale anaerobic digester at Bailonggang municipal wastewater treatment plant (WWTP) (Shanghai, China). It had TS of 4.1% (w/w) and VS of 52.3% of TS. The main characteristics (average data plus standard deviations in duplicate tests) of substrates and inoculums are listed in Table 1. The collected FW was heated to 35°C before daily feeding.

Reactors and operation

Three identical reactors (numbered R1, R1 and R3), with liquid working volume of 6.0 L, were equipped with helix-type stirrers to provide sufficient mixing for substrates. The rotation speed was set at a rate of 60 rpm (rotations per minute) with 9 min stirring and 1 min break, continuously. Daily feeding was carried out by pushing semi-fluid substrate through the feeding piston. Since the digestate of FW in each reactor was completely fluid, daily draw-off was easily carried out by opening the discharge valve.

On the first day of the experiments, 6.0 L seed sludge was added to each reactor, which was operated semi-continuously (once-a-day draw-off and feeding) under single phase mesophilic conditions (35°C). The reactors were purged with N_2 for 10 min in order to provide anaerobic conditions. During the start-up period, the OLR was increased stepwise with high-solids FW before the TS content of the substrate in each reactor did not reach its designed TS level. Once the TS of the substrate in each reactor approached its designed level, the feeding FW was diluted to its designed TS level (5%, 15% and 20%, respectively) with de-ionized water before feeding. Each reactor was operated for five SRTs at 20 days SRT. For a full understanding of the microbial community structures in anaerobic fermentation reactors with different TS contents, the anaerobically digested FW samples were taken on Day 100 when the systems could be deemed to have reached their steady state operation (determined by constant methane yield and VS reduction) after running for more than 3 months. The fermentation substrate samples in the reactors were taken every three days during the operation period of the fifth SRT for reactor performance analysis.

DNA extraction, PCR and Pyrosequencing

To analyze the bacterial and archaeal communities in mesophilic anaerobic digesters with different feeding TS levels, 0.5 g of sample in reactor operated for 100 d was used for DNA extraction using a Fast DNA Spin Kit (QBIOgene, Carlsbad, CA, USA) following the manufacturer's instructions. For each sample, two independent PCR reactions were conducted using the primer pairs of 27F (5′-AGAGTTTGATCCTGGCTCAG-3′) and 533R (5′-TTACCGCGGCTGCTGGCAC-3′) for bacteria and 344F (5′-ACGGGGYGCAGCAGGCGCGA-3′) and 915R (5′-GTGCTCCCCCGCCAATTCCT-3′) for archaea [4]. To achieve the sample multiplexing during pyrosequencing, barcodes were incorporated in the 5′end of reverse primers 553R and 915R. All PCR reactions were carried out in a 25 uL mixture containing 0.5 uL of each primer at 30 mmolL^{-1}, 1.5 uL of template DNA (10 ng), and 22.5 uL of Platinum PCR SuperMix (Invitrogen, Shanghai, China). The PCR amplification program contained an initial denature at 95°C for 5 min, followed by 25 cycles of denaturing at 95°C for 30 s, annealing at 55°C for 30 s, and extension at 72°C for 30 s, followed by a final extension at 72°C for 5 min. The thermal cycling for archaea was similar to that for bacteria except that the annealing temperature was 57°C. After amplification, the PCR products were purified and quantified, and an equal amount of the PCR product was combined in a single tube to be run on a Roche GS FLX 454 Pyrosequencing machine at Majorbio Bio-Pharm Technology Co., Ltd., Shanghai, China.

Analysis of Pyrosequencing-derived Data

After sequencing completed, all sequence reads were quality checked using Mothur software [13]. Raw sequence reads were filtered before subsequence analyses to minimize the effect of random sequencing errors. The sequence reads that did not contain the correct primer sequence after the initial quality check (primer sequences were subsequently removed), were shorter than 200 bp, contained one or more ambiguous base(s), or checked as chimeric artifact were eliminated. Finally, the high-quality

Table 1. Characteristics of the substrates and inoculums.

Parameters	FW[a] 1 (days 1–30)	FW 2 (days 31–60)	FW 3 (days 61–90)	FW 4 (days 91–120)	Inoculums
TS[b] (%, w/w)	26.5±0.6	27.8±1.1	27.3±1.2	26.8±1.2	4.1±0.1
VS[c]/TS (%)	94.7±3.9	92.2±3.7	93.4±4.6	93.9±4.2	52.3±2.4
pH	4.72±0.21	4.64±0.11	4.79±0.24	4.87±0.23	7.9±0.3
C/N (w/w)	13.4±0.6	14.2±0.7	13.9±0.4	13.6±0.6	-
TAN[d] (mg/L)	538±24	546±19	534±25	543±19	299±13

–Not determined.
[a]FW: food waste.
[b]TS: total solids.
[c]VS: volatile solids.
[d]TAN: total ammonia nitrogen.

sequences after filtering were assigned to samples according to barcodes. Sequences were aligned in according with SILVA alignment [14]. Mothur was also used to conduct rarefaction curve, abundance base coverage estimator (ACE), richness (Chao), Shannon diversity, Simpson diversity indices and Good's coverage analysis, assign sequences to operational taxonomic units (OTUs, 97% similarity) using furthest neighbor approach. For taxonomy-based analysis, the SILVA database project (http://www.arb-silva.de) was used as a repository for aligned rRNA sequences. The sequences have been deposited into the NCBI short read archive (SRA) under the accession number SRX484115 for bacteria and SRX485028 for archaea.

Analytical methods

Volumes of produced biogas were measured by wet gas meters every day. The methane content of the biogas was measured by a gas chromatograph (GC) (Agilent Technologies 6890N, CA, USA) with a thermal conductivity detector equipped with Hayseq Q mesh and Molsieve 5A columns. For the analysis of volatile fatty acid (VFA), the fermentation mixtures withdrawn from digesters were centrifuged at 10, 000 ×g for 10 min, and then the supernatants were immediately filtered through 0.45 um cellulose nitrate membrane fiber paper. The filtrate was collected in a 1.5 ml gas chromatograpgy (GC) vial and acidified by formic acid to adjust the pH to approximately 2.0, and then analyzed using a gas chromatograph (GC, Agilent 7820) with a flame ionization detector (FID) and equipped with a 52 CB column (30 m×0.32 mm×0.25 mm). The concentration of total VFA was calculated as the sum of the measured acetic, propionic, n-butyric iso-butyric, n-valeric, and iso-valeric acids. Metrohm 774 pH-meter was used in all pH measurements. The TS, VS, total alkalinity (TA) and total ammonia-nitrogen (TAN) were measured according to Standard Methods [15]. Free ammonia-nitrogen (FAN) was calculated in the same way as described by Østergaard [16]. The degradation or removal level based on VS (i.e., VS reduction) was calculated by the same formula as reported previously [3]. All experimental analyses were performed in triplicate. The data on performances of each reactor were expressed as mean±standard deviation of the samples.

Results and Discussion

Effect of TS content on anaerobic digestion performance

Table 2 summarizes the values of the main parameters indicating system stability (pH, VFA, TA) and potential inhibitory chemicals (TAN and FAN) for three reactors operated at different TS contents, and the performance data were the average values of the last five samples during the operation period of the fifth SRT after the system reached steady state (determined by constant methane yield and VS reduction).

For each semi-continuously experiment with a good anaerobic digestion performance (between 5% and 20% TS), there was no accumulation of VFA and low pH. The concentration values of VFA showed increasing trend with increasing TS contents. Under mesophilic semi-dry anaerobic digestion of sorted organic fraction of municipal solid waste (OFMSW), Li et al., also observed an increasing trend of the VFA concentrations with TS contents increasing (for TS contents of 11.0%, 13.5% and 16.0%), the maximum VFA value was 4.2 g L^{-1}, 6.8 g L^{-1} and 22.4 g L^{-1}, respectively [17]. In this study, higher VFA concentrations were obtained in the reactors with higher TS contents, which could be explained by the fact that more organic matter was hydrolyzed and transformed to VFA in the reactors. High VFA levels and almost steady VS reduction (Table 3) in reactors indicated that the acidogenic activity was not influenced significantly. In addition, the reactor stability was maintained and the digestion occurred normally because a constant pH was maintained for each reactor. The average pH value was about 7.39, 7.68 and 7.82 at 5%, 15% and 20% TS, respectively. These pH values were within the permissible range for AD 6.5–8.5 but not with the optimal range 6.8–7.4 [18]. As we all know, the increase of VFA concentration contributes to the decrease of pH. However, low pH value was not observed in R3 in which the VFA concentration was highest. It could be explained by the fact that high buffering capacity was observed in high-solids anaerobic system at TS 20%, for which the total alkalinity value of 13.8 g CaCO$_3$/L was detected.

It was known that ammonia nitrogen concentration (especially free ammonia concentration) was an important factor influencing the stability of anaerobic digestion system. The TAN and FAN concentrations in three reactors at steady state were also observed. They showed a similar trend to that of above parameters with increasing TS contents. However, the maximum FAN value was just 163 mg/L. It has been reported that the FAN at concentrations above 200–1100 mg/L can inhibit the anaerobic system [19]. Therefore, the effect of FAN concentration on the system stability was probably negligible for the three reactors with TS contents ranged from 5% to 20%.

Biogas generation and methane efficiency of different reactors are shown in Table 3. The average daily cumulative biogas (based on added VS) of R1-5%, R2-15% and R3-20% accounted to 700, 760 and 870 ml and 370, 410 and 480 ml methane content, respectively. Hence, both of biogas production and methane content showed increasing trend with increasing TS contents. This result was in contrast with a previous work [2], in which the

Table 2. Summary of performance parameters on system stability and inhibition in three reactors.

reactor	SRT[a]	OLR[b] (Kg VS m^{-3}d^{-1})	pH	TA[c] (g/L)	TAN[d] (g/L)	FAN[e] (mg/L)	VFA[f] (g/L)	
							Total	Acetic
R1(5%)	20	2.35	7.39±0.08	3.8±0.1	0.40±0.01	11±0.4	0.12±0.01	0.11±0.01
R2(15%)	20	7.01	7.68±0.06	10.9±0.3	1.31±0.15	66±2.5	0.53±0.02	0.43±0.01
R3(20%)	20	9.41	7.82±0.09	13.8±0.2	1.92±0.04	163±8.0	0.94±0.01	0.64±0.02

aSRT: solid retention time.
bOLR: organic loading rate.
cTA: total alkalinity.
dTAN: total ammonia nitrogen.
eFAN: free ammonia nitrogen.
fVFA: volatile fatty acid.

Table 3. Performance parameters of three reactors with different total solids contents.

Reactor	SRT (d)	OLR (Kg VS m^{-3}d^{-1})	Y$_{biogas}$[a] (LBiogas gVS^{-1} added)	CH$_4$ (%)	Y$_{methane}$[b] (L CH$_4$ g VS^{-1} added)	VS$_r$[c] (%)	SBP[d] (L Biogas gVS^{-1} removed)	SMP[e] (LCH$_4$ gVS^{-1} removed)	BP[f] (Biogas L^{-1}d^{-1})	MP[g] (LCH$_4$ L^{-1}d^{-1})
R1(5%)	20	2.35	0.70±0.02	52.5±2.1	0.37±0.01	80.1±2.4	0.88±0.02	0.46±0.01	1.65±0.06	0.87±0.03
R2(15%)	20	7.01	0.76±0.01	54.2±2.7	0.41±0.01	82.4±2.2	0.92±0.05	0.50±0.01	5.36±0.2	2.90±0.07
R3(20%)	20	9.41	0.87±0.02	55.1±2.6	0.48±0.01	85.6±2.6	1.01±0.04	0.56±0.02	8.21±0.34	4.52±0.05

aY$_{biogas}$: biogas yield.
bY$_{methane}$: methane yield.
cVS$_r$: VS reduction.
dSBP: specific biogas production rate based on removed VS.
eSMP: specific methane production rate based on removed VS.
fBP: volumetric biogas production rate.
gMP: volumetric methane production rate.

reactors with smaller TS contents showed higher biogas production and methane percentage in the batch anaerobic digestion of FW. It was suggested that the increasing of feeding TS contents lower than 20% has positive effect on the methane production. A maximum methane content of 55.1% in R3 agreed with the previous study on anaerobic digestion of FW [1], but was lower than in another reference [20], which was probably due to the differences in substrate composition. In addition, it could also be observed that reactors with higher TS contents showed higher volumetric biogas and methane production rate. It is well known that FW is a high degradable substrate for anaerobic digestion [21]. For reactors R1-R3 at a fixed 20 days SRT, increased feeding TS content of FW meant higher applied OLR and larger proportion of easily degradable substrate for microorganisms, which results in higher volumetric biogas yield and methane production rate. As showed in Table 3, higher VS reduction was observed in the anaerobic digesters with higher TS contents. The reasons for this important result obtained were investigated from the microbiology aspect in the following chapters. The specific biogas and methane product rate based on removed VS increased slightly. The highest specific biogas production rate determined on removed VS was 1.01 L gVS^{-1} removed in R3, which was higher than corresponding data obtained in a previous study [1].

Overall analysis of pyrosequencing

The latest developed 454 high-throughput pyrosequencing that can generate huge amounts of DNA reads is widely employed to investigate the bacterial and archaeal community structures and dynamics in various environmental samples. To investigate the compositions of microbial populations involved in the fermentative reactors with different TS contents, a total of 9571, 7769 and 5598 trimmed bacterial 16S rRNA gene sequences and 5245, 4654 and 4432 trimmed archaeal 16S rRNA gene sequences were recovered from samples R1, R2 and R3 (Table S1), respectively. The sequences were grouped into OTUs at a distance level of 3% to estimate the phylogenetic diversities of microbial communities. The OTUs number identified by bacterial sequences in R1 was the largest among three samples. The bacterial community richness levels can also reflected using ACE, Chao, Shannon and Simpson diversity indices (Table S1), which also revealed that the R1-5% had the highest bacterial diversity among three samples. However, the number of archaeal OTUs in R2 was the largest. The rarefaction curves of three samples generated at 3% cutoff for bacterial and archaeal communities are shown in the Figure S1 (Supporting information), demonstrating clearly that the bacterial community richness of R1 and the archaeal community richness of R2 was the highest among these samples, respectively. However, none of the curves approached a plateau, suggesting that this sequencing depth was still not enough to cover the whole microbial diversity and further sequencing would have resulted in more OTUs for each sample. Pyrosequencing analysis of environmental samples can obtain much more sequences and OTUs than conventional cloning and sequencing methods [11,12]. The bacterial (or archaeal) PCR amplicons from anaerobic digester were grouped into only 238–514 (or 8–26) OTUs according to the clone library in a previous publication [22]. To the authors's knowledge, this was the first study using pyrosequencing technology to characterize the microbial communities in anaerobic digesters with different TS contents. It can be found that compared with traditional clone library, 454 high-throughput pyrosequencing could be a powerful tool to elucidate the microbial community structures and diversities in anaerobic reactors treating food waste with different TS contents.

Effect of total solids content on functional bacterial populations involved in food waste hydrolysis and acidification

Large numbers of bacterial populations are involved in the hydrolysis and acidification processes of anaerobic fermentation for food waste. The distribution of sequences at the phylum level in each sample is shown in Figure 1. There are seven phyla with relative abundance of higher than 0.5% in at least one sample. From the phylum assignment results, it can be seen that most bacterial sequences in the anaerobic digester treating food waste were distributed among three major phyla: *Chloroflexi*, *Bacteroidetes* and *Firmicutes*, the total relative abundances of them accounted for 96.13%, 95.61% and 81.35% in R1, R2 and R3, respectively, along with other phyla at minor predominance. Similarly, *Bacteroidetes*, *Firmicutes*, *Chloroflexi*, *Synergistetes*, and *Actinobacteria* were reported to the major populations at phylum level in the mesophilic anaerobic digester treating food waste [4]. The dominance of *Bacteroidetes*, *Firmicutes* and *Chloroflexi* was also found in other previous studies [7,23]. In addition, R3 with 20% feeding TS content had high relative abundance of *Spirochaetes* (8.09%), *Tenericutes* (6.86%) and *Proteobacteria* (2.16%).

Although most bacteria in reactors were affiliated to these dominant phyla, the relative abundances of these phyla in each reactor were different and each digester had its own characteristic bacterial community composition. The proportion of phylum *Chloroflexi* in each reactor was the highest in this study. This was in good accordance with previous reports that *Choroflexi* populations were abundant in anaerobic digesters, as determined by membrane hybridization [7], FISH [23] and 16S rRNA gene clone analysis [23,24]. Rivière et al., also found large proportions (25–45%) of *Chloroflexi* sequences in municipal WWTP sludge samples [22]. An important trend is the small proportion of *Choroflexi* at the highest TS content: 31% for the 20% TS, compared to 58% with the 15% TS and 65% at the 5% TS. The proliferation of *Choroflexi* (formerly known as Green Nonsulfur Bacteria), a well known scavenger biomass-derived organic carbon such as soluble microbial products (SMP), supports a greater influence of difficult-to-biodegrade organic materials from the input substrates and from endogenous dacay of the anaerobic biomass [22,25]. For R1-R3 at a fixed SRT, increased feeding TS of FW meant higher applied OLR and larger amount of easily degradable substrate per unit volume for microorganisms, which resulted in a smaller relative abundance of phylum *Choroflexi*.

On the other hand, the *Bacteroidetes* population was enriched in the reactors with higher TS contents (from 18.2% at the 5% TS to 26.40% at the 15% TS and 36.33% at the 20% TS). The phylum *Bacteroidetes* are proteolytic bacteria and were probably involved in the degradation of various proteins used for anaerobic digestion studies [22,25]. The majority of proteolytic microorganisms are able to metabolize amino acids to produce VFA such as acetate, propionate and succinate and NH_3 [22]. Interestingly, their selective enrichment at high TS contents seems to be in consistent with the observation of high protein-input rate and VFA production in the reactors with higher TS contents (Table 2). This result indicated the importance of the *Bacteroidetes* performing protein hydrolysis. However, the changing trend of relative abundance of the phylum *Firmicutes* was not obvious with increasing TS contents. The average value of *Firmicutes* proportion was 12% in three reactors. *Firmicutes* are well-known to be acetogenic and syntrophic bacteria that can degrade VFA, such as butyrate and its analogs. The prevalence of organisms belonging to *Firmicutes* suggested that these products are readily available due to the prior fermentation of these simple VFA and played a critical

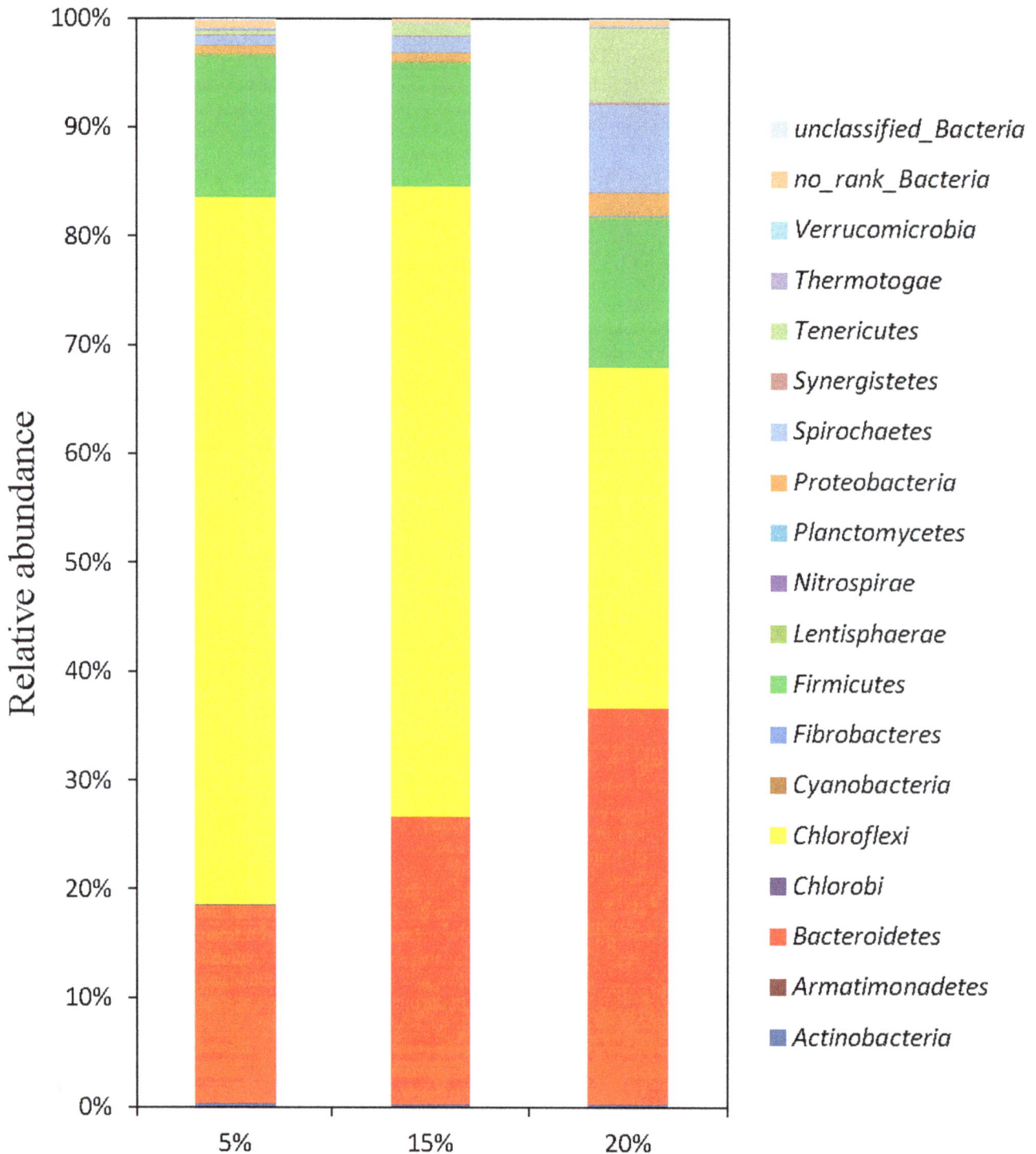

Figure 1. Taxonomic compositions of bacterial communities at phyla level in each sample retrieved from pyrosequencing.

role in anaerobic digestion of FW, especially on the production of acetic acid, an essential step for methane production by acetoclastic methanogenic microorganisms. In addition, the relative abundances of other phyla including *Proteobacteria*, *Spirochaetes* and *Tenericutes* obviously increased with the feeding TS contents increasing. It has been suggested that they might play important roles in the degradation of FW. *Proteobacteria* are also involved in the first step of the degradation of organic wastes and

they are important consumers of propionate, butyrate, and acetate [23]. *Spirochaetes* are reported to ferment carbohydrates or amino acids into, mainly, acetate, H_2 and CO_2 [8] and *Tenericutes* was found to be related with lignin utilization [26].

In order to further compare the difference of bacterial communities in anaerobic digesters with different feeding TS contents, it is preferable to deconstruct the sequencing date at the subdivision level. Therefore, the relative abundance of each genus

Table 4. Taxonomic composition of bacterial communities at the genus level for the sequences retrieved from each sample.

Phylum	Genus	5% Relative abundance	15%	20%
	Bacteroidales	0.43%	0.40%	1.54%
	Bacteroides	0.54%	0.27%	0.82%
	Barnesiella	2.83%	0.08%	0.00%
	Marinilabiaceae	0.37%	0.22%	0.61%
Bacteroidetes	Parabacteroides	0.18%	0.06%	0.34%
	Petrimonas	0.45%	0.37%	1.11%
	Proteiniphilum	1.15%	2.59%	4.29%
	Rikenellaceae	11.16%	21.70%	26.58%
	Sphingobacteriales	0.48%	0.17%	0.52%
Chloroflexi	Anaerolineaceae	64.99%	58.03%	31.37%
	Anaerobranca	0.11%	0.36%	1.43%
	Christensenellaceae	0.30%	0.14%	0.18%
	Clostridiales	1.73%	4.00%	3.88%
	Erysipelotrichaceae	7.17%	1.85%	2.41%
Firmicutes	Fastidiosipila	0.30%	0.79%	1.09%
	Gelria	0.85%	0.85%	0.45%
	Lachnospiraceae	0.13%	0.72%	0.39%
	Lutispora	0.09%	0.06%	0.36%
	Ruminococcaceae	1.11%	0.75%	0.86%
Proteobacteria	Novosphingobium	0.21%	0.14%	0.77%
	Rhizobiales	0.07%	0.06%	0.36%
Spirochaetes	Spirochaeta	0.28%	0.26%	0.73%
	Spirochaetes	0.46%	1.08%	6.98%
Tenericutes	Acholeplasma	0.40%	1.12%	6.75%
	Minor group	4.21%	3.93%	6.20%

in three samples was calculated. The sequence distributions at genus level in each sample are shown in Table 4. A total of 17 genera were detected among which 7 genera with relative abundance of higher than 0.5% in at least one sample were screened as the abundant genera. Other genera were grouped into the minors. As mentioned in the previous section, lower proportions of population from the phylum *Choroflexi* were markedly detected in the reactors with higher TS contents. All sequences classified to phylum *Choroflexi* in three reactors were assigned to genus *Anaerolineaceae* (Figure 1 and Table 4) and class *Anaerolineae* at class level (previous known as "subphylum I" [24]) (Table S2), and the relative abundance of genus *Anaerolineaceae* decreased with increasing TS contents. Because all the characterized species of the class *Anaerolineae* are anaerobic bacteria that decompose carbohydrates via fermentation [27], the genus *Anaerolineaceae* seemed to be involved in carbohydrate decomposition in anaerobic digestion of FW. Similarly, in the previous studies, it was found that all the *Choroflexi* sequences obtained from the up-flow anaerobic sludge blanket reactors treating various food-processing and high-strength organic wastewaters belong to the class *Anaerolineae* [27] and *Anaerolineaceae* group was dominant in phylum *Choroflexi* with its maximum proportion of 8.9% at the 58 days in mesophilic anaerobic digestion of FW [4].

Concerning *Bacteroidetes*, another very abundant phylum which increased with the increasing TS contents, the subdivisions at genus level were multiple and many genera were mainly presented in three anaerobic reactors. *Rikenellaceae* spp. and *Proteiniphilum* spp. were the mostly major genera within this dominant phylum and the changing trends of the relative abundances of these two genera were the same as that of the *Bacteroidetes*. *Rikenellaceae* spp. showed a remarkable proportion from 11% to 27%. The genus *Rikenellaceae* could utilize lactate as substrate in the fermentation processes, and acetate and propionate are the main end-products [28]. *Proteiniphilum*, a relatively new genus showed an unusual ability to grow well at 20–45°C and pH 6.0–9.7. The strains were proteolytic and yeast extract, peptone and l-arginine could be used as carbon and energy sources. Acetic acid and NH_3 were produced after utilizing these substrates [29]. The predominance of *Proteiniphilum* was also obtained in other anaerobic digesters by using a meta-analysis approach [9]. Other genera in this phylum with individual proportion higher or lower than 0.5% might also have played important roles in FW degradation. Regarding to *Firmicutes*, the generic distributions were also distinct with genera *Clostridiales* and *Erysipelotrichaceae* as the main groups in three anaerobic reactors (Table 4). The latter was especially notable in sample R1-5% with relative abundance of 7.13%. Moreover, the proportion of the reigning genera *Spirochaetes* within the abundant phylum

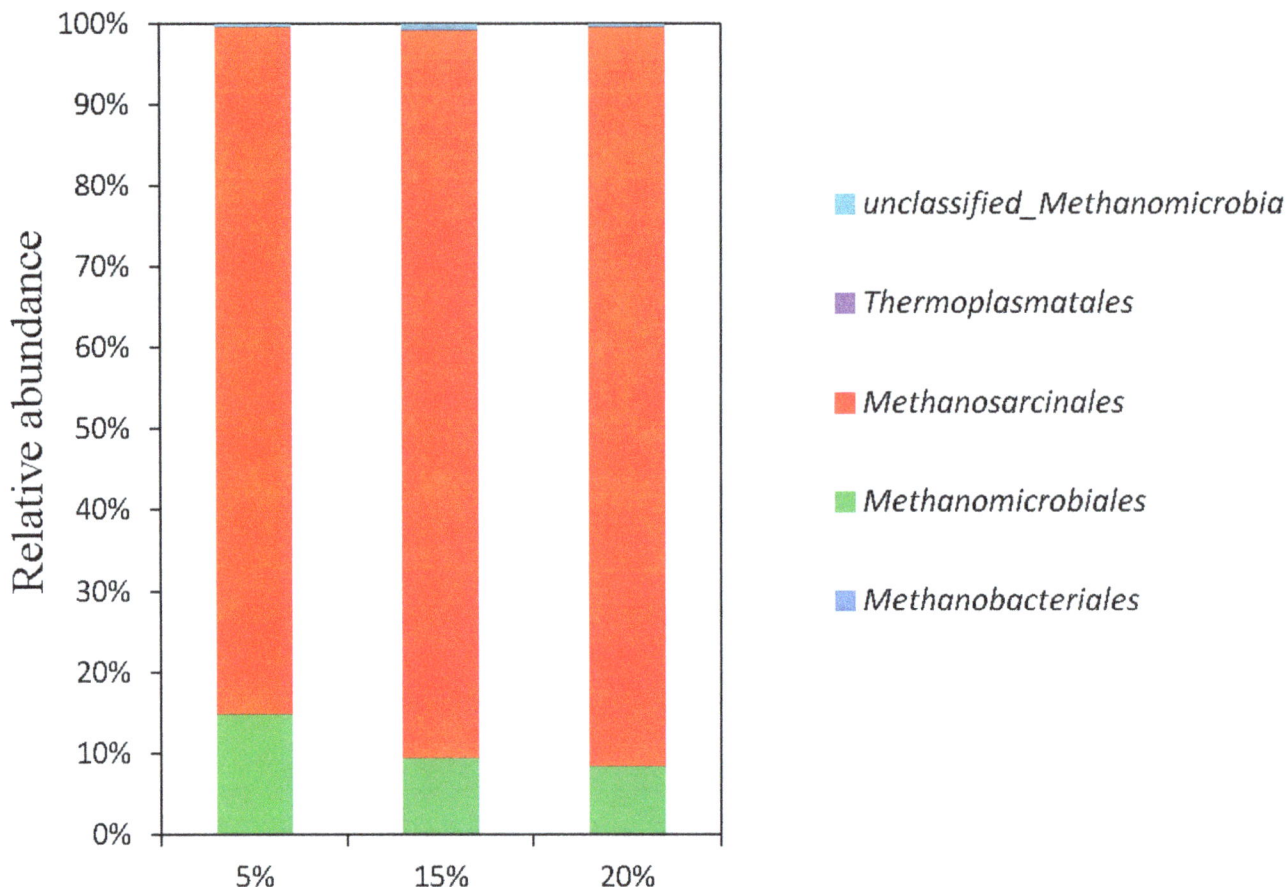

Figure 2. Taxonomic compositions of methanogens at order level in each sample retrieved from pyrosequencing.

Spirochaetes and *Acholeplasma* within the *Tenericutes* increased obviously with TS contents increasing. From the analyses made above, it can be seen that the changing patterns of main microbial population abundances were closely related to the performance variations with TS contents increasing, especially for VS reduction. The increasing degradation of organic matter to precursors for methanogenesis was jointly accomplished by the compatible collaborations of these microorganisms which played their respective roles in one of several trophic levels including hydrolysis, fermentation and acetogenesis.

Effect of total solids content on functional archaeal populations involved in food waste methanogenesis

The diversities of archaeal populations in three anaerobic digesters were also revealed by high-throughput pyrosequencing target 16S rRNA gene segments. All species richness estimators including ACE, Chao Shannon and Simpson indices are shown in Table S1. The Good's coverage estimated at least 97% coverage at a similarity of 97%, indicating good coverage of archaeal community. Two hydrogen-utilizing methanogenic groups, *Methanobacteriales* and *Methanomicrobiales*, and acetoclastic methanogenic order *Methanosarcinales* were detected in three reactors. The sum relative abundances of these three methanogenic groups accounted for 99.64%, 99.19% and 99.62% of total archaeal sequences in R1, R2, R3, respectively. However, *Methanococcales* was not detected in any DNA samples in this study (Figure 2). This result was in accordance with previous work characterizing the microbial community shifts in anaerobic

digestion of secondary sludge [30]. The relative abundances of sequencing data were also analyzed more specifically at genus level (Table. S3). It was showed that the phylogenetic diversity of methanogens was much lower than that of the bacterial community due partly to the inherent phylogenetic low diversity of methanogens.

As shown in Table S3, there was no large gap in terms of methanogens diversity and distinct discrimination in the taxonomic compositions at genus level. Most of methanogens were assigned to the genus *Methanosarcina* (accounting for 84.4%, 89.5% and 90.9% of total archaeal sequences in R1, R2, R3, respectively), indicating that acetoclastic methanogens played important roles in anaerobic digestion of FW and acetoclastic methanogenesis was the principal pathway of methane production. The low-solids anaerobic digester R1 was secondly dominated by hydrogenotrophic *Methanoculleus* while another hydrogenotrophic methanogens *Methanomicrobiales* was the second most detected group in anaerobic digesters R2 and R3.

Methanosarcina, a typical member of acetoclastic methanogens, have been often reported as the dominant methanogens in AD [31]. The ability of genus *Methanosarcina* having high growth rates and forming irregular cell clumps makes them more tolerant to changing in pH and high concentrations of toxic ionic agents [32]. The genus *Methanosarcina* produce methane from acetate, although some species are more versatile and can also utilize H_2/CO_2, methylated amines and methanol. In addition, *Methanosarcina* spp. are able to use both the acetoclastic and the hydrogenotroph methanogenesis pathways, making them more tolerant to specific inhibitors of the acetoclastic pathway compared to *Methanosaeta* spp. Therefore, anaerobic digester dominantly based on *Methanosaricna* spp. could potentially achieve stable methanogenesis [33], as their special morphological characteristics and flexibility in metabolism.

Besides, the changing patterns of the proportions of three major genera with TS contents increasing were different. The relative proportion of the genus *Methanosaricna* slightly increased from 84% to 90.9% with the TS content increased from 5% in R1 to 20% in R3. On the basis of stable operation, increased feeding TS contents of FW meant higher applied OLR and more VS for microorganisms, which resulted in higher VFA concentrations. In this study, it was observed in Table 2 that the acetate concentration increased with increasing TS contents. It is suggested that higher acetate concentrations would favor the growth of *Methanosarcina* [33]. Therefore, higher concentrations of VFA (especially acetate) and, by extension, at higher OLR caused by the anaerobic systems with higher TS contents induced the selective proliferation of *Methanosarcina*.

The relative abundance of genus *Methanoculleus* obvious decreased from 7.63 to 2.91% with increasing TS contents, indicating that hydrogenotrophic methanogenesis by *Methanoculleus* contributed less to the methane production in high-solids AD than it did in low-solids AD. It has been reported that *Methanoculleus* methanogens had been widely distributed with large proportion in various thermophilic ananerobic digesters [34] and their population ratio seems to be affected by HRT, OLR, or the concentration of VFA. In this study, similar result was obtained that the dominance of *Methanoculleus* declined in the mesophilic anaerobic digesters with TS content increasing resulting in the increase of OLR and the concentration of VFA. Summarily, the changing of microbial communities in mesophilic anaerobic digestion of FW was responsible for the different performances of the reactors with the increasing TS contents. The results obtained in this study expand our knowledge about the role of the TS content on the behavior of the microbial community structure involved in the anaerobic digestion degradation of solids, from low-solids to high-solids technology, and hence to provide valuable information to optimize fermentation process to favor efficient breakdown of food waste.

Conclusions

Three stable processes were achieved for AD of food waste with TS contents increasing from 5% to 20%. Better performances, mainly including VS reduction and methane yield and significant shifts in bacterial community, were obtained with the increasing TS contents. The relative abundance of phylum *Chloroflexi* decreased while other functional bacteria increased. The genus *Methanosarcina* absolutely dominated in archaeal communities in three reactors and the relative abundance of this group showed increasing trend with TS contents increasing. These results revealed the effect of the TS content on the performance parameters and the behavior of the microbial community involved in the AD of food waste from wet to dry technologies.

Supporting Information

Figure S1 Rarefaction cures of bacterial (A) and archaeal (B) sequences from the fermentation reactors with different total solids contents.

Table S1 Bacterial and archaeal richness and diversity indices for three reactors. All values were calculated at a distance level of 3%.

Table S2 Taxonomic composition of bacterial communities at the class level for the sequences retrieved from each samples.

Table S3 Taxonomic composition of archaeal communities at the genus level for the sequences retrieved from each samples.

Author Contributions

Conceived and designed the experiments: XD. Performed the experiments: JY. Analyzed the data: BD. Contributed reagents/materials/analysis tools: JJ. Contributed to the writing of the manuscript: JY.

References

1. Dai X, Duan N, Dong B, Dai L (2013) High-solids anaerobic co-digestion of sewage sludge and food waste in comparison with mono digestions: Stability and performance. Waste Manage. 33: 308–316.
2. Forster-Carneiro T, Pérez M, Romero L (2008) Influence of total solid and inoculum contents on performance of anaerobic reactors treating food waste. Bioresour Technol. 99: 6994–7002.
3. Duan N, Dong B, Wu B, Dai X (2012) High-solid anaerobic digestion of sewage sludge under mesophilic conditions: feasibility study. Bioresour Technol. 104: 150–156.
4. Guo X, Wang C, Sun F, Zhu W, Wu W (2014) A comparison of microbial characteristics between the thermophilic and mesophilic anaerobic digesters exposed to elevated food waste loadings. Bioresour Technol. 152: 420–428.
5. Pavan P, Battistoni P, Mata-Alvarez J (2000) Performance of thermophilic semi-dry anaerobic digestion process changing the feed biodegradability. Water Sci Technol. 41: 75–81.
6. Abbassi-Guendouz A, Brockmann D, Trably E, Dumas C, Delgenès JP, et al. (2012) Total solids content drives high solid anaerobic digestion via mass transfer limitation. Bioresour Technol. 111: 55–61.
7. Chouari R, Le PD, Daegelen P, Ginestet P, Weissenbach J, et al. (2005) Novel predominant archaeal and bacterial groups revealed by molecular analysis of an anaerobic sludge digester. Environ Microbiol. 7: 1104–1115.
8. Fernández A, Huang S, Seston S, Xing J, Hickey R, et al. (1999) How stable is stable? Function versus community composition. Applied Environ Microb. 65: 3697–3704.
9. Nelson MC, Morrison M, Yu Z (2011) A meta-analysis of the microbial diversity observed in anaerobic digesters. Bioresour Technol. 102: 3730–3739.
10. Pinto AJ, Xi C, Raskin L (2012) Bacterial community structure in the drinking water microbiome is governed by filtration processes. Environ Sci Technol. 46: 8851–8859.
11. Kwon S, Moon E, Kim TS, Hong S, Park HD (2011) Pyrosequencing demonstrated complex microbial communities in a membrane filtration system for a drinking water treatment plant. Microbes Environ. 26: 149–155.
12. Roesch L, Fulthorpe RR, Riva A, Casella G, Hadwin AK, et al. (2007) Pyrosequencing enumerates and contrasts soil microbial diversity. ISME J. 1: 283–290.
13. Schloss PD, Westcott SL, Ryabin T, Hall JR, Hartmann M, et al. (2009) Introducing mothur: open-source, platform-independent, community-supported software for describing and comparing microbial communities. Appl Environ Microb. 75: 7537–7541.
14. Quast C, Pruesse E, Yilmaz P, Gerken J, Schweer T, et al. (2013) The SILVA ribosomal RNA gene database project: improved data processing and web-based tools. Nucleic acids res. 41: 590–596.
15. APHA (American Public Health Association) (1995) Standard Methods for the Examination of Water and Wastewater, 19th ed, Washington, DC, USA.
16. Østergaard N (1985) Biogasproduktion i det thermofile temperaturinterval: Kemiteknik, Teknologisk Institut.
17. Li R, Chen S, Li X (2010) Biogas production from anaerobic co-digestion of food waste with dairy manure in a two-phase digestion system. Appl Environ Microb. 160: 643–654.
18. Malina J, Pohland JF, Frederick G (1992) Design of anaerobic processes for the treatment of industrial and municipal wastes: CRC Press 7.
19. Hansen KH, Angelidaki I, Ahring BK (1998) Anaerobic digestion of swine manure: inhibition by ammonia. Water Res. 32: 5–12

20. Lou XF, Nair J, Ho G (2012) Field performance of small scale anaerobic digesters treating food waste. Energy Sustain Dev. 16: 509–514.

21. Heo NH, Park SC, Kang H (2004) Effects of mixture ratio and hydraulic retention time on single-stage anaerobic co-digestion of food waste and waste activated sludge. J Environ Sci Heal A. 39: 1739–1756.

22. Rivière D, Desvignes V, Pelletier E, Chaussonnerie S, Guermazi S, et al. (2009) Towards the definition of a core of microorganisms involved in anaerobic digestion of sludge. ISME J. 3: 700–714.

23. Ariesyady HD, Ito T, Okabe S (2007) Functional bacterial and archaeal community structures of major trophic groups in a full-scale anaerobic sludge digester. Water Res. 4: 1554–1568.

24. Yamada T, Sekiguchi Y (2009) Cultivation of uncultured Chloroflexi subphyla: significance and ecophysiology of formerly uncultured Chloroflexi 'subphylum I' with natural and biotechnological relevance. Microbes Environ. 24: 205–216.

25. Kindaichi T, Ito T, Okabe S (2004) Ecophysiological interaction between nitrifying bacteria and heterotrophic bacteria in autotrophic nitrifying biofilms as determined by microautoradiography-fluorescence in situ hybridization. Appl Environ Microb. 70: 1641–1650.

26. Boucias DG, Cai Y, Sun Y, Lietze VU, Sen R, et al. (2013) The hindgut lumen prokaryotic microbiota of the termite Reticulitermes flavipes and its responses to dietary lignocellulose composition. Mol EcoL. 22:1836–1853.

27. Narihiro T, Terada T, Kikuchi K, Iguchi A, Ikeda M, et al. (2008) Comparative analysis of bacterial and archaeal communities in methanogenic sludge granules from upflow anaerobic sludge blanket reactors treating various food-processing, high-strength organic wastewaters. Microbes Environ. 24: 88–98.

28. Su Y, Li B, Zhu WY (2012) Fecal microbiota of piglets prefer utilizing dl-lactate mixture as compared to d-lactate and l-lactate in vitro. Anaerobe. 19: 27–33.

29. Chen S, Dong X (2005) *Proteiniphilum acetatigenes* gen nov, sp nov, from a UASB reactor treating brewery wastewater. Int J Syst Evol Micr. 55: 2257–2261.

30. Shin SG, Lee S, Lee C, Hwang K, Hwang S (2010) Qualitative and quantitative assessment of microbial community in batch anaerobic digestion of secondary sludge. Bioresour Technol. 101: 9461–9470.

31. Demirel B, Scherer P (2008) The roles of acetotrophic and hydrogenotrophic methanogens during anaerobic conversion of biomass to methane: a review. Rev Environ Sci Biotechnol. 7: 173–190.

32. Conklin A, Stensel HD, Ferguson J (2006) Growth kinetics and competition between Methanosarcina and Methanosaeta in mesophilic anaerobic digestion. Water Environ Res. 78: 486–496.

33. Vrieze JD, Hennebel T, Boon N, Verstraete W (2012) Methanosarcina: the rediscovered methanogen for heavy duty biomethanation. Bioresour Technol. 112: 1–9.

34. Bourque JS, Guiot S, Tartakovsky B (2008) Methane production in an UASB reactor operated under periodic mesophilic-thermohilic conditions. Biotechnol Bioeng. 100: 1115–1121.

PERMISSIONS

LIST OF CONTRIBUTORS

Willow Hallgren, Udaya Bhaskar Gunturu and Adam Schlosser
The MIT Joint Program on the Science and Policy of Global Change, Massachusetts Institute of Technology, Cambridge, Massachusetts, United States of America

Rui Carvalho and David K. Arrowsmith
School of Mathematical Sciences, Queen Mary University of London, London, United Kingdom

Lubos Buzna
University of Zilina, Univerzitna 8215/1, Zilina, Slovakia

Flavio Bono
European Laboratory for Structural Assessment, Institute for the Protection and Security of the Citizen (IPSC), Joint Research Centre, Ispra(VA), Italy

Marcelo Masera
Energy Security Unit, Institute for Energy and Transport, Joint Research Centre, Petten, The Netherlands

Dirk Helbing
ETH Zurich, Zurich, Switzerland
Risk Center, ETH Zurich, Swiss Federal Institute of Technology, Zurich, Switzerland

David K. Y. Lim, Sourabh Garg, Eugene S. B. Zhang, Skye R. Thomas-Hall, Holger Schuhmann, Yan Li and Peer M. Schenk
School of Agriculture and Food Sciences, The University of Queensland, Brisbane, Queensland, Australia

Matthew Timmins
School of Agriculture and Food Sciences, The University of Queensland, Brisbane, Queensland, Australia
ARC Centre of Excellence in Plant Energy Biology, Centre for Metabolomics, School of Chemistry and Biochemistry, The University of Western Australia, Crawley, Western Australia, Australia

Naomi Hayashida and Noboru Takamura
Department of Global Health, Medical and Welfare, Nagasaki University Graduate School of Biomedical Sciences, Nagasaki, Japan

Shunichi Yamashita
Department of Radiation Medical Science, Nagasaki University Graduate School of Biomedical Sciences, Nagasaki, Japan

Hitoshi Yamaguchi
Department of Ecomaterials Science, Nagasaki University Graduate School of Engineering, Nagasaki, Japan
Department of Microbiology, Semey State Medical Academy, Semey, the Republic of Kazakhstan

Rimi Tsuchiya
Nagasaki University School of Medicine, Nagasaki, Japan

Jumpei Takahashi
Center for International Collaborative Research, Nagasaki University, Nagasaki, Japan

Alexander Kazlovsky
Department of Pediatrics, Gomel State Medical University, Gomel, the Republic of Belarus

Marat Urazalin and Tolebay Rakhypbekov
Department of Microbiology, Semey State Medical Academy, Semey, the Republic of Kazakhstan

Yasuyuki Taira
Department of Global Health, Medical and Welfare, Nagasaki University Graduate School of Biomedical Sciences, Nagasaki, Japan
Nagasaki Prefectural Institute for Environmental Research and Public Health, Omura, Japan

Tong Zhang and Zilin Song
College of Forestry and the Research Center of Recycle Agricultural Engineering and Technology of Shaanxi Province, Northwest A&F University, Yangling, Shaanxi, People's Republic of China

Guangxin Ren, Yongzhong Feng, Xinhui Han, Gaihe Yang and Linlin Liu
College of Agronomy and the Research Center of Recycle Agricultural Engineering and Technology of Shaanxi Province, Northwest A&F University, Yangling, Shaanxi, People's Republic of China

Sebastian Jaenicke, Christina Ander, Thomas Bekel, Regina Bisdorf, Felix Tille, Martha Zakrzewski and Alexander Goesmann
Computational Genomics, Center for Biotechnology (CeBiTec), Bielefeld University, Bielefeld, Germany

Marcus Dröge and Olaf Kaiser
Roche Diagnostics GmbH, Penzberg, Germany

Karl-Heinz Gartemann
Department of Genetechnology/Microbiology, Bielefeld University, Bielefeld, Germany

Sebastian Jünemann
Computational Genomics, Center for Biotechnology (CeBiTec), Bielefeld University, Bielefeld, Germany
Department of Periodontology, University Hospital Münster, Münster, Germany

Lutz Krause
Division of Genetics and Population Health, Queensland Institute of Medical Research, Herston, Australia

Alfred Pühler and Andreas Schlüter
Institute for Genome Research and Systems Biology, Center for Biotechnology (CeBiTec), Bielefeld University, Bielefeld, Germany

Hongguang Yu, Qiaoying Wang, Zhiwei Wang, Jinxing Ma and Zhichao Wu
State Key Laboratory of Pollution Control and Resource Reuse, School of Environmental Science and Engineering, Tongji University, Shanghai, PR China

Erkan Sahinkaya
Istanbul Medeniyet University, Bioengineering Department, Kadıköy, Istanbul, Turkey

Yongli Li
Laboratory of Polyméres, Biopolyméres and Surfaces, UMR 6270, University of Rouen-CNRS-INSA, Boulevard Maurice de Broglie, Mont-Saint-Aignan, France

Bernadetta Rina Hastilestari, Sebastian Guretzki and Jutta Papenbrock
Institute of Botany, Gottfried Wilhelm Leibniz University Hannover, Hannover, Germany

Marina Mudersbach, Filip Tomala, Hartmut Vogt and Bettina Biskupek-Korell
Technology of Renewable Resources, University of Applied Sciences Hannover, Hannover, Germany

Patrick Van Damme
Department of Plant Production, Laboratory for Tropical and Subtropical Agriculture and Ethnobotany, Ghent University, Ghent, Belgium
Institute of Tropics and Subtropics, Czech University of Life Sciences Prague, Prague, Czech Republic

Zilin Song, Xiaofeng Liu, Yuexiang Yuan and Yinzhang Liao
Chengdu Institute of Biology, Chinese Academy of Science, Chengdu, Sichuan, PR China

GaiheYang and Zhiying Yan
Research Center of Recycle Agricultural Engineering Technology of Shaanxi Province, Northwest A&F University, Yangling, Shaanxi, PR China

Jibin Zhang, Ziniu Yu, Ziduo Liu and Sen Yang
State Key Laboratory of Agricultural Microbiology, National Engineering Research Centre of Microbial Pesticides, College of Life Science and Technology, Huazhong Agricultural University, Wuhan, Hubei, People's Republic of China

Qing Li
State Key Laboratory of Agricultural Microbiology, National Engineering Research Centre of Microbial Pesticides, College of Life Science and Technology, Huazhong Agricultural University, Wuhan, Hubei, People's Republic of China
College of Science, Huazhong Agricultural University, Wuhan, Hubei, People's Republic of China

Qinglan Zeng
Department of Biological Engineering, Xianning Vocational Technical College, Xianning, China

Puneet Dwivedi
Warnell School of Forestry and Natural Resources, University of Georgia, Athens, Georgia, United States of America

Madhu Khanna
Energy Biosciences Institute, University of Illinois at Urbana-Champaign, Urbana, Illinois, United States of America

Xiaotian Han and Zhiming Yu
Key Laboratory of Marine Ecology and Environmental Sciences, Institute of Oceanology, Chinese Academy of Sciences, Qingdao, China

Liyan He
Key Laboratory of Marine Ecology and Environmental Sciences, Institute of Oceanology, Chinese Academy of Sciences, Qingdao, China
University of Chinese Academy of Sciences, Beijing, China

Robert I. McDonald and Jimmie Powell
Worldwide Office, The Nature Conservancy, Arlington, Virginia, United States of America

Julian D. Olden
School of Aquatic & Fishery Sciences, University of Washington, Seattle, Washington, United States of America

Jeffrey J. Opperman
Freshwater Focal Area Program, The Nature Conservancy, Chargin Falls, Ohio, United States of America

William M. Miller
Department of Chemical and Biological Engineering, Northwestern University, Evanston, Illinois, United States of America

Joseph Fargione
North America Region, The Nature Conservancy, Minneapolis, Minnesota, United States of America

Carmen Revenga
Marine Focal Area Program, The Nature Conservancy, Arlington, Virginia, United States of America

Jonathan V. Higgins
Freshwater Focal Area Program, The Nature Conservancy, Chicago, Illinois, United States of America

Yangkai Duan, Zhi Zhu, Ke Cai, Xiaoming Tan and Xuefeng Lu
Key Laboratory of Biofuels, Qingdao Institute of Bioenergy and Bioprocess Technology, Chinese Academy of Sciences, Qingdao, China

Paul Illmer, Christoph Reitschuler, Andreas Otto Wagner, Thomas Schwarzenauer and Philipp Lins
University Innsbruck, Institute of Microbiology, Innsbruck, Austria

Chaochun Wei
Department of Bioinformatics and Biostatistics, School of Life Sciences and Biotechnology, Shanghai Jiao Tong University, Shanghai, China
Shanghai Center for Bioinformation Technology, Shanghai, China

Peng Jia and Lei Liu
Key Laboratory of Systems Biology, Shanghai Institutes for Biological Sciences, Chinese Academy of Sciences, Shanghai, China
Graduate School of the Chinese Academy of Sciences, Shanghai, China
Shanghai Center for Bioinformation Technology, Shanghai, China

Liming Xuan
Shanghai Center for Bioinformation Technology, Shanghai, China
Department of Biochemistry and Molecular Biology, School of Bioengineering, East China University of Science and Technology, Shanghai, China

Xiaojiao Wang, Fang Li and Gaihe Yang
College of Agronomy, Northwest A&F University, Yangling, Shaanxi, People's Republic of China

Xingang Lu
School of Chemical Engineering, Northwest University, Xian, Shaanxi, People's Republic of China

Ana T. Marcos and David Cánovas
Department of Genetics, Faculty of Biology, University of Seville, Seville, Spain

Almudena Escobar-Niño
Department of Genetics, Faculty of Biology, University of Seville, Seville, Spain
Department of Microbiology and Parasitology, Faculty of Pharmacy, University of Seville, Seville, Spain

Encarnación Mellado
Department of Microbiology and Parasitology, Faculty of Pharmacy, University of Seville, Seville, Spain

Carlos Luna and Diego Luna
Department of Organic Chemistry, University of Córdoba, Córdoba, Spain

Gregory J. O. Martin, David R. A. Hill, Ian L. D. Olmstead, Amanda Bergamin, Sandra E. Kentish and Peter J. Scales
Department of Chemical and Biomolecular Engineering, The University of Melbourne, Parkville, Victoria, Australia

Daniel A. Dias
Metabolomics Australia, The School of Botany, The University of Melbourne, Parkville, Victoria, Australia

Melanie J. Shears
Metabolomics Australia, The School of Botany, The University of Melbourne, Parkville, Victoria, Australia
Apicolipid Group, Laboratoire Adaption et Pathogenie des Microorganismes UMR5163, CNRS, University of Grenoble I, La Tronche, France

Cyrille Y. Botté
Apicolipid Group, Laboratoire Adaption et Pathogenie des Microorganismes UMR5163, CNRS, University of Grenoble I, La Tronche, France

Damien L. Callahan
Metabolomics Australia, The School of Botany, The University of Melbourne, Parkville, Victoria, Australia
Centre for Chemistry and Biotechnology, School of Life and Environmental Sciences, Deakin University, Burwood, Victoria, Australia

Yan Yang, Chuang Wen and Shuli Wang
Jiangsu Key Laboratory of Oil-Gas Storage and Transportation Technology, Changzhou University, Changzhou, Jiangsu Province, China

Yuqing Feng
Computational Informatics, Commonwealth Scientific and Industrial Research Organization, Melbourne, The State of Victoria, Australia
Richard W. Hammack
National Energy Technology Laboratory, Pittsburgh, Pennsylvania, United States of America

Arvind Murali Mohan and Kelvin B. Gregory
National Energy Technology Laboratory, Pittsburgh, Pennsylvania, United States of America
Department of Civil and Environmental Engineering, Carnegie Mellon University, Pittsburgh, Pennsylvania, United States of America

Daniel Lipus
National Energy Technology Laboratory, Pittsburgh, Pennsylvania, United States of America

Department of Civil and Environmental Engineering, University of Pittsburgh, Pittsburgh, Pennsylvania, United States of America

Kyle J. Bibby
National Energy Technology Laboratory, Pittsburgh, Pennsylvania, United States of America
Department of Civil and Environmental Engineering, University of Pittsburgh, Pittsburgh, Pennsylvania, United States of America
Department of Computational and Systems Biology, University of Pittsburgh Medical School, Pittsburgh, Pennsylvania, United States of America

Jing Yi, Bin Dong, Jingwei Jin and Xiaohu Dai
National Engineering Research Center for Urban Pollution Control, College of Environmental Science and Engineering, Tongji University, Shanghai, People's Republic of China

Index

www.ingramcontent.com/pod-product-compliance
Lightning Source LLC
Chambersburg PA
CBHW080535200326
41458CB00012B/4445